TECHNICAL
COMMUNICATION
PROCESS AND PRODUCT

TECHNICAL COMMUNICATION

PROCESS AND PRODUCT

SEVENTH EDITION

SHARON J. GERSON
DeVry University (Retired)

STEVEN M. GERSON
Johnson County Community College

Prentice Hall
Boston   Columbus   Indianapolis   New York   San Francisco   Upper Saddle River
Amsterdam   Cape Town   Dubai   London   Madrid   Milan   Munich   Paris   Montreal
Toronto   Delhi   Mexico City   Sao Paulo   Sydney   Hong Kong   Seoul   Singapore   Taipei   Tokyo

Editorial Director: Vernon Anthony
Executive Editor: Gary Bauer
Development Editor: Linda Cupp
Editorial Assistant: Tanika Henderson
Director of Marketing: David Gessel
Marketing Manager: Stacey Martinez
Marketing Assistant: Les Roberts
Senior Managing Editor: JoEllen Gohr
Project Manager: Christina Taylor

Senior Operations Specialist: Pat Tonneman
Senior Art Director: Diane Ernsberger
Full-Service Project Management: Kelli Jauron,
    S4Carlisle Publishing Services
Composition: S4Carlisle Publishing Services
Printer/Binder: R.R. Donnelly/Willard
Cover Printer: Lehigh-Phoenix Color/
    Hagerstown
Text Font: 10/12 Sabon

Microsoft® and Windows® are registered trademarks of the Microsoft Corporation in the U.S.A. and other countries. Screen shots and icons reprinted with permission from the Microsoft Corporation. This book is not sponsored or endorsed by or affiliated with the Microsoft Corporation.

Many of the designations by manufacturers and seller to distinguish their products are claimed as trademarks. Where those designations appear in this book, and the publisher was aware of a trademark claim, the designations have been printed in initial caps or all caps.

Cataloging-in-Publication Data is available from the Library of Congress.

10 9 8 7 6 5 4 3 2 1

Prentice Hall
is an imprint of

www.pearsonhighered.com

ISBN 10:   0-13-137734-5
ISBN 13: 978-0-13-137734-9

FOR OUR DAUGHTERS, STACY AND STEFANI

ABOUT THE AUTHORS

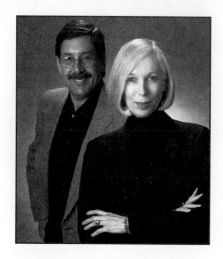

SHARON J. AND STEVEN M. GERSON are dedicated career professionals who have a combined total of over 75 years teaching experience at the college and university level. They have taught technical writing, business writing, professional writing, and technical communication to thousands of students, attended and presented at dozens of conferences, written numerous articles, and published several textbooks, including *Technical Writing: Process and Product* (seventh edition), *The Red Bridge Reader* (third edition, co-authored by Kin Norman), *Writing That Works: A Teacher's Guide to Technical Writing* (second edition), *Workplace Communication: Process and Product* (first edition), and *Workplace Writing: Planning, Packaging, and Perfecting Communication* (first edition).

In addition to their academic work, Sharon and Steve are involved in business and industry through their business, Steve Gerson Consulting. In this business, they have worked for companies such as Sprint, AlliedSignal–Honeywell, General Electric, JCPenney, Avon, the Missouri Department of Transportation, H&R Block, Mid America Regional Council, and Commerce Bank. Their work for these businesses includes writing, editing, and proofreading many different types of technical documents, such as proposals, marketing collateral, reports, and instructions.

Steve also has presented hundreds of hands-on workshops on technical/business writing, business grammar in the workplace, oral presentations in the workplace, and business etiquette. Over 10,000 business and governmental employees have benefited from these workshops. For the past decade, Steve has worked closely with K–12 teachers. He has presented many well-attended, interactive workshops to give teachers useful tips about technical writing in the classroom.

Both Steve and Sharon have been awarded for teaching excellence and are listed in *Who's Who Among America's Teachers*. Steve is a Society for Technical Communication Fellow. In 2003, Steve was named Kansas Professor of the Year by the Council for the Advancement and Study of Education.

Their wealth of experience and knowledge has been gathered for you in this seventh edition of *Technical Communication: Process and Product*.

PREFACE

Welcome to the seventh edition of *Technical Communication: Process and Product*. This reader-friendly textbook combines easy-to-follow instructions for producing all forms of technical communication with interesting scenarios and examples featuring real people facing communication challenges on the job.

For over 30 years, the writing process—prewriting, writing, and rewriting—has been the standard for teaching students how to write effectively. Many of us were taught in K–12 how to write based on this process, or we were introduced to the writing process in college. In this textbook, we build upon this process with a unique approach that applies the writing process to both oral and written communication.

Each chapter features a unit entitled **The Writing Process at Work** with a business person facing a communication challenge. The person uses the writing process to create effective technical communication.

- In the prewriting stage, we show how the writer determines goals, audience, and communication channel. Then we show a planning technique the writer uses to gather data for the correspondence.
- In the writing stage, we show how the writer organizes his or her rough draft and formats the text for ease of access. We also show how the writer receives suggestions for revision.
- In the rewriting stage, we show how the rough draft has been revised. This stage illustrates the finished product: a written document that communicates successfully with its intended audience.

Throughout the textbook, we also feature real-world business writers in **Spotlights** showing how they confront communication challenges at work.

The Writing Process in Action

The writing process allows us to accomplish key goals, including the following:

- Provide criteria for different communication channels
- Give examples of different types of communication
- Teach an effective process to help students succeed in the classroom and in their careers
- Show the importance of revision and how it can be achieved
- Expose students to real people facing real workplace communication challenges
- Reveal the thought processes and the approach to writing and revising followed by businesspeople
- Bring to life the principles of technical communication

Addressing the Needs of the Evolving Workplace

Today's workplace is constantly evolving. This text addresses the important topical issues encountered in workplace communication today including the following:

- Multiculturalism and the global economy
- Social media and its impact on technical communication
- Criteria for effective workplace use of blogging, FaceBook, YouTube videos, instant messages, text messages, and Twitter
- Communication in dispersed team settings
- Strategies for ethical workplace communication
- Collaboration and teamwork especially with Wikis, Google Docs, Google Sites, Google Wave, and other technological tools
- Persuasive communication techniques for meeting the needs of the audience
- Up-to-date job search information
- Oral communication and the workplace
- Updated documentation information in Appendix B including Council of Science Editors (CSE) style format

New to the 7th Edition

To provide you up-to-date information about technical communication, this 7th edition offers the following new additions:

- New chapter covering communication via social media, including blogging, YouTube, Twitter, Facebook, Virtual World, and more
- Updated information on the use of Wikis and other groupware including Google Docs and Google Wave for collaboration
- Updated information on multicultural concerns
- Updated information on text messages
- New types of letters, including
  - Job acceptance letter
  - Confirmation letter
  - Recommendation letter
- New samples of short, informal reports
- Updated information on online help
- Updated information in Appendix B on APA, CSE, and MLA documentation style sheets
- New information on standard operating procedures
- New information about listening skills
- New information about ethics in the workplace, focusing on boilerplate content, copyright, questions to ask when confronted by an ethical dilemma at work, and end-of-chapter ethics Problem-Solving Think Pieces
- Technology Tips updated to Word 2010

The Technical Communication Learning System

Design and Instructional Aids

To ensure readability and enjoyment of this 7th edition, we have created a colorful new layout and design. In addition, each chapter in the text contains a variety of helpful pedagogical features that enhance the narrative including the following:

FREQUENTLY ASKED QUESTION boxes provide answers to questions people ask about topical issues.

> **FAQs:** Using the Writing Process on the Job
>
> **Q:** Do writers actually follow a process when they compose correspondence?
>
> **A:** Most good writers follow a process. It's like plotting your route before a trip. Sure, you can get in a car without a map, head west (or east or north or south) and find the destination without getting lost, but mapping the route before a trip assures that you won't get lost and waste time.
>
> There's no one way to plot your destination. Prewriting might entail only a quick outline, a few brief notes that list the topics you plan to cover and the order in which you'll cover them. This way, you will know where you're going before you get there.
>
> In addition to creating both brief and sometimes much more detailed outlines, an important part of

TECHNOLOGY TIP boxes show students how to use Microsoft 2010.

> **TECHNOLOGY TIPS**
>
> Using Microsoft Word 2010 for Rewriting
>
> Word processing programs help you rewrite your document in many ways:
>
> - **Spell check.** When you misspell a word, often spell check will underline the error in red (as shown in the following example with "grammer" incorrectly spelled). Spell check, unfortunately, will not catch all errors. If you use a word like to instead of too, spell check will not "no" the difference (of course, that should be "know" but spell check did not mark the error). Microsoft Word 2010's Review tab also provides you access to proofreading help and allows you to make comments and track changes.

CHECKLISTS guide students through the revision stage of their writing.

> **AUDIENCE CHECKLIST**
>
> ____ 1. What is my audience's level of understanding regarding the subject matter?
> - High tech
> - Low tech
> - Lay
> - Multiple
>
> ____ 2. Given my audience's level of understanding, have I written accordingly?
> - Have I defined my acronyms, abbreviations, and jargon?
>
> - Negative
> - Positive
> - Noncommittal
> - Uninformed
>
> ____ 7. Is my audience in a position of authority to act according to my wishes? If not, who will make the decision?
>
> ____ 8. What are my audience's personality traits?
> - Slow to act
> - Eager

BEFORE AND AFTER EXAMPLES demonstrate to students the importance of revision.

> | BEFORE | AFTER |
> |---|---|
> | Biased | Unbiased |
> | Current tax regulations allow a head of household to deduct for the support of his children. | Current tax regulations allow a head of household to deduct for child support. |
> | The manager is responsible for the productivity of his department; the foreman is responsible for the work of his linemen. | Management is responsible for departmental productivity. Supervisors are responsible for their personnel. |
> | The administrative assistant brought her boss his coffee. | The administrative assistant brought the boss's coffee. |
> | The teacher must be sure her lesson plans are filed. | The teacher must file all lesson plans. |

DOCUMENT EXAMPLES WRITTEN FOR DIFFERENT AUDIENCES provide examples of documents written for lay, low-tech, and high-tech audiences.

> FIGURE 4.2   E-mail Message to Low-Tech Audience
>
> Date: June 15, 2012
> To: Security Personnel
> From: Fatai Sayanti, Facilities Manager
> Subject: Security of Heating, Ventilating, and Air Conditioning (HVAC)
>
> When you are patrolling the facilities and grounds, be on the alert for suspicious substances, such as fine powders, residues, fog, mist, oily liquids, or unusual odors. Not every liquid or odor is necessarily suspicious. Just focus on those with unexplainable origins. Also watch for discomfort in our employees, such as two or more people experiencing difficulty breathing. These signs could be evidence of dangerous chemicals in our HVAC system.
>
> If you discover problems, follow these procedures:
> • DO NOT touch the substance.
>
> This e-mail to low-tech readers deals with a very important subject (security in the face of potential terrorism). The writer presents only what the audience needs to know in order to complete their job responsibilities. The writer conveys this content simply, using short sentences, a list, and easy-to-understand words. HVAC is defined.

WEB RESOURCES direct students to additional samples, interactive activities, and exercises.

Web Resources

at www.pearsonhighered .com/gerson For more information about research and documen-

TOPICAL CHAPTER REFERENCES provide links to related topics in other chapters.

Long Reports

See Chapter 17 for more discussion of long, formal reports.

Building Technical Communication Skills
End-of-Chapter Individual and Collaborative Activities

The end-of-chapter activities provide students with opportunities to apply chapter principles. In addition, because students learn in a variety of ways, we provide activities that engage students through memorization, critical thinking, collaboration, writing, oral presentations, and research.

CHAPTER HIGHLIGHTS summarize key points discussed in the chapter.

CHAPTER HIGHLIGHTS

1. You can research a topic either using secondary sources, such as books, periodicals, or the Internet; or primary sources, such as interviews or questionnaires.
2. You need to consider the audience's level of technical knowledge when you write a research paper.
3. Primary and secondary research can be conducted to assist you in developing content for reports.

CASE STUDIES present real-world scenarios and on-the-job communication challenges.

CASE STUDIES

1. The City of Oak Springs, Iowa, needs to improve a ten-mile stretch of Ironhorse Avenue that runs east and west through the town. The winding, two-lane road was built in 1976. Since then, the city has grown with several businesses and new housing along this road, and the road no longer is sufficient for the increased traffic load. The road must be expanded to four lanes and straightened. To do so, new sewer lines, sidewalks, easements, esplanades, and lighting must be added. However, this construction will impact current homeowners (whose land, through eminent domain, can be expropriated without the owner's consent). In addition, part of the

INDIVIDUAL AND TEAM PROJECTS provide students an opportunity to apply principles discussed in the chapter drawing from personal experience, research, or collaboration.

INDIVIDUAL AND TEAM PROJECTS
Avoiding Sexist Language

1. Revise the following sentences to avoid sexist language.
   a. All the software development specialists and their wives attended the conference.
   b. The foremen met to discuss techniques for handling union grievances.
   c. Every technician must keep accurate records for his monthly activity reports.
   d. The president of the corporation, a woman, met with her sales staff.

PROBLEM-SOLVING THINK PIECES help students practice critical thinking skills.

PROBLEM-SOLVING THINK PIECES

1. Many communities have recycling projects that allow residents to recycle paper products, cans, and plastic. Not all businesses recycle, however. Research the benefits of recycling, determine how a business or businesses could implement a corporate recycling plan, and write a report recommending action based on your research.
2. In today's global economy, understanding and accommodating multiculturalism and cross-culturalism in business is important. Research the following:

WEB WORKSHOPS allow students to have an opportunity for discovery of topics related to technical communication beyond the classroom.

WEB WORKSHOP

1. USA.gov allows you to research a wide variety of topics, such as education and jobs, benefits and grants, consumer protection, environment and energy, science and technology, and public safety and health. This Web site also provides information on breaking news. Access USA.gov and research a topic relevant to your career goals. Write an e-mail message or report to your instructor summarizing your findings.

2. Go to an online news magazine, such as *Slate*, *Time*, or *U.S. News Online*. Type a topic of interest in the magazine's search engine. Research this topic and make a brief oral

---

QUIZ QUESTIONS allow students to test their knowledge of the chapter's content.

QUIZ QUESTIONS

1. In what places can you locate research information?
2. Explain why you need to consider audience when you write a research report.
3. What is the difference between primary and secondary research?
4. What is the appropriate writing style of a research report?
5. What is the difference between a quote and a paraphrase?

Quick Reference Technology Tips for Microsoft Word 2010

Throughout the book are Technology Tips that provide helpful information about how to use Microsoft Word 2010 features to improve the effectiveness of your communication. Below is a listing of the tips by chapter.

Chapter 2—Using Microsoft Word for Rewriting

Chapter 3—Using Microsoft Word to Check the Readability Level of your Text

Chapter 5—Using Microsoft Word for Documentation

Chapter 6—Using Memos and Letter Templates in Microsoft Word

Chapter 8—Using Resume Templates in Microsoft Word

Chapter 10—Creating Graphics in Microsoft Word (Pie Charts, Bar Charts, Line Graphs, etc.)

Chapter 11—Using Microsoft Word for Flier and Brochure Design Templates

Chapter 12—How to Make Callouts Using Microsoft Word

Chapter 13—Using Screen Captures to Visually Depict Steps

Chapter 14—Using Microsoft Word to Creae a Web Site

Chapter 16—Creating Headers and Footers in Microsoft Word

Chapter 17—Creating a Hierarchy of Headings Using Microsoft Word

Chapter 19—Adding Hypertext Links to a PowerPoint Presentation

Using Microsoft Word 2010 for Rewriting

Word processing programs help you rewrite your document in many ways:

- **Spell check.** When you misspell a word, often spell check will underline the error in red (as shown in the following example with "grammer" incorrectly spelled). Spell check, unfortunately, will not catch all errors. If you use a word like to instead of too, spell check will not "no" the difference (of course, that should be "know" but spell check did not mark the error). Microsoft Word 2010's **Review** tab also provides you access to proofreading help and allows you to make comments and track changes.

- **Grammar check.** Word processors also can help you catch grammar errors. Grammar check underlines errors in green. When you right-click on the underlined text, the word processing package will provide an optional correction.

Online Resources to Enhance Technical Communication

To access our Web site, go to **www.pearsonhighered.com/gerson**.

We are especially excited about the wealth of new material on the Web site including cases, exercises, activities, and documents that have been developed for each chapter. Online materials for each chapter in the textbook include the following:

- **Chapter Learning Objectives**—Overview of major chapter concepts.
- **Writing Process Exercises**—Prewriting, Writing, and Rewriting assignments.
- **Interactive Editing and Revision Exercises**—Interactive documents allow students to see poorly done and corrected versions of documents with additional assignable document revision exercises.

- **Communication Cases**—Students encounter real-world situations with writing assignments.
- **Activities and Exercises**—Activities specific to a variety of technical and career fields allow students to practice producing communication relevant to their interests.
- **Collaboration Exercise**—Assignments designed to provide practice writing and communicating in teams.
- **Web Resources**—Links to helpful online resources related to chapter content.
- **Document Library**—Additional documents.
- **Chapter Quizzes**—Self-grading, multiple-choice quizzes help students master chapter concepts and prepare for tests.

Do You Need Help Refining Grammar, Mechanics, Punctuation Skills?

My WritingLab www.mywritinglab.com

MyWritingLab is an online learning system that provides better writing practice through diagnostic assessment and progressive exercises to move students from literal comprehension to critical thinking and writing. With this better practice model, students develop the skills needed to become better writers.

A MyWritingLab access code can be packaged with the textbook or can be purchased online at www.pearsonhighered.com.

Your One-Stop Source for Technical Communication Resources

MyTechcommLab for Gerson, *Technical Communication: Process and Product*, 7th Edition

This comprehensive resource can be packaged at no additional cost with purchase of a new text. MyTechCommLab provides a wide array of multimedia tools, all in one place, and all designed specifically for workplace writers.

- **Over 90 Model Documents,** most with interactive activities and annotations selected from a variety of professions and purposes (letters, memos, career correspondence, proposals, reports, instructions and procedures, descriptions and definitions, websites, and presentations). Mytechcommlab also contains **50 Interactive Documents** with rollover annotations highlighting purpose, audience, design, and other critical topics.
- **Grammar, Mechanics, and Writing Help:** If you need more practice in basic grammar and usage, MyTechCommLab's grammar diagnostics will generate a study plan linked to the thousands of test items in ExerciseZone, with results tracked by Pearson's exclusive GradeTracker.
- **Document Design Resources:** A **Writing Process Tutorial** leads you through each stage of the writing process—from prewriting to final formatting. A new **tutorial on Writing Formal Reports** offers step-by-step guidance for creating one of the most common document types in technical communication and working with sources. **Activities and Case Studies** provide over 65 exercises, all rooted in technical communication and many document-based, including new case studies on usability. An **online reference library of e-books** includes pdf files for books on Visual Communication and Workplace Literacy.

- **Pearson's MySearchLab: Research, Grammar, and Writing Tips** access gives you research tips, access to the EBSCO document database, writing and assessment and instruction and access to the Longman Online Handbook for Writers.
- **E-book with Online Reference Sources:** You can choose to purchase a version of Mytechcommlab that includes an ebook with links embedded in the pages to all online resources.

To order *Technical Communication: Process and Product*, 7th Edition with the generic MyTechCommLab access code, order ISBN: 0-13-262777-9.

To order *Technical Communication: Process and Product*, 7th Edition with MyTech CommLab with ebook in CourseCompass access code, order ISBN: 0-13-261750-1.

Standalone access codes for either MyTechCommLab with or without e-book can be purchased online at www.pearsonhighered.com.

Instructor's Resources

New to the 7th edition is an expanded Instructor's Manual loaded with helpful teaching notes for your classroom. We have also added additional sample answers to case studies and other end-of-chapter assignments.

- **Instructor's Manual** The Instructor's Manual is loaded with helpful teaching notes for your classroom. Included in the manual are answers to the chapter quiz questions, answers to case studies and other end of chapter assignments, and instructor notes for assignments.
- **MyTest**
- **PowerPoint Lecture Presentation Package**
- **Online Newsletter:** In addition to this textbook, the instructor's manual, and the online components of *Technical Communication: Process and Product*, you can have up-to-date information about the field of technical communication in our quarterly newsletters. Contact Prentice-Hall at www.pearsonhighered.com/gerson to obtain a free subscription to this enlightening newsletter.

The Instructor's Manual, MyTest, and PPt Package can be downloaded from our Instructor's Resource Center. To access supplementary materials online, instructors need to request an instructor access code. Go to **www.pearsonhighered.com/irc**, where you can register for an instructor access code. Within 48 hours of registering, you will receive a confirming e-mail including an instructor access code. Once you have received your code, locate your text in the online catalog and click on the Instructor Resources button on the left side of the catalog product page. Select a supplement and a log-in page will appear. Once you have logged in, you can access instructor material for all Pearson textbooks.

Acknowledgments

We would like to thank the following reviewers and focus group participants for their helpful comments on this and past editions of the textbook:

Floyd Brigdon, Trinity Valley Community College

Elizabeth Christensen, Sinclair Community College

Amy Tipton Cortner, Caldwell Community College and Technical Institute

Natalie Daley, Linn-Benton Community College

Myra G. Day, North Carolina State
Rosemary Day, Central New Mexico Community College
Treven Edwards, SUNY-Pittsburgh
Paul Fattaruso, University of Denver
Bart Ganzert, Forsyth Technical Community College
Cynthia Gillispie-Johnson, North Carolina A&T State University
Joshua Hamling, New Mexico State University and Dona Ana Community College
Joyce Harp, New Mexico State University
Arthur Khaw, Kirkwood Community College
Liz Kleinfeld, Red Rocks Community College
Angela W. Lamb, Robeson Community College
Lindsay Lewan, Arapahoe Community College
Caroline Mains, Palo Alto College
Josie Mills, Arapahoe Community College
Heather Milton, University of California, Davis
Sharon Mouss, Oklahoma State University, Okmulgee College
Kevin Nebergall, Kirkwood Community College
Joseph Nocera, Jefferson Community College
Nancy Roberts, Griffin Technical College
Sherry Rosenthal, Community College of Southern Nevada
Brian Still, Texas Tech University
Joyce C. Staples, Patrick Henry Community College
Alan Tessaro, Spartanburg Community College
Chris Thaiss, University of California, Davis
Marc Wilson, Ivy Tech Community College
Esther J. Winter, Central Community College
JB Zwilling, Allen County Community College

We also would like to thank the following people who contributed activities, exercises, and documents to the Companion Web site: May Beth Van Ness, Terra Community College; Linda Gray, Oral Roberts University; Catharine Schauer, Visiting Professor, Embry-Riddle Aeronautical University; Malanie Rosen Brown, St. Johns River Community College; and Connie Cerniglia, Guilford Technical Community College. And thank you to Bart Ganzert for preparing the MyTest.

We would especially like to thank Executive Editor Gary Bauer, Developmental Editor Linda Cupp, and Project Manager Christina Taylor for their efforts, patience, and creativity in helping us bring out this seventh edition. We also want to thank Kelli Jauron of S4Carlisle Publishing Services for her copy editing and project assistance.

Sharon J. Gerson   *Steven M. Gerson*

BRIEF CONTENTS

CONTENTS

TECHNICAL COMMUNICATION

PROCESS AND PRODUCT

CHAPTER ONE

An Introduction to Technical Communication

COMMUNICATION AT WORK

In the Gulfview scenario, employees in diverse locations reveal the importance of technical communication.

Gulfview Architectural and Engineering Services is based in Gulfview, Texas, with office sites in ten U.S. cities and five locations throughout the world. Gulfview hopes to build a power plant in Saudi Arabia. To accomplish this task, a team of employees is working on two continents. The project requires that all team members be involved in numerous communication challenges.

Proposal. First, one team, consisting of engineers, architects, marketing specialists, accountants, lawyers, and technical communicators, put together a proposal. In this proposal, they focused on the services they could offer, the expertise of their workforce, the price they would charge for the construction, and a timeline for their work. Despite many competitors, Gulfview won the account.

E-mail, Text Messages, and Instant Messages. The construction would take Gulfview approximately two years. During that time, Gulfview personnel had to communicate with their Saudi contractors on a daily basis. E-mail, text messages, and instant messages answered this need. The team members communicated with each other by writing approximately 50 e-mail messages a day. In these transmittals, the team members

Objectives

When you complete this chapter, you will be able to

1. Define and understand technical communication
2. Use many different channels of oral and written technical communication
3. Understand the importance of technical communication
4. Recognize the importance of teamwork in technical communication
5. Use groupware for collaborative projects
6. Deal with challenges to effective teamwork
7. Resolve conflicts in collaborative projects
8. Apply the checklist to team activities

focused on construction permits, negotiated costs with vendors, changed construction plans, and asked questions and received answers. They used text messages and instant messages for quick updates and to build rapport with co-workers.

Intranet Web Site and Corporate Blog. To help all parties involved (those in Saudi Arabia as well as Gulfview employees throughout the United States), Gulfview's information technology department built an intranet site and a blog geared specifically toward the power plant project. This firewall-protected site, open to Gulfview employees and external vendors associated with the project, helped all construction personnel submit online forms, get corporate updates, and access answers to frequently asked questions. Many of these FAQs were managed through online help screens with pull-down menus. The blog allowed employees to provide work journals, web logs in which they could comment on construction challenges and get feedback from other employees working with similar issues.

Letters. To secure and revise construction permits, Gulfview personnel had to write formal letters to government officials in Saudi Arabia. In addition, Gulfview employees had to write letters to vendors, asking for quotes.

Reports. Finally, all of the employees involved in the power plant project had to report on their activities. To encourage collaboration and improve the quality of the company's writing, management created a corporate wiki where participants could write the following:

- Progress reports providing updates on the project's status
- Incident reports when job-related accidents and injuries occurred
- Feasibility reports to recommend changes to the project's plan or scope
- Meeting minutes following the many team meetings

Like all companies engaged in job-related projects, Gulfview Architectural and Engineering Services spent much of its time communicating with a diverse audience. The challenges they faced involved teamwork, multicultural and multilingual concerns, a vast array of communication technologies, and a variety of communication channels.

Check out our quarterly newsletters at www.pearsonhighered.com/gerson for dot.com updates, new case studies, insights from business professionals, grammar exercises, and facts about technical communication.

What is Technical Communication?

Technical communication is oral and written communication for and about business and industry. Technical communication focuses on products and services—how to manufacture, market, manage, deliver, and use them.

Technical communication is composed primarily in the work environment for supervisors, colleagues, subordinates, vendors, and customers. As either a professional technical communicator, an employee at a company, or a consumer, you can expect to write the following types of correspondence for the following reasons (and many more):

- As a computer information systems (CIS) employee, you work at a 1-800 hotline helpdesk. A call comes from a concerned customer. Your job is to answer that client's questions and follow up with a one-page *e-mail* documenting the problem and your responses.

- As a social media manager, you market your company by using *blogs* and *tweets*, maintaining a personal *Web site*, and having a presence on *Facebook, MySpace, LinkedIn, Delicious, FriendFeed*, and *YouTube*. Using social media allows you to make business contacts, find new clients, manage the company's brand, and hear and communicate with the expansive constituency of the online community.

- You are a technical communicator, working in engineering, biomedical equipment manufacturing, the automotive industry, computer software development, or a variety of other job areas. Your job is to write *user manuals* to explain the steps for building a piece of equipment, performing preventative maintenance, or for shipping and handling procedures.

- As a trust officer in a bank, one of your jobs is to make *proposals* to potential clients. To do so, you must write a 20- to 30-page proposal about your bank's services.

- You are a customer. You ordered an automotive part from a national manufacturer. Unfortunately, the part was shipped to you five days later than promised, it arrived broken, and you were charged more than the agreed-upon price. You need to write a *letter of complaint*.

- As the manager of a medical records reporting department, one of your major responsibilities is ensuring that your staff's training is up to date. After all,

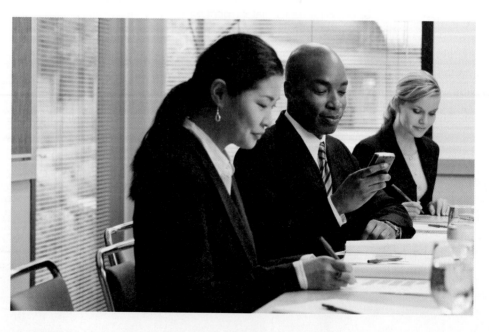

insurance rules and regulations keep changing. To document your department's compliance, you must write a monthly *progress report* to upper-level management.

- You are a webmaster. Your job is creating a corporate *Web site*, complete with *online help screens* and *instructional videos*. The Web site gives clients information about your locations, pricing, products and services, mission statement, and job openings. The help screens provide easy-to-access answers for both customer and employee questions. The instructional videos show customers and employees step-by-step procedures.

- As an entrepreneur, you are opening your own computer-maintenance service (or services for HVAC repair, deck rebuilding, home construction, lawn care, or automotive maintenance). To market your company, you will need to write *fliers*, *brochures*, *sales letters*, and post a *YouTube video* about your business.

- You have just graduated from college (or you have just been laid off). It's time to get a job. You need to write a *resume* and a *letter of application* to show corporations what assets you will bring to their company.

Communication Channels

Technical communication takes many different forms. Not only will you communicate both orally and in writing, but also you will rely on various types of correspondence and technology, dependent upon the audience, purpose, and situation. To communicate successfully in the workplace, you must adapt to many different channels of communication.

Table 1.1 gives you examples of different communication channels, both oral and written.

Table 1.2 illustrates how different writers and speakers might use various channels to communicate effectively to both internal and external audiences. Internal audiences consist of the co-workers, subordinates, and supervisors in your workplace; external audiences consist of vendors, customers, and other workplace professionals.

Many communication channels overlap in terms of purpose and audience. If you are requesting information from a vendor, for example, you could write a letter, send an e-mail message, or make a telephone call. However, in other instances, communication channels are more exclusive. You would not want to communicate bad news—such as layoffs, loss of benefits, or corporate closings—to employees by way of mass e-mail messages or

TABLE 1.1 Communication Channels

Written Communication Channels	Oral Communication Channels
<ul><li>E-mail</li><li>Memos</li><li>Letters</li><li>Reports</li><li>Proposals</li><li>Fliers</li><li>Brochures</li><li>Faxes</li><li>Web sites</li><li>Instant messages and text messages</li><li>Blogs</li><li>YouTube</li><li>Facebook</li><li>Twitter</li><li>Job information (resumes, letters of application, follow-up letters, interviews)</li></ul>	<ul><li>Leading meetings</li><li>Conducting interviews</li><li>Making sales calls</li><li>Managing others</li><li>Participating in teleconferences and videoconferences</li><li>Facilitating training sessions</li><li>Participating in collaborative team projects</li><li>Providing customer service</li><li>Making telephone calls</li><li>Leaving voice mail messages</li><li>Making presentations at conferences or to civic organizations</li><li>Participating in interpersonal communication at work</li><li>Conducting performance reviews</li></ul>

TABLE 1.2 Communication Channels—Audience and Purpose

Writers/ Speakers	Type of Communication and Communication Channel	Purposes	Internal or External
Human resources/training	Instructions—either hard copy or online	Help employees and staff perform tasks	Internal
Marketing personnel	Brochures, sales letters, blogs, and phone calls	Promote new products or services	External
Customers	Inquiry or complaint letters or phone calls	Ask about/complain about products or services	External
Quality assurance	Investigative, incident, or progress reports	Report to regulatory agencies about events	External
Vendors	Newsletters, phone calls, and e-mail	Update clients on new prices, products, or services	External
Management	Oral presentations to department staff	Update employees on mergers, acquisitions, layoffs, raises, or site relocations	Internal
Everyone (management, employees, clients, vendors, governmental agencies)	E-mail, blogging, Web sites, instant messages, text messages, and other forms of social media	All conceivable purposes	Internal and external

televised reports. In these instances, face-to-face meetings would be more appropriate. A key to successful technical communication is choosing the right channel.

To clarify the use of different technical communication channels, look at Figure 1.1. In a survey of approximately 120 companies employing over 8 million people, the National Commission on Writing for America's Families, Schools, and Colleges found that employees "almost always" use many forms of writing, including e-mail messages, PowerPoint, memos, letters, and reports ("Writing: A Ticket to Work" 11).

FIGURE 1.1 Channels "Almost Always" Used in Workplace Communication

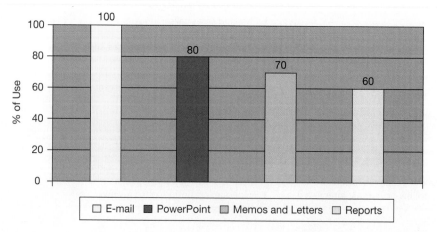

The Importance of Technical Communication

To succeed on the job, you need to write and speak effectively to others—constantly. Gaston Caperton, president of the National Commission on Writing for America's Families, Schools, and Colleges, says, "Writing is a fundamental professional skill. Most of the new jobs in the years ahead will emphasize writing. If students want professional work in service firms, in banking, finance, insurance, and real estate, they must know how to communicate on paper clearly and concisely" ("Writing Skills").

The National Commission on Writing concluded that "in today's workplace writing is a 'threshold skill' for hiring and promotion among salaried . . . employees. Survey results indicate that writing is a ticket to professional opportunity, while poorly written job applications are a figurative kiss of death" ("Writing: A Ticket to Work" 3). Technical communication is a significant factor in your work experience for several reasons.

Business

Technical communication is not a frill or an occasional endeavor. It is a major component of the work environment. Through technical correspondence, employees

- Maintain good customer-client relations (follow-up letters)
- Ensure that work is accomplished on time (directive memos or e-mail)
- Provide documentation that work has been completed (progress reports)
- Generate income (sales letters, brochures, fliers)
- Provide quick updates or answers to work-related questions (instant messages and text messages)
- Keep machinery working (user manuals)
- Ensure that correct equipment is purchased (technical descriptions)
- Participate in teleconferences or videoconferences (oral communication)
- Get a job (resumes)
- Define terminology (online help screens)
- Inform the world about your company's products and services (Web sites and blogs)
- Allow opportunities for collaboration (wikis)

Time

In addition to serving valuable purposes on the job, getting a job, or meeting your needs as a customer, communication is important because it is time consuming. Just imagine how much of your time at work will be spent communicating with others.

Across professions such as engineering, finance, accounting, management, marketing and sales, and computer programming, approximately 80 percent of workers spend more than 30 percent of their time writing. Figure 1.2 shows that 50 percent of the respondents to a survey spend between 11 and 30 percent of their time writing. Another 31 percent of respondents spend between 31 and 50 percent of their time writing (Miller 10).

Generally speaking, new hires might spend less time writing on the job. As a supervisor, you will spend more time directing your subordinates through written correspondence. Though you will spend a great deal of time writing on the job, you will spend even more time communicating in other ways. Calculate the time you will spend verbally communicating in meetings, on the phone with vendors and clients, walking to and from the elevator with your co-workers, and on collaborative work teams while discussing how to complete a project. You will use oral communication skills when speaking to a customer in your office or in the showroom. You will need to convince your boss to let you miss two workdays while you coach your child's soccer team at an out-of-town tournament. You will need to communicate face-to-face constantly with co-workers. You also will need to use effective oral

FIGURE 1.2   Amount of Time Spent Writing on the job

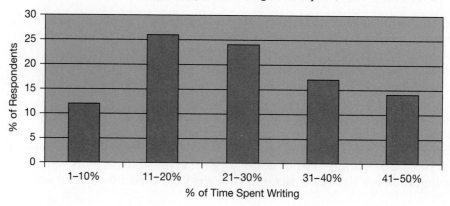

communication when you represent your company at the local speaker's bureau. When you are not writing at work, you will be speaking, listening, and reading much of the time.

Money

You have heard it before—time is money. Here are four simple ways of looking at the cost of your technical communication:

- **Cost of correspondence.** A recent study by Dartnell's Institute of Business Research says that the "average cost of producing and mailing a letter is $19.92" ("Business Identity"). That amount factors in the time it takes a worker to write the letter as well as the cost of the paper, printing, and stamp. If one letter costs almost $20, imagine how much an entire company's correspondence might cost annually, including every employee's e-mail, letters, memos, and reports.

- **Percentage of salary.** Consider how much of your salary is being paid for your communication skills. Let's say you make $35,000 a year. If you are spending 30 percent of your time writing (as many employees do), then your company is paying you approximately $10,500 just to write. That does not include the additional time you spend on oral communication.

  If you are not communicating effectively on the job, then you are asking your bosses to pay you a lot of money for substandard work. Your time spent communicating, both in writing and orally, is part of your salary—and part of your company's expenditures.

- **Cost of training.** Corporations spend money to improve their employees' writing skills. The National Commission on Writing reported that "more than 40 percent of responding firms offer or require training for salaried employees with writing deficiencies. "We're likely to send out 200 to 300 people annually for skills upgrade courses like "business writing" or "technical writing," said one respondent." Based on survey responses, the Commission estimates that remedying deficiencies in writing costs American corporations as much as $3.1 billion annually" ("Writing Skills").

- **Generating income.** Your communication skills do more than just cost the company money; these talents can earn money for both you and the company. A well-written sales letter, flier, brochure, proposal, or Web site can generate corporate income. Effectively written newsletters to clients and stakeholders can keep customers happy and bring in new clients. Good written communication is not just part of your salary—it helps pay your wages.

Interpersonal Communication

A major component of a successful company is the environment it develops, the tone it expresses, and the atmosphere it creates. Successful companies know that effective communication, both written and oral, creates a better workplace. These "soft skills" make customers want to shop with you and employees work for you.

Your technical communication reflects something about you. E-mail messages, Facebook sites, corporate blogs, YouTube videos, or telephone skills are a photograph of you and your company. If you communicate well, you are telling your audience that you can think logically and express your thoughts clearly. When your writing is grammatically correct, or when your telephone tone of voice is calm and knowledgeable, you seem professional to your audience. Technical communication is an extension of your interpersonal communication skills. Co-workers or customers will judge your competence and professionalism based on what you say and how you say it.

The Importance of Teamwork

Companies have found that teamwork enhances productivity. Teammates help and learn from each other by providing checks and balances. Through teamwork, employees can develop open lines of communication to ensure that projects are completed successfully.

Collaboration

In business and industry, many user manuals, reports, proposals, PowerPoint presentations, and Web sites are team written. Teams consist of engineers, graphic artists, marketing specialists, and corporate employees in legal, delivery, production, sales, accounting, and management. These collaborative team projects extend beyond the company. A corporate team also will work with subcontractors from other corporations. The collaborative efforts include communicating with companies in other cities and countries through teleconferences, faxes, and e-mail. Modern technical communication requires the participation of "communities of practice": formal and informal networks of people who collaborate on projects based on common goals, interests, initiatives, and activities (Fisher and Bennion 278).

The National Association of Colleges and Employers lists the "Top Ten Qualities/Skills Employers Want" (see Table 1.3). Notice how interpersonal, teamwork, oral communication,

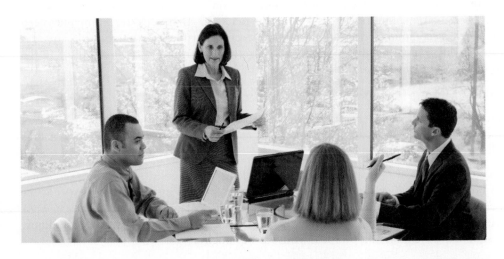

TABLE 1.3 Top Ten Qualities/Skills Employers Want

Skill	Rating
1. Communication skills (verbal and written)	4.7
2. Honesty/integrity	4.7
3. Teamwork skills (work well with others)	4.6
4. Strong work ethic	4.5
5. Analytical skills	4.4
6. Flexibility/adaptability	4.4
7. Interpersonal skills (relate well to others)	4.4
8. Motivation/initiative	4.4
9. Computer skills	4.3
10. Detail oriented	4.1

(5-point scale, where 1 = not at all important and 5 = extremely important)

Source: National Association of Colleges and Employers: Job Outlook 2006 Student Version

and written communication skills take precedence over other skills. Collaboration ranks so highly for employers because teamwork helps employees accomplish the following goals:

- Brainstorm for new ideas and consensus
- Make decisions
- Solve problems
- Determine team roles
- Assign tasks
- Complete work by team-determined deadlines

The Problems with "Silo Building"

Working well with others requires collaboration versus "silo building." The silo has become a metaphor for departments and employees who behave as if they have no responsibilities outside their areas. They build bunkers around themselves, failing to collaborate with others. In addition, they act as if no other department's concerns or opinions are valuable.

Such stand-alone departments or people isolate themselves from the company as a whole and become inaccessible to other departments. Their limited perspective creates problems. Poor accessibility and poor communication "can cause duplicate efforts, discourage cooperation, and stifle cross-pollination of ideas." To be effective, companies need "open lines of communication within and between departments" (Hughes 9). The successful employee must be able to work collaboratively with others to share ideas. In the workplace, teamwork is essential.

Why Teamwork Is Important

Teamwork benefits employees, corporations, and consumers. By allowing all constituents a voice in project development, teamwork helps to create effective workplaces and ensures product integrity.

Diversity of Opinion.   When you look at problems individually, you tend to see issues from limited perspectives—*yours*. In contrast, teams offer many points of view. For instance, if a team has members from accounting, public relations, customer service, engineering, and information technology, then that diverse group can offer diverse opinions. You should always look at a problem from various angles.

Using a Variety of Communication Channels to Achieve Collaboration

Rob Studin is the Executive Director of Financial Advisory Services for Lincoln Financial Advisors (LFA). With a home office in Philadelphia, LFA's 3,000 advisors and employees provide fee-based financial planning for clients, including "Estate Planning, Investment Planning, Retirement Planning Strategies, and Business Owner Planning" (http://www.lfg.com/).

LFA uses four electronic oral communication channels to ensure a consistent, collaborative workforce: teleconferences, videoconferences, webinars, and LFA's Virtual University.

- **Teleconferences.** Rob, who works in LFA's Birmingham office, has six key managers who work in San Francisco, Salt Lake City, Cleveland, Columbus, Rochester, and Baltimore. To communicate with dispersed team members, Rob states, "We have a conference call just to touch base. Sometimes we have a formal agenda, and sometimes I just ask, 'What's going on?' A casual, weekly teleconference allows us to stay up to date on issues facing us individually or as a group. We collectively understand that six heads are better than one for problem solving."

- **Videoconferences.** You can't communicate effectively with 3,000 people on the telephone. While teleconferencing works well for Rob and smaller groups, when LFA needs to communicate to all of its employees and/or advisors about corporate-wide issues that affect policy, budget, personnel, and strategic planning, face-to-face meetings might be the optimum solution. However, transporting 3,000 people to a central location is neither time-efficient nor cost-effective. A three-hour meeting might require two days of travel plus hotel, food, and air fares. To save time and money, LFA uses videoconferences.

- **Webinars.** Videoconferences create at least three challenges for LFA employees. First, to participate in a videoconference, the employees must be in a room fitted with LFA's companywide videoconference system. Second, if many people are in the audience and seated at a distance from the TV screen, visibility/readability can be an obstacle. Finally, videoconferences don't allow the audience any hands-on opportunity to practice new skills. Webinars solve these corporate communication challenges. To participate in an online seminar, employees only need a computer.

- **LFA's Virtual University.** Most of LFA's webinars are synchronous. Employees are asked to log on at a given time while a webinar host runs the training program. Inevitably, however, an employee can't participate in the webinar when it is initially presented. LFA has solved this problem. Their Virtual University offers asynchronous "training on demand." All training videoconferences and all webinars are recorded and archived. By accessing LFA's password-protected www.LFAplanner.com site, advisors and employees can retrieve training materials at their convenience.

Rob says that he spends approximately 50 percent of his work time communicating via e-mail messages, telephone calls, and teleconferences. For efficiency, cost savings, and consistent communication to a geographically dispersed workforce, LFA has found that multiple, electronic channels help team members achieve their communication goals.

Checks and Balances. Diversity of opinion also provides the added benefit of checks and balances. Rarely should one individual or one department determine outcomes. When a team consists of members from different disciplines, those members can say, "Wait a minute. Your idea will negatively impact my department. We better stop and reconsider."

Broad-Based Understanding. If decisions are made in a silo, by a small group of like-minded individuals, then these conclusions might surprise others in the company. Surprises are rarely good. You always want buy-in from the majority of your stakeholders. An excellent way to achieve this is through team projects. When multiple points of view are shared, a company benefits from broad-based knowledge. Improved communication allows people to see the bigger picture.

Empowerment. Collaboration gives people from varied disciplines an opportunity to provide their input. When groups are involved in the decision-making process, they have a stake in the project. This allows for better morale and productivity.

Team Building. Everyone in a company should have the same goals—corporate success, customer satisfaction, and quality production. Team projects encourage shared visions, a better work environment, a greater sense of collegiality, and improved performance. Employees can say, "We are all in this together, working toward a common goal."

Diverse Teams . . . Dispersed Teams in a Global Economy

Collaborative projects will depend on diverse team members and dispersed team members.

Diversity

See Chapter 4 for more discussion of diversity in the workplace.

Diverse Teams

Teams will be diverse, consisting of people from different areas of expertise. Your teams will be made up of engineers, graphic artists, accountants, technical communicators, financial advisors, human resource employees, and others. In addition, the team will consist of people who are different ages, genders, cultures, and races.

Dispersed Teams

In a global economy, members of a team project might not be able to work together face to face. Team members might be located across time and space. They could work in different cities, states, time zones, countries, or different shifts. For example, you might work for your company in New York, while members of your team work for the company at other sites in Chicago, Denver, and Los Angeles. This challenge to collaboration is compounded when you also must team with employees at your company's sites in India, Mexico, France, and Japan. According to a recent report, 41 percent of employees at the top international corporations live outside the borders of their company's home country (Nesbitt and Bagley-Woodward 25). As companies "get comfortable with new technologies," these companies will become mobile enterprises. "Forrester [Research] defines a mobile enterprise as one that can connect and control suppliers, partners, employees, assets, products, and customers from any location" ("Achieving the Promise").

Using Collaborative Software (Groupware) in Virtual Teams

When employees are dispersed geographically, getting all team members together would be costly in terms of time and money. Companies solve this problem by forming virtual, remote teams that collaborate using electronic communication tools called groupware. Groupware consists of software and hardware that helps companies reduce travel costs, allows for telecommuting, and facilitates communication for employees located in different cities and countries.

Groupware includes the following types of software (Nesbitt and Bagley-Woodward 28):

- Electronic conferencing tools such as webinars, listservs, chat systems, message and discussion boards, videoconferences, and teleconferences.
- Electronic management tools, such as Digital Dashboards, that schedule, track, and chart the steps in a project. Another example of an electronic management tool is Microsoft Outlook's electronic calendaring. This allows you to send a meeting request to dispersed team members, check the availability of meeting attendees, reschedule meetings electronically, forward meeting requests, and cancel a meeting—without ever visiting with your team member face-to-face.
- Electronic communication tools for writing and sending documents. These include tools like instant messages, e-mail, blogs, intranets and extranets, and wikis.
- Social media for communication of text and videos through Facebook, YouTube, Twitter, virtual worlds (Second Life), and more.

Collaborative Writing Tools

Social Media

See Chapter 7 for more discussion of social media.

Wikis.   What's a wiki? A wiki "is a website that allows the visitors . . . to easily add, remove, and otherwise edit and change available content, and typically without the need for registration. This ease of interaction and operation makes a wiki an effective tool for mass collaborative authoring" ("Wiki"). In addition, wikis let collaborative writers track "the history of a document as it is revised." Whenever a team member edits text in the "wiki," that new text becomes the current version, while older versions are stored" (Mader "Using Wiki").

Who's Using Wikis?   Many companies use wikis for collaborative writing projects. Yahoo, for example, uses a wiki. Eric Baldeschwieler, director of software development of Yahoo!, says, "Our development team includes hundreds of people in various locations all over the world, so web collaboration is VERY important to us." Cmed runs pharmaceutical clinical trials and develops new technology. In this heavily regulated environment, wikis "improved communication and increased the quantity (and through peer review, the quality) of documentation." At Cingular Wireless, "project managers have been encouraged to utilize the site for any issues needing collaborating efforts in lieu of emails." Disney uses a wiki "for an engineering team that was rearchitecting the Go.com portal. During this time we found TWiki to be a very effective means of posting and maintaining development specs and notes as well as pointers to resources." Texas Instrument's "India design centre" uses wikis to manage all project-specific information, such as documenting ideas, plans, and status; sharing information with other teams across various worksites; and updating information and content to team members ("TWiki").

How Can You Use a Wiki?   In your dispersed teams, whether virtual, remote, or mobile, you might use wikis in the following ways to create collaborative documents:

- **Create websites.** Wikis help team members easily add pages, insert graphics, create hyperlinks, and add navigation.
- **Project development with peer review.** A wiki makes it easy for team members to write, revise, and submit projects, since all three activities can take place in the wiki.
- **Group authoring.** Wikis allow group members to build and edit a document. This creates a sense of community within a group, allows group members to build on each other's work, and provides immediate, asynchronous access to all versions of a document.
- **Track group projects.** Each wiki page lets you track how group members are developing their contributions. The wiki also lets you give feedback and suggest editorial changes.

Benefits of Using Wikis.   Often in the business world, people are only familiar with wikis through Wikipedia, the open-source, online encyclopedia. Because it is an open source, anyone can write and revise content. The notion that something is an open source can frighten some business professionals. However, a wiki offers many benefits to collaboration in the workplace, including the following:

1. **Involve all team members in the writing process.** When many individuals brainstorm ideas and participate in writing, revising, and proofreading, this can lead to an improved document.
2. **Minimize the need for face-to-face meetings.** An asynchronous wiki lets team members work on a project anytime and anywhere. Rather than having to attend a meeting, team members can receive "updates from the wiki via RSS or e-mail" (Mader "Your Wiki"). Because a wiki allows for detailed revisions and discussions, meetings can be shorter and concentrate only on the topics that would benefit from face-to-face communication.
3. **Limit excessive and lengthy e-mail messages.** When content is available for threaded discussion within the wiki, the need for a constant flow of e-mail messages about the project diminishes.

4. **Transmit the finished product to end users.** The content in the wiki can be shared with customers or other individuals via Word documents or PDF files.

5. **Organize the team's work and maintain privacy.** A wiki allows you and your team to have in one location all of the material related to a project. In the wiki you can designate which content is viewable by team members. You also can share designated information with anyone in the company (Mader "Your Wiki").

Figures 1.3, 1.4, and 1.5 are illustrations of a wiki, edited text in a wiki, and a record of edited versions.

FIGURE 1.3   Wiki Page

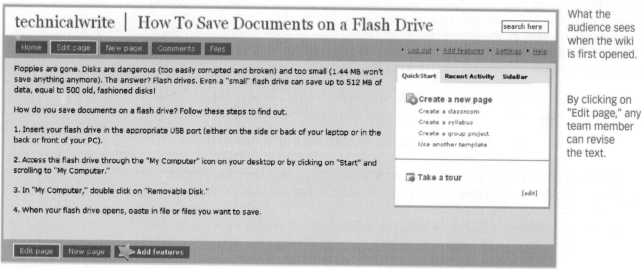

Source: Courtesy of PBWiki

What the audience sees when the wiki is first opened.

By clicking on "Edit page," any team member can revise the text.

FIGURE 1.4   Wiki in "Edit" Mode

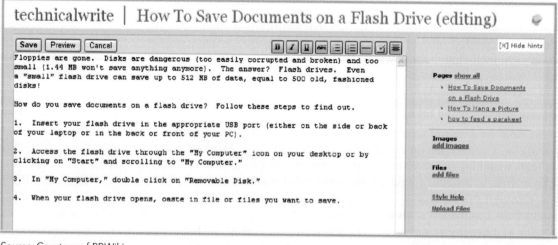

Source: Courtesy of PBWiki

The "editing" screen allows team members to add, delete, and enhance text (by bolding, italicizing, underlining, adding bullets, etc.) just as in a Word document.

FIGURE 1.5 Record of Edited Versions

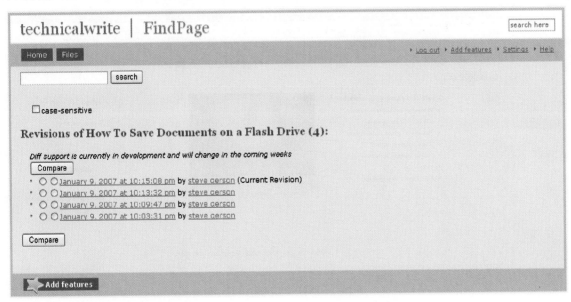

Source: Courtesy of PBWiki

This screen allows team members to compare all versions of the text, thus seeing what has been added, deleted, or enhanced.

Google Docs, Google Sites, and Google Wave. Other collaborative writing tools you can use easily are Google Docs, Google Sites, and Google Wave. These online tools, free to anyone with a Web browser and an Internet connection, are useful for document sharing, collaborating on group projects, and publishing to the Internet.

Google Docs. Using Google Docs, you and group members can edit Word documents, rich text format (RTF), and hypertext markup language (HTML) files. Teams can be at any location on their computers and work on one document simultaneously. Changes made by one writer will be seen by all team members instantly.

Google Docs (www.docs.google.com) provides you these benefits:

- Upload and save files
- Edit and view a document
- Show changes in real time
- View a document's revision history
- Return to earlier versions
- Add new team members or delete writers
- Choose who can access your site
- Post documents to a blog or publish a document to a Web page

See Figure 1.6 for an example of Google Docs applications.

Google Sites. Google defines this collaborative writing tool as "a free and easy way to create and share webpages." Through Google Sites, you can "create rich web pages easily, collect all your info in one place, and control who can view and edit" (www.google.com/sites). See Figures 1.7, 1.8, and 1.9 for samples of Google Sites screen captures.

FIGURE 1.6   Google Docs Homepage

From the Google Docs home page, you can create a secure, collaborative workspace.

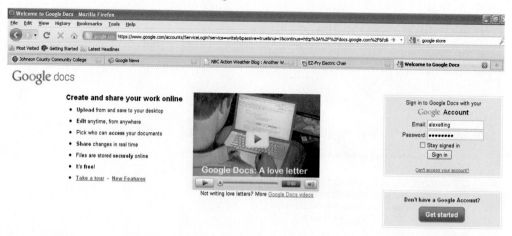

Source: Courtesy of Alex Elting

Google Wave.   Google Wave is a "mashup," a Web development application that combines "two or more external sources to create a new service" ("Mashup"). Google Wave allows for "instant messaging, blogs, forums, tweets, social media, file-sharing, [and] e-mail" (Reddick). With combined services, Google Wave is perfect "for real-time communication and collaboration. A wave can be both a conversation and a document where people can discuss and work together using richly formatted text, photos, videos, maps, and more" ("About Google Wave").

Google Wave offers these benefits for collaboration:

- **Shared content.** Participants on a team project can reply to parts of a document, edit content, add participants, and "playback" content by rewinding "the wave to

FIGURE 1.7   Google Sites Customized Homepage

Google Sites lets you customize a design for your collaborative writing site.

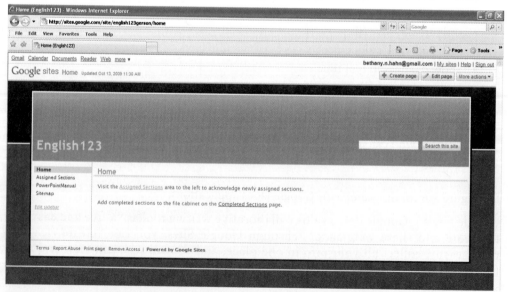

Source: Courtesy of Bethany Hahn

FIGURE 1.8 Google Sites File Cabinet

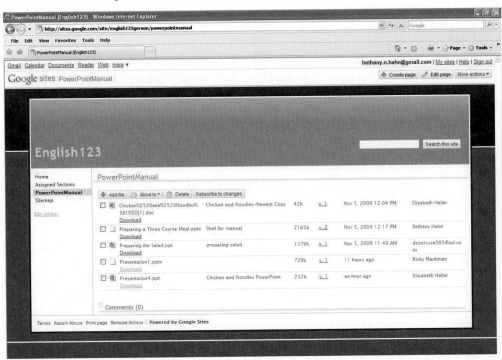

Google Sites lets you create a file cabinet into which you can download files, note which version of text you're working on, date the text, and show who has authored a document.

Source: Courtesy of Bethany Hahn

FIGURE 1.9 Google Sites Task Screen

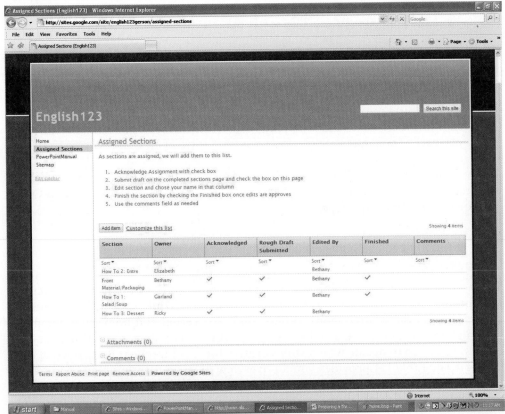

Google Sites lets you assign tasks. It also lets you record when work has been submitted and edited.

Source: Courtesy of Bethany Hahn

see who said what and when" (Trapani). In this sense, Google Wave allows for wiki functionality.

- **Live transmission.** You can see what a team member is typing, letter by letter. Members of your team can see edits and can converse in real time.
- **One-stop media.** With its mashup capabilities, Google Wave allows collaborators to brainstorm, converse, view graphics, draft content, revise, archive, and review feedback in one place.
- **Autocorrects.** Wave can spell-check text.
- **Drag and drop.** Instead of adding attachments, you can drag and drop files into Wave, including graphics.
- **Personalization.** On Wave's navigation bar, you can include a list of your contacts and team members, complete with their photographs. This feature is similar to Facebook and personalizes Wave's e-mail and instant messaging capabilities.

Employees are finding value to Google Wave. One consultant from the Center for Disease Control (CDC) concluded that Google Wave could improve the delivery of H1N1 vaccines. Located at the Atlanta Department of Health and Human Services, the consultant works with a team of individuals in Washington, DC, and Dallas. Without Google Wave, he has had to use e-mail, intranet sites, shared drives, and face-to-face meetings to update plans and report cases relating to the H1N1 virus. Doing so leads to "duplicated work and wasted time." In contrast, with Google Wave, the consultant knows that he could improve the timeliness and efficiency of documentation and communication (Trapani).

An air traffic coordinator at the Philadelphia International Airport concluded that Google Wave could help reduce departure delays and maintain air traffic efficiency. Through the mashup of text and graphics, the Philadelphia International Airport can use Google Wave as follows:

- Google Spreadsheet can communicate real-time departure information, weather concerns, construction updates, and snow removal plans.
- Wave can embed real-time weather videos and graphics.
- Chat features can create robust communication capabilities, allowing the airport staff to discuss issues and archive the conversations for documentation and playback (Trapani).

In addition to Google Documents, Google Sites, and Google Wave, other collaborative sites include WetPaint, PBWiki, Wikispaces, Access Grid, eGroupWare, Group-Office, Mindquarry, TikiWiki, and Trac.

Choosing the Appropriate Channel for Collaboration

Technologies for collaboration are not equal. Some allow for more interactivity than others. Some are more synchronous than others. While some technologies depend solely on the written word, others allow for verbal and nonverbal communication among team members. When deciding which technology to use for your team efforts, you need to decide which collaborative tools aid frequent and effective communication. Table 1.4 provides you an overview of various technology channels for collaboration.

TABLE 1.4 Technology Channels for Collaboration

Collaboration Channel	Synchronous or Asynchronous	Written, Verbal, or Nonverbal	Interaction
Teleconference	Synchronous. Team members can have real-time, spontaneous communication.	All participants can be seen and heard. They can write, read, speak, and convey nonverbal cues (frowns, laughter, gestures, etc.).	All team members can participate equally with feedback and suggestions.
Webinar	Synchronous. Participants can engage in real-time, spontaneous communication.	All participants can talk and share written communication. Nonverbal communication is limited unless participants have web cameras.	All team members can participate equally with feedback and suggestions.
E-mail, Text Message, Instant Message, and Twitter	Near synchronous. Participants can engage in near real-time communication if everyone is online or has his or her cell phone turned on.	All participants can share written communication. Nonverbal communication is limited to emoticons.	Participation is dependent on the team members' willingness to send and receive messages.
Blogs and Wikis	Asynchronous. Team members can access a blog or wiki at their own convenience.	All participants can share written communication. Nonverbal communication is limited to emoticons.	Participation is dependent on the team members' willingness to respond to the blog or wiki posting.

Source: Adapted from Robidoux and Hewett

Challenges to Effective Teamwork

Any collaborative activity is challenging to manage: Team members do not show up for class or work; one student or employee monopolizes the activity while another individual snoozes; people exert varying amounts of enthusiasm and ability; personalities clash. Some people fight over everything. Occasionally, when a boss participates on a team, employees fear speaking openly. Some team members will not stay focused on the subject. One team member will not complete assignments ("Individual's and Teams' Roles and Responsibilities"). Group dynamics are difficult and can lead to performance gaps.

Human Performance Improvement

Human Performance Improvement (HPI) focuses on "root cause analysis" to assess and overcome the barriers inherent in teamwork. To close performance gaps, HPI analyzes the following possible causes for collaborative breakdowns:

1. **Knowledge.** Perhaps employees do not know how to perform a task. They have never acquired the knowledge or do not understand which skills are needed to complete the specific job. Varying skills of team members can impede the group's progress.

2. **Resources.** Think of these possibilities: Tools are broken or missing; the department is out of funds; you do not have enough personnel to do the job; the raw material needed for the job is below par; you ordered one piece of machinery but were shipped something different; you needed 100 items but have only 50 in stock. To complete a project, you often have to solve problems with resources.

3. **Processes.** For teams to succeed in collaborative projects, everyone must know his or her responsibilities. Who reports to whom? How will these reports be handled (orally, in writing)? Who does what job? Are responsibilities shared equally? Structure, of some sort, is needed to avoid chaos, lost time, inefficiency, hurt feelings, and many other challenges to teamwork. To achieve successful collaboration, the team should set and maintain effective procedures.

4. **Information.** A team needs up-to-date and accurate information to function well. If required database information is late or incorrect, then the team will falter. If the information is too high tech for some of the team members, a lack of understanding may undermine the team effort.

5. **Support.** To succeed in any project, a team needs support, such as financial, attitudinal, or managerial. When managers from different departments are fighting "turf wars" over ownership of a project, teams cannot succeed. Teams need enough money for staffing, personnel, or equipment.

6. **Wellness.** Another consideration involves the team's health and well-being. People get sick or miss work for health reasons. People have car accidents. If a teammate must miss work for a day or an extended period, this will negatively impact the team's productivity. Stress and absences can lead to arguments, missed deadlines, erratic work schedules, and poor quality.

7. **Acknowledgment.** For teams to succeed, management needs to recognize the benefits and results of collaboration. Successful workplaces reward staff with promotions, merit raises, bonuses, and awards.

8. **Trust.** A team is only as good as the trust each member has for the team effort. Team members who trust each other are open in their brainstorming, complimentary of each other's efforts, and work effectively toward a common goal.

9. **Leadership.** A team needs direction. Successful teams work well when their team leaders provide clear deadlines, responsibilities, task goals, and fair-minded leadership.

10. **Efficient communication.** If everyone on the team is overwhelmed with e-mail messages, text messages, tweets, or instant messages, no work will get done. Too much information is as bad as no information. To help keep team members on track, avoid overloading them with excessive communication.

HPI Intervention Techniques. After assessing root causes that challenge a team's success, HPI creates intervention options. These might include the following:

- Improved compensation packages
- Employee recognition programs
- Revised performance appraisals
- Improved employee training
- Simulations
- Mentoring or coaching
- Restructured work environments to enhance ergonomics
- Safety implementations
- Strategic planning changes
- Improved communication channels
- Health and wellness options—lectures, on-site fitness consultants, incentives for weight loss, and therapist and social worker interventions

People need help in order to work effectively with each other. A progressive company recognizes these challenges and steps in to help.

Conflict Resolution in Collaborative Projects

To ensure that team members work well together and that projects are completed successfully, consider these approaches to conflict resolution:

1. **Choose a team leader.** Sometimes, team leaders are chosen by management; sometimes, team leaders emerge from the group by consensus. However this person gains the position, he or she can serve many valuable purposes. The team leader becomes "point person," the individual whom all can turn to for assistance. He or she can solve problems, seek additional resources, or organize the team effort. For instance, a team leader can give the team direction, interface with management, and/or act as the team's mediator.

2. **Set guidelines.** One reason that conflicts occur is because people do not know what to expect or what is expected of them. On the other hand, if expectations are clear, then several major sources of conflict can be resolved. For example, one simple conflict might be related to time. A team member could be unaware of when the meeting will end and schedules another meeting. If that team member then has to leave the first meeting early, disrupting the team's progress, this can cause a conflict.

   To solve this problem, set guidelines. Hold an initial meeting (online or teleconferenced for remote, virtual teams) to define goals and establish guidelines, establish project milestones, or create schedules for synchronous dialogues. Communicate to all team members (before the meeting via e-mail or early in a project) how long the project will last. Also, clarify the team's goals, the chain of command (if one exists), and each team member's responsibilities. Use text messages, instant messages, or Twitter messages for frequent updates.

3. **Ensure that all team members have compatible hardware and software.** This is especially important for virtual, remote teams. To communicate successfully, all team members need access to the same e-mail platform. Some members cannot use Yahoo or MSN or Hotmail while others use Outlook. This would cause communication challenges if software is incompatible. The problem is further heightened when video or telephone equipment is different.

4. **Encourage equal discussion and involvement.** A team's success demands that everyone participate. A team leader should encourage involvement and discussion. All team members should be mutually accountable for team results, including planning, writing, editing, proofreading, and packaging the finished project. Be sure that everyone is allowed a chance to give input by participating orally, in writing, or in a wiki.

   Conflicts also arise when one person monopolizes the work. If one person speaks excessively, others will feel left out and disregarded. A team leader should ensure equal participation. He or she should call on others for their opinions and ask for additional input from the team. In addition, team leaders must limit an overly aggressive team member's participation by saying, "Thanks, John, for your comments. Now, let's see what others have to say." Or, "Wait one second, John. I'll come back to you after we've heard from a few others."

5. **Discourage taking sides.** Discussion is necessary, but conflict will arise if team members take sides. An "us against them" mentality will harm the team effort. You can avoid this pitfall by seeking consensus, tabling issues, or creating subcommittees.

6. **Seek consensus.** Not every member of the team needs to agree on a course of action. However, a team cannot go forward without majority approval. To achieve consensus, your job as team leader is to listen to everyone's opinion, seek compromise, and value diversity. Conflict can be resolved by allowing everyone a chance to speak. Once everyone has spoken, then take a vote.

7. **Table topics when necessary.** If an issue is so controversial that it cannot be agreed upon, take a time-out. Tell the team, "Let's break for a few minutes. Then we can reconvene with fresh perspectives." Maybe you need to table the topic for the next meeting. Sometimes, conflicts need a cooling-off period.

8. **Create subcommittees.** If a topic cannot be resolved, teammates are at odds, or sides are being taken, create a subcommittee to resolve the conflict. Let a smaller group tackle the issue and report back to the larger team.

9. **Find the good in the bad.** Occasionally, one team member comes to a meeting with an agenda. This person does not agree with the way things have been handled in the past or the way things are being handled presently. You do not agree, nor do other team members. However, you cannot resolve this issue simply by saying, "That's not how we do things." A disgruntled team member will not accept such a limited viewpoint.

    As team leader, seek compromise. Let the challenging team member speak. Discuss each of the points of dissension. Allow for input from the team. Some of the ideas might have more merit than you originally assumed.

10. **Deal with people individually.** From time to time, a team member will cause problems for the group. The teammate might speak out of turn or say inappropriate things. These could include off-color or off-topic comments. A team member might cause problems for the group by habitually showing up late, missing meetings, or monopolizing discussions.

    To handle these conflicts, avoid pointing a finger of blame at this person during the meeting. Do not react aggressively or impatiently. Doing so will lead to several problems:

    - Your reaction might call more attention to this person. Sometimes people come to meetings late or speak out in a group *just* to get attention. If you react, you might give the individual exactly what he or she wants.
    - Your reaction might embarrass this person.
    - Your reaction might make you look unprofessional.
    - Your reaction might deter others from speaking out. You want an open environment, allowing for a free exchange of ideas.

    Speak to any offending team members individually. This could be accomplished at a later date, in your office, or during a coffee break. Speaking to the person later and individually might defuse the conflict.

11. **Stay calm.** Act professionally when dealing with conflict. To resolve conflicts, speak slowly, keep your voice steady and quiet, and stay seated (rising will look too aggressive). You also might want to take notes. This will provide you with a record of the discussion.

CHECKLIST FOR COLLABORATION

1. Have you chosen a team leader (or has a team leader been assigned)?
2. Do all participants understand the team's goal and their individual responsibilities?
3. Does the team have a schedule, complete with milestones and target due dates?
4. Does the team have compatible hardware and software for their wiki site or teleconference equipment?
5. In planning the team's project, did you seek consensus?
6. Have all participants been allowed to express themselves?
7. If conflicts occurred, did you table topics for later discussion or additional research?
8. Did you encourage diversity of opinion?
9. Have you avoided confronting people in public, choosing to meet with individuals privately to discuss concerns?
10. Have you remained calm in your interaction with the team?

CHAPTER HIGHLIGHTS

1. Technical communication is written for and about business and industry and focuses on products and services.
2. You will use various channels of communication, dependent upon the audience, purpose, and situation.
3. Technical communication costs a company both time and money, so employees must strive to write effectively.
4. The top five skills employers want include communication skills, honesty, interpersonal skills, a strong work ethic, and teamwork.
5. Avoid "silo building," isolating yourself on the job.
6. Working in teams allows you to see issues from several points of view.
7. Human Performance Improvement (HPI) solves problems inherent in teamwork.
8. Teams face problems including varied knowledge levels, differing motives, and insufficient resources.
9. Conflict resolution strategies are essential to a team's success.
10. To resolve conflicts in a team, you should set guidelines, encourage all to participate, and avoid taking sides.

CASE STUDY

You are the team leader of a work project at Gulfview Architectural and Engineering Services. The team has been involved in this project for a year. During the year, the team has met weekly, every Wednesday at 8:00 A.M. It is now time to assess the team's successes and areas needing improvement.

Your goal will be to recommend changes as needed before the team begins its second year on this project. You have encountered the following problems:

- One team member, Caroline Jensen, misses meetings regularly. In fact, she has missed at least one meeting a month during the past year. Occasionally, she missed two or three in a row. You have met with Caroline to discuss the problem. She says she has had child care issues that have forced her to use the company's flextime option, allowing her to come to work later than usual, at 9:00 A.M.

- Another team member, Guy Stapleton, tends to talk a lot during the meetings. He has good things to say, but he speaks his mind very loudly and interrupts others as they are speaking. He also elaborates on his points in great detail, even when the point has been made. If Guy isn't talking, he's texting.

- A third team member, Sharon Mitchell, almost never provides her input during the meetings. She will e-mail comments later or talk to people during breaks. Her comments are valid and on-topic, but not everyone gets to hear what she says.

- A fourth team member, Craig Mabrito, is very impatient during the meetings. This is evident from his verbal and nonverbal communication. He grunts, slouches, drums on the table, gets up to walk around while others are speaking, and answers his cell phone.

- A fifth employee, Julie Jones, is overly aggressive. She is confrontational, both verbally and physically. Julie points her finger at people when she speaks, raises her voice to drown out others as they speak, and uses sarcasm as a weapon. Julie also crowds people, standing very close to them when speaking.

Assignment

How will you handle these challenges? Try this approach:

- Analyze the problem(s). To do so, brainstorm. What gaps might exist causing these problems?
- Invent or envision solutions. How would you solve the problems? Consider Human Performance Improvement issues, as discussed in this chapter.
- Plan your approach. To do so, establish verifiable measures of success (including time frames and quantifiable actions).

Write an e-mail to your instructor sharing your findings.

INDIVIDUAL AND TEAM PROJECTS

Teamwork—Business and Industry Expectations

Individually or in small groups, visit local banks, hospitals, police or fire stations, city offices, service organizations, manufacturing companies, engineering companies, or architectural firms. Once you and your teammates have visited these sites, asked your questions (see the following assignments), and completed your research, share your findings using one of the following methods:

- **Oral:** As a team, give a three- to five-minute briefing to share with your colleagues the results of your research.
- **Oral:** Invite employee representatives from other work environments to share with your class their responses to your questions.
- **Written:** Write a team e-mail message, letter, or report about your findings.

1. Ask employees at the sites you visit if, how, and how often they are involved in team projects. In your team, assess your findings and report your discoveries.
2. Ask employees at the sites you visit about the challenges they face with conflict resolution. In your team, assess your findings and report your discoveries.
3. Use the Internet and/or your library to research companies that rely on teamwork. Focus on which industries these companies represent and the goals of their team projects. You also could consider the challenges they encounter, their means of resolving conflicts, the numbers of individuals on each team, and whether the teams are diverse. Then report these findings to your professor or classmates, either orally or in writing.
4. Visit the Society for Technical Communication (STC) Web site to learn about its membership. See which industries employ technical communicators and determine these writers' job responsibilities. Also, learn which colleges and universities have programs in technical communication and what the programs entail. What else can you learn about technical communication from the STC Web site?
5. Research major publications of technical communication, such as *Intercom, Technical Communication*, and *The Journal of Scientific and Technical Communication*. What topics do the articles in these journals focus on?

PROBLEM-SOLVING THINK PIECES

To understand and practice conflict resolution, complete the following assignments.

1. Attend a meeting. This could be at your church, synagogue, or mosque; a city council meeting; your school, college, or university's board of trustees meeting; or a meeting at your place of employment. Was the meeting successful? Did it have room for improvement? To help answer these questions, use the following Conflict Resolution in Team Meetings Matrix. Then report your findings to your professor or classmates, either orally or in writing. Write an e-mail message, memo, or report, for example.

CONFLICT RESOLUTION IN TEAM MEETINGS MATRIX			
Goals	Yes	No	Comments
1. Were meeting guidelines clear?			
2. Did the meeting facilitator encourage equal discussion and involvement?			
3. Were the meeting's attendees discouraged from taking sides?			
4. Did the meeting facilitator seek consensus?			
5. Were topics tabled if necessary?			
6. Were subcommittees created if necessary?			
7. Did the meeting facilitator find the good in the bad?			
8. Did the meeting facilitator deal with people individually?			
9. Did the meeting's facilitator stay calm?			

2. Have you been involved in a team project at work or at school? Perhaps you and your classmates grouped to write a proposal, research Web sites, create a Web site, or perform mock job interviews. Maybe you were involved in a team project for another class. Did the team work well together? If so, analyze how and why the team succeeded. If the team did not function effectively, why not? Analyze the gaps between what should have been and what was. To help you with this analysis, use the following Human Performance Index Matrix. Then, report your findings to your instructor or classmates either orally or in writing. Write an e-mail message, memo, or report, for example.

HUMAN PERFORMANCE INDEX MATRIX			
Potential Gaps	Yes	No	Comments
1. Did teammates have equal and appropriate levels of knowledge to complete the task?			
2. Did teammates have equal and appropriate levels of motivation to complete the task?			
3. Did the team have sufficient resources to complete the task?			
4. Did teammates understand their roles in the process needed to complete the task?			
5. Did the team have sufficient and up-to-date information to complete the task?			
6. Did the team have sufficient support to complete the task?			
7. Did wellness issues affect the team's success?			

WEB WORKSHOP

1. How important is technical communication in the workplace? Go online to research this topic. Find five Web sites that discuss the importance of communication in the workplace, and report your discoveries to your teacher and/or class. To do so, write a brief report, memo, or e-mail message. You could also report your information orally.

2. Create a class wiki for collaborative writing. To do so, consider using one of the following sites:
   - [http://www.wikispaces.com/]
   - [http://pbworks.com/]
   - [http://www.wetpaint.com/]
   - http://docs.google.com

3. As a technical communicator, you have access to many different types of collaborative software. Research various types of groupware to find different products and to determine the strengths and weaknesses of each.

QUIZ QUESTIONS

1. Define technical communication.
2. What are five channels of technical communication?
3. List three reasons why technical communication is important in business.
4. What is the percentage of time employees spend writing?
5. What are the top five skills employers want?
6. Define silo building.
7. Explain why business depends on teamwork to ensure quality.
8. List five causes for collaborative breakdown, according to Human Performance Improvement (HPI).
9. List four HPI intervention options to help a team solve its problems.
10. List three things a team leader can do to ensure successful teamwork.

CHAPTER TWO

The Communication Process: Prewriting, Writing, and Rewriting

COMMUNICATION AT WORK

In the following scenario, Creative International uses the writing process to create effective communication with its clients.

Connie Jones (president), Mary Michelson (project director), and Lori Smith (director of sales and marketing) have made Creative International a cutting-edge company. Creative International works with organizations to define strategic communication goals. A key to their success is following a process "from the beginning to the end of a communication project." They prewrite, write, and rewrite.

Prewriting

- **Initial client contact.** Through telephone calls, e-mail messages, networking, their Facebook site, or a preliminary meeting, Creative gathers data to discover the client's needs. In this phase, the Creative team interviews the end users and observes them at work.

- **Clarification meeting.** Meeting face-to-face with an upper-level decision maker, the Creative team collects information about the end user's needs. Connie, Mary, and Lori don't just say, "Sure, we can do that job." Instead, they ask probing questions, such as "Why do you need that?" "Why do you want that?" "What do you want to communicate to your audience?"

Objectives

When you complete this chapter, you will be able to

1. Understand the writing process including prewriting, writing, and rewriting

2. Prewrite to examine purposes and goals, determine audience, gather data, and choose the communication channel

3. Write to organize your information, provide visuals, and format content

4. Test for usability of your technical communication

5. Rewrite by adding or deleting information, simplifying terms, moving or reformatting content, changing style of writing, and correcting

6. Apply the checklist to your technical communication

- **Proposal creation.** Following the initial meeting, Creative writes a schedule, complete with project plan, scope, and description of the deliverables.

Writing

- **Drafting through collaboration.** The Creative team creates a wiki to help them write collaboratively.

- **Design, development, production, and pilot testing.** Creative creates text, graphics, and audio and video training modules for final beta testing. This rough draft verifies that the product works the way everyone expects it to.

Rewriting

- **Editing.** The Creative writing team (with input from other co-workers and the end user) revises their initial drafts by making revisions through their wiki.

This includes adding details, deleting unnecessary content, and correcting errors.

- **Evaluation and maintenance.** Through end-user analysis, commentary on their blog, usability testing, and customer measurement, Creative ensures that the performance needs are met and that training materials are current and valid.

Creative International refers to its "process map" from the beginning to the end of a project. They use process for marketing, for internal communication, and for project planning and management. The writing process that Creative follows is recursive. It includes constant sign-offs and change orders. With input from all parties, during prewriting, writing, and rewriting, Creative provides its customers "communication that provides custom solutions."

> Check out our quarterly newsletters at www.pearsonhighered.com/gerson for dot.com updates, new case studies, insights from business professionals, grammar exercises, and facts about technical communication.

The Writing Process: An Overview

Technical communication is a major part of your daily work experience. It takes time to write the correspondence, and your writing has an impact on those around you. A well-written report, e-mail message, user manual, Web site, or blog gets the job done and makes you look good. Poorly written correspondence wastes time and creates a negative image of you and your company.

However, recognizing the importance of technical communication does not ensure that your correspondence will be well written. How do you effectively write the memo, letter, blog entry, or report? How do you successfully produce the finished product? To produce successful technical communication, you need to approach writing as a process. The process approach to writing has the following sequence:

1. **Prewrite.** Before you can write your document, you must have something to say. Prewriting allows you to spend quality time, prior to writing the correspondence, generating information, considering the needs of the audience, and choosing the communication channel.
2. **Write.** Once you have gathered your data and determined your objectives, the next step is to state them. You need to draft your document. To do so, you should organize the draft so that it's easy to read, contains appropriate visual aids, and is formatted for easy access.
3. **Rewrite.** The final step, and one that is essential to successful communication, is to rewrite your draft. This step requires that you revise the rough draft. Revision allows you to test for usability and to perfect your memo, Web site, letter, report, or any document.

The writing process is dynamic, with the three parts—prewriting, writing, and rewriting—often occurring simultaneously. You may revisit any of these parts of the process at various times as you draft your document. The writing process is illustrated in Figure 2.1.

FIGURE 2.1 The Writing Process

The Writing Process

Prewriting	Writing	Rewriting

• Determine whether your audience is internal or external.
• Write to inform, instruct, persuade, and build trust.
• Choose the correct communication channel for your audience and purpose.
• Gather your data.
	• Organize your content using modes such as problem/solution, cause/effect, comparison, argument/persuasion, analysis, chronology, etc.
	• Use figures and tables to clarify content.
	• Format the content for ease of access.
		• Test for usability.
		• Revise your draft by
		  • Adding details
		  • Deleting wordiness
		  • Simplifying words
		  • Enhancing the tone
		  • Reformatting your text
		  • Proofreading and correcting errors

Prewriting

Prewriting, the first stage of the process, allows you to plan your communication. If you do not know where you are going in the correspondence, you will never get there, and your audience will not get there with you. Through prewriting, you accomplish many objectives, including

- Examining purposes
- Determining goals
- Considering audience
- Deciding what action you want your audience to perform
- Gathering data
- Determining the most effective communication channel

Examine Your Purposes

Before you write the document, you need to know why you are communicating. Are you planning to write because you have chosen to do so of your own accord or because you have been asked to do so by your boss? In other words, is your motivation external or internal?

External Motivation. If someone else has requested the correspondence, then your motivation is external. Your boss, for example, expects you to write a monthly status report, a performance appraisal of your subordinate, or an e-mail report suggesting solutions to a current problem. Perhaps a vendor has requested that you write a letter documenting due dates, or a customer asks that you respond to a letter of complaint. In all of these instances, someone else has asked you to communicate.

Internal Motivation. If you have decided to write on your own accord, then your motivation is internal. For example, you need information to perform your job more effectively, so you write a letter of inquiry. You need to meet with colleagues to plan a job, so you write an e-mail message calling a meeting and setting an agenda. Perhaps you recognize a problem in your work environment, so you create a questionnaire and transmit it via the company blog. Then, analyzing your findings, you call a meeting to report your findings. In all of these instances, you initiate the communication.

Determine Your Goals

Once you have examined why you are planning to communicate, the next step is to determine your goals in the correspondence or presentation. You might be communicating to

- Persuade an audience to accept your point of view
- Instruct an audience by directing actions
- Inform an audience of facts, concerns, or questions you might have
- Build trust and rapport by managing work relationships

These goals can overlap, of course. You might want to inform by providing an instruction. You might want to persuade by informing. You might want to build trust by persuading. Still, it is worthwhile looking at each of these goals individually to clarify their distinctions.

Communicating to Persuade. If your goal in writing is to change others' opinions or a company's policies, you need to be persuasive. For example, you might want to write a proposal, a brochure, or a flier to sell a product or a service. Maybe you will write your annual progress report to justify a raise or a promotion. As a customer, you might want to write a letter of complaint or an e-mail message about poor service. Your goal in each of these cases is to persuade an audience to accept your point of view.

Communicating to Instruct.   Instructions will play a large role in your technical communication activities. As a manager, for example, you often will need to direct action. Your job demands that you tell employees under your supervision what to do. You might need to write an e-mail providing instructions for correctly following procedures. These could include steps for filling out employee forms, researching documents in your company's intranet data bank, using new software, or writing reports according to the company's new standards.

As an employee, you also will provide instructions. Your boss might ask you to create an instructional YouTube video. As a computer information specialist, maybe you work the 1-800 hotline for customer concerns. When a customer calls about a computer problem, your job would be to give instructions for correcting the problem. You either will provide a written instruction in a follow-up e-mail or a verbal instruction while on the phone.

Communicating to Inform.   Often, you will write letters, reports, and e-mails merely to inform. In an e-mail message, for instance, you may invite your staff to an upcoming meeting. A trip report will inform your supervisor what conference presentations you attended or what your prospective client's needs are. A letter of inquiry will inform a vendor about questions you might have regarding her services. Maybe you will be asked to write a corporate blog entry or a Web site link informing your co-workers about the company's picnic, personnel birthdays, or new stock options available to employees. In these situations, your goal is not to instruct or persuade but to share information objectively.

Communicating to Build Trust.   Building rapport is a very important component of your communication challenge. As a manager or employee, your job is not merely to "dump data" in your written communication. You also need to realize that you are communicating with co-workers, people with whom you will work every day. To maintain a successful work environment, you want to achieve the correct, positive tone in your writing. This might require nothing more than writing a tweet, saying "Thanks for the information," or "You've done a great job reporting your findings." A positive tone shows approval for work accomplished and recognition of the audience's time.

Recognizing the goals for your correspondence makes a difference. Determining your goals allows you to provide the appropriate tone and scope of detail in your communication. In contrast, failure to assess your goals can cause communication breakdowns.

Consider Your Audience

Audience

See Chapter 4 for more discussion of audience recognition and involvement.

Multicultural Audiences

See Chapter 4 for more discussion of multiculturalism.

What you say and how you say it is greatly determined by your audience. Are you writing up to management, down to subordinates, or laterally to co-workers? Are you speaking to a high-tech audience (experts in your field), a low-tech audience (people with some knowledge about your field), or a lay audience (customers or people outside your work environment)? Face it—you will not write the same way to your boss as you would to your subordinates. You will not speak the same way to a customer as you would to a team member. You must provide different information to a multicultural audience than you would to individuals with the same language and cultural expectations. You must consider issues of diversity when you communicate.

Gather Your Data

Once you know why you are writing and who your audience is, the next step is deciding what to say. You have to gather data. The page or screen remains blank until you fill it with content. Your communication, therefore, will consider personnel, dates, actions required, locations, costs, methods for implementing suggestions, and so forth. As the writer, it is your obligation to flesh out the details. After all, until you tell your readers what you want to tell them, they do not know.

There are many ways to gather data. In this chapter, and throughout the textbook, we provide options for gathering information. These planning techniques include the following:

- Answering the reporter's questions
- Mind mapping
- Brainstorming or listing
- Outlining
- Storyboarding
- Creating organization charts
- Flowcharting
- Researching

Research Techniques

See Chapter 5 for more discussion of research and documentation.

Each prewriting technique is discussed in greater detail in Table 2.1. Table 2.2 lists Web sites for online research.

Determine How the Content Will Be Provided— The Communication Channel

After you have determined your audience, goals, and content, the last stage in prewriting is to decide which communication channel will best convey your message. Will you write an instant message or a text message, letter, memo, report, e-mail, Web site, blog entry, YouTube video, proposal, instructional procedure, flier, or brochure or will you make an oral presentation?

In Table 2.3, you can review the many channels or methods you may use for communicating your content.

Single Sourcing.   Maybe you will create content that will be used in a variety of communication channels simultaneously. Single sourcing is the act of "producing documents designed to be recombined and reused across projects and various media" (Carter 317). In a constantly changing marketplace, you will need to communicate your content to many different audiences using a variety of communication channels. For instance, you might need to market your product or service using the Internet, a flier, a brochure, newsletters, and a sales letter. You might need to write hard-copy user manuals and develop online help screens. To ensure that content is reusable, the best approach would be to write a "single source of text" that will "generate multiple documents for different media" (Albers 337).

DOT.COM UPDATES

For more information about single sourcing, check out the following links:

- "STC Single Sourcing SIG" http://www.stcsig.org/ss/
- "Single Sourcing: An Introduction" http://www.cherryleaf.com/ news_single_sourcing.htm
- "Single Sourcing and Its Applicability to Small Projects" http://www.techscribe .co.uk/ techw/single _source.htm

TABLE 2.1 Prewriting Techniques

Answering the reporter's questions
By answering *who*, *what*, *when*, *where*, *why*, and *how*, you create the content of your correspondence.

Sample Reporter's Questions	
Who	Joe Kingsberry, Sales Rep
What	Need to know what our discount is if we buy in quantities • What the guarantees are • If service is provided on-site • If the installers are certified and bonded • If Acme provides 24-hour shipping
When	Need the information by July 9 to meet our proposal deadline
Where	Acme Radiators 11245 Armour Blvd. Oklahoma City, Oklahoma 45233 Jkings@acmerad.com
Why	As requested by my boss, John, to help us provide more information to prospective customers
How	Either communicate with a letter or an e-mail. I can write an e-mail inquiry to save time, but I must tell Joe to respond in a letter with his signature to verify the information he provides.

Mind mapping
Envision a wheel. At the center is your topic. Radiating from this center, like spokes of the wheel, are different ideas about the topic. Mind mapping allows you to look at your topic from multiple perspectives and then cluster the similar ideas.

Sample Mind Mapping

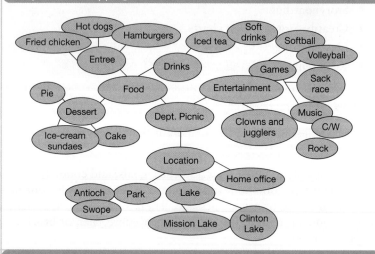

Brainstorming or listing
Performing either individually or with a group, you can randomly suggest ideas (brainstorming) and then make a list of these suggestions. This method, which works for almost all kinds of communication, is especially valuable for team projects.

Sample Listing—Improving Employee Morale

• Before meetings, ask employees for agenda items (that way, they can feel empowered)
• Consider flex-time
• Review employee benefits packages
• Hold yearly awards ceremony for best attendance, highest performance, most cold calls, lowest customer complaints, etc.
• Offer employee sharing for unused personal days/sick leave days
• Roll over personal days to next calendar year
• Include employees in decision-making process
• Add more personal days (as a tradeoff for anticipated lower employee raises)

TABLE 2.1 (Continued)

Outlining

This traditional method of gathering and organizing information allows you to break a topic into major and minor components. This is a wonderful all-purpose planning tool.

Sample Topic Outline

1.0 The Writing Process
   1.1 Prewriting
      • Planning Techniques
   1.2 Writing
      • All-Purpose Organizational Template
      • Organizational Techniques
   1.3 Rewriting
2.0 Criteria for Effective Technical Communication
   2.1 Clarity
   2.2 Conciseness
   2.3 Document Design
   2.4 Audience Recognition
   2.5 Accuracy

Storyboarding

Storyboarding is a visual planning technique that lets you graphically sketch each page or screen of your text. This allows you to see what your document might look like.

Sample Brochure Storyboard

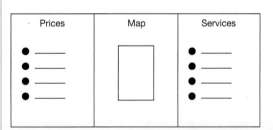

Creating organization charts

This graphic allows you to see the overall organization of a document as well as the subdivisions to be discussed.

Sample Organization Chart

(Continued)

TABLE 2.1   Prewriting Techniques (Continued)

Flowcharting	Sample Flowchart
Flowcharting is another visual technique for gathering data. Because flowcharting organizes content chronologically, it is especially useful for instructions. For example: Stop/Start = ◯ Step = ▭ Decision = ◇	

TABLE 2.2   Internet Search Engines

Purpose	Sites
Popular online search engines	Yahoo.com, Google.com, Bing.com, Ask.com
Meta-search engines (multithreaded engines that search several major engines at once)	MetaCrawler.com, Clusty.com, Dogpile.com
Specialty search engines	Findlaw.com focuses on legal resources. Webmd.com lets you access health and medical sites. MedlinePlus.gov provides consumer information. GoPubMed.com provides biomedical resources. Scirus.com provides scientific resources. ZDNet.com, EarthLink.net, and BusinessWeek.com provide resources for business.
Broad academic searches	Librarians' Internet Index to the Internet (http://lii.org), Infomine.ucr.edu, GoogleScholar.com, FindArticle.com
Government search sites	USA.gov, Google's Uncle Sam
International search sites	Search Engine Colossus, Abyz News Links (international newspapers and magazines), and Worldpress.org (international perspectives on the United States)
Directory of search engines	Yahoo's Web Search, Search Engine Collosus

TABLE 2.3 Options for Providing Content Through Different Communication Channels

Communication Channels	Benefits and Challenges
E-mail messages, instant messages (IM), text messages (TM)	**Benefits:** These types of electronic communication are quick and can almost be synchronous. You can have a real-time, electronic chat with one or more readers. Though e-mail messages should be short (20 or so lines of text), you can attach documents, Web links, graphics, and sound and movie files. **Challenges:** E-mail, instant messages, and text messages tend to be less formal than other types of communication. E-mail might not be private (a company's network administrators can access your electronic communication). IM and TM have size limitations.
Letters	**Benefits:** Typed on official corporate letterhead stationery, letters are formal correspondence to readers outside your company. **Challenges:** Letters are time consuming because they must be mailed physically. Although you can enclose documents, this might demand costly or bulky envelopes.
Memos	**Benefits:** Memos—internal correspondence to one or several co-workers—allow for greater privacy than e-mail (e-mail messages are archived and can be observed by administrators within a company). Even though most memos are limited in length (one or two pages), you can attach or enclose documents. **Challenges:** Memos are both more time consuming than e-mail and less formal than letters.
Reports and proposals	**Benefits:** Reports, internal and external, are usually very formal. They can range in length from one page to hundreds of pages (proposals and annual corporate reports to stakeholders, for example). Because of their length, reports are appropriate for detailed information. **Challenges:** They can be time consuming to write.
Brochures and fliers	**Benefits:** Brochures and fliers are appropriate for informational and promotional communication to large audiences. **Challenges:** Most brochures are limited to six or so panels, the equivalent of a back and front hard copy. Similarly, a flier is usually only one page long. Thus, in-depth coverage of a topic will not occur.
Web sites and blogs	**Benefits:** A Web site or blog can provide informal and public communication to the entire world—anytime, anyplace. Web sites essentially have unlimited size, so you can provide lots of information, and the content can be updated instantaneously by Web designers. A Web site can include links to other sites, animation, graphics, and color. **Challenges:** Audiences need access to the Internet. Web sites and blogs could divulge sensitive corporate communication.
Oral presentations	**Benefits:** Oral presentations can be both formal and informal and communicate directly with the audience. **Challenges:** Many people are afraid to make an oral presentation. Sometimes audience members can be poor listeners.
Social media	**Benefits:** Social media such as Twitter, Facebook, and YouTube allow you to communicate rapidly with many people. You can communicate in a friendly, informal way with your audience. **Challenges:** Social media can pose confidentiality issues and consume too much on-the-job time. Twitter allows only 140 characters of text.
Videos	**Benefits:** Instructional videos avoid language barriers and provide clear actions for people to follow. Videos can be entertaining. **Challenges:** Creating a precise script with detailed directions for the video camera operator can be challenging, time consuming, and expensive.

Writing

Writing lets you package your data. Once you have gathered your data, determined your objectives, recognized your audience, and chosen the channel of communication, the next step is writing the document. You need to write a draft so that your readers can follow your train of thought readily and easily access your data. Writing the draft lets you organize your thoughts in some logical, easy-to-follow sequence. Writers usually know where they are going, but readers do not have this same insight. When readers pick up your document, they can read only one line at a time. They know what you are saying at the moment, but they don't know what your goals are. They can only hope that in your writing, you will lead them along logically and not get them lost with unnecessary data or illogical arguments.

Organization

To avoid confusing or misleading your audience, you need to organize your thoughts. As with prewriting, you have many organizational options. In Chapter 3, we discuss the following traditional methods of organization:

- Space (spatial organization)
- Chronology
- Importance
  - Comparison/contrast
  - Problem/solution

These organizational methods are not exclusive. Many of them can be used simultaneously within your technical communication to help your reader understand your content.

Formatting

You also must format your text to allow for ease of access. In addition to organizing your ideas, you need to consider how the text looks on the page or screen. If you give your readers a massive wall of words, they will file your document for future reading and look for the nearest exit. An unbroken page or screen of text is not reader friendly. To invite your readers into the document, to make them want to read the technical communication, you need to highlight key points and break up monotonous-looking text. You need to ensure that your information is accessible.

Rewriting

Rewriting lets you perfect your writing. After you have prewritten (to gather data, organize thoughts, and understand audience) and written your draft, your final step is to rewrite. Revision requires that you look over your draft to determine its usability and correctness.

Document Design

See Chapter 9 for more discussion of document design and formatting techniques.

Usability Testing

A report, instruction, or Web site is only good if your audience can understand the content and use the information. Usability testing helps you determine the success of your draft. Through usability testing, you decide what works in the draft and what needs to be rewritten. Thus, usability focuses on the following key factors:

- **Retrievability.** Can the user find specific information quickly and easily?
- **Readability.** Can the user read and comprehend information quickly and easily?
- **Accuracy.** Is the information complete and correct?
- **User satisfaction.** Does the document present information in a way that is easy to learn and remember? (Dorazio)

USABILITY EVALUATION CHECKLIST	
Audience Recognition	1. Are technical terms defined? 2. Are examples used at the reader's level of understanding to explain difficult steps or concepts? 3. Are graphics used appropriately to help the audience understand the procedure or concept? 4. Are tone and word usage appropriate for the intended audience?
Development	1. Are steps or concepts precisely developed? 2. Is all required information provided including hazards, technical descriptions, process analyses, warranties, accessories, and required equipment or tools? 3. Is irrelevant or rarely needed information omitted?

Ease of Use	Can readers easily find what they want through
	    • A table of contents
	    • A navigation bar
	    • A glossary
	    • Hierarchical headings
	    • Headers or footers
	    • An index
	    • Links
	    • FAQs

Conciseness	1. Are words limited to one to two syllables when possible? 2. Are sentences limited to 10 to 15 words when possible? 3. Are paragraphs limited to four to six lines of text when possible?
Consistency	1. Is a consistent hierarchy of headings used? 2. Are graphics presented consistently (same location, same use of figure titles and numbers, similar sizes, similar style of graphics, etc.)? 3. Does word usage mean the same throughout the text (technical terms, hazard notations, etc.)? 4. Is the same style of numbering used throughout?
Document Design	1. Do graphics help depict how to perform steps? 2. Is white space used to make information accessible? 3. Does color emphasize hazards, key terms, or important information? 4. Do headings, subheadings, and talking headings aid clarity and help the readers navigate the text?

Revision Techniques

After testing your document for usability, revise your text by using the following revision techniques:

- Add missing detail for clarity.
- Delete dead words and phrases for conciseness.
- Simplify unnecessarily complex words and phrases to allow for easier understanding.
- Move around information (cut and paste) to ensure that your most important ideas are emphasized.
- Reformat (using highlighting techniques) to ensure reader-friendly ease of access.
- Enhance the tone and style of the text.
- Correct errors to ensure accurate grammar and content.

We discuss each of these points in greater detail throughout the text.

How Important Is Proofreading?

Do employees in the workplace really care about grammar and mechanics? Is proofreading only important to teachers? Proofreading is absolutely important. Incorrect documentation costs companies money. "A misplaced decimal point resulted in one company paying . . . $120,000 in taxes on a piece of industrial equipment, instead of the $1,200 the firm rightfully owed" (Rizzo A1).

A Chicago-based company purchased an industrial sander for $54,589.62. Unfortunately, when listing the purchase on their year-end taxes, the company reported the purchase price as $5,458,962. This misplaced comma and decimal point equaled a difference of over $5 million. The issue went to court, costing even more money. A single mark of punctuation can be important (Rizzo A6).

Table 2.4 shows the importance of proofreading. The National Commission on Writing highlights what employers and employees consider to be essential skills in technical communication ("Writing: A Powerful Message from State Government" 19).

Revision is possibly the most important stage in the writing process. If you prewrite effectively (gathering data, determining objectives, and recognizing audience) and write an effective draft, you are off to a great start. However, if you then fail to rewrite your text, you run the risk of having wasted the time you spent prewriting and

TABLE 2.4 Essential Skills in Technical Communication

Skills	Extremely Important
Accuracy	87.8%
Spelling, grammar, punctuation	71.4%
Clarity	69.4%
Documentation/support	61.2%
Logic	55.1%
Conciseness	42.9%
Visual appeal	12.2%

Using Microsoft Word 2010 for Rewriting

Word processing programs help you rewrite your document in many ways:

- **Spell check.** When you misspell a word, often spell check will underline the error in red (as shown in the following example with "grammer" incorrectly spelled). Spell check, unfortunately, will not catch all errors. If you use a word like *to* instead of *too*, spell check will not "no" the difference (of course, that should be "know" but spell check did not mark the error). Microsoft Word 2010's **Review** tab also provides you access to proofreading help and allows you to make comments and track changes.

- **Grammar check.** Word processors also can help you catch grammar errors. Grammar check underlines errors in green. When you right-click on the underlined text, the word processing package will provide an optional correction.

Grammer error look like this, for example.

- **Add/Delete.** Word processing makes adding new content and deleting unneeded text very easy. All you need to do is place your cursor where you want to add/delete. Then, to add, you type. To delete, you hit the Backspace key or the Delete key.
- **Move.** The Copy, Cut, and Paste features of word processing allow you to move text with ease.
- **Enhance/Reformat.** In addition to changing the tone of your text, you also can enhance the visual appeal of your document at a keystroke. From the **Home** tab on your toolbar, you can choose from the Word 2010 Ribbon and include bullets, italics, boldface, font changes, numbered lists, etc.

writing. Rewriting is the stage in which you make sure that everything is correct. Failure to do so not only can cause confusion for your readers but also can destroy your credibility.

The process approach to writing—including prewriting, writing, and rewriting (including usability testing)—can help you communicate successfully in any work environment or writing situation. In fact, the greatest benefit of process is that it is generic. Process is not designed for any one profession or type of correspondence. No author of a technical communication book can anticipate exactly where you will work, what type of documents you will be required to write, or what your supervisors will expect in your

Prewriting

1. Have you determined the purpose for communicating?
2. Is your goal to inform, instruct, persuade, or build trust?
3. Is your audience high-tech, low-tech, or lay?
4. Have you gathered data through prewriting techniques or researching?
5. Have you chosen the correct communication channel?

Writing

6. Have you organized the information (using modes such as spatial, chronology, comparison/contrast, etc.)?
7. Is the content formatted for readability?

Rewriting

8. Have you tested for usability?
9. Have you revised by adding, deleting, simplifying, enhancing tone, and reformatting?
10. Have you proofread for accuracy?

writing. However, we *can* give you a methodology for tackling any communication activity. Writing as a process will help you write any kind of oral or written communication, for any boss, in any work situation.

The Writing Process at Work

Following is an e-mail report written using the process approach to writing. The document was produced in the workplace by John, an information technologist writing to his supervisor. John needed to create a user manual and instructional video for the

SPOTLIGHT

An Editor's Use of the Writing Process

Candice Millard, an editor at *Wireless World* and *National Geographic* magazines and author of *The River of Doubt* sees herself as "the reader's advocate." As an editor, her job is to make sure that readers understand the text she is editing. Candice states, "Because authors invest enormous time and effort in their writing, they often become so attached to their work that they get lost in the details." Candice's job is to be the detached eye, the objective reader's point of view.

In the writing process, Candice's authors are in charge of the prewriting and writing. Candice's job focuses on the final stage of the writing process—rewriting. To accomplish this goal, she offers these editorial hints:

- **Start big and get small.** To achieve editorial objectivity, Candice says that she must "step back, get the big picture, and then work toward the details." She asks herself, "What's superfluous, what's confusing, what questions do I have that haven't been answered, and where is clarity needed?"

- **Flesh out the details.** Candice's next job is to "fill in the holes." Sometimes, her biggest challenge is defining her authors' highly technical terminology. When authors depend on jargon, Candice "translates the terms."

- **Slice and dice.** The opposite goal to adding detail for clarity is deleting text. Candice also must "edit out contradictions, weed the irrelevancies, remove the distractions, and excise whatever might ring false."

- **Practice diplomacy.** Candice's authors are all experts in their fields. When these authors write about what they love, every word they use is precious to them; every detail is crucial. Candice, on the other hand, must cut their text to fit space limitations and to meet a reader's limited attention span. Cutting makes "people bristle." Therefore, when editing, she considers how her comments will affect the writer. She "engages in conversation, points out what's good, tempers the bad with the positive, and then weaves in required changes."

As an editor, Candice knows that she isn't just dealing with words; she is working with people.

FIGURE 2.2   Using Reporter's Questions to Prewrite

Who—my boss

What—write a progress report about our new user manual and instructional video

Where—the manual will be available online (www.vCard.com/customerhelp), saved to disc, and packaged with the product; the instructional video will be saved on disc and accessible on our Web site

When—to accompany the release of our new product (November 2012)

Why—in response to boss's request for an update on how to improve customer service

How—I worked with SMEs and a panel of customers to test usability to create the user manual and instructional video

installation of a video card, with added functions such as video capture, MPEG-2 and MPEG-4 decoding, light pen, TV output, and the ability to connect multiple monitors. His boss asked for a follow-up status report about the creation of the manual and the instructional video.

Prewriting

John worked on a user manual and video for several weeks. After receiving an e-mail request for a follow-up status report from his boss, John used the reporter's questions to prewrite for his report. John's prewriting (shown in Figure 2.2) tells us *who* he is writing to; *what* types of documents he worked on; *where* the user manual will be accessible to the customer; *when* the new product will be released; *why* the boss requested the update; *how* John gathered data for his manual and tested for usability.

   With data gathered and objectives determined, John was ready to write.

Writing

First, John wrote a rough draft, as shown in Figure 2.3. In this draft, John took the word "rough" seriously and did not consider document design, grammar, or sentence structure. When you draft, do not worry about errors or how the correspondence looks. It is meant to be rough, to free you from worry about making errors. You can correct errors when you revise.

Rewriting

No writing is ever perfect. Every memo, letter, report, or e-mail message can be improved. John asked a colleague to review his rough draft and suggest revisions. Figure 2.4 shows the rough draft with suggestions for improvement.

   After John read his colleague's comments, he revised the report and sent it to his boss. Figure 2.5 shows you his finished product. Once the manager received the report, he commended John for a job well done. The manager found the report easy to read because of the use of white space, headings, listed and bulleted information, and smaller paragraphs. The manager also noted the grammatically correct and well-constructed sentences. When you approach writing as a step-by-step process (prewriting, writing, and rewriting), your results usually are positive—and you will receive positive feedback from your supervisors.

FIGURE 2.3 Rough Draft

REPORT ON MANUAL

Recently, you asked us to develop an instruction manual and video for installation of our vCard 9600 pro series video card. Our focus was to create a manual that was aesthetically pleasing and easy to follow for all levels of computer literate people. All in efforts to increase our customer service satisfaction percentage.

We feel we successfully created a manual that can be beneficial to anyone who uses this product. Our installation guide is an extremely user friendly, providing an easy-to-follow, step-by-step instruction to the installation of our 9600 pro and can be used as a basis for developing manuals for all of our products. We feel we should develop detailed installation manuals for every product and service we provide, this will increase customer service satisfaction and relieve our technical support department's hold time on calls. We're sure you agree that Extreme PCs should not only be known for our great products, but for our great customer service as well. Since the release of this manual, our customer satisfaction percentage has increased greater than our competitors. We encountered two problems in writing our manual. Using computer jargon when describing installation. Our use of high-tech terminology led to reader confusion, as documented by customer complaints. Screen captures should parallel the content. We had problems with the instructional video.

The manual was a complete success. However, we could improve future manuals as follows: More instruction manuals detailing installation on all our products. We're certain that our customer satisfaction will continue to increase if all of our products are packaged with manuals or if manuals are provide online for easy download. To manage the challenge we encountered with high-tech terminology, all manuals will now include glossaries to define challenging word usage. SMEs will help with this. More time to work on the manuals. We had so many ideas during the development process, that we ran out of time to implement them all. With this first manual created, however, we can use it as a template. This will save time for future manual development. We'd also like to see an increase in staffing to take on such a massive project. Plus, we'd like to increase our tech support department to reduce hold time and ensure customer satisfaction levels.

Each company you work for over the course of your career will have its own unique approach to writing memos, letters, reports, e-mail messages, proposals, or Web sites. Your employers will want you to do it their way. Company requirements vary. Different jobs and fields of employment require different types of correspondence. However, you will succeed in tackling any writing task if you have a consistent approach to writing. A process approach to writing will allow you to write any correspondence effectively.

FIGURE 2.4   Rough Draft with Suggested Revisions

REPORT ON MANUAL

Recently, you asked us to develop an instruction manual and video for installation of our vCard 9600 pro series video card. Our focus was to create a manual that was aesthetically pleasing and easy to follow for all levels of computer literate people. All in efforts to increase our customer service satisfaction percentage.

We feel we successfully created a manual that can be beneficial to anyone who uses this product. Our installation guide is an extremely user friendly, providing an easy-to-follow, step-by-step instruction to the installation of our 9600 pro and can be used as a basis for developing manuals for all of our products. We feel we should develop detailed installation manuals for every product and service we provide, this will increase customer service satisfaction and relieve our technical support department's hold time on calls. We're sure you agree that Extreme PCs should not only be known for our great products, but for our great customer service as well. Since the release of this manual, our customer satisfaction percentage has increased greater than our competitors. We encountered two problems in writing our manual. Using computer jargon when describing installation. Our use of high-tech terminology led to reader confusion, as documented by customer complaints. Screen captures should parallel the content. We had problems with the instructional video.

The manual was a complete success. However, we could improve future manuals as follows: More instruction manuals detailing installation on all our products. We're certain that our customer satisfaction will continue to increase if all of our products are packaged with manuals or if manuals are provide online for easy download. To manage the challenge we encountered with high-tech terminology, all manuals will now include glossaries to define challenging word usage. SMEs will help with this. More time to work on the manuals. We had so many ideas during the development process, that we ran out of time to implement them all. With this first manual created, however, we can use it as a template. This will save time for future manual development. We'd also like to see an increase in staffing to take on such a massive project. Plus, we'd like to increase our tech support department to reduce hold time and ensure customer satisfaction levels.

John, don't put the subject line in all caps. What about the manual? You need additional information about the video.

Change the word "recently" to a specific date, clarify who you worked with on this project, and remind the reader exactly why you're writing this report.

Note that the last line in paragraph one is a fragment (incomplete sentence). You have similar grammar errors elsewhere.

No one will read this paragraph or your last paragraph, John. They're both too bulky. Can you add white space, add headings, and use some bulleted lists?

Define "SMEs" John.

Our boss asked for a status report about the vCard project. Your conclusion does not emphasize what was learned from the project. A status report of the vCard project will help him decide how to set up the next company project.

Your conclusion needs some "punch" for emphasis.

FIGURE 2.5   Finished Product

Subject: Report on Video Card Installation Manual and Instructional Video

What we hoped to achieve

On Friday, November 2, 2012, you asked our team to develop an instruction manual and video for installation of our vCard 9600 pro series video card. Our focus was to create a hard-copy manual and an instructional video that were aesthetically pleasing and easy to follow for all levels of computer literate people. We hoped to increase our customer service satisfaction percentage. We completed the user manual and video in time for the vCard 9600 rollout. This report will bring you up to date on our achievements.

What we accomplished

We successfully created a manual and a video that can be beneficial to anyone who uses our product. Our installation guide is user friendly, providing an easy-to-follow, step-by-step instruction to the installation of our 9600 pro. The manual and video are available online at www.vCard.com/customerhelp. Plus, we saved the text and accompanying video to disc and packaged them with the product.

We believe that these end-user help aids can serve as a basis for developing manuals for all our products. Our instructional video is easy to use for the consumer and can be a model for all future instructional videos. Working with our customer usability panel and SMEs (subject matter experts) proved to be an invaluable resource and a great way to achieve customer service satisfaction.

Work remaining to be done

We feel we should develop detailed installation manuals and instructional videos for every product and service we provide. This will increase customer service satisfaction and relieve our technical support department's hold time on calls. Since the release of this manual and video, our customer satisfaction percentage has increased by approximately 12 percent over our competitors, as validated by independent survey results.

Problems encountered

We encountered two problems in writing our manual:

1. We used computer jargon when describing installation. Our use of high-tech terminology led to reader confusion, as documented by customer complaints.
2. We did not include enough screen captures to help our target audience. This led to complaints about lack of understanding.

We encountered two problems when scripting our video:

1. The instructional steps were not detailed enough for the video camera operator to shoot.
2. Our "actors" had insufficient time for learning lines.

Suggestions for future projects

The manual and video were successful after we made revisions based on comments from our usability panel. However, we could improve future manuals and videos as follows:

- **More Manuals.** More instruction manuals detailing installation of our products need to be created. We're certain that our customer satisfaction will continue to increase if all products are packaged with manuals or if manuals are provided online for easy download.

FIGURE 2.5 (Continued)

- **More Instructional Videos.** Most if not all of our products could benefit from instructional videos. We could create videos in different languages to be sold in our international market. YouTube videos would be appealing to many of our customers.

- **Glossaries.** To manage the challenge we encountered with high-tech terminology, all manuals will now include glossaries to define challenging word usage. Our in-house SMEs can help with definitions of terms.

- **Time.** We need more time to work on the manuals and the videos. We had so many ideas during the development process that we ran out of time to implement them all. With this first manual and video created, however, we can use them as templates. This will save time for future manual and video development.

- **Personnel.** We'd also like to see an increase in staffing to take on such a massive project. Plus, we'd like to increase our tech support department to reduce hold time and ensure customer satisfaction levels. My team believes that hiring only two more help desk employees could successfully handle this need.

If I can answer any questions, let me know. Thank you for giving me the opportunity to work on this project.

CHAPTER HIGHLIGHTS

1. Writing effectively is a challenge for many people. Following the process approach to writing will help you meet this challenge.
2. Prewriting helps you determine goals, consider audience, gather data, examine purposes, and determine the communication channel.
3. Prewriting techniques will help you get started. Try answering reporter's questions, mind mapping, brainstorming or listing, outlining, storyboarding, creating organization charts, flowcharting, or researching.
4. When you prewrite, decide whether you are communicating to persuade, instruct, inform, or build trust.
5. To begin writing a rough draft, organize your material, consider the layout and design of the communication, and add visual aids such as tables and figures.
6. You can communicate content through e-mail messages, instant messages, blogging, letters, memos, reports, brochures, proposals, Web sites, and PowerPoint presentations.
7. Perfect your text by testing for usability.
8. Rewrite your document by adding, deleting, simplifying, moving, reformatting, enhancing, and correcting.
9. Proofreading is an essential part of the rewriting step in the writing process. Lack of proofreading causes businesses to lose money.
10. Accuracy is an essential skill in business according to the National Commission on Writing.

CASE STUDIES

1. You are the co-chair of the "Mother's Weekend" at your sorority, fraternity, or other school organization. Using mind mapping and listing, brainstorm the activities, menus, locations, decorations, dates, and fees for this weekend's festivities. Brainstorm the pros and cons of hosting the weekend at your sorority or fraternity house or at a hotel or restaurant.

Assignment

Write an outline showing the decisions you've made regarding the topics above. Then, write a short memo or e-mail to your organization's executive board sharing your findings.

2. You work for the Oneg, Oregon, City Planning Department. Your boss, Carol Haley, has received complaints recently from citizens concerned about a wastewater facility being built in their neighborhood. The homeowners are worried about odors, chemical runoff in nearby Tomahawk Creek, decreases in home values, and a generally diminished quality of life in the neighborhood. The wastewater facility will be built. Despite the citizens' concerns, City Planning has decided that Oneg needs and will profit from the plant. Nonetheless, you must respond to these complaints, acting upon the citizens' issues.

For odor abatement, the wastewater management company plans to control fumes and particulate matter through the use of cross flow and wet scrubbers, thermal oxidizers, absorption materials, and bio-filters. Many of the concerns regarding runoff and home values can be solved through improved land management and ecological restoration. By planting more reeds, bushes, and trees in the green space between the homes and the proposed plant, runoff can be absorbed more efficiently, and green barriers will improve home values. Finally, you have learned that the wastewater company wants to be a good neighbor. To do so, it plans to become actively involved in the community by building more parks, playgrounds, hike/bike trails, and stocking the nearby pond.

Assignment

In small teams or as individuals, decide which communication channel you will use to share your findings with the community. Will you use a Web site, blog entry, tweet, letter, e-mail, oral communication, and so forth? Brainstorm the pros and cons of each channel. Then, write an e-mail to your teacher explaining which channel you chose and why.

3. Electronic City is a retailer of DVDs, televisions, DVD players with MP3 playback/JPEG viewer, computer systems, cameras, telephones, fax machines, printers, and more. Electronic City needs to create a Web site to market its products and services. The content for this Web site should include the following:

Prices	Store hours	Warranties	Service agreements
Job opportunities	Installation fees	Extended holiday hours	Discounts
Technical support	Product information	Special holiday sales	Delivery fees

Assignment

Review the list of Web site topics for Electronic City. Using an organizational chart, decide how to group these topics. Which will be major links on the Web site's navigation bar? Which will be topics of discussion within each of the major links? Once you have organized the links, sketch the Web site by creating a storyboard.

4. You are the special events planner in the marketing department at Thrill-a-Minute Entertainment Theme Park. You and your project team need to plan the grand opening of the theme park's newest sensation ride—The Horror—a wooden roller coaster that boasts a 10 g drop. What activities should your team plan to market and introduce this special event?

Assignment

Using a planning technique discussed in this chapter, gather ideas for a day-long event to introduce The Horror. Report your findings as follows:

- In the brief report to your marketing department boss, explain why you are writing, give options for the event, clarify which techniques you used to gather ideas, and sum up by recommending what you think are the best marketing approaches.
- Write an e-mail to your teacher providing options for the event and explaining which techniques you used to gather data.
- Give an oral presentation in class providing options for the event and explaining which techniques you used to gather data.
- Create a posting on the corporate blog about this special event. Send an e-mail to your teacher with the text for the blog entry, and explain how you gathered data.
- Create a link on the company Web site about the day-long event introducing the newest ride. Send an e-mail to your teacher with the text to be included in the link, and explain how you gathered data.

INDIVIDUAL AND TEAM PROJECTS

1. To practice prewriting, take one of the following topics. Then, using the suggested prewriting technique, gather data.
   a. **Reporter's questions.** To gather data for your resume, list answers to the reporter's questions for two recent jobs you have held and for your past and present educational experiences.
   b. **Mind mapping.** Create a mind map for your options for obtaining college financial aid.
   c. **Brainstorming or listing.** List five reasons why you have selected your degree program or why you have chosen the school you are attending.
   d. **Outlining.** Outline your reasons for liking or disliking a current or previous job.
   e. **Storyboarding.** If you have a personal Web site, use storyboarding to depict graphically the various screens. If you do not have such a site, use storyboarding to depict graphically what your site's screens would include.
   f. **Creating organizational charts.** What is the hierarchy of leadership or management at your job or college organization (fraternity, sorority, club, or team)? To depict graphically who is in charge of what and who reports to whom, create an organizational chart.
   g. **Flowcharting.** Create a flowchart of the steps you followed to register for classes, buy a car, or seek employment.
   h. **Researching.** Go online or find a hard copy of the *Occupational Outlook Handbook*. Then, research a career field that interests you. Reading the handbook, find out the nature of the work, working conditions, employment opportunities, educational requirements, and pay scale.

2. Using the techniques illustrated in this chapter, edit, correct, and rewrite the following flawed memo.

Date: April 3, 2012
To: William Huddleston
From: Julie Schopper
Subject: Trainning Classes

Bill, our recent training budget has increased beyond our projections. We need to solve this problem. My project team has come up with several suggestions, you need to review these and then get back to us with your input. Here is what we have come up with.

We could reduce the number of training classes, fire several trainers, but increase the number of participants allowed per class. Thus we would keep the same amount of income from participants but save a significant amount of money due to the reduction of trainer salaries and benefits. The downside might be less effective training, once the trainer to participant ratio is increased. As another option, we could outsource our training. This way we could fire all our trainers which would mean that we would save money on benefits and salaries, as well as offer the same number of training sessions, which would keep our trainer to participant ratio low.

What do you think. We need your feedback before we can do anything so even if your busy, get on this right away. Please write me as soon as you can.

PROBLEM-SOLVING THINK PIECE

In an interview, a company benefits manager said that she spent over 50 percent of her workday on communication issues. These included the following:

- Consulting with staff, answering their questions about retirement, health insurance, and payroll deductions
- Meeting weekly with human resources (HR) colleagues
- Collaborating with project team members
- Preparing and writing quarterly reports to HR supervisors
- Teleconferencing with third-party insurance vendors regarding new services and/or costs
- E-mailing supervisors and staff, in response to questions
- Updating information about benefits on the corporate blog
- Creating a new link on the company Web site about changes to insurance and benefits
- Calling and responding to telephone calls
- Faxing information as requested
- Responding to and sending instant messages and text messages
- Writing letters to vendors and staff to document services

Though she had to use various methods of both written and oral communication, the communication channels each have benefits and drawbacks. E-mailing, for example, has pluses and minuses (convenience over depth of discussion, perhaps). Think about each of the communication options above. Using the following table, list the benefits of each particular type of communication versus the drawbacks.

Communication Channels	Benefits	Drawbacks	Possible Solutions
One-on-one discussions			
Group meetings			
Collaborative projects			
Written reports			
Teleconferences			
E-mail			
Blogging			
Web site			
Phone calls			
Faxes			
IM/TM			
Letters			

WEB WORKSHOP

1. Proofreading is a key component of successful technical communication. Access the following Web sites and read what these sites suggest as editing/proofreading hints. Compare the content in the Web sites to your approaches to proofreading and editing. Write an e-mail message or memo summarizing your findings.
   - Literacy Education Online:
     http://leo.stcloudstate.edu/acadwrite/genproofed.html
   - Purdue University's Online Writing Lab:
     http://owl.english.purdue.edu/handouts/general/gl_edit.html
   - University of North Carolina:
     http://www.unc.edu/depts/wcweb/handouts/proofread.html
2. The Society for Technical Communication provides a link to professional articles about usability testing: http://www.stcsig.org/usability/. Read any of the articles found in this Web site and report on your findings in an e-mail message or memo to your professor.

QUIZ QUESTIONS

1. What are the three main parts of the writing process?
2. What are four ways you can provide technical communication content?
3. What can you achieve by prewriting?
4. What is the difference between external and internal motivation?
5. Why should you consider your audience before you begin writing?
6. What are four different prewriting techniques?
7. Why do you consider format when you write a business document?
8. What are four rewriting techniques?
9. What happens when you fail to revise accurately?
10. How can the writing process help ensure that you become a successful writer?
11. What are four goals of technical communication?
12. What are three search engines?
13. How do reporter's questions differ from mind mapping?
14. What is usability testing?
15. How can software help you to rewrite your documents?

CHAPTER THREE

Clarity, Conciseness, Accuracy, and Ethics in Technical Communication

COMMUNICATION AT WORK

In the CompToday scenario, employees address the needs of an audience by writing text that is clear, concise, accurate, and ethical.

CompToday is a computer hardware company located in Boston, Massachusetts. Like all publicly traded companies, CompToday must meet the Sarbanes-Oxley Act, passed by Congress in response to accounting scandals following the Enron, Tyco, and WorldCom illegalities. This act makes all corporate executives responsible for their companies' accounting practices. Penalties for failing to abide by this act are severe, including heavy fines as well as the potential for prison sentences.

CompToday, like many other companies, realizes that maintaining ethics and legalities in business is not just a matter of being a good citizen in the community. Maintaining ethics in the work environment demands a great deal of written communication. A key component of

Objectives

When you complete this chapter, you will be able to

1. Achieve clarity by providing details and answering reporter's questions

2. Delete wordiness in sentences and paragraphs to achieve conciseness

3. Simplify word usage

4. Learn proofreading techniques to ensure accuracy

5. Organize your thoughts to help readers better understand documents

6. Communicate ethically orally, in writing, and through graphics

7. Evaluate your technical communication for clarity, conciseness, accuracy, organization, and ethics using a checklist

the Sarbanes-Oxley Act, combating ethical problems in the workplace, is the need for extensive documentation, including the following:

- Written policies provided for all corporate employees, clarifying in low-tech terms the intent of the act as well as its standards

- Clear instructional procedures (hard copy and online) guiding employees in the steps they must follow to abide by Sarbanes-Oxley

- Concise online help screens providing answers to FAQs (frequently asked questions) relevant to Sarbanes-Oxley

- Accurate monthly reports documenting key financial activities, such as accounts payables, cash disbursement, purchases, and earnings

- Annual reports submitted to and signed by outside auditors

To achieve this daunting goal, CompToday's chief executive officer William Huddleston has reassigned a number of technical writers from his corporate communication department. They had been working on other projects, such as the company's Web site, instructional manuals, and online help screens. Now, they are tasked with new responsibilities—creating the documentation required by CompToday to meet the Sarbanes-Oxley requirements.

This hurts the company's bottom-line profit margin. After all, new writers must be hired to complete unfinished writing projects while the reassigned technical writers work toward compliance with the act. Maintaining an ethical workplace can be costly. However, practicing good ethics and abiding by the law benefits everyone—employees, stockholders, clients, and vendors.

Source: Adapted from Harkness 16–18.

Check out our quarterly newsletters at www.pearsonhighered.com/gerson for dot.com updates, new case studies, insights from business professionals, grammar exercises, and facts about technical communication.

Achieving Clarity in Technical Communication

The ultimate goal of good technical communication is clarity. If you write a letter, report, e-mail message, user manual, blog entry, or Web site that is unclear to your readers, problems can occur. Unclear technical communication can lead to missed deadlines, damaged equipment, inaccurate procedures, incorrectly filled orders, or danger to the end user. To avoid these problems and many others caused by unclear communication, write to achieve clarity.

Provide Specific Detail

One way to achieve clarity is by supplying specific, quantified information. If you write using vague, abstract adjectives or adverbs, such as *some* or *recently*, your readers will interpret these words in different ways. The adverb *recently* will mean 30 minutes ago to one reader, yesterday to another, and last week to a third reader. This adverb, therefore, is not clear. The same applies to an adjective like *some*. You write, "I need some information about the budget." Your readers can only guess what you mean by *some*. Do you want the desired budget increase for 2012 or the budget expenditures for 2011?

Look at the following example of vague writing caused by imprecise, unclear adjectives. (Vague words are underlined.)

BEFORE

Our <u>latest</u> attempt at molding preform protectors has led to <u>some</u> positive results. We spent <u>several</u> hours in Dept. 15 trying different machine settings and techniques. <u>Several</u> good parts were molded using two different sheet thicknesses. Here's a summary of the findings.

First, we tried the <u>thick</u> sheet material. At 240°F, this thickness worked well. Next, we tried the <u>thinner</u> sheet material. The <u>thinner</u> material is less forgiving, but after a <u>few</u> adjustments we were making good parts. Still, the <u>thin</u> material caused the most handling problems.

The engineer who wrote this report realized that it was unclear. To solve the problem, she rewrote the report, quantifying the vague adjectives.

AFTER

During the week of 10/4/12, we spent approximately 12 hours in Dept. 15 trying different machine settings, techniques, and thicknesses to mold preform mold protectors. Here is a report on our findings:

- 0.030" Thick Sheet
  At 240°F, this thickness worked well.
- 0.015" Thick Sheet
  This material is less forgiving, but after decreasing the heat to 200°F, we could produce good parts. Still, material at 0.015" causes handling problems.

Your goal as a technical communicator is to express yourself clearly. To do so, state your exact meaning through specific, quantified word usage (measurements, dates, monetary amounts, and so forth).

Answer the Reporter's Questions

A second way to write clearly is to answer the reporter's questions—who, what, when, where, why, and how. The best way we can emphasize the importance of answering these reporter's questions is by sharing with you the following e-mail message, written by a highly placed executive, to a newly hired employee.

E-mail Messages

See Chapter 6 for more discussion of e-mail.

BEFORE

Date: November 16, 2012
To: Staff
From: Earl Eddings, Manager
Subject: Research

Please be prepared to plan a presentation on research. Make sure the information is very detailed. Thanks.

That's the entire e-mail. The questions are, "What doesn't the newly hired employee know?" "What additional information would that employee need to do the job?" "What needs clarifying?"

To achieve successful communication, the writer needs to answer reporter's questions. *What* is the subject of the presentation and the research? *What* exactly is the reader supposed to do? Will the reader of this e-mail make a presentation, plan a presentation, or prepare to plan a presentation? *Who* is the audience? The word "staff" is too encompassing. Will all of the staff be involved in this project? *Why* is this presentation being made? That is, what is the rationale or motivation for this presentation? *When* will the presentation be made? *How* much detail is "very detailed"? *Where* will the presentation take place?

In contrast, the "After" e-mail message below achieves clarity by answering reporter's questions.

AFTER

Date: November 16, 2012   *What*
To: Melissa Hider
From: Earl Eddings   *Who*
Subject: Research for Homeland Security Presentation

Please make a presentation on homeland security for the Weston City Council. This meeting is planned for November 20, 2012, in Conference Room C, from 8:00 A.M.–5:00 P.M.
                              When   *Where*

We have a budget of $6,000,000. Thus, to use these funds effectively, our city must be up to date on the following concerns:

- Bomb-detection options
- Citizen preparedness   *Why*
- Defense personnel training
How • Crisis management

Use PowerPoint software to make your presentation. With your help, I know Weston and the Kansas City metro area will benefit.

Use Easily Understandable Words

Another key to clarity is using words that your readers can understand easily. Avoid obscure words and be careful when you use acronyms, abbreviations, and jargon.

Avoid Obscure Words.   A good rule of thumb is to *write to express, not to impress; write to communicate, not to confuse.* If your reader must use a dictionary, you are not writing clearly. Read the following unclear example.

EXAMPLE ▶

Words like "nonduplicatable" and phrases like "execute without abnormal termination" are hard to understand. For clarity, write "do not duplicate secure messages" and "JCL system testing will ensure that applications continue to work."

> The following rules are to be used when determining whether or not to duplicate messages:
>
> - Do not duplicate nonduplicatable messages.
> - A message is considered nonduplicatable if it has already been duplicated.
>
> Your job duties will be to ensure that distributed application modifications will execute without abnormal termination through the creation of production JCL system testing.

Following is a list of difficult, out-of-date terms and the modern alternatives:

OBSCURE WORDS	ALTERNATIVE WORDS
aforementioned	discussed above
initial	first
in lieu of	instead of
accede	agree
as per your request	as you requested
issuance	send
this is to advise you	I'd like you to know
subsequent	later
inasmuch as	because
ascertain	find out
pursuant to	after
forward	mail
cognizant	know
endeavor	try
remittance	pay
disclose	show
attached herewith	attached
pertain to	about
supersede	replace
obtain	get

Define Acronyms, Abbreviations, and Jargon.   In addition to obscure, unclear words, a similar obstacle to readers is created by acronyms, abbreviations, and jargon.

We have all become familiar with common acronyms such as *scuba* (<u>s</u>elf-<u>c</u>ontained <u>u</u>nderwater <u>b</u>reathing <u>a</u>pparatus), *radar* (<u>ra</u>dio <u>d</u>etecting <u>a</u>nd <u>r</u>anging), *NASA* (<u>N</u>ational <u>A</u>eronautics and <u>S</u>pace <u>A</u>dministration), *FICA* (<u>F</u>ederal <u>I</u>nsurance <u>C</u>ontributions <u>A</u>ct), and *MADD* (<u>M</u>others <u>A</u>gainst <u>D</u>runk <u>D</u>riving)—single words created from the first letters of multiple words. We are comfortable with abbreviations like *FBI* (<u>F</u>ederal <u>B</u>ureau of <u>I</u>nvestigation), *JFK* (<u>J</u>ohn <u>F</u>. <u>K</u>ennedy), *NFL* (<u>N</u>ational <u>F</u>ootball <u>L</u>eague), *IBM* (<u>I</u>nternational <u>B</u>usiness <u>M</u>achines), and *LA* (<u>L</u>os <u>A</u>ngeles). Some jargon (in-house language) has become so common that we reject it as a cliché. Baseball jargon is a good example. It is hard to tolerate sportscasters who speak baseball jargon, describing line drives as "frozen ropes" and fast balls as "heaters."

More often than not, acronyms, abbreviations, and jargon cause problems, not because they are too common but because no one understands them. Your technical communication loses clarity if you depend on them. You have to decide when to use acronyms, abbreviations, and jargon and how to use them effectively. One simple rule is to define your terms. You can do so either parenthetically or in a glossary. Rather than just writing *CIA*, write *CIA* (*cash in advance*). Such parenthetical definitions, which are only used once per correspondence, don't take a lot of time and won't offend your readers. Instead, the result will be clarity.

If you use many potentially confusing acronyms or abbreviations, or if you need to use a great deal of technical jargon, then parenthetical definitions might be too cumbersome. In this case, supply a separate glossary.

A *glossary* is an alphabetized list of terms, followed by their definitions, as in Figure 3.1.

Acronyms and Abbreviations

See Chapter 4 for more discussion about defining acronyms and abbreviations.

Use Verbs in the Active Voice Versus the Passive Voice.   Using passive voice verb construction can cause wordiness and a lack of clarity.

BEFORE

It has been decided that Joan Smith will head our Metrology Department.

The preceding sentence is written in the *passive voice* (the primary focus of the sentence, *Joan Smith*, is acted on rather than initiating the action). Passive voice causes two problems:

1. Passive constructions are often unclear. In the preceding sentence, *who* decided that Joan Smith will head the department? To solve this problem and to achieve clarity, replace the vague indefinite pronoun *it* with a precise noun as in the "After" example.
2. Passive constructions are often wordy. Passive sentences always require helping verbs (such as *has been*). In the "After" revised sentence, the helping verb "has been" disappears.

AFTER

Kin Norman decided that Joan Smith will head our Metrology Department.

The revision is written in the *active voice*. When you use the active voice, your subject *(Kin Norman)* initiates the action.

FIGURE 3.1   Glossary

CPA	Certified public accountant
FICA	Federal Insurance Contributions Act (Social Security taxes)
Franchise	Official establishment of a corporation's existence
Gross pay	Pay before deductions
Line of credit	Amount of money that can be borrowed
Net pay	Pay after all deductions
Profit and loss statement	Report showing all income and expenses for a specified time

Another common problem with passive voice construction concerns prepositions. Look at the following "Before" example.

BEFORE

Overtime is favored by hourly workers.

Again, this sentence, written in the passive voice, uses a helping verb *(is)*, has the doer *(hourly workers)* acted on rather than initiating the action, and also includes a preposition *(by)*. The sentence is wordy. Revised, the sentence reads as follows:

AFTER

Hourly workers favor overtime.

By omitting the helping verb *is* and deleting the preposition *by*, the sentence is less wordy and more precise. Occasionally, you can use the passive voice when an individual is less important than an inanimate object or when the individual is unnamed.

EXAMPLE ▶

> The tax preparation software can be learned easily even when the user is new to our CPA firm.

Passive voice is created when an inanimate object ("overtime") takes precedence over people ("workers").

Active voice is created when people take precedence over inanimate objects.

The individual, unnamed "user" is not as important as the inanimate object, "software."

SPOTLIGHT

The Importance of Clarity, Conciseness, and Accuracy

How important is clarity, conciseness, and accuracy in technical communication? For answers to this question, look at the varied communication channels used by **Joel Blobaum,** Community Relations Manager at the Missouri Department of Transportation (MoDOT). MoDOT, which plans, builds, inspects, and maintains over 32,000 miles of roadway, is the sixth largest state highway system in the nation.

As community relations manager, Joel conducts public meetings and writes newsletters, news releases, and daily work zone reports. His audience includes approximately 1,200 current and retired MoDOT employees and thousands of external readers, consisting of citizens, public works employees, and local and state elected officials.

- Elected officials base legislative decisions and funding allocations on Joel's documentation.
- Business consultants and vendors depend on MoDOT news for job and bidding opportunities.
- Community members engage in meetings and surveys to provide feedback for MoDOT construction plans.

In his hard-copy, electronic, and online communication (Facebook, Twitter, MoDOT's blog, YouTube, podcasts, and Widgets), Joel must be clear, concise, and accurate. Why? Look at the following clear and concise documentation from Joel's daily "Road Zones Today" updates, provided online in MoDOT's Web site:

I-70: Shoulder-repair crews will work in the left lane of westbound I-70 between I-470 and Lee's Summit Road from 9 a.m. to 3 p.m. Friday. Bridge-maintenance crews will close the right lane of westbound I-70 between Grain Valley and west of Adams Dairy Parkway from 9 a.m. to 3 p.m. Monday.

Route 291: Pavement-repair crews will work on southbound Route 291 between Kentucky Road and Route 78 from 9 a.m. to 2 p.m. weekdays through March 16, and on northbound Route 291 between Route 78 and Route 24 from 9 a.m. to 2 p.m. weekdays March 12-30.

I-35 at Route 69: Bridge-maintenance crews will close the right lane of northbound I-35 between Route 69 and north of I-435 from 9 a.m. to 3 p.m. Friday.

In each of the examples above, Joel writes short paragraphs, provides specific details, and answers the reporter's questions (who, what, when, where, why, and how). His audience can access this information quickly and use the news to plot their driving routes around the city. Thus, Joel's content must be accurate. If his information is incorrect, and people get snarled in traffic, bad things can occur—accidents, delays, and poor public relations.

Simplifying Words, Sentences, and Paragraphs for Conciseness

A second major goal in technical communication is conciseness, providing detail in fewer words. Conciseness is important for at least three reasons.

Conciseness Saves Time

Technical communication in the workplace is time consuming. American workers spend on average 30 percent of their time writing. Therefore, conciseness in writing can help save some of this time.

Conciseness Aids Clarity

Concise writing can aid comprehension. If you dump an enormous number of words on your readers, they might give up before finishing your correspondence or skip and skim so much that they miss a key concept. Wordy writing will lead your readers to think, "Oh no! I'll never be able to finish that. Maybe I can skim through it. I'll probably get enough information that way." Conciseness, on the other hand, makes your writing more appealing to your readers. They'll think, "Oh, that's not too bad; I can read it easily." If they can read your correspondence easily, they will read it with greater interest and involvement. This, of course, will aid their comprehension and increase readability—the audience's ability to easily read the text.

Technology Demands Conciseness

Technology is impacting the size of your technical communication. The size of the screen and the character limitations placed on the length of the communication by Twitter and instant messages make the difference. Thus, when you write, you need to consider the way in which technology limits your space. Today, more and more, effective technical communication must be concise enough to *fit in a box*.

Notice how the size of the "box" containing the following communication affects the way you package your content (see Figure 3.2 and Table 3.1).

DOT.COM UPDATES

For more information about active versus passive voice, check out the following link:

- http://www.plainlanguage.gov/whatisPL/govmandates/memo.cfm. President Clinton's "Presidential Memorandum on Plain Language" mandates plain language in government documents, focusing specifically on clarity and conciseness.

FIGURE 3.2 The Shrinking Size of Technical Communication

TABLE 3.1 Size Characteristics of Communication Channels

Communication Channel	Overall Size of the Screen	Lines per Page or Screen	Characters per Line
Hard-copy paper	8½ × 11 inches	55	70–80
E-mail screen	4 × 6 inches	20–22	60–70
PDA screen	2 × 2 inches	10–15	30–35
Cell phone screen	2 × 1 inch	4–6	10–15

Resumes. A one-page resume is standard. One hard-copy page measures 8½ × 11 inches. That is a box. In fact, this box (one page of text) allows for only about 55 lines of text, and each line of text allows for only about 70 to 80 characters. (A "character" is every letter, punctuation mark, or space.)

E-mail Messages. In contrast, much of today's written communication in the workplace is accomplished through e-mail messages. In Chapter 1, we quote statistics stating that close to 100 percent of business employees report using e-mail every day ("Writing: A Ticket to Work").

Because e-mail screens tend to be smaller (around only 4 × 6 inches) than the typical computer screen, e-mails should be brief and concise. Yes, you can scroll an e-mail message endlessly, but no one wants to do that. In fact, the reason that readers like a one-page resume is that one page allows for what is known as the "W-Y-S-I-W-Y-G" factor ("What You See Is What You Get"). Readers like to see what they will be getting in the correspondence. In contrast, if you make your readers scroll endlessly in e-mail, they do not see what they get. This causes problems. Therefore, a good e-mail message should fit in the box, letting the reader see the entire content at one glance. It should be limited to about 20 lines of text.

Mobile Messaging. The technological impact is even more dramatic when you consider the screen size for handheld, mobile messaging equipment such as PDAs, cell phones, and combination products. The Verizon enV™ screen display allows for 11 lines of text. The Palm Treo Smartphone™ and Blackberry have screen sizes that are about 2 × 2 inches. The Apple iPhone's display screen is only 3½ inches wide. In "Serving the Electronic Reader," Linda E. Moore says, "More and more e-readers are accessing documents using cell phones, PDAs, and other wireless devices" (17). Technical communicators need "to figure out how to create content that works on a four-line cell phone screen" (Perlin 4–8).

Online Help

See Chapter 14 for more discussion of online help.

Online Help Screens. Another example of technical communication that must fit in a box is online help screens. If you work in information technology or computer sciences, you might need to create or access online help screens. If you do so, then your text will be limited by the size of these screens.

PowerPoint

See Chapter 19 for more discussion of PowerPoint.

Microsoft PowerPoint. You also must fit your technical communication within a box (or boxes) when you use Microsoft PowerPoint software.

Twitter and Instant Messages. A further size limitation created by technology is evident in tweets and instant messages. A tweet (the name for messages sent via Twitter) is limited to 140 characters. Instant messages are limited to 160 characters.

Technology requires that you write concisely. If you are filling out an online form, you will be limited by the size of the form's fields. If you are writing correspondence that will be viewed as a PowerPoint presentation, an online help screen, or e-mail sent to someone's PDA or cell phone, you need to limit the size of your text. Therefore, as you write, consider the impact of technology and write concisely.

Limit Paragraph Length

Let's look at some poor writing—writing that is wordy, time consuming to read, and not easily comprehensible.

BEFORE

Please prepare to supply a readout of your findings and recommendations to the officer of the Southwest Group at the completion of your study period. As we discussed, the undertaking of this project implies no currently known incidences of impropriety in the Southwest Group, nor is it designed specifically to find any. Rather, it is to assure ourselves of sufficient caution, control, and impartiality when dealing with an area laden with such potential vulnerability. I am confident that we will be better served as a company as a result of this effort.

This "Before" paragraph is not easy to understand. The difficulty is caused by the lengthy paragraph, words, and sentences. An excessively long paragraph is ineffective. In a long paragraph, you force your reader to wade through many words and digest large amounts of information. This hinders comprehension. In contrast, short, manageable paragraphs invite reading and help your reader understand your content.

A paragraph in a technical document should consist of no more than four to six typed lines, or no more than 50 words. Sometimes you can accomplish these goals by cutting your paragraphs in half; find a logical place to stop a paragraph and then start a new one. Even the previous poorly written example can be improved in this way.

AFTER

Please prepare to supply a readout of your findings and recommendations to the officer of the Southwest Group at the completion of your study period. As we discussed, the undertaking of this project implies no currently known incidences of impropriety in the Southwest Group, nor is it designed specifically to find any.

Rather, it is to assure ourselves of sufficient caution, control, and impartiality when dealing with an area laden with such potential vulnerability. I am confident that we will be better served as a company as a result of this effort.

The writing is still difficult to understand, but at least it is a bit more manageable. You can read the first paragraph, stop, and consider its implications. Then, once you have grasped its intent, you can read the next paragraph and try to tackle its content. The paragraph break gives you some room to breathe.

Limit Word and Sentence Length

In addition to the length of the example paragraph, the writing is flawed because the paragraph is filled with excessively long words and sentences. This writer has created an impenetrable wall of haze—the writing is foggy. In fact, we can determine how foggy this prose is by assessing it according to Robert Gunning's fog index.

Use the Fog Index. The following is Gunning's mathematical way of determining foggy writing:

1. Count the number of words in successive sentences. Once you reach approximately 100 words, divide these words by the number of sentences. This will give you an average number of words per sentence.

2. Count the number of long words within the sentences you have just reviewed. Long words are those with three or more syllables. You cannot count (a) proper names, like Leonardo Dicaprio, Julio Iglesias, Helio Castroneves, or LaDainian Tomlinson; (b) long words that are created by combining shorter words, such as *chairperson* or *firefighter*; or (c) three-syllable verbs created by *–ed* or *–es* endings, such as *united* or *arranges*. Discounting these exceptions, count the remaining multisyllabic words. (A good example of a multisyllabic word is the word *mul-ti-syl-lab-ic.*)

3. To determine the fog index, add the number of words per sentence and the number of long words. Then multiply your total by 0.4.

Given this system, let's see how the original difficult, "Before" paragraph scores. The paragraph is composed of 92 words in four sentences. Thus, the average number of words per sentence is 23. The paragraph contains 16 multisyllabic words (*recommendations, officer, completion, period, undertaking, currently, incidences, impropriety, specifically, sufficient, impartiality, area, potential, vulnerability, confident,* and *company*).

$$
\begin{array}{ll}
23 & \text{(words per sentence)} \\
+\ 16 & \text{(multisyllabic words)} \\
\hline
39 & \text{(total)} \\
\times\ 0.4 & \text{(fog factor)} \\
\hline
=15.6 & \text{(fog index)}
\end{array}
$$

What does a fog index of 15.6 mean? Look at Table 3.2. It shows that the paragraph is written at a level midway between college junior and senior, definitely above the danger line.

A fog index of 15.6 enters the danger zone for three reasons:

- Approximately 28 percent of Americans graduate from college ("College Degree"). Thus, if you are writing at a college level, you could be alienating approximately 72 percent of your audience.
- College graduates do not read well. According to the National Assessment of Adult Literacy, "The average American college graduate's literacy in English declined

TABLE 3.2 Fog Index and Reading Level

	Fog Index	By Grade	By Magazine
	17	College graduate	No popular magazine scores this high.
	16	College senior	
	15	College junior	
	14	College sophomore	
Danger Line	13	College freshman	
	12	High school senior	*Atlantic Monthly*
	11	High school junior	*Time* and *Newsweek*
	10	High school sophomore	*Reader's Digest*
	9	High school freshman	*Good Housekeeping*
	8	Eighth grade	*Ladies' Home Journal*
	7	Seventh grade	Modern romances
	6	Sixth grade	Comics

significantly over the past decade." Only 31 percent of college graduates could read lengthy, complex English texts and draw complicated inferences (Dillon).

- Reading habits are changing. Many of today's readers "power browse"—they skim rather than read text in depth. Much of this, according to studies, is due to the emergence of social media. Look at the following statistics which highlight the change in reading habits of average college students (Self 11):

Number of books read in a year	8
Number of Web pages read in a year	2,300
Number of Facebook profiles viewed in a year	1,281
Number of pages of college assignments written in a year	42
Number of e-mail messages written in a year	500

Given these facts, many businesses ask their employees to write at a sixth- to eighth-grade level. To accomplish this, you would have to strive for an average of approximately 15 words per sentence and no more than 5 multisyllabic words per 100 words.

$$15 \text{ (words per sentence)} + 5 \text{ (multisyllabic words)} = 20 \times 0.4 \text{ (fog factor)}$$
$$= 8.0 \text{ (fog index level)}$$

You cannot always avoid multisyllabic words. Scientists would find it impossible to write if they could never use words like *electromagnetism*, *nitroglycerine*, *telemetry*, or *trinitrotoluene*. The purpose of a fog index is to make you aware that long words and sentences can create reading problems. In addition, long words with multiple syllables add to the character count in a tweet or instant message. Therefore, although you cannot always avoid long words, you should be careful when using them. Similarly, you cannot always avoid lengthy sentences. However, try not to rely on sentences over 15 words long. Vary your sentence lengths, relying mostly on sentences less than 15 words long.

Use the Meat Cleaver Method of Revision. One way to limit the number of words per sentence is to cut the sentence in half or thirds. The following sentence, which contains 43 words, is too long.

BEFORE

To maintain proper stock balances of respirators and canister elements and to ensure the identification of physical limitations that may negate an individual's previous fit-test, a GBC-16 Respirator Request and Issue Record will need to be submitted for each respirator requested for use.

AFTER

Please submit a GBC-16 Respirator Request and Issue Record for each requested respirator. We then can maintain proper respirator and canister element stock balances. We also can identify physical limitations that may negate an individual's previous fit-test.

Using the meat cleaver approach makes this sentence more concise and easier to understand. The "Before" sentence, now rewritten as three sentences, is easier to read.

Avoid Shun Words. In the preceding examples, the "Before" sentence contained 43 words; the revised "After" version is composed of three sentences totaling 37 words. Where did the missing six words go?

One way to write more concisely is to shun words ending in *–tion* or *–sion*—words ending in a *shun* sound. For example, the original sentence reads "to ensure the identifica*tion* of physical limita*tions*." To revise this, you could write, "identify physical challenges." That's three words versus seven in the original version. Deleting four words in this case reduces wordiness by over 50 percent. "Shun" words are almost always unnecessarily wordy.

Using Microsoft Word 2010 to Check the Readability Level of Your Text

Microsoft Word 2010 allows you to check the "readability" level of your writing. Doing so will let you know the following:

- How many words you've used
- How many sentences you've written
- How many words per sentence
- If you've used passive voice constructions
- The grade level of your writing

1. Click on the **File** tab, and then click on **Options**.

2. From the **Word Options** pop up, click on **Proofing**.

3. Click on the **Show readability statistics** button and then **OK**.

Once you have enabled this readability feature, open a file and click **ABC Spelling & Grammar**.

(Continued)

When Word has finished checking the spelling and grammar, you will see a display similar to the one below.

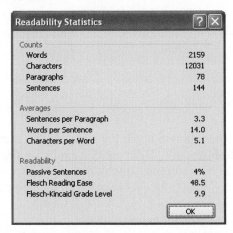

In this example, the text consisted of 2159 words, 78 paragraphs, and 144 sentences. This averaged out to 14 words per sentence, equaling about 10th grade level writing. Four percent of the sentences were written in passive voice.

Let's try another example. Instead of writing "I want you to take into considera*tion* the following," you could write "consider the following." That is three words versus nine, a 66.6 percent savings in word count.

Look at the following "shun" words and their concise versions:

SHUN WORDS	CONCISE VERSIONS
came to the conclu*sion*	concluded (or decided)
with the excep*tion* of	except for (or but)
make revi*sions*	revise (or change, edit, fix)
investiga*tion* of the	investigate
consider implementa*tion*	implement
utiliza*tion* of	use

Avoid Camouflaged Words.   Camouflaged words are similar to "shun" words. In both instances, a key word is buried in the middle of surrounding words (usually helper verbs or unneeded prepositions). For example, in the phrase "with the exception of," the key word "except" is camouflaged behind the unneeded *with*, *the*, *–tion*, and *of*. Once we prune away these unneeded words, the key word "except" is left, making the sentence less wordy.

Camouflaged words are common. Here are some examples and their concise versions:

CAMOUFLAGED WORDS	CONCISE VERSIONS
make an *amend*ment to	amend (or change)
make an *adjust*ment of	adjust
have a *meet*ing	meet
*thank*ing *you* in advance	thank you
for the purpose of *discuss*ing	discuss
arrive at an *agree*ment	agree
at a *later* moment	later

Avoid the Expletive Pattern.   Another way to write more concisely is to avoid the following expletives:

- *there* is, are, was, were, will be
- *it* is, was

Both these expletives (*there* and *it*) lead to wordy sentences. For example, consider the following sentence.

EXAMPLE ▶

> There are three people who will work for Acme.

This sentence can be revised to read, "Three people will work for Acme." The original sentence contains nine words; the revision has six. Deleting the expletive *there* omits three words. Deleting three words in one sentence might not seem like an achievement; however, if you can delete three words from every sentence, the benefits add up.

The expletive *it* creates similar wordiness, as in the following sentence.

EXAMPLE ▶

> It has been decided that ten engineers will be hired.

By deleting the expletive *it*, the sentence reads, "Ten engineers will be hired." The original sentence contained ten words; the revision has only five.

Omit Redundancies.   Redundancies are words that say the same thing. Conciseness is achieved by saying something once rather than twice. For example, in each of the following instances, the boldface words are redundant:

during **the year of** 2012
  (2012 is a year; the words *the year of* are redundant.)
in **the month of** December
  (As in the preceding example, *the month of* is redundant.)
needless to say
  (If it's *needless to say*, why say it?)
The computer will cost **the sum of** $1,000.
  (One thousand dollars *is* a sum.)
the results **so far achieved** prove
  (A result, by definition, is something that has been achieved.)
our **regular** monthly status reports require
  (Monthly status reports must occur every month; regularity is a prerequisite.)
We collaborated **together** on the project.
  (One can't collaborate alone.)
the **other** alternative is to
  (Every alternative presumes that some option exists.)
This is a **new** innovation.
  (All innovations are new.)
the consensus **of opinion** is to
  (The word *consensus* implies opinion.)

Avoid Wordy Phrases.   Sentences may be wordy not because you have been redundant or because you have used "shun" words, camouflaged words, or expletives. Sometimes sentences are wordy simply because you've used wordy phrases.

Here are examples of wordy phrases and their concise revisions:

WORDY PHRASES	CONCISE REVISIONS
in order to purchase	to buy
at a rapid rate	fast (or state the exact speed)
it is evident that	evidently
with regard to	about
in the first place	first
a great number of times	often (or state the number of times)
despite the fact that	although
is of the opinion that	thinks
due to the fact that	because
am in receipt of	received
enclosed please find	enclosed is
as soon as possible	by 11:30 A.M.
in accordance with	according to
in the near future	soon (or provide the date)
at this present writing	now (or provide the date)
in the likely event that	if
rendered completely inoperative	broken

Achieving Accuracy in Technical Communication

Clarity and conciseness are primary objectives of effective technical communication. However, if your writing is clear and concise but incorrect—grammatically or contextually—then you have misled your audience and destroyed your credibility. To be effective, your technical communication must be *accurate*.

Accuracy in technical communication requires that you *proofread* your text. The examples of inaccurate technical communication below are caused by poor proofreading (we have underlined the errors to highlight them).

First City Federal Savings and Loan
1223 Main
Oak Park, Montana

October 12, 2012

Mr. and Mrs. David Harper
2447 N. Purdom
Oak Park, Montana

Dear Mr. and Mrs. <u>Purdom</u>:

Note that the savings and loan incorrectly typed the customer's street rather than the last name.

> National Bank
> 1800 Commerce Street
> Houston, TX
>
> September 9, 2012
>
> Adler's Dog and <u>Oat</u> Shop
> 8893 Southside
> Bellaire, TX
>
> Dear <u>Sr.</u>:
> In response to your request, your account with us has been <u>close</u> out. We <u>are</u> <u>submitted</u> a check in the amount of $468.72 (your existing balance). If you have any questions, please <u>fill</u> free to <u>contact</u> us.

In addition to all the other errors, it should be "Dog and <u>Cat</u> Shop." The errors make the writer look incompetent.

To ensure accurate writing, use the following proofreading tips:

1. **Let someone else read it.** We miss errors in our own writing for two reasons. First, we make the error because we don't know any better. Second, we read what we think we wrote, not what we actually wrote. Another reader might help you catch errors.

2. **Use the gestation approach.** Let your correspondence sit for a while. Then, when you read it, you'll be more objective.

3. **Read backwards.** You can't do this for content. You should read backwards only to slow yourself down and to focus on one word at a time to catch typographical errors.

4. **Read one line at a time or print it out.** Use a ruler or scroll down your PC screen to isolate one line of text. Again, this slows you down for proofing. You also can proofread more effectively by printing the document and then reading it for errors, line by line.

5. **Read long words syllable by syllable.** How is the word *responsiblity* misspelled? You can catch this error if you read it one syllable at a time (re-spon-si-bil-i-ty).

6. **Use technology.** Computer spell checks are useful for catching most errors. They might miss proper names, homonyms (*their, they're,* or *there*) or incorrectly used words, such as *device* to mean *devise*.

7. **Check figures, scientific and technical equations, and abbreviations.** If you mean $400,000, don't write $40,000. Double-check any number or calculations. If you mean to say *HCl* (hydrochloric acid), don't write *HC* (a hydrocarbon).

8. **Read it out loud.** Sometimes we can hear errors that we cannot see. For example, we know that *a outline* is incorrect. It just sounds wrong. *An outline* sounds better and is correct.

9. **Try scattershot proofing.** Let your eyes roam around the page at random. Sometimes errors look wrong at a glance. If you wander around the page randomly reading, you often can isolate an error just by stumbling on it.

10. **Use a dictionary.** If you are uncertain, look it up.

If you commit errors in your technical communication, your readers will think one of two things about you and your company: (1) They will conclude that you are unprofessional, or (2) they will think that you are lazy. In either situation, you lose. Errors create a negative impression at best; at worst, a typographical error relaying false figures, calculations, amounts, equations, or scientific or medical data can be disastrous.

Organizing Technical Communication

If you are clear, concise, and accurate, but no one can follow your train of thought because your text rambles, you still haven't communicated effectively. Successful technical communication also must be well organized. No one method of organization always works. Following are five patterns of organization that you can use to help clarify content.

Spatial

If you are writing a technical specification to describe the parts of a machine or a plot of ground, you might want to organize your text spatially. You would describe what you see as it appears in space—left to right, top to bottom, inside to outside, clockwise, or by direction (north, south, east, or west). These spatial sequences help your readers visualize what you see and, therefore, better understand the physical qualities of the subject matter. They can envision the layout of the land you describe or the placement of each component within the machine.

For example, let's say you are a contractor describing how you will refinish a basement. Your text reads as follows:

Organization

See Chapter 12 for more discussion of spatial and chronological organization in technical descriptions and process analyses.

> At the basement's north wall, I will build a window seat 7' long by 2' wide by 2' high. To the right of this seat, on the east wall, I will build a desk 4' high by 5' long by 3' wide. On the south wall, to the left of the door, I will build an entertainment unit the height of the wall including four, 4' high by 4' wide by 2' deep shelving compartments. The west wall will contain no built-ins. You can use this space to display pictures and to place furniture.

◀ EXAMPLE

Note how this text is written clockwise, uses points of the compass to orient the reader, and includes the transitional phrases "to the right" and "to the left" to help the reader visualize what you will build. That's spatial organization.

Chronological

Whereas you would use spatial organization to describe a place or a piece of equipment, you would use chronology to document time, steps in a process analysis, or the steps in an instruction. For example, an emergency medical technician (EMT) reporting services provided during an emergency call would document those activities chronologically.

> At 1:15 P.M., we arrived at the site and assessed the patient's condition, taking vitals (pulse, respiration, etc.). At 1:17 P.M., after stabilizing the patient, we contacted the hospital and relayed the vitals. By 1:20 P.M., the patient was on an IV drip and en route to the hospital. Our vehicle arrived at the hospital at 1:35 P.M. and hospital staff took over the patient's care.

◀ EXAMPLE

Chronology also would be used to document steps in an instruction. No times would be provided as in the EMT report. In contrast, the numbered steps would denote the chronological sequence a reader must follow.

Importance

Your page of text is like real estate. Certain areas of the page are more important than others—location, location, location. If you bury key data on the bottom of a page, your reader might not see the information. In contrast, content placed approximately one-third from the top of the page and two-thirds from the bottom (eye level) gets more attention. The same applies to a bulleted list of points. Readers will focus their attention on the first several points more than on the last few.

Knowing this, you can decide which ideas you want to emphasize and then place that information on the page accordingly. Organize your ideas by importance. Place the more important ideas above the less important ones.

The following "Before" agenda is incorrectly organized.

BEFORE

Agenda

- Miscellaneous ideas
- Questions from the audience
- Refreshments
- Location, date, and time
- Main topic
- Guest speakers

Which of these points is most important? Certainly the first two items are not important. A better list would be organized by importance, as in the following "After" example.

AFTER

Agenda

- Main topic
- Guest speakers
- Location, date, and time
- Refreshments
- Questions from the audience
- Miscellaneous ideas

Comparison/Contrast

Many times in business you will need to document options and ways in which you surpass a competitor. These require that you organize your text by comparison/contrast. You compare similarities and contrast differences. For example, if you are writing a sales brochure, you might want to present your potential client alternatives regarding services, personnel, timetables, and fee structures. Table 3.3 shows how comparison/contrast provides the client options for cost and features. Each housing option is comparable in that the developer provides a four-bedroom, three-bath home with fully equipped kitchen. However, the homes contrast regarding garage size, deck availability, and basement. These options then

TABLE 3.3 Housing Costs

Item	Features	Cost
The Broadmoor	4 bedrooms, 3½ baths 2-car garage fully equipped kitchen	$300,000
The Aspen	4 bedrooms, 3½ baths finished basement 3-car garage fully equipped kitchen	$340,000
The Regency	4 bedrooms, 3½ baths patio/deck finished basement with ½ bath 3-car garage fully equipped kitchen	$380,000

affect the cost. The table, organized according to comparison/contrast, helps the reader understand these distinctions.

Problem/Solution

Proposals and sales letter are often problem/solution oriented. When you write a proposal, for instance, you are proposing a solution to an existing problem. If your proposal focuses on new facilities, your reader's current building must be flawed. If your proposal focuses on new procedures, your reader's current approach to doing business must need improvement. Similarly, if your sales letter promotes a new product, your customers will purchase it only if their current product is inferior.

To clarify the value of your product or service, therefore, you should emphasize the readers' need (their problem) and show how your product is the solution. Note how the following summary from a proposal is organized according to problem/solution.

> Your city's 20-year-old wastewater treatment plant does not meet EPA requirements for toxic waste removal or ozone depletion regulations. This endangers your community and lessens property values in its neighborhoods.
>
> Anderson and Sons Engineering Company has a national reputation for upgrading wastewater treatment plants. Our staff of qualified engineers will work in partnership with your city's planning commission to modernize your facilities and protect your community's values.

◀ **EXAMPLE**

The summary's first paragraph identifies the problem. The second paragraph promotes the solution. The problem/solution organization clarifies the writer's intent.

Recognizing the Importance of Ethical Communication

In addition to achieving clarity, conciseness, and accuracy in technical communication, ethics are a final important objective. What is ethical communication? Ethical communication is an honest expression of ideas about any product or service and considers the needs of both the consumer and the business.

Here is the scenario. You are a biomedical equipment salesperson. Your company has created a new piece of equipment to be marketed worldwide. Part of the sales literature that your boss tells you to share with potential customers contains the following sentence.

> NOTE: Our product has been tested for defects and safety by trained technicians.

◀ **EXAMPLE**

When read literally, this sentence is true. The product has been tested, and the technicians are trained. However, you know that the product has been tested for only 24 hours by technicians trained on site without knowledge of international regulations. As a loyal employee, are you required to do as your boss requests? Even though the statement is not completely true, can you legally and ethically include it in your sales literature?

The answer to both questions is no! You have an ethical responsibility to write the truth. Your customers expect it, and it is in the best interests of your company. Equally important, including the sentence in your sales literature is illegal. Though the sentence is essentially true, it implies something that is false. Readers will assume that the product has been *thoroughly* tested by technicians who have been *correctly* trained. Thus, the sentence deceives the readers. Such comments are "actionable under law" if they lead to false impressions (Wilson WE-68). If you fail to properly disclose information, including dangers, warnings, cautions, or notes like the sentence in question, then your company is legally liable.

Knowing this, however, does not make writing easy. Ethical dilemmas exist in corporations. The question is, what should you do when confronted with such problems? One way to solve this dilemma is by checking your actions against these three concerns: legal, practical, and ethical. For example, if you plan to write operating instructions for a mechanism, will your text be

1. *Legal,* focusing on liability, negligence, and consumer protection laws?
2. *Practical,* because dishonest technical communication backfires and can cause the company to lose sales or to suffer legal expenses?
3. *Ethical,* written to promote customer welfare and avoid deceiving the end user? (Bremer et al. 76–77)

These are not necessarily three separate issues. Each interacts with the other. Our laws are based on ethics and practical applications. Ethical considerations in technical communication are so important that we provide additional coverage of this topic in the following chapters:

- Chapter 5, Research and Documentation
- Chapter 7, Building Online Communities
- Chapter 8, Communicating Effectively to Get a Job
- Chapter 10, Using Visual Aids in Technical Communication
- Chapter 11, Communicating to Persuade
- Chapter 13, Instructions, User Manuals, and Standard Operating Procedures
- Chapter 14, Online Help and Web Sites
- Chapter 17, Long, Formal Reports
- Chapter 18, Proposals
- Chapter 19, Oral Communication

Legalities

When asked to write text that profits the company but deceives the customer, you might question where your loyalties lie. After all, the boss pays the bills, but your customers might also be your next-door neighbors. Such conflicts exist and challenge all employees. What do you do? You should trust your instincts and trust the laws.

Laws are written to protect the customer, the company, and you—the employee. Corporate, employee, and customer rights are covered by many laws and organizations. Your company must adhere legally to laws such as the Family Educational Rights and Privacy Act (FERPA), the Family and Medical Leave Act (FMLA), and the Health Insurance Portability and Accountability Act (HIPAA). Your company must meet the legal expectations of the Equal Employment Opportunity Commission (EEOC) or of the Occupational Safety and Health Administration (OSHA). If you believe you are being asked to do something illegal that will harm your community, seek legal counsel.

Practicalities

Even though it might appear to be in the best interests of the company to hide potentially damaging information from customers, such is not the case. First, as a technical communicator your goal is candor. That means you must be truthful, stating the facts. It also means you must not lie, keeping silent about facts that are potentially dangerous (Girill 178–179). Second, practically speaking, the best business approach is *good* business. The goal of a company is not just making a profit, but making money the right way—"good ethics is good business" (Guy). What good is it to earn money from a customer who will never buy from you again or who will sue for reparation? That is not practical.

One simple example of the importance of practicality in technical writing is your resume. Let's say that a job posting requires the applicant to be OSHA certified in the handling of hazardous materials. You don't have this certification, but you want the job. To meet the company's requirements, you state in your resume that you are OSHA certified. Your actions are not only unethical because you have lied on the resume, but also your assertion is impractical. If the company hires you based on your falsehood, and you commit an OSHA violation that leads to a customer illness or injury, the company can be held liable. You most likely will lose your job. Where's the practicality in harming a client, hurting the company, and losing your employment?

Ethicalities

Here's the ultimate dilemma. Defining ethical standards has been challenging for professional organizations. As Shirley A. Anderson-Hancock, manager for the Society for Technical Communication (STC) Ethical Guidelines Committee, states, "Everyone who participates in discussions of ethical issues may offer a different perspective, and many viewpoints may be legitimate" (6). Due to the difficulty of clearly defining what is and what is not ethical, the STC struggled for three years before publishing its guidelines (see Figure 3.3).

These ethical principles, as the STC Ethical Guidelines Committee admits, had to be "sufficiently broad to apply to different working environments" (Anderson-Hancock 6).

One way to clarify these necessarily broad standards is by looking at the Society for Technical Communication's "Code for Communicators," which reads as follows:

Society for Technical Communication's Code for Communicators

As a technical communicator, I am the bridge between those who create ideas and those who use them. Because I recognize that the quality of my services directly affects how well ideas are understood, I am committed to excellence in performance and the highest standards of ethical behavior.

I value the worth of the ideas I am transmitting and the cost of developing and communicating those ideas. I also value the time and effort spent by those who read or see or hear my communication.

I therefore recognize my responsibility to communicate technical information truthfully, clearly, and economically.

My commitment to professional excellence and ethical behavior means that I will

- Use language and visuals with precision.
- Prefer simple, direct expression of ideas.
- Satisfy the audience's need for information, not my own need for self-expression.
- Hold myself responsible for how well my audience understands my message.
- Respect the work of colleagues, knowing that a communication problem may have more than one solution.
- Strive continually to improve my professional competence.
- Promote a climate that encourages the exercise of professional judgment and that attracts talented individuals to careers in technical communication.

("Code for Communicators")

FIGURE 3.3   Ethical Principles for Technical Communicators

Ethical Principles for Technical Communicators

As technical communicators, we observe the following ethical principles in our professional activities.

Legality

We observe the laws and regulations governing our profession. We meet the terms of contracts we undertake. We ensure that all terms are consistent with laws and regulations locally and globally, as applicable, and with STC ethical principles.

Honesty

We seek to promote the public good in our activities. To the best of our ability, we provide truthful and accurate communications. We also dedicate ourselves to conciseness, clarity, coherence, and creativity, striving to meet the needs of those who use our products and services. We alert our clients and employers when we believe that material is ambiguous. Before using another person's work, we obtain permission. We attribute authorship of material and ideas only to those who have made an original and substantive contribution. We do not perform work outside our job scope during hours compensated by clients or employers, except with their permission; nor do we use their facilities, equipment, or supplies without their approval. When we advertise our services, we do so truthfully.

Confidentiality

We respect the confidentiality of our clients, employers, and professional organizations. We disclose business-sensitive information only with their consent or when legally required to do so. We obtain releases from clients and employers before including any business-sensitive materials in our portfolios or commercial demonstrations or before using such materials for another client or employer.

Quality

We endeavor to produce excellence in our communication products. We negotiate realistic agreements with clients and employers or schedules, budgets, and deliverables during project planning. Then we strive to fulfill our obligations in a timely, responsible manner.

Fairness

We respect cultural variety and other aspects of diversity in our clients, employers, development teams, and audiences. We serve the business interests of our clients and employers as long as they are consistent with the public good. Whenever possible, we avoid conflicts of interest in fulfilling our professional responsibilities and activities. If we discern a conflict of interest, we disclose it to those concerned and obtain their approval before proceeding.

Professionalism

We evaluate communication products and services constructively and tactfully, and seek definitive assessments of our own professional performance. We advance technical communication through our integrity and excellence in performing each task we undertake. Additionally, we assist other persons in our profession through mentoring, networking, and instruction. We also pursue professional self-improvement, especially through courses and conferences.

Approved by the STC Board of Directors
September 1998

(STC "Ethical Principles for Technical Communicators")

Guidelines for Achieving Ethical Standards

To explain how you can achieve ethical communication, we have expanded on each of the seven bulleted points mentioned in the STC "Code for Communicators."

Use Language and Visuals with Precision.   In a survey we conducted comparing technical writers and teachers of technical communication, we discovered an amazing finding: Professional technical communicators rate grammar and mechanics higher than teachers do (Gerson and Gerson). On a 5-point scale (5 equaling "very important"), writers rated grammar and mechanics 4.67, whereas teachers rated grammar and mechanics only 3.54. That equals a difference of 1.13, which represents a 22.6 percent divergence of opinion.

Given these numbers, would we be precise in writing "Teachers do not take grammar and mechanics as seriously as writers do"? The numbers accurately depict a difference of opinion, and the 22.6 percent divergence is substantial. However, these figures do not assert that teachers ignore grammar. To say so is imprecise and would constitute an ethical failure to present data accurately. Even though writers are expected to highlight their client's values and downplay their client's shortcomings, technical communicators ethically cannot skew numbers to accomplish these goals (Bowman and Walzer). Information must be presented accurately, and writers must use language precisely.

Precision also is required when you use visuals to convey information. Look at Figures 3.4 and 3.5. Both figures show that sales have risen. However, the "Before" example in Figure 3.4 shows a meteoric rise in sales over ten years without any variations. The "After" example in Figure 3.5 more accurately shows the highs and lows of sales revenue year to year. As with language, the writer is ethically responsible for presenting visual information precisely.

Prefer Simple, Direct Expression of Ideas.   While writing an instructional manual, you might be asked to use legal language to help your company avoid legal problems (Bowman and Walzer). Here is a typically obtuse warranty that legally protects your company.

BEFORE

FIGURE 3.4   Imprecise Depiction of Sales Growth

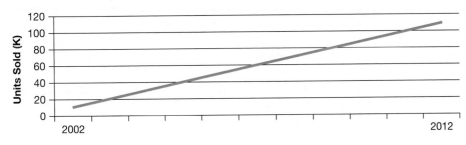

AFTER

FIGURE 3.5   Precise Depiction of Sales Growth

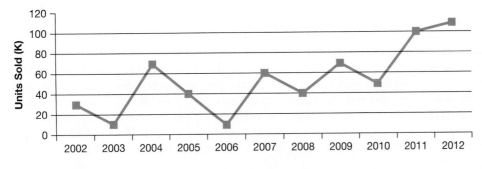

> Acme's liability for damages from any cause whatsoever, including fundamental breach, arising out of this Statement of Limited Warranty, or for any other claim related to this product, shall be limited to the greater of $10,000 or the amount paid for this product at the time of the original purchase, and shall not apply to claims for personal injury or damages to personal property caused by Acme's negligence, and in no event shall Acme be liable for any damages caused by your failure to perform your responsibilities under this Statement of Limited Warranty, or for loss of profits, lost savings, or other consequential damages, or for any third-party claims.

Can your audience easily grasp this warranty's lengthy sentence structure and difficult-to-understand words? No, the readers will become frustrated and will fail to recognize what is covered under such a warranty. Although you are following your boss's directions, you are not fulfilling one of your ethical requirements to use simple and direct language.

Technical communication can be legally binding *and* easy to understand. The first requirement does not negate the second. As a successful technical communicator who values "the time and effort spent by those who read or see or hear [your] communication," you want to aid communication by using simple words and direct expression of ideas whenever possible ("Code for Communicators").

Satisfy the Audience's Need for Information, Not My Own Need for Self-Expression. The importance of simple language also encompasses the third point in STC's "Code for Communicators": satisfying "the audience's need for information, not my own need for self-expression." The previously mentioned warranty might be considered poetic in its sentence structure and sophisticated in its word usage. It does not communicate what the audience needs, however. Such elaborate and convoluted writing will satisfy only one's need for self-expression: That is not the goal of effective technical communication.

Electronic communication, including e-mail messages, blogging, instant messages, text messages, and tweets, present another opportunity for unethical behavior. Because electronic communication is so easy to use, many company employees use it too frequently. They use electronic communication tools for business purposes, but they also send e-mail, text and instant messages, and tweets when writing to family and friends. Employees know not to abuse their company's metered mail by using corporate envelopes and stamps to send in their gas bills or write thank-you notes to Aunt Rose. These same employees, however, will use company-owned electronic communication for personal reasons and on company time. They are not satisfying their business colleagues' need for information; instead, they are using company-owned systems and equipment for their own self-expression (Hartman and Nantz 60). That is unethical.

Hold Myself Responsible for How Well My Audience Understands My Message. As a technical communicator, you must place the audience first. When you write a precise proposal or instruction, using simple words and syntax, you can take credit for helping the reader understand your text. In contrast, you must also accept responsibility if the reader fails to understand your text.

Ethical standards for successful communication require the writer to always remember the readers—the real people who read manuals to put together their children's toys, who read corporate annual reports to understand how their stocks are doing, and who read proposals to determine whether to purchase a service. An ethical writer remembers that these real people will be frustrated by complex instructions, confused by inaccessible stock reports, and misled by inaccurate proposals. It is unethical to forget that your writing can frustrate people and even endanger them. It is unethical to use your writing to mislead or to confuse.

In contrast, as an ethical writer, you need to treat these readers like neighbors, people you care about, and then write accordingly. Take the time to check your facts, present your

FIGURE 3.6 Illustration of a
Warning to Prevent Liability Problems

information precisely, and communicate clearly so that your readers—your friends, co-workers, and clients—are safe and satisfied. You are responsible for your message (Barker).

Liability Laws.   Keeping your audience safe is a key concern for technical communicators. Liability laws cover injuries, such as accidents or exposure to hazardous materials, caused by faulty construction or dangerous products. An ethical technical communicator must warn the audience about dangers. Doing so requires the following:

- Provide clear instructions for use of the product.
- Warn users about potential risks.
- Ensure that the warning is clearly and concisely stated.
- Ensure that the warning is brought to the user's attention through headings, color, and illustrations.

Figure 3.6 is an illustration of an effective way to communicate potential product liability issues.

Respect the Work of Colleagues.   Each of the ethical considerations already discussed relates to your clients, the readers of your technical communication. This fifth point differs in that it relates to co-workers or professional colleagues.

Much of today's technical communication takes place online, electronically. Electronic communication creates two unique reasons for an increased focus on ethics. As the technical communicator, you must ethically consider confidentiality and courtesy when working with the Internet or any private electronic network.

> **Ethically Communicating Online**
>
> See Chapter 6 Routine Correspondence, Chapter 7 Building Online Communities, Chapter 14 Online Help and Web Sites for more discussion about ethics.

- **Confidentiality.** The 1974 Privacy Act allows "individuals to control information about themselves and to prevent its use without consent" (Turner 59). The Electronic Communication Privacy Act of 1986, which applies all federal wiretap laws to electronic communication, states that e-mail messages can be disclosed only "with the consent of the senders or recipients" (Turner 60). However, both of these laws can be abused easily on the Internet. First, neither law specifically defines "consent." Data such as your credit records can be accessed without your knowledge by anyone with the right hardware and software. Your confidentiality can be breached easily.

  Second, the 1986 Electronic Communication Privacy Act fails to define "the sender." In the workplace, a company owns the e-mail system, blog sites, Web sites, and your business phone or PDA, just as it owns other more tangible items such as desks, computers, and file cabinets. Companies have been held liable for electronic messages sent by employees. Thus, many corporations consider the contents of one's e-mail, blog entry, or text message company property, not the property of the employee. With ownership comes the right to inspect an employee's messages. Although you might write an electronic message assuming

Courtesy in E-mail, Instant Messages, Text Messages, and Social Media

See Chapters 6 and 7 for more discussion of courtesy in electronic communication.

that your thoughts are confidential, this message can be monitored without your knowledge (Hartman and Nantz 61).

These instances might be legal, but are they ethical? Even though a company can eavesdrop on your e-mail or access your life's history through databases, that doesn't mean it should. As an employee or corporate manager, you should respect another's right to confidentiality. Ethically, you should avoid the temptation to read someone else's e-mail or to access data about an individual without consent.

- **Courtesy.** As already noted, electronic communication is not as private as you might believe. Whatever you write in your e-mail, text message, or tweet can be read by others. Given this reality, you should be very careful about what you say. Specifically, you don't want to offend co-workers by "flaming" (writing discourteous messages). Before you criticize a co-worker's ideas or ability, and before you castigate your employer, remember that common courtesy, respect for others, is ethical (Adams et al. 328).

Strive Continually to Improve My Professional Competence. Promote a Climate that Encourages the Exercise of Professional Judgment. We've combined these last two tenets from the STC's Code for Communicators because together they sum up all of the preceding ethical considerations. Professional competence, for example, includes one's ability to avoid plagiarizing. Competence should also include professional courtesy, respect for another's right to confidentiality, a sense of responsibility for one's work, and the ability to write clearly and precisely. Ultimately, your competence is dependent upon professional judgment. Do you know the difference between right and wrong? Recognizing the distinction is what ethics is all about.

The Ethics of Intellectual Property Laws

The creator of a unique product, term, image, or invention is protected by intellectual property rights. Types of intellectual property include copyrights and trademarks.

Copyright Laws. Every English teacher you have ever had told you to avoid *plagiarism* (stealing another writer's words, ideas, or images). Plagiarism, unfortunately, is a problem in the workplace. If you were an unethical technical communicator, you could access text and images from many sources and easily print them as your own. An ethical communicator will not fall prey to such temptation. If you have not written it, give the other author credit. Words are like any other possession. Taking words and ideas without attributing your source through a footnote or parenthetical citation is wrong. You should respect copyright laws.

"Every element of a Web page—text, graphics, and HTML code—is protected by U.S. and international copyright laws, whether or not a formal copyright application has been submitted" (LeVie 20–21). The U.S. Copyright Office's "Circular 66, Copyright Registration for Online Works" states the following:

> Copyright protects original authorship fixed in tangible form. For works transmitted online, the copyrightable authorship may consist of text, artwork, music, audiovisual material (including any sounds), sound recordings, etc. Copyright does not protect ideas, procedures, systems, or methods of operation. Under U.S. law, copyright protection subsists from the time the work is fixed in any tangible medium of expression from which it can be perceived, reproduced, or otherwise communicated, either directly or with the aid of a machine or device.

If you and your company "borrow" from an existing Internet site, thus infringing upon that site's copyright, you can be assessed actual or statutory damages. The Copyright

DOT.COM UPDATES

For more information about online copyright laws, check out the following links:

- http://www.bitlaw.com/index.html. **BitLaw** provides information about Internet patent, copyright, trademark, and legal issues.
- http://www.copyright.gov/. The United States Copyright Office.

Act allows the owner of Web material to stop infringement and obtain damages and attorney fees. In addition to financial damages, your company, if violating intellectual property laws, could lose customers, damage its reputation, and lose future capital investments.

Trademark Laws.   In addition to laws protecting copyrighted words and phrases, the same holds true for a company's name; logo, symbol, design, or image that identifies the company; or combinations of these elements. In these instances, you must respect trademark laws. A company can protect its intellectual property by branding products or services with the following trademark symbols:

- ™ (unregistered trademark)
- SM (unregistered service mark)
- ® (registered trademark)

If you copy images or text designated as trademarks by these symbols, the individual or company from whom you have taken this information can sue for trademark infringement.

Boilerplate Content.   However, avoiding intellectual property theft and plagiarism in the workplace is not always clear cut. According to Jessica Reyman, "communicators commonly perform a variety . . . of composing activities that could be considered plagiarism in the context of the classroom" but are not necessarily plagiarism in the workplace (61). For example, workplace communicators often use boilerplate information and templates—text that has been used before. They rely on "existing designs and layouts"; write collaboratively; and cut, paste, and repurpose corporate content (61). Do you need to cite content that has already been used in your company? The answer is yes . . . and no.

If you are writing a proposal to a new client, and you draw from your company's library of existing proposals for content, that is acceptable use of boilerplate content and templates. You do not need to cite the source of your information. However, if you find a proposal from some other company on the Internet, and use their content or images, that is plagiarism and unethical behavior. You must cite the source of this original information or get the company's approval.

Guidelines for Protecting Intellectual Property

To protect your rights and those of others, you should

- Assume that information is covered under copyright and trademark protection laws unless proven otherwise.
- Obtain permission for use from the original creator of graphics or text.
- Cite the source of your information.
- Create your own graphics and text.
- Copyright any information you create.
- Place a copyright notice at the bottom of your Web site (Johnson 17).

Ten Questions to Ask When Confronting an Ethical Dilemma

You have encountered a situation at work which does not have a clear cut solution. You have first asked yourself whether or not the situation is legal, practical, or ethical, but you still do not have an answer. Like many employees, you might feel isolated when confronted by apparently unethical behavior. What should you do next? Consider these questions:

1. Does your company have a policy regarding the situation?
2. Have you discussed the situation with your boss?

3. Have you met with the director of human resources for guidance?

4. Does your company have an ombudsman who provides support or training?

5. What do your colleagues say about the situation?

6. Do you belong to a professional organization with published policies regarding ethical behavior?

7. Do you have a moral authority with whom you can confer?

8. How has your organization dealt with similar situations in the past?

9. Does your company have a way to communicate anonymously about ethical issues, such as a drop box or secure e-mail address?

10. Has your company offered training about ethical behavior in the workplace?

Strategies for Making Ethical Decisions

When confronted with ethical challenges, try following these writing strategies (Guy 165):

Define the problem. Is the dilemma legal, practical, ethical, or a combination of all three?

Determine your audience. Who will be affected by the problem (clients, co-workers, management)? What is their involvement, what are their individual needs, and what is your responsibility—either to the company or to the community?

Maximize values; minimize problems. Ethical dilemmas always involve options. Your challenge is to select the option that promotes the greatest worth for all stakeholders involved. You won't be able to avoid all problems. The best you can hope for is to minimize those problems for both your company and your readers while you maximize the benefits for the same stakeholders.

Consider the big picture. Don't just focus on short-term benefits when making your ethical decisions. Don't just consider how much money the company will make now or how easy the text will be to write now. Focus on long-term consequences as well. Will what you write please your readers so that they will be clients for years to come? Will what you write have a long-term positive impact on the economy or the environment?

Write your text. Implement the decision by writing your memo, letter, proposal, manual, or report. When you write your text, remember to

- Use precise language and visuals.
- Use simple words and sentences.
- Satisfy the audience's need for information, not your own need for self-expression.
- Take responsibility for your content, remembering that real people will follow your instructions or make decisions based on your text.
- Respect your colleagues' confidentiality, be courteous, and abide by copyright laws.
- Promote professionalism and good judgment.

CHAPTER HIGHLIGHTS

1. If your technical communication is unclear, your readers may misunderstand you and then do a job wrong, damage equipment, injure themselves, or contact you for further explanations.
2. Use details to ensure reader understanding. Whenever possible, specify and quantify your information.
3. Answering *who, what, when, where, why,* and *how* (the reporter's questions) helps you determine which details to include.
4. For some audiences, you should avoid acronyms, abbreviations, and jargon.
5. Avoid words that are not commonly used (legalisms, outdated terms, etc.).
6. Write to express, not to impress—to communicate, not to confuse.
7. Avoid passive voice constructions, which tend to lengthen sentences and confuse readers.
8. Writing concisely helps save time for you and your readers.
9. Shorter paragraphs are easier to read, so they hold your reader's attention.
10. When possible, use short, simple words (always considering your reader's level of technical knowledge).
11. Apply readability formulas to determine your text's degree of difficulty.
12. Proofreading is essential to effective technical communication.
13. Well-organized documents are easy to follow.
14. Different organizational patterns—spatial, chronological, importance, comparison/contrast, and problem/solution—can help you explain material.
15. Consider whether or not your technical communication is legal, practical, and ethical.

APPLY YOUR KNOWLEDGE

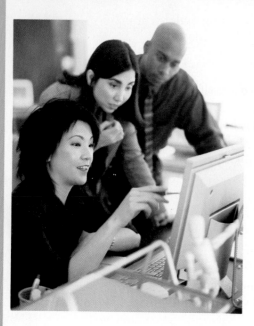

CASE STUDIES

1. CompToday, a computer hardware company, must abide by the Sarbanes-Oxley Act, passed by Congress in response to accounting scandals. This act specifically mandates the following related to documentation standards:

   - **Section 103: Auditing, Quality Control, and Independence Standards and Rules.** Companies must "prepare, and maintain for a period of not less than 7 years, audit work papers, and other information related to any audit report, in sufficient detail to support the conclusions reached in such report."

   - **Section 401(a): Disclosures in Periodic Reports; Disclosures Required.** "Each annual and quarterly financial report . . . must be presented so as not to contain an untrue statement or omit to state a material fact necessary in order to make the pro forma financial information not misleading."

   Beverly Warden, technical documentation specialist at CompToday, is responsible for managing the Sarbanes-Oxley reports. She is being confronted by the following ethical issues:

   a. To help Beverly prepare the first annual report, her chief financial officer (CFO) has given her six months of audits (January through June). These prove that the company is meeting its accounting responsibilities. However, Beverly's report covers the entire year, including July through December. Section 103 states that the report must provide "sufficient detail to support the conclusions reached in [the] report."

Assignment

Are the company's first six months of audits sufficient? If Beverly writes a report stating that her company is in compliance, is she abiding by her Society for Technical Communication Ethical Principles, which state that a technical writer's work is "consistent with laws and regulations"? What are her ethical, practical, and legal responsibilities?

   - Share your findings in an oral presentation.
   - Write a letter, memo, report, or e-mail to your professor stating your opinion regarding this issue.

   b. Beverly's CFO also has told her that during the year, the company fired an outside accounting firm and hired a new one to audit the company books. The first firm expressed concerns about several bookkeeping practices. The newly hired firm, providing a second opinion after reviewing the books, concluded that all bookkeeping practices were acceptable. The CFO sees no reason to mention the first firm.

Assignment

Section 401(a) states that reports must contain no untrue statements or omit to state a material fact. The STC Ethical Principles also say that writing must be truthful and accurate, to the best of a writer's ability. Beverly can report factually that the new accounting firm finds no bookkeeping errors. Should she also report the first accounting firm's assessment? Is that a material fact? If she omits any mention of the first accounting firm, as her boss suggests, is she meeting both her STC technical writer's responsibilities and the needs of Sarbanes-Oxley? What are her ethical, practical, and legal responsibilities?

   - Share your findings in an oral presentation.
   - Write a letter, memo, report, or e-mail to your professor stating your opinion regarding this issue.

2. The Blue Valley Wastewater Treatment Plant processes water running to and from Frog Creek, a water reservoir that passes through the North Upton community. This water is usually characterized by low alkalinity (generally, <30 mg/l), low hardness (generally, <40 mg/l), and minimal water discoloration. Inorganic fertilizer nutrients (phosphorus and nitrogen) are also generally low, with limited algae growth.

Despite the normal low readings, algae-related tastes and odors occur occasionally. Although threshold odors range from 3 to 6, they have risen to 10 in summer months. Alkalinity rises to <50 mg/l, hardness to <60 mg/l, and discoloration intensifies. Taste and odor problems can be controlled by powdered activated carbon (PAC); nutrient-related algae growth can be controlled by filtrated ammonia. Both options are costly.

The odors and tastes are disturbing North Upton residents. The odors are especially bothersome to outdoor enthusiasts who use the trails bordering Frog Creek for biking and hiking. Residents also worry about the impact of increased alkalinity on fish and turtles, many of which are dying, further creating an odor nuisance. Frog Creek is treasured for its wildlife and recreational opportunities.

The Blue Valley Wastewater Treatment Plant is under no legal obligation to solve these problems. The alkalinity, hardness, color, and nutrient readings are all within regulated legal ranges. However, community residents are insistent that their voices are heard and that restorative steps are taken to improve the environment around their homes.

This is an ethical and practical dilemma. In response to this dilemma, divide into small groups to write one of the following documents:

- You are Blue Valley Wastewater Treatment Plant's director of public relations. Write a letter to the City Commission stating your plant's point of view. (We discuss routine correspondence in Chapter 6.)

- You are North Upton community's resident representative. Write a letter to the City Commission stating your community's point of view.

- You are an employee for the Blue Valley Wastewater Treatment Plant. Write an e-mail to the plant director suggesting ways in which the plant could solve this environmental and community relations problem.

- You are the Blue Valley Wastewater Treatment Plant's director. Write an e-mail to the engineering supervisor (your subordinate) stating how to solve this environmental and community relations problem.

These documents could be organized by the problem/solution or comparison/contrast methods. Whichever method you choose, consider the strategies for making ethical decisions, discussed in this chapter.

INDIVIDUAL AND TEAM PROJECTS

Achieving Clarity

The following sentences are unclear. They will be interpreted differently by different readers. Revise these sentences—replace the vague, impressionistic words with more specific information.

1. We need this information as soon as possible.
2. The machinery will replace a flawed piece of equipment in our department.
3. Failure to purchase this will have a negative impact.
4. Weather problems in the area resulted in damage to the computer systems.
5. The most recent occurrences were caused by insufficient personnel.

Avoiding Obscure Words

Obscure words make the following sentences difficult to understand. Improve the sentences by revising the difficult words and making them more easily understood.

1. As Very Large Scale Integration (VLSI) continues to develop, a proliferation of specialized circuit simulators will be utilized.

2. As you requested at the commencement of the year, I am forwarding my regular quarterly missive.

3. Though Randolph was cognizant of his responsibility to advise you of any employment aberrations, he failed to abide by this mandate.

4. Herewith is an explanation of our rationale for proffering services, pursuant to your request.

5. Please be advised that the sale constitutes a successful closure.

Using the Active Voice Versus the Passive Voice

Use of the passive voice often leads to vague, wordy sentences. Revise the following sentences by writing them in the active voice.

1. Implementation of this procedure is to be carried out by the Accounting Department.

2. Benefits derived by attending the conference were twofold.

3. The information was demonstrated and explained in great detail by the training supervisor.

4. Discussions were held with representatives from Allied, who supplied analytical equipment for automatic upgrades.

5. Also attended was the symposium on polymerization.

Limiting Paragraph Length to Achieve Conciseness

You can achieve clarity and conciseness if you limit the length of your paragraphs. An excessively long paragraph (beyond six typed lines) requires too much work for your reader.

Revise the following paragraph to make it more reader friendly.

As you know, we use electronics to process freight and documentation. We are in the process of having terminals placed in the export departments of some of our major customers around the country so they may keep track of all their shipments within our system. I would like to propose a similar tracking mechanism for your company. We could handle all of your export traffic from your locations around the country and monitor these exports with a terminal located in your home office. This could have many advantages for you. You could generate an export invoice in your export department, which could be transmitted via the computer to our office. You could trace your shipments more readily. This would allow you to determine rating fees more accurately. Finally, your accounting department would benefit. All in all, your export operations would achieve greater efficiency.

Reducing Word Length to Achieve Conciseness

Multisyllabic words can create long sentences. To limit sentence length, limit word length. Find shorter words to replace the following words:

1. advise
2. anticipate
3. ascertain
4. cooperate
5. determine

6. endeavor
7. inconvenience
8. indicate
9. initially
10. presently

Reducing Sentence Length to Achieve Conciseness

Each of the following sentences is too long. Revise them using the techniques suggested in this chapter: Use the meat cleaver approach; avoid "shun" words, camouflaged words, and expletives; omit redundancies; and delete wordy phrases.

1. In regard to the progress reports, they should be absolutely complete by the fifteenth of each month.
2. I wonder if you would be so kind as to answer a few questions about your proposal.
3. I am in receipt of your e-mail requesting an increase in pay and am of the opinion that it is not merited at this time due to the fact that you have worked here for only one month.
4. On two different occasions, I have made an investigation of your residence, and I believe that your sump pump might result in damage to your neighbor's adjacent property. I have come to the conclusion that you must take action to rectify this potential dilemma, or your neighbor might seek to sue you in a court of law.
5. If there are any questions that you might have, please feel free to contact me by phone.

Using Organization

1. **Spatial**
   a. Using spatial organization, write a paragraph describing your classroom, office, work environment, dorm room, apartment, or any room in your house.
   b. Using spatial organization, write an advertisement describing the interior of a car, the exterior of a mechanism or tool, a piece of clothing, or an iPod.
2. **Chronological.** Organizing your text chronologically, write a report documenting your drive to school or work, your activities accomplished in class or at work, your discoveries at a conference or on vacation, or your activities at a sporting event.
3. **Importance.** A fashion merchandising retailer asked her buyers to purchase a new line of clothing. In her e-mail, she provided them the following list to help them accomplish their task. Reorganize the list by importance, and justify your decisions.

Date: January 15, 2012
To: Buyers
From: Sharon Baker
Subject: Clothing Purchases

It is time again for our spring purchases. This year, let's consider a new line of clothing. When you go to the clothing market, focus on the following:

- Colors
- Materials
- Our customers' buying habits
- Price versus markup potential
- Quantity discounts
- Wholesaler delivery schedules

Good luck. Your purchases at the market are what make our annual sales successful.

---

4. **Comparison/contrast.** Visit two auto dealerships, two clothing stores, two restaurants, two bookstores, two prospective employers, two colleges, and so on. Based on your discoveries, write a report using comparison/contrast to make a value judgment. Which of the two cars would you buy, which of the two restaurants would you frequent, and at which of the two bookstores would you purchase CDs?

5. **Problem/solution.** You work for Acme Electronics as an electrical engineer. Dejuan Simpson, your boss, informs you that your department's electronic scales are measuring tolerances inaccurately. You are asked to study the problem and determine solutions. In your study, you find that one scale (ID #1893) is measuring within 90 percent of tolerance; another scale (ID #1887) is measuring within 75 percent of tolerance; a third scale (ID #1890) is measuring within 60 percent of tolerance; a final scale (ID #1885) is measuring within 80 percent of tolerance. Standards suggest that 80 percent is acceptable. To solve this problem, the company could purchase new scales ($2,000 per scale); reduce the vibration on the scales by mounting them to the floor ($1,500 per scale); or reduce the vibration around the scales by enclosing the scales in plexiglass boxes ($1,000 per scale).

   Write an e-mail to your boss detailing your findings (the problems) and suggesting the solutions.

PROBLEM-SOLVING THINK PIECES

Considering Ethics

1. The Society for Technical Communication constantly is trying to redefine its Ethical Principles. This is especially challenging as new communication channels, such as YouTube, Facebook, Twitter, and other social media emerge. As a class, how would you define the word *ethics?* Brainstorm new definitions as they apply to technical communication and come to a class consensus. Then in small groups, based on your class's definition of *ethics*, do the following:

   • List five or more examples of ethical responsibilities you believe technical communicators should have when writing letters, reports, proposals, Web sites, e-mail messages, social media, or instructions (other than those already discussed in this chapter).

   • List five or more examples of ethical responsibilities you believe employees should have when using social media in the workplace. Consider ethical lapses you've seen with the use of instant messages, text messages, blog entries, tweets, Facebook sites, or YouTube videos.

   • List five or more examples of failures to abide by ethical responsibilities you have seen either in writing or in other types of media (recordings, movies, television programs, newspapers, magazines, news reports, etc.).

   • Technical communication is factual, as are newspaper, magazine, radio, and television news reports. In contrast, recordings, movies, and television programs are art forms. Do the same ethical considerations apply for all types of media? If there are differences, explain your answer.

2. Bring to class examples of ethically flawed communication. These could include poorly written warranties (which are too difficult to understand) or dangers, warnings, and cautions (which do not clearly identify the potential for harm). You might find misleading annual reports or unethical advertisements. You might have seen videos on YouTube that do not ethically represent a company. Poor examples could even include graphics that are visually misleading. Then in small groups, rewrite these types of communication or redraw the graphics to make them ethical.

3. Mary Madero has been employed at Commercial Savings and Loans for many years. She frequently communicates via e-mail with colleagues in the bank and at other savings and loans organizations. Recently, Mary was elected president of the ASLTS (the Association for Savings and Loans Teller Supervisors). As ASLTS president, she is involved in fundraising for charity, community service outreach projects, organization picnics, and the ASLTS monthly bowling league.

Sometimes she gets e-mail during the workday that relates to her new position as president. In addition, at home, she must answer ASLTS e-mail communication. In the evening and on weekends, she uses her company laptop for additional work related to her position as teller supervisor at the savings and loan.

Assignment

Is it ethical for Mary to use her company's e-mail account, or should she use her personal e-mail account for ASLTS business? Is it ethical for Mary to use the company laptop, or should she be using her own laptop for ASLTS business? Justify your answer based on information provided in this chapter.

4. Nicole Stefani is a public relations employee for ImageSkill, a corporate "make-over" company that works to improve a client's image. One of ImageSkill's new clients, the city of Greenfield, has asked for help. To learn about the city, Nicole planned to interview the city's residents, business owners, and employees.

   Prior to the interviews, Nicole clarified her goals to the city participants. They all were made aware that their comments eventually would be used for reporting purposes. During the interviews, to ensure accuracy, Nicole either taped what the people said or took very thorough notes. Before writing her rough draft, Nicole checked with her sources of information to be sure that their quotes were correct.

   Nicole contacted one individual, Burt Knoblauch, for verification. Upon hearing what Nicole planned to quote, Burt first denied having said it. Nicole, however, had his comment on tape and played it for him. Burt told Nicole that he did not want her to use his comments because his boss and his neighbors might be offended. Nicole reminded Burt that she had told him and all participants that their comments would be used in a follow-up report. She even had signed release forms for Burt and others.

   What should Nicole do? She knows for a fact that her quote is accurate. In addition, Burt's quote is essential to her report, since his comments highlight a key problem facing the city. Still, Burt has told her that using the quote could cause him problems at work and in his neighborhood.

Assignment

What is Nicole's ethical responsibility in this situation? Write an e-mail message to your professor explaining your reasons for whatever decision you make. Justify your answer based on information provided in this chapter.

5. Ta'isha Roberts contacted her company's computer technician to complain about her company laptop. She told the technician that the laptop wasn't working effectively. When her husband used the laptop to make a PowerPoint presentation at their son's elementary school, the 2003 PowerPoint slides he had created on Ta'isha's laptop were not compatible with the school's 2007 PowerPoint capabilities. Ta'isha wants the technician to fix the laptop. However, the technician realizes that Ta'isha is using her corporate-issued laptop for purposes other than work.

Assignment

Is it ethical for Ta'isha's husband to use her company laptop? Is it the technician's ethical responsibility to report this use of company equipment to his boss? Justify your answer in an e-mail message to your professor based on information provided in this chapter.

WEB WORKSHOP

1. Many governmental organizations and technical professions ask employees to follow the standards for "plain language," which focus on clarity and conciseness. To learn more about plain language, visit http://www.plainlanguage.gov. This site provides a definition of plain language, governmental mandates, and before and after comparisons. Click on any of the links in this site and report your findings either orally or in an e-mail message.

2. The Internet plays a major role in technical communication. You might create Internet, extranet, or intranet sites; communicate through online blogs, YouTube, or Facebook; or consider downloading information and graphics from the Internet. What is legal, and what isn't? To learn more about online copyright laws, visit BitLaw (http://www.bitlaw.com/index.html) or the United States Copyright Office (http://www.copyright.gov/). Research issues related to the Internet and report your findings either orally or in writing.

3. Communicating ethically is both the legal and moral responsibility of a technical communicator. To learn more about ethical codes of conduct, visit either of the following sites:
   - http://www.ethicscentre.ca/EN/resources/Online%20Ethics%20Center%20Codes%20Of%20Ethics%20And%20Conduct.htm. This site provides ethical codes from societies for engineers and scientists.
   - http://www.iit.edu/departments/csep/. The Center for the Study of Ethics in the Professions focuses on ethics in education.
   - Research a code of conduct from your major field of interest and report your findings either orally or in writing.

QUIZ QUESTIONS

1. Why is clarity the ultimate goal of technical communication?
2. How can you achieve clarity?
3. How will answering the reporter's questions help you achieve clarity?
4. Why should you avoid using acronyms, abbreviations, and jargon in your technical communication?
5. In what two places in a business document can you define a term?
6. Why should you strive for conciseness in your technical communication?
7. Why is a long paragraph ineffective in your business documents?
8. What is a fog index?
9. Why should you use shorter sentences?
10. What are three causes of wordy sentences?
11. What are four proofreading tips?
12. When will you organize a document spatially?
13. Why would you organize chronologically?
14. What is the effect of organization by importance?
15. When is comparison/contrast used effectively in a business document?
16. How can you use problem/solution in a document?
17. What do you consider when you focus on the legal aspect of a document?
18. What do you consider when you focus on the practical aspect of a document?
19. What do you consider when you focus on the ethical aspect of a document?
20. What are three things you consider when you follow the guidelines for ethical standards?

CHAPTER FOUR

Communicating Effectively to Your Audience in a Multicultural World

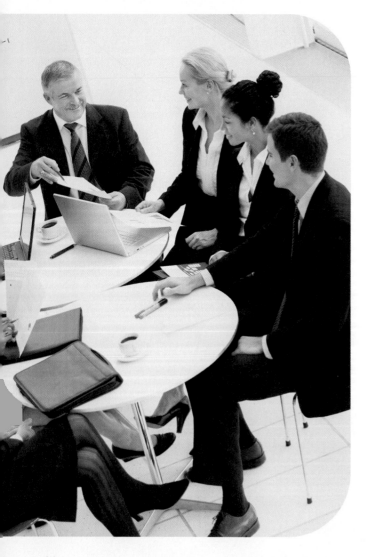

COMMUNICATION AT WORK

In this scenario, Home and Business Mortgage works to meet the company's commitment to diversity through its technical communication.

Home and Business Mortgage (HBM), a Phoenix, Arizona, company, prides itself on being a good neighbor, an asset to the community. Its goal is to ensure that the company meets all governmental personnel regulations, including those required by the Equal Employment Opportunity Commission (EEOC), the Family Medical Leave Act (FMLA), and the Americans with Disabilities Act (ADA). HBM is committed to achieving diversity in its workplace.

In fact, the company is so focused on its appreciation of and commitment to diversity that HBM's technical communicators have created the following:

- Bilingual brochures, newsletters, and annual reports
- Brochures geared toward different stakeholders—senior citizens (age 55 and up), African-American, Hispanic, and Native American homebuyers, women, and young singles (18–25)

Objectives

When you complete this chapter, you will be able to

1. Achieve audience recognition
2. Distinguish among high-tech, low-tech, lay, and multiple audiences
3. Define terms for different audience levels
4. Consider audience personality traits
5. Avoid biased language
6. Recognize diversity including multiculturalism and cross-culturalism
7. Follow the guidelines for effective multicultural communication
8. Avoid sexist language
9. Achieve audience involvement
10. Evaluate your audience using the audience checklist

- A Web site with photographs depicting Phoenix's diverse population

HBM's Web site is unique in another important way. It has sought to abide by W3C WAI mandates. The World Wide Web Consortium (W3C) hosts the Web Accessibility Initiative (WAI). This initiative states that "the power of the Web is its universality. Access by everyone regardless of disability is an essential aspect." HBM's webmasters, in agreement with the WAI, realize that graphic-laden Web sites, as well as those with audio components, are not accessible to people with disabilities, such as hearing and visual impairment.

To meet this WAI platform, HBM's webmasters have achieved the following at the company's Web site:

- Provided links to access information in different languages
- Avoided online audio
- Limited online video and used the Alt attribute to describe each visual's function
- Increased the font of Web text to 20-point versus the standard 12- to 14-point font
- Avoided all designer fonts, like cursive, which are hard to read online
- Avoided red and green for color emphasis because these colors are hard to read for audiences with visual impairments
- Avoided frames that not only fail to load on all computers but also create readability challenges online

HBM is honored to serve all of its customers, regardless of race, age, religion, or physical challenges. One way to achieve the company's commitment to diversity is through its technical communication.

Check out our quarterly newsletters at www.pearsonhighered.com/gerson for dot.com updates, new case studies, insights from business professionals, grammar exercises, and facts about technical communication.

When you write a memo, e-mail message, blog entry, Web site, instant message, tweet, let-ter, or report, someone reads it. That individual or group of readers is your audience. To compose effective technical communication, you should achieve audience recognition and audience involvement.

Audience Recognition

In the business world, you will never write or speak in a vacuum. When you write, give an oral presentation, convene a meeting, communicate with customers in a salesroom, or make a speech at a conference, consider the following questions:

- Who is your audience?
- What does this reader or listener know?
- What does this reader or listener not know?
- What must you write or say to ensure that your audience understands your point?
- How do you communicate to more than one person (multiple audiences)?
- What is this person's position in relation to your job title?
- What is this person's attitude toward your topic?
- What diversity issues (gender, sexual orientation, cultural, multicultural) must you consider?

If you do not know the answers to these questions, your communication might contain jargon or acronyms the reader will not understand. The tone of the e-mail may be inappropriate for management (too dictatorial) or for your subordinates (too relaxed). Your verbal communication might not consider your audience's unique culture and lan-guage. To communicate successfully, you must recognize your audience's level of under-standing. You also should factor in your audience's unique personality and traits, which could impact how successfully you communicate.

See Table 4.1 for a summary of audience variables to consider when communicating.

Knowledge of Subject Matter

What does your audience know about the subject matter? Does the person work closely with you on the project? That would make the audience a high-tech peer. Does the audience have general knowledge of the subject matter but a different area of expertise? That would make the audience a low-tech peer. Is the audience totally uninvolved in the subject matter? That

TABLE 4.1 Audience Variables

Knowledge of Subject Matter
- High tech
- Low tech
- Lay
- Multiple

Personality Traits
- Audience's perceived personality
- Audience's attitude or position regarding the topic
- Audience's preference regarding style
- Audience's response to the topic

Issues of Diversity
- Age
- Gender
- Race and/or religion
- Sexual orientation
- Language and/or culture of origin—multicultural or cross-cultural

would make the person a lay audience. Finally, could your audience be a combination of these types? Then, you would be communicating with an audience with multiple levels of expertise.

High-Tech Audience.   High-tech readers work in your field of expertise. They might work directly with you in your department, or they might work in a similar capacity for another company. Wherever they work, they are your colleagues because they share your educational background, work experience, or level of understanding.

If you are a computer programmer, for example, another computer programmer who is working on the same system is your high-tech peer. If you are an environmental engineer working with hazardous wastes, other environmental engineers focusing on the same concerns are your high-tech peers.

Once you recognize that your reader is high tech, what does this tell you? High-tech readers have the following characteristics:

- They are experts in the field you are writing about. If you write an e-mail message to a department colleague about a project you two are working on, your associate is a high-tech peer. If you write a letter to a vendor requesting specifications for a system she markets, that reader is a high-tech expert. If you write a proposal geared toward your colleagues, they are high tech.

- Because their work experience or education is comparable to yours, high-tech readers share your level of understanding. Therefore, they will understand high-tech jargon, acronyms, and abbreviations. You do not have to explain to an electronic technician, for example, what MHz means. Defining megahertz for this high-tech reader would be unnecessary. We recently read a procedure written by a manager to her engineers explaining how to write departmental technical reports. In the procedure, the boss specified that the engineers should use "metric SI units in their reports." Because the boss recognized that her audience was high tech, she did not have to define SI as "System International."

- High-tech readers require minimal detail regarding standard procedures or scientific, mathematical, or technical theories. Two biomedical technologists, when discussing a metabolic condition in which the body cannot buffer changes in pH, do not need to define pH or acidosis (the terms being discussed). Although this jargon is unintelligible to most of us, to high-tech peers no explanations are required.

- High-tech colleagues read to discover new technical knowledge or for updates regarding the status of a project.

- High-tech readers need little background information regarding a project's history or objectives unless the specific subject matter of the correspondence is new to them. If, for example, you are writing a status report to your supervisor, who has been involved in a project since its inception, then you will not need to flesh out the history of the project. On the other hand, if a new supervisor is hired or if you are updating a colleague new to your department, even though these readers are high tech, you will need to provide background data.

If you work in an environment in which you write only to high-tech peers, you are a rare and lucky individual. Writing to high-tech readers is rather easy because you can use acronyms, abbreviations, jargon, complex graphics, and so on. You usually do not have to define terms or provide background information.

Figure 4.1 is an e-mail message to high-tech HVAC employees, informing them of the capabilities of a new HVAC security system. In contrast, Figure 4.2 discusses the same topic for a low-tech audience.

Low-Tech Audience.   Writing in the workplace, however, rarely allows you to communicate only with high-tech readers. You also will write to low-tech readers. Low-tech readers include your co-workers in other departments. Low-tech readers also might include your

FIGURE 4.1   E-mail Message to High-Tech Audience

This e-mail, written to high-tech readers, uses terminology without explanation or definition. Its tone is businesslike and directive.

Date: June 15, 2012

To: HVAC Personnel

From: Fatai Sayanti, Facilities Manager

Subject: Security of HVAC

Due to increased concerns about the possibility of terrorist threats, our company is installing an interconnected wireless HVAC system and wireless security system. These two systems will be interconnected through a common wireless technology, using the same frequency modulation and protocols.

Wireless Technology

The HVAC system uses wireless control systems, including RF, IR, and protocols such as 802.1 Ix, Bluetooth, and IRDA$_5$. We will install one wireless thermostat with a transceiver for controlling the HVAC system.

Improved HVAC Capabilities

This new security system will include the following:

- Wireless CO sensors and wireless smoke or fire detectors
- Wireless control keypads for communicating with and controlling the security system
- IR sensors to detect the presence of employees in rooms
- Room controls to regulate the temperature of unoccupied rooms
- GUIs to provide programmable thermostats
- An air quality sensor configured to monitor toxicity levels and presence of foreign substances

Next week, you will receive training for installation and monitoring of this new security system.

bosses, subordinates, or colleagues who work for other companies. For instance, if you are a biomedical equipment technician, the accountant or personnel director or graphic artists in your company are low-tech peers. These individuals have worked around your company's equipment and, therefore, are familiar with your technology. However, they do not understand the intricacies of this technology.

Your bosses are often low tech because they no longer work closely with the equipment. Although they were technicians at one time, as they moved into management, they moved further and further away from technology. Your subordinates might be low tech because their levels of education or work experience are less than yours.

Although your colleagues at other companies might have your level of education or work experience, they could be low tech if they are not familiar with your company's procedures or in-house jargon, acronyms, and abbreviations. When you write e-mail, letters, and reports to low-tech readers, remember that they share the following characteristics:

- Low-tech readers are familiar with the technology you are writing about, but their job responsibilities are peripheral to the subject matter. They either work in another department, manage you, work under your supervision, or work outside your company.

- Because low-tech readers are familiar with your subject matter, they understand some abbreviations, jargon, and technical concepts. To ensure that readers understand your content, therefore, define your terms. An abbreviation like BCA can't stand alone. Define it parenthetically: BCA (burst cutting area).

- Abbreviations, acronyms, and technical jargon like "burst cutting area" sometimes need follow-up explanations. You should write, "BCA (burst cutting area) is a circular section located near the center of a DVD disc where manufacturing information can be inscribed in a barcode."

- In addition, technical concepts must be defined for low-tech readers. For example, whereas high-tech readers understand the function of pressure transducers, a low-tech reader needs further information, such as "Pressure transducers: Solid-state components sense proximal pressure in the patient tubing circuit. The transducers convert this pressure value into a proportional voltage for the control system."

- Since the low-tech reader is not in your normal writing "loop"—that is, not someone to whom you write often regarding your field of expertise—you need to provide more background information. When you submit a status report to upper-level management, for example, you cannot just begin with work accomplished. You need to explain why you are working on the project (objectives, history), who is involved (other personnel), when the project began and its scheduled end date, and how you are accomplishing your goals. Low-tech readers understand the basic concepts of your work, but they have not been involved in it daily. Fill them in on past history.

Figure 4.2 is an e-mail to low-tech security personnel, informing them of procedures to follow if they discover suspicious substances on site.

FIGURE 4.2   E-mail Message to Low-Tech Audience

Date: June 15, 2012

To: Security Personnel

From: Fatai Sayanti, Facilities Manager

Subject: Security of Heating, Ventilating, and Air Conditioning (HVAC)

When you are patrolling the facilities and grounds, be on the alert for suspicious substances, such as fine powders, residues, fog, mist, oily liquids, or unusual odors. Not every liquid or odor is necessarily suspicious. Just focus on those with unexplainable origins. Also watch for discomfort in our employees, such as two or more people experiencing difficulty breathing. These signs could be evidence of dangerous chemicals in our HVAC system.

If you discover problems, follow these procedures:

- DO NOT touch the substance.

- Evacuate all employees from the affected area.

- Contact our HVAC personnel and ask them to shut down all systems to avoid spreading contamination.

- Secure the area.

- Report the situation to management, including the location, number of affected employees, and a description of the substance.

As first responders, your alertness and professionalism are essential to ensure the safety of our 5,000 employees.

This e-mail to low-tech readers deals with a very important subject (security in the face of potential terrorism). The writer presents only what the audience needs to know in order to complete their job responsibilities. The writer conveys this content simply, using short sentences, a list, and easy-to-understand words. HVAC is defined.

Lay Audience. Customers and clients who neither work for your company nor have any knowledge about your field of expertise are the lay audience. If you work in network communications for a cable company, for example, and you write an e-mail to a client regarding a problem with the high-speed Internet connection, your audience is a lay reader. If your field of expertise is biomedical equipment and you write a user manual for the patient, you are writing to a lay audience. If you are an automotive technician writing a service report for a customer, that customer is a lay reader.

Although you understand your technology, your reader, who uses the high-speed Internet service or the medical equipment or the car, is not an expert in the field. These readers are using your equipment or require your services, but the technology you are writing about is not within their daily realm of experience.

This makes writing for a lay audience difficult. It is easy to write to a high-tech reader who thoroughly understands the technology you are discussing. However, writing to a lay audience totally outside your field of expertise is demanding. To write successfully to a lay audience, remember that these readers share the following characteristics:

- Lay readers are unfamiliar with your subject matter. They do not understand your terminology. Therefore, you should write simply. That's not to say that you should insult your lay reader with a remedial discussion or with a patronizing tone. You must, however, explain your topic clearly through precise word usage, depth of detail, and simple graphics.
- Because your lay readers do not understand your terminology or work environment, they won't understand your in-house jargon, abbreviations, or acronyms. Avoid high-tech terms or define them thoroughly.
- Lay readers will need background information. If you leap into a discussion about a procedure without explaining to your lay readers why they should perform each step, they will not understand the causes or rationale. High-tech and possibly even low-tech readers might not need such explanations, but the lay reader needs clarity.

For example, look at the maintenance procedure, which is part of a user manual provided for the purchaser of an audio recorder (Figure 4.3).

Next, look at the procedure for cable preparation from an installation manual directed toward high-tech readers (Figure 4.4).

A lay reader would have no idea what is going on in Figure 4.4. Why are we performing this act? What's the purpose? What will be achieved? Perhaps these questions are irrelevant for the technician, who knows the background or rationale, but a lay audience will be confused without an introductory overview. In addition, a lay audience will not know the terms *gasket*, *cable dielectric*, or *center conductor*. These terms are high tech. What dimensions are we supposed to achieve? The high-tech reader, who is familiar with this operation, knows what is meant by proper length but not a lay audience. Finally, when is a clamp seated "properly in gasket"? Again, a lay audience needs these points clarified.

Composing effective technical communication requires that you recognize the differences among high-tech, low-tech, and lay audiences. If you incorrectly assume that all readers are experts in your field, you will create problems for yourself as well as for your readers. If you write using high-tech terms to low-tech or lay audiences, your readers will be confused and anxious. You will waste time on the phone or answering e-mail messages clarifying the points that you did not make clear in the technical document.

Multiple Audiences. Correspondence is not always sent to just one type of audience. Your correspondence may have an audience with multiple levels of expertise. Sometimes you write to high-tech, low-tech, and lay audiences simultaneously. For example, when writing a report, most people assume that their supervisor will be the only reader. This

FIGURE 4.3 Instructions for a Lay Reader

MAINTENANCE

Cleaning the Head Section

The heads, capstan, and pinch rollers get dirty easily. If this head section becomes dirty, the high-frequency sound will not be reproduced and the stereo balance will be impaired. This hurts your system's sound quality.

To avoid these problems, clean your system's head section regularly by following these simple steps:

1. Push the STOP/EJECT button to open the cassette door.
2. Dip a cleansing swab into the cleaning fluid.
3. Wipe the heads, capstan, and pinch rollers with the swab.
4. Allow 30 seconds to dry.

Note:

- Do not hold screwdrivers, metal objects, or magnets close to the heads.
- When demagnetizing the heads, be sure the unit's POWER switch is in the OFF position.

The introductory paragraph clarifies for the lay reader why he or she should clean the head section, instead of incorrectly assuming that the reader will understand without being told. This procedure also uses a simple graphic to clearly depict what action is required.

FIGURE 4.4 Instructions for a High-Tech Audience

1. Place nut and gasket over cable and cut jacket to dimension shown.
2. Comb out braid and fold out. Cut cable dielectric to dimension shown. Tin center conductor.
3. Pull braid wires forward and taper toward center conductor. Place clamp over braid and push back against cable jacket.
4. Fold back braid wires as shown, trim to proper length (D), and form over clamp as shown. Solder contact to center conductor.
5. Insert cable and parts into connector body. Make sure sharp edge of clamp seats properly in gasket. Tighten nut.

This example provides no background about why to perform the action, no clarity about how to perform the steps, and no graphics to help the reader visualize the procedure. This procedure is written for a high-tech audience.

FIGURE 4.5 Examples of Possible Audiences

might not be the case, however. The supervisor could send a copy of your report to the manager, who could then submit the same report to the executive officer. Similarly, your supervisor might send the report to your colleagues or to your subordinates. Your report might be sent out to other lateral departments. Figure 4.5 shows the possibilities.

Writing correspondence for multiple readers with different levels of understanding and reasons for reading creates a challenge. When you add the necessity of using a tone that will be appropriate for all of these varied readers, the writing challenge becomes even greater.

How do you meet such a challenge? The first key to success is recognizing that multiple audiences exist and that they share the following characteristics:

- Your intended audience will not necessarily be your only readers. Others might receive copies of your correspondence.
- Some of the multiple readers will be unfamiliar with the subject matter. Proper background data (objectives, overviews) will clarify the history of the report for these readers. In a letter, report, e-mail message, Web site, or blog, this background information cannot be too elaborate. Often, all you do is provide a reference line, subject line, or link suggesting where the readers can find out more about the subject matter if they wish—"Reference: Operations Procedure 321 dated 9/21/12." In longer reports, background data will appear in the summary or abstract, as well as in the report's introduction.
- Multiple readers have diverse understandings of your technology. This requires that you define jargon, abbreviations, and acronyms.

Communication Channels
See Chapters 7, 14, and 17 for more discussion of blogging, Web sites, and long reports.

Writing for Future Audiences

Time creates another challenge regarding audience recognition. Technical communication usually is archived for future reference. You might write an e-mail, report, proposal, or instruction using terms or focusing on content that your primary audience understands . . . at the moment.

However, months (or years) later when the correspondence is retrieved from the files, will your readers still be familiar with the topic? Will you still have the same readers? Many people, some of whom didn't even work for your company at the time of the original writing, will read your correspondence. These future readers need clarity, background information, and terms defined.

As a technical communicator, you need to know that your correspondence is not just written for the moment. It is written for future readers. Depending on what you have written, these readers could include

- Judges and juries who depend on past reports to decide cases
- New employees at a company who retrieve archived documents to learn how to complete current tasks
- Accountants and auditors who look at past records to find what work was accomplished at a prior date
- New managers or supervisors who need to familiarize themselves with workplace procedures

Defining Terms for Audiences

Based on whether or not the current or future audience is high tech, low tech, lay, or multiple, you have to decide when to use acronyms, abbreviations, and technical terminology and how to use them effectively. One simple rule for low-tech, lay, and multiple audiences is to define your terms. You can do so as follows:

- Parenthetically
- In a sentence
- In an extended paragraph
- In a glossary
- As online help with a pop-up definition

Defining Terms Parenthetically.   Rather than just writing *CIA*, write *CIA* (*cash in advance*) or *cash in advance (CIA)*. Such parenthetical definitions, which are only used once per correspondence, don't take a lot of time and won't offend your readers. Instead, the result will be clarity. If you use many potentially confusing acronyms or abbreviations, or if you need to use a great deal of technical jargon, then parenthetical definitions might be too cumbersome. In this case, supply a separate glossary.

Defining Your Terms in a Sentence.   If you provide a sentence definition, include the following: **Term** + **Type** + **Distinguishing characteristics**. For example, using a sentence to define *HTTP*, you would write the following:

Term   Type

Hypertext Transfer Protocol is a set of rules for the secure transfer of files, including **text, graphics, videos, and sound on the** Internet.

◀ EXAMPLE

Distinguishing
characteristics

Using Extended Definitions of One or More Paragraphs.   When you need to provide an extended definition of a paragraph or more, in addition to providing the term, type, and distinguishing characteristics, also consider including examples, procedures, and descriptions. Look at the following paragraph definition of a video card.

EXAMPLE ▶

A video card (graphics accelerator card, display adapter, or graphics card) is a type of printed circuit board that can be inserted into an expansion slot of a computer's motherboard. The video card creates and outputs images to a display. Functions of a video card can include video capture, MPEG-2 and MPEG-4 decoding, TV output, or the ability to connect multiple monitors.

Using a Glossary.   If you have not defined your terms parenthetically, in a sentence, or in a paragraph, you should use a glossary. A glossary is an alphabetized list of terms placed after your conclusion/recommendation.

EXAMPLE ▶

GLOSSARY

ADC	analog to digital converter
Analog	a variable signal continuous in both time and amplitude
DAC	digital to analog converter
LAN	local area network
PSTN	Public Switched Telephone Network—analog phone systems
ROI	return on investment
Voice signal	allows users to access phone options with a single voice command
VoIP	voice over Internet protocol
WAN	wide area network

Defining Terms

See Chapters 13 and 14 for more discussion of defining terms in manuals, online help, and Web sites.

Providing Pop-ups and Links with Definitions.   Use pop-up screens for online help. (See Figure 4.6.)

Other ways to define terms include the following:

- Place the definition in the front matter or in an appendix of a document (such as a manual).
- Use endnotes in a report with other researched material.

FIGURE 4.6   Online Help Screen with Pop-up Definition

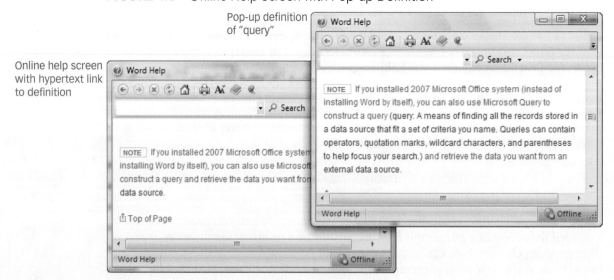

Pop-up definition of "query"

Online help screen with hypertext link to definition

TABLE 4.2 How to Communicate to Different Audience Levels

Audience Level	How to Communicate	Sample
High Tech	Use jargon, acronym, or abbreviation alone.	The wastewater is being treated for DBPs.
Low Tech	Use jargon, acronym, or abbreviation with a parenthetical definition.	The wastewater is being treated for DBPs (disinfection by-products).
Lay	Use jargon, acronym, or abbreviation with a parenthetical definition *and* a brief explanation or extended definition.	The wastewater is being treated for DBPs (disinfection by-products), such as acid, methane, chlorine, and ammonia.

Table 4.2 shows techniques for communicating with audiences who have different levels of technical knowledge. These various techniques for defining terms are neither foolproof nor mandatory. You are always the final judge of how much information to provide your readers. However, if your readers fail to understand your content, then you have failed to communicate.

Audience Personality Traits

In addition to considering an audience's level of knowledge, a technical communicator should determine aspects of an audience's personality. By considering personality traits, the writer then can write appropriately using the proper tone, visual aids, and writing style. This will help the writer meet the audience's needs. Of equal importance, by recognizing the audience's personality traits, the writer can more effectively get the desired response from the reader.

Let us start with a caveat. You obviously cannot always know your audience's personality traits. This would be a difficult goal to achieve when working with customers you've never met, a dispersed team of co-workers located in different cities, members of an audience at a large conference, or the widely diverse audience that reads your Web site or blog postings. However, many times you will write or speak to an audience that you know. This could include customers you have served in the past, vendors you often work with, as well as co-workers in your office and your immediate supervisors. To communicate successfully with an audience whom you know, you want to factor in your knowledge of their personalities, attitudes, and preferences.

For example, we work with a company that offers its employees a pay incentive. The company will pay employees to identify a problem in the workplace and then suggest solutions to this problem. The goal is to improve the company's services, efficiency, budget, and profits. To earn this monetary award, the employees must write a report analyzing the perceived problem and proposing the solutions, create PowerPoint slides regarding the topic, and make a brief oral presentation to their supervisors.

We have found from interviews that employees who successfully earn the financial incentive prewrite for their report and prepare for their presentation by first assessing their audience. This includes determining their audience's level of knowledge, but it also involves assessing their audience's personality. An audience's personality traits and possible responses to a topic of discussion can include issues such as those shown in Table 4.3. Reviewing these considerations will help you determine the type of content you want to provide, how persuasive you need to be, and what tone and style your technical communication must achieve.

TABLE 4.3 Audience Personality Traits and Responses

Audience's perceived personality
• Slow to act?
• Eager?
• Receptive?
• Questioning?
• Organized?
• Disorganized?
• Oppositional?
Audience's attitude or position regarding the topic
• Negative?
• Positive?
• Noncommittal?
• Informed?
• Uninformed?
Audience's preference regarding style
• Will he, she, or they accept contractions?
• Should you use first names, last names, or position titles?
• Is short and to-the-point better?
• Is long and detailed better?
Desired audience response
• Do you want the audience to consider this idea and pass on a recommendation?
• Do you want the audience to act now?
• Do you want the audience to confirm what has been suggested?
• Do you want the audience to reject some options but act on others?
• Do you want the audience to file this information for future reference?

Biased Language—Issues of Diversity

Your audience will not be composed of people just like you. In contrast, your readers or listeners will more than likely be diverse. Diversity includes "gender, race/ethnicity, religion, age, sexual orientation, class, physical and mental characteristics, language, family issues, [and] departmental diversity" (Grimes and Richard 8).

Think of it this way: You work for a city government, and your audience is the citizenry. Who comprises your city's populace? Or, you work in a hospital. Who will visit your facility? Or, you work in a retail establishment. Who will shop there? Or, you work in a company, any company. Who are your co-workers? They are people of many different interests, levels of knowledge, and backgrounds—and they are all valuable to your business.

Why should you be concerned about a diverse audience?

1. **Diversity is protected by the law.** Prejudicial behavior and discrimination on the job will not be tolerated. "Non-compliance with Equal Opportunity or Affirmative Action legislation can result in fines and/or loss of contracts" (McInnes).

2. **Respecting diversity is the right thing to do.** People should be treated equally, regardless of their age, gender, sexual orientation, culture, or religion.

3. **Diversity is good for business.** A workplace where employees feel valued and respected results in more motivation and commitment to the company ("What Is the 'Business Case' for Diversity?"). Clients and customers prefer shopping in an environment devoid of prejudice. In addition, people from diverse backgrounds spend money and buy products. "To ensure that . . . products and services are designed to appeal to this diverse customer base, 'smart' companies are hiring people from those walks of life—for their specialized insights and knowledge" (McInnes).

4. **A diverse workforce keeps companies competitive.** Talent does not come in one color, nationality, or belief system. Instead, talent is "represented by people from a vast array of backgrounds and life experiences. Competitive companies cannot allow discriminatory preferences and practices to impede them from attracting the best available talent" (McInnes). Diversity management is such an important concern that American businesses spend millions of dollars training employees to understand multiculturalism and diversity (Jordan 1).

This is further verified in a Society for Human Resource Management Survey of Diversity Initiatives which stated that "84 percent of human resource professionals at Fortune 500 companies say their top-level executives think diversity management is important" ("What Is the 'Business Case' for Diversity?").

DOT.COM UPDATES

Go to http://www.shrm.org/diversity/ (the Society for Human Resource Management) for additional information about diversity in the workplace, including definitions, links, and articles about diversity in the news.

Web Resources

www.pearsonhighered.com/gerson For more information about issues of diversity, visit our companion website.

FAQs: Multiculturalism and Technical Communication in a Global Economy

Q: Why is "multiculturalism" important in technical communication? Doesn't everyone in the United States speak English?

A: No, not everyone in the United States speaks English. Look at these facts:

- 53 million Americans or 20 percent of the population speak a language other than English at home.
- 32 million Americans speak Spanish in the home.
- 2.3 million Americans speak Chinese in the home.
- 1.4 million Americans speak French in the home (Progressive Policy Institute). See figure 4.7 for a comparison of languages other than English spoken at home.

FIGURE 4.7 Languages Other than English Spoken at Home

Q: OK, but, shouldn't we write for the majority of our audience?

A: Your audience will be diverse—and the "majority" in the United States is changing.

- "America's population has become increasingly diverse as Hispanics, African Americans, Asians and people from many other segments rapidly contribute to the cultural richness of our population."
- Hispanics are expected to number 51.4 million by 2011.
- Asians, African Americans, and Hispanics accounted for more than 70 percent of population growth, and Hispanics accounted for more than 50 percent since 2000.
- By 2011 the population in the three largest ethnic groups will be more than 107 million and Hispanics will represent nearly half of that population.
- America has 113.5 million households. Of this total, nearly 13 million are Hispanic, nearly 11.5 million African American, and nearly 4 million Asian. (Melgoza)

Multiculturalism

Multicultural Communication

The global economy also impacts the importance of diversity in technical communication. Your company will market its products or services worldwide. International business requires multicultural communication, the sharing of written and oral information between businesspeople from many different countries (Nethery). Who is doing business internationally? Almost everyone! According to one recent study, "67% of professionals communicate internationally" (Miller).

The Global Economy. Read Table 4.4 to see how production and communication at various major corporations are affected by the global economy.

The Challenges of Multicultural Communication. The Internet, e-mail, Facebook, and YouTube affect global communication and global commerce constantly. With these technologies, companies can market their products internationally and communicate with multicultural clients and co-workers at a keystroke. An international market is great for companies because a global economy increases sales opportunities. However, international commerce also creates written and oral communication challenges. Companies that work internationally must communicate with all of their employees and clients; therefore, communication must be multilingual.

For example, Medtronic, a leading medical technology company, does business in 120 countries. Many of those countries mandate that product documentation be written in the local language. To meet these countries' demands, Medtronic translates its manuals into 11 languages: French, Italian, German, Spanish, Swedish, Dutch, Danish, Greek, Portuguese, Japanese, and Chinese (Walmer 230).

Multilingual reports create unique communication challenges:

- Will each language version be identical in content and readability?
- Will the first language version suggest advantages to investors over translations?
- Are all translations carefully prepared according to tone, style, and content?
- Is each translation tested for usability and accuracy? (Courtis and Hassan 395).

TABLE 4.4 Global Communication

Company	Impact
Black & Veatch	Black & Veatch, a global engineering company, has locations in Argentina, Australia, Botswana, Brazil, China, the Czech Republic, Egypt, Germany, Hong Kong, India, Indonesia, Korea, Kuwait, Malaysia, Mexico, Philippines, Poland, Saudi Arabia, Singapore, South Africa, Swaziland, Taiwan, Thailand, Turkey, United Arab Emirates, United Kingdom, United States, Vietnam, Zambia, and Zimbabwe.
Cerner	As a global supplier of healthcare solutions, Cerner works with more than 8,000 clients worldwide and has locations in the United States, Australia, Canada, France, Germany, Hong Kong, India, Ireland, Kingdom of Saudi Arabia, Malaysia, Singapore, Spain, United Arab Emirates, and the United Kingdom.
Coca Cola	Coca Cola produces 400 different brands in 200 countries. Seventy-five percent of their income is generated from outside the United States (Coca-Cola Company).
Hallmark Cards	Margaret Keating, vice president-operations for Hallmark Cards Inc., oversees Hallmark's North American manufacturing and distribution, graphics, global procurement, and global operations. She says, "Communication is critical with a group [of 6,000 employees] geographically dispersed." "She spends a lot of time making communication clear, concise and compelling. That it has to be translated into different languages makes efficient communication even more important." Technical communication at Hallmark "'has to make sense to the audience, and the audiences are very different'" (Cardarella).

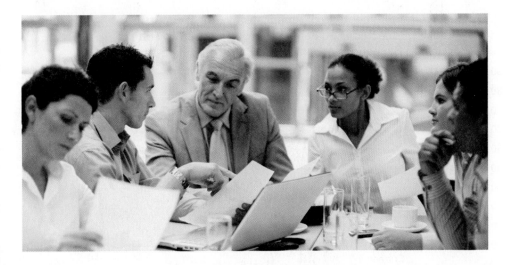

Multicultural Team Projects.   What about international, multilingual project work teams? If, for example, your U.S. company is planning to build a power plant in China, you will work with Chinese engineers, construction workers, financial planners, regulatory officials, and others. To do so effectively, you will need to understand that country's

- Verbal and nonverbal communication norms
- Management styles
- Decision-making procedures
- Sense of time and place
- Local values, beliefs, and attitudes

Our natural instinct is to evaluate people and situations according to our sense of values, our cultural perspectives. That is called ethnocentrism—a belief that one's own culture represents the norm. Such is not the case. The world's citizenry does not share the same perspectives, beliefs, values, political systems, social orders, languages, or habits. Successful technical communication takes into consideration language differences, nonverbal communication differences, and cultural differences.

Due to the multicultural makeup of your audience, you must ensure that your writing, speaking, and nonverbal communication skills accommodate language barriers and cultural customs. The classic example of one company's failure to recognize the importance of translation concerns a car that was named Nova. In English, *nova* is defined as a star that spectacularly flares up. In contrast, *no va* in Spanish is translated as "no go," a poor advertisement for an automobile.

Communicating Globally . . . in Your Neighborhood

Cross-Cultural Communication.   Multiculturalism will affect you not only when you communicate globally, but also when you are confronted with communication challenges in your own city and state. Many experts agree that "'the challenge posed by the increasing cultural diversity of the U.S. workforce is perhaps the most pressing challenge of our times'" (Grimes and Richard 9). Another term for this challenge is *cross-cultural communication,* writing and speaking between businesspeople of two or more different cultures within the same country (Nethery). How big a challenge is this? Look at the following statistics regarding America's melting pot, as shown in Figure 4.8.

One challenge presented by the increasingly multicultural nature of our society and workplace is language. Language barriers are an especially challenging situation for hospitals, police and fire personnel, and governmental agencies where failure to communicate effectively can have dangerous repercussions.

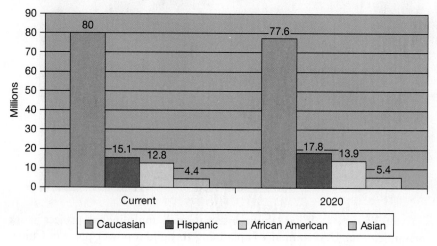

FIGURE 4.8 Current Percentage of Americans by Ethnic Group (U.S. Census Bureau) vs. Projected 2020 Percentage of Americans by Ethnic Group ("Population Projections")

SPOTLIGHT

Why do some corporations provide foreign language training for their employees?

Phil Wegman, Program Director of Skills Enhancement for the Center for Business and Technology, sighed deeply and said, "I receive two to three calls *every day* from companies, desperate for Spanish language training. They need to teach their supervisors how to communicate more effectively with customers as well as with employees for whom English is a second language."

Phil's client base covers literally hundreds of fields, including dental and medical, criminal justice, public safety, transit, government, and education, as well as industries such as hospitality management (hotels and restaurants), construction, casinos, manufacturing, warehousing, banking, retail, childcare, and accounting.

Police, nurses, physicians, and paramedics, for example, have asked Phil to provide them occupational language training that is "work specific and real life." Then, when an emergency situation occurs, the safety personnel can ask, in Spanish, French, German, or Chinese, "where does it hurt?"

McDonalds, Hardees, and Burger King supervisors need to know enough Spanish to be able to say to their staff, "here's how you ring up this sale," "here's how you make this meal," or "here's how you clean this piece of equipment." Of equal importance, these companies have asked Phil to teach their supervisors how to answer customer questions or to take customer orders. Many of these fast-food restaurants have on staff a worker who is fluent in a second language. However, the pace is so hectic in fast food that the restaurants can't always pull that one employee off the job to handle the situation. Thus, *all* supervisors need enough language skills to provide direct commands, make simple statements, or answer common questions in a foreign language.

Sometimes, the needs are immediate . . . and life threatening. Phil says "when a construction or manufacturing company is rolling out a new piece of equipment, where the learning curve is enormous, language can be a huge barrier. Maybe a company has either employees for whom English is a second language, or that company does business internationally. Fluency becomes mandatory to ensure productivity, quality, and personal safety." If the new machinery is operated incorrectly, due to a failure to understand the English instructions, the end products might not meet standards and, of greater concern, injuries can occur.

There is one more key component to language knowledge, Phil states. "When a company teaches its employees a different language, the company honors that culture. The company tells its client base, 'we're trying to learn who you are and what makes you unique.'" The company is learning cross-cultural information to minimize barriers; the company is saying to its staff and customers, "we respect who you are and we want to work with you. That's just good business."

One hospital reported that it has "13 staff members supplying Spanish, Arabic and Somali translations." However, this hospital's successful use of translators to help with doctor-patient communication is rare. "Only about 14 percent of U.S. hospitals provide training for volunteer translators." Most hospitals depend on the patient's relatives. "In one case studied, an 11-year-old sibling translated. The child made 58 mistakes" (Sanchez A4). Imagine how such errors can negatively impact healthcare and medical records.

The communication challenges are not only evident for employees in healthcare and other community infrastructures. Industries as diverse as banking, hospitality (restaurants and hotels), construction, agriculture, meat production and packing, and insurance also face difficulties when communicating with clients and employees for whom English is a second language.

Guidelines for Effective Multicultural Communication

To achieve effective multicultural technical communication, follow the guidelines in this section.

Define Acronyms and Abbreviations

Acronyms and abbreviations cause most readers a problem. Although you and your immediate colleagues might understand such high-tech usage, many readers won't. This is especially true when your audience is not native to the United States. For example, corporate employees often abbreviate the job title "system manager" as *sysmgr*. However, in German, the title "system manager" is called the *system leiter*; in French, it's *le responsable*. The abbreviation *sysmgr.* would make no sense in either of these countries (Swenson WE-193).

Avoid Jargon and Idioms

The same dilemma applies to *jargon* and *idioms*, words and phrases that are common expressions in English but that could be meaningless outside our borders. Every day in the United States, we use *on the other hand* as a transitional phrase and *in the black* or *in the red* to denote financial status. What will these idioms mean in a global market? Following is a list of idioms that are commonly used in America but which will not translate well in other countries:

ballpark figure	eye on the prize
belt-tightening	face the music
bite the bullet	guesstimate
brownie points	jump through hoops
buck stops here	pass the buck
crunch time	pull the plug
dog-eat-dog world	shape up or ship out
drum up business	through the roof

Distinguish Between Nouns and Verbs

Many words in English act as both nouns and verbs. This is especially true with computer terms, such as *file*, *scroll*, *paste*, *boot*, *code*, and *help*. If your text will be translated, make sure that your reader can tell whether you're using the word as a noun or a verb (Rains).

DOT.COM UPDATES

Check out these sites related to the importance of multiculturalism, cross-culturalism, and diversity:

- Diversity in the Workplace (http://www.business.com/ directory/human_resources/ diversity/) provides information about affirmative action, diversity training, minority-owned businesses, and related cultural issues.
- International Business Multiculturalism and the Internet (http://www.infotoday .com/searcher/oct01/fink .htm) provides an article and access to various websites related to the global economy.
- American Society of Mechanical Engineers (http://www.asme.org/ Communities/Diversity/) provides information about the importance of diversity in the field of engineering.

Watch for Cultural Biases/Expectations

Your text will include words and graphics. As a technical communicator, you need to realize that many colors and images that connote one thing in the United States will have different meanings elsewhere. For example, the idioms "in the red" and "in the black" will not necessarily communicate your intent when they are translated. Even worse, the colors black and red have different meanings in different cultures. Red in the United States connotes danger; therefore, "in the red" suggests a financial problem. In China, however, the word *red* has a positive connotation, which would skew your intended meaning. The word *black* often implies death and danger, yet "in the black" suggests financial stability. Such contradictions could confuse readers in various countries.

Animals represent another multicultural challenge. In the United States, we say you're a "turkey" if you make a mistake, but success will make you "soar like an eagle." The same meanings don't translate in other cultures. Take the friendly piggy bank, for example. It represents a perfect image for savings accounts in the United States, but pork is a negative symbol in the Mideast. If you are "cowed" by your competition in the United States, you lose. Cows, in contrast, represent a positive and sacred image in India.

Be Careful When Using Slash Marks

Does the slash mark mean "and," "or" or both "and/or"? The word *and* means "both," but the word *or* means "one or the other," not necessarily both. If your text will be translated to another language, will the translator know what you meant by using a slash mark? To avoid this problem, determine what you want to say and then say it.

Avoid Humor and Puns

Humor is not universal. In the United States, people talk about regional humor. If a joke is good in the South but not in the North, how could that same joke be effective overseas? Microsoft's software package Excel is promoted by a logo that looks like an *X* superimposed over an *L*. This visual pun works in the United States because we pronounce the letters *X* and *L* just as we would the names of the software package. If your readers are not familiar with English, however, they might miss this clever sound-alike image (Horton 686).

Realize That Translations May Take More or Less Space

Paper Size.   If your writing will be conveyed not on paper but on disk or on the Internet, you must consider software's line-length and screen-length restrictions.

TABLE 4.5 Translations Increase Word Length

English Word	Translations into Other Languages
Print	*Impression* (French)
File	*Archivo* (Spanish)
View	*Visualizzare* (Italian)
Help	*Assistance* (French)
E-mail	*Courriel* (Swiss)

For example, a page of hard-copy text in the United States will consist of approximately 55 lines that average 80 characters per line. How could this present a problem? The standard sheet of paper in the United States measures 8½ × 11 inches. In contrast, the norm in Europe for standard-sized paper is A4—210 × 297 millimeters, or 8.27 × 11.69 inches.

Why is the size of paper important? Here is the problem: If you format your text and graphics for an 8½ × 11-inch piece of paper in Atlanta, travel to London, download the files on a computer there, and hit Print, you might find that what you get is not what you hoped for. The line breaks will not be the same. You will not be able to three-hole punch and bind the text. The margins will be off, and so will the spacing on your flowchart or table (Scott 20–21).

Web Sites. An even greater problem occurs when you are writing for the Internet. On a Web site, you will provide a navigation bar and several frames. Why is this a problem? The word count of a document written in English will expand more than 30 percent when translated into some European languages. In Table 4.5, notice how English words become longer when translated into other languages (Horton 691).

The Swiss government is trying to curb what it defines as "the encroachment of English." To do so, the government's French Linguistics Service asked its citizens to avoid using the word *spam*, instead opting for *courier de masse non sollicite*, meaning "unsolicited bulk mail" ("Swiss Fight Encroachment of English" A16).

Whereas the above examples add length to a document, "Chinese character writing captures the subject matter more compactly than English. Consequently, the length of the Chinese translation is usually about 60% of that needed by the English version" (Courtis and Hassan 397). You must consider length to accommodate translations.

Avoid Figurative Language

Many of us use sports images to figuratively illustrate our points. We "tackle" a chore; in business, a "good defense is the best offense"; we "huddle" to make decisions; if a sale isn't made, you might have "booted" the job; if a sale is made, you "hit a home run." Each of these sports images might mean something to native speakers, but they may not communicate worldwide. Instead, say what you mean, using precise words (Weiss 14).

Be Careful with Numbers, Measurements, Dates, and Times

Numbers and Measurements. If your text uses measurements, you are probably using standard American inches, feet, and yards. However, most of the world measures in metrics. Thus, if you write 18 high × 20 wide × 30 deep, what are the measurements? There is a huge difference between 18 × 20 × 30 inches and 18 × 20 × 30 millimeters.

Dates. In the United States, we tend to abbreviate dates as MM/DD/YY: 05/03/12. In the United Kingdom, however, this could be perceived as March 5, 2012, instead of May 3, 2012. See Table 4.6 for additional examples.

TABLE 4.6 Different Ways of Understanding and Writing the U.S. Date 05/03/12

Country	Date
United States	May 3, 2012
United Kingdom	March 5, 2012
France	5 mars 2012
Germany	5. Marz 2012
Sweden	12–05–03
Italy	5.3.12

TABLE 4.7 Different Ways of Writing the U.S. Time 5:15 P.M.

Country	Time
United States	5:15 P.M.
France	17:15
Germany	17.15
Quebec, Canada	17 h 15

Time. Time is another challenge. Table 4.7 shows how different countries write times.

In addition to different ways of writing time, you must also remember that even within the United States, 1:00 P.M. does not mean the same thing to everyone. Is that central time, Pacific time, mountain time, or eastern time?

Another challenge with time occurs when we incorrectly assume that everyone everywhere abides by the same work hours. In the United States, the average workweek is 40 hours, and the typical workday is from 8:00 A.M. to 5:00 P.M. However, this is not the norm globally. French laws have reduced the workweek to 35 hours. Many Middle Eastern countries close work on Friday, the beginning of Sabbath. Offices in parts of southern Europe shut down for a traditional two-hour lunch (noon to 2:00 P.M.) (St. Amant "Communication" 28). Therefore, if you write an e-mail telling a co-worker in Spain or Jordan that you will call at 2:00 P.M. his or her time, that could be an inappropriate time for your audience.

Finally, even simple words like *today, yesterday,* or *tomorrow* can cause problems. Japan is 14 hours ahead of U.S. eastern standard time. Thus, if you need a report "tomorrow," do you mean tomorrow—the next day *your time*—or tomorrow—two days from *your reader's time*? To solve these problems, determine your audience and make changes accordingly. That might mean

- Writing out the date completely (January 12, 2012)
- Telling the reader what standard of measurement you will use ("This document provides all measurements in metrics.")
- Telling the reader what scheme of time presentation you will use ("This document relates time using a 24-hour clock rather than a 12-hour clock.")
- Using multiple formats ("Let's meet at 2:30 P.M./14:30.")
- Avoiding vague words like *today, tomorrow,* or *yesterday*
- Recognizing that people have different work schedules globally

Use Stylized Graphics to Represent People

A photograph or realistic drawing of people will probably offend someone and create a cultural conflict. You want to avoid depicting race, skin color, hairstyles, and even gender.

To solve this problem, avoid shades of skin color, choosing instead pure white or black to represent generic skin. Use simple, abstract, even stick figures to represent people. Stylize hands so they are neither male nor female—and show a right hand rather than a left hand, if possible (a left hand is perceived as "unclean" in some countries) (Flint 241). Recognizing the importance of the global marketplace is smart business and a wise move on the part of the technical communicator.

Following is an example of poor communication in an e-mail message for a multilingual audience.

BEFORE

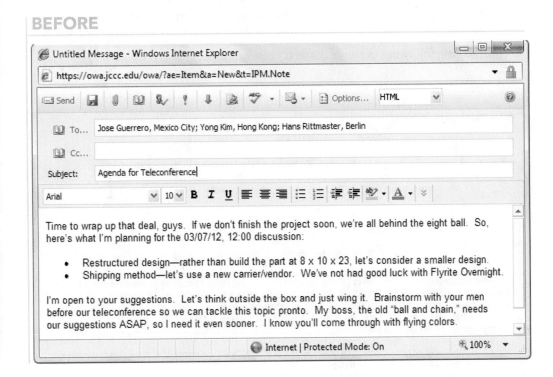

This e-mail fails for many reasons:

1. **Date.** Does "03/07/12" mean March 7, 2012 or July 3, 2012?
2. **Time.** "12:00" not only fails to specify the time zone, but also creates the question whether midnight or noon is intended.
3. **Figurative language.** "Wrap up that deal," "behind the eight ball," "think outside the box," "just wing it," "brainstorm," "tackle the topic," "ball and chain," and "flying colors" all are idiomatic phrases. Though these phrases might communicate within the United States, they may not translate internationally.
4. **Informal tone.** Informality is common in most stateside businesses, but this is not true in all countries. Germany and Japan, for example, are far more formal in their technical communication expectations. Thus, "guys" and the contractions could cause problems creating incorrect tone.
5. **Sexist language.** "Guys" and "men" should be avoided when writing any correspondence. The implication is that only men will be involved in the discussions.

6. **Measurements.** "8 × 10 × 23" is confusing. Is the writer discussing inches, feet, or meters?

7. **Slash marks.** "New carrier/vendor" could mean either "new carrier *or* vendor," or "new carrier *and* vendor." Which is it?

8. **Undefined abbreviations or acronyms.** "ASAP" must be defined. Though all readers within the United States might understand this to mean "as soon as possible," you cannot assume the same level of understanding internationally.

9. **Cultural sensitivity.** In the United States, we might see humor in referring to a boss as "the old ball and chain" (though this is highly doubtful). In Japan, however, where saving face is a cultural norm, potentially offending a superior would be a problem. Finally, "pronto" is cowboy slang. To communicate internationally, you must consider each country's cultural norms—and be careful to avoid offense or confusion.

In contrast, the following example corrects these communication problems.

AFTER

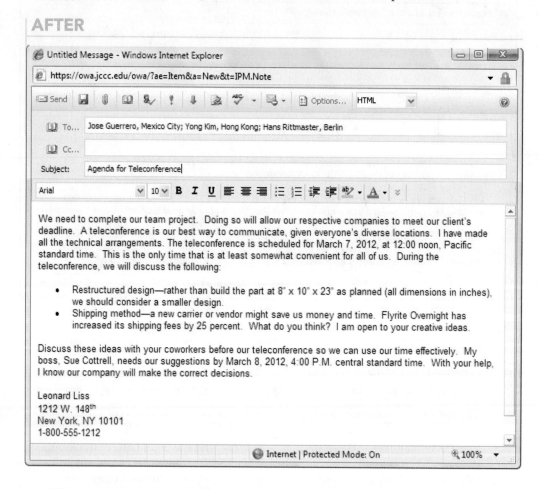

Avoiding Biased Language

E-mail

See Chapter 6 for more discussion of e-mail messages.

In addition to recognizing your audience's level of knowledge, roles, and cultural diversity, you also must consider your audience's age, physical limitations, and gender.

Ageist Language

A word like "elderly" could imply feebleness. The words "Old folks" create a negative image. To avoid these biases, write "people over seventy" or "retirees." Better yet, avoid reference to age.

BEFORE	AFTER
Professor Jones, an elderly teacher at State University, is publishing a textbook despite his age.	Professor Jones, a State University teacher, is publishing a textbook.

Biased Language about People with Disabilities

The word "handicap" creates a negative image. "Disability" is generally preferred. However, any euphemism can be offensive. You should avoid reference to a person's disabilities. If you need to refer to a physical problem, do so without negative characterizations.

BEFORE	AFTER
Debbie Brown, a blind market researcher, won "Employee of the Month."	Debbie Brown, a market researcher, won "Employee of the Month."
The AIDS victim changed insurance carriers.	The AIDS patient changed insurance carriers.
John suffers from diabetes.	John is diabetic.
Sheila is confined to a wheelchair.	Sheila uses a wheelchair.

Sexist Language

Many of your readers will be women. This does not constitute a separate audience category. Women readers will be high tech or low tech, management or subordinate. Thus, you don't need to evaluate a woman's level of understanding or position in the chain of command any differently than you do for readers in general.

Recognize, however, that women constitute at least half the workforce. As such, when you write, you should avoid *sexist language*, which is offensive to all readers. Let's focus specifically on ways in which sexism is expressed and techniques for avoiding this problem. Sexism creates problems through omission, unequal treatment, and stereotyping, as well as through word choice.

Web Resources

www.pearsonhighered .com/gerson For more information about sexist language, visit our companion website.

Ignoring Women or Treating Them as Secondary. When your writing ignores women or refers to them as secondary, you are expressing sexist sentiments. The following are examples of biased comments and their nonsexist alternatives.

BEFORE	AFTER
Biased	Unbiased
The CEO of Yahoo is a woman named Carol A. Bartz.	The CEO of Yahoo is Carol A. Bartz.
When setting up his experiment, the researcher must always check for errors.	When setting up experiments, the researcher must always check for errors.
As we acquired scientific knowledge, men began to examine long-held ideas more critically.	As we acquired scientific knowledge, people began to examine long-held ideas more critically.

Modifiers that describe women in physical terms not applied to men treat women unequally.

BEFORE	AFTER
Biased	Unbiased
The poor women could no longer go on; the exhausted men . . .	The exhausted men and women could no longer go on.
Mrs. Acton, a gorgeous blonde, is Joe Granger's assistant.	Jan Acton is Joe Granger's assistant.

Stereotyping. If your writing implies that only men do one kind of job and only women do another kind of job, you are stereotyping. For example, if all management jobs are held by men and all subordinate positions are held by women, this is sexist stereotyping.

BEFORE	AFTER
Biased	Unbiased
Current tax regulations allow a head of household to deduct for the support of his children.	Current tax regulations allow a head of household to deduct for child support.
The manager is responsible for the productivity of his department; the foreman is responsible for the work of his linemen.	Management is responsible for departmental productivity. Supervisors are responsible for their personnel.
The administrative assistant brought her boss his coffee.	The administrative assistant brought the boss's coffee.
The teacher must be sure her lesson plans are filed.	The teacher must file all lesson plans.

Sexist language disappears when you use pronouns and nouns that treat all people equally.

Pronouns.   Pronouns such as *he, him,* or *his* are masculine. Sometimes you read disclaimers by manufacturers stating that although these masculine pronouns are used, they are not intended to be sexist. They're only used for convenience. This is an unacceptable statement. When *he, him,* and *his* are used, a masculine image is created, whether or not such companies want to admit it.

To avoid this sexist image, avoid masculine pronouns. Instead, use the plural, generic *they* or *their.* You also can use *he or she* and *his or her.* Sometimes you can solve the problem by omitting all pronouns.

BEFORE	AFTER
Biased	Unbiased
Sometimes the doctor calls on his patients in their homes.	Sometimes the doctor calls on patients in their homes.
The typical child does his homework after school.	Most children do their homework after school.
A good lawyer will make sure that his clients are aware of their rights.	A good lawyer will make sure that clients are aware of their rights.

Gender-Tagged Nouns.   Use nouns that are nonsexist. To achieve this, avoid nouns that exclude women and denote that only men are involved.

BEFORE	AFTER
Gender-tagged, Biased Nouns	Unbiased
mankind	people
manpower	workers, personnel
the common man	the average citizen
wise men	leaders
businessmen	businesspeople
policemen	police officers
firemen	firefighters
foreman	supervisor
chairman	chairperson, chair
stewardess/steward	flight attendant
waitress/waiter	server

Consider the following examples of sexist and nonsexist writing in a letter advertising landscaping services. In the first letter, the sexist writing is boldfaced, whereas in the second example the sexist terms are replaced with nonsexist boldfaced words.

Sexist Writing

DiBono and Sons Landscaping
1349 Elm
Oakland Hills, IA 20981

April 2, 2012

Owner
TurboCharge, Inc.
2691 Sommers Rd.
Iron Horse, IA 20992

Dear **Sir:**

Interested in beautifying your company property, with no care or worries? We have been in business for over 25 years helping **men** just like you. Here's what we can offer:

1. Shrub Care—your **workmen** will no longer need to water shrubs. We'll take care of that with a sprinkler system geared toward your business's unique needs.
2. Seasonal System Checks—don't worry about your **foreman** having to ask **his** workers to turn on the sprinkler system in the spring or turn it off in the winter. Our **repairmen** take care of that for you as part of our contract.
3. New Annual Plantings—we plant rose bushes and daffodils that are so pretty, **your secretary will want to leave her desk and pick a bunch for her** office. Of course, you might want to do that for **her** yourself ☺.
4. Trees—we don't just plant annuals. We can add shade to your entire property. Just think how beautiful your parking lot will look with elegant elms, oaks, and maples. Then, when you have that company picnic, your **employees and their wives** will be surrounded by nature.

Call today for prices. Let our skilled **craftsmen** work for you!

Sincerely,

Richard DiBono

Nonsexist Revision

DiBono and Sons Landscaping
1349 Elm
Oakland Hills, IA 20981

April 2, 2012

Owner
TurboCharge, Inc.
2691 Sommers Rd.
Iron Horse, IA 20992

Subject: Sales Information

Interested in beautifying your company property, with no care or worries? We have been in business for over 25 years helping **business owners** just like you. Here's what we can offer:

- Shrub Care—your **employees** will no longer need to water shrubs. We'll take care of that with a sprinkler system geared toward your business's unique needs.
- Seasonal System Checks—don't worry about your **supervisor** having to **ask his or her** workers to turn on the sprinkler system in the spring or turn it off in the winter. Our **staff** takes care of that for you as part of our contract.
- New Annual Plantings—we plant rose bushes and daffodils that are so pretty, **all of your employees** will want vase-filled flowers on **their** desks.
- Trees—we don't just plant annuals. We can add shade to your entire property. Just think how beautiful your parking lot will look with elegant elms, oaks, and maples. Then, when you have that company picnic, your **employees** and **their families and friends** will be surrounded by nature.

Call today for prices. Let our skilled **experts** work for you!

Sincerely,

Richard DiBono

Audience Involvement

In addition to audience recognition, effective technical communication demands audience involvement. You not only need to know whom you are writing to in your technical correspondence (audience recognition), but also you need to involve your readers—draw them into your writing and keep them interested. Achieving audience involvement requires that you strive for personalized tone and reader benefit.

Personalized Tone

Companies do not write to companies; people write to people. Remember when you write your e-mail, letter, report, or procedure, another person will read it, so you want to achieve a personalized, collaborative tone to involve your reader. Personalized tone is even more important in social media (Facebook, YouTube, Twitter, blogs, and so on) since this communication channel seeks to build community. Personalization can be achieved by using pronouns and people's names.

Pronouns. The best way to personalize correspondence is through pronoun usage. If you omit pronouns in your technical communication, the text will read as if it has been computer generated, absent of human contact. When you use pronouns in your technical communication, you humanize the text. You reveal that the memo, letter, report, or procedure is written by people, for people.

The hierarchy of pronoun usage is as follows:

Pronoun	Focus
You	
Your	The reader
We	
Us	The team
Our	
I	
Me	The ego
My	

The first group of pronouns, *you* and *your*, is preferred. When you use *you* or *your*, you are speaking directly to your reader(s) on a one-to-one basis. The readers, whoever they are, read the word *you* or *your* and see themselves in the pronoun, envisioning that they are being spoken to, focused on, and singled out. In other words, *you* or *your* appeals to the reader's sense of self-worth. We all like to be thought of as special and worthy of the writer's attention. By focusing on the pronouns *you* or *your* in technical communication, you make your readers feel special and involved in the written communication.

The second group of pronouns (*we, us,* and *our*) uses team words to connote group involvement. These pronouns are valuable when writing to multiple audiences or when writing to subordinates. *We, us,* or *our* implies to the readers that "we're all in this together." Such a team concept helps motivate by making the readers feel an integral part of the whole.

The third group (*I, me,* and *my*) denotes the writer's involvement. If overused, however, these pronouns can connote egocentricity ("All I care about is *me, me, me*"). Because of this potential danger, emphasize *you* and *your* and downplay *I, me,* and *my*. Strive for a two-to-one ratio. For every *I, me,* or *my,* double your use of *you* or *your*. We are not saying that you should avoid using first-person pronouns. You cannot write without involving yourself through *I* or *my*. We are just saying that an overuse of these first-person pronouns creates an egocentric image, whereas an abundance of *you* orientation achieves audience involvement.

Let's look at an automobile manufacturer's user manual, which omits pronouns and, thus, reads as if it is computer generated.

BEFORE

Claims Procedure

To obtain service under the Emissions Performance Warranty, take the vehicle to the company dealer as soon as possible after it fails an I/M test along with documentation showing that the vehicle failed an EPA-approved emissions test.

Without any pronouns, the version has no personality. It's dull and stiff, and it reads as if no human beings are involved.

Compare this flat, dry, dehumanized example with the more personalized version.

AFTER

Claims Procedure

How do you get service under the Emissions Performance Warranty? To get service under this warranty, take your car to the dealer as soon as possible after it has failed an EPA-approved test. Be sure to bring along the document that shows your car failed the test.

When pronouns are added in this version, the writing involves the reader on a personal level.

Many writers believe that the first version is more professional. However, professionalism does not require that you write without personality or deny the existence of your reader. The pronoun-based writing in the second version is more friendly. Friendliness and humanism are positive attributes in technical communication. The second version also ensures reader involvement. As a professional, you want your readers (customers, clients, colleagues) to be involved, for without these readers no transaction occurs.

Names.   Another way to personalize and achieve audience involvement is through use of names—incorporating the reader's name in your technical communication. By doing so, you create a friendly reading environment in which you speak directly to your audience. In the following example, the writer uses the reader's name and pronouns to personalize the text and to achieve audience involvement.

◀ EXAMPLE

Pfeiffer Consulting

February 12, 2012

Aleia Shanmugan
Home Health Care
1238 Roe
Phoenix, AZ 34133

Subject: Proposed Crisis Communication Seminar

Thank you, Ms. Shanmugan, for inviting us to familiarize you and your colleagues with techniques for communicating in crisis situations. We are looking forward to working with you. Following is a schedule for the seminar:

Date: March 19, 2012
Time: 8:00 A.M.–5:00 P.M.

Site: Cedar Inn, Room C

If I can provide additional information, please let me know. You can contact me at gbpfeiffer@pfeifferco.com.

Glenn Pfeiffer, CEO

When do you use first names versus last names? The decision is based on your closeness to or familiarity with your reader. If you know your reader well, have worked with him or her for a while, and know your reader will not be offended, then use the first name. However, if you do not know the reader well, have not worked with this reader before, or worry that the reader might be offended, then use the surname instead. In either instance, calling a reader by name will involve that reader and personalize your writing.

Reader Benefit

A final way to achieve audience involvement is to motivate your readers by giving them what they want or need. We're not saying that you should make false promises. Instead, you should show your audience how they will benefit from your technical communication. You can do this in two ways: Explain the benefit and use positive words.

Explain the Benefit. Until you tell your readers how they'll benefit, they may not know. Therefore, in your letter, e-mail, report, or manual, state the benefit clearly. You can do this anywhere, but you're probably wise to place the statement of benefit either early (first paragraph or abstract/summary) or late (last paragraph or conclusion). Placing the benefit early in the writing will interest readers and help ensure that they continue to read. Placing the benefit at the end could provide a motivational close, leaving readers with a positive impression.

For example, when writing a procedure, you want your readers to know that by performing the steps appropriately, they will avoid mechanism breakdowns, maintain tolerance requirements, achieve a proper fit, or ensure successful equipment operation. By following the procedures, they will reap a benefit.

EXAMPLE ▶

This instruction conveys reader benefit.

> ### Instructions for Poured Foundations
>
> A poured foundation will provide a level surface for mounting both the pump and motor. Carefully aligned equipment will provide you a longer and more easily maintained operation.

E-mail messages can also develop reader benefit to involve audiences, as in the following example.

EXAMPLE ▶

The benefit is evident in the positive words—*improved, strengthen, reduce, timely, accurate, happier, pride,* and *contributions.* These words motivate readers by suggesting how they will benefit from their actions.

Reader benefit is emphasized in the last paragraph. The pronouns *we* and *our* in the last sentence involve the readers in the activity. A team concept is implied.

> To: AMcWard
>
> Subject: Improved Billing Control Procedures
>
> Three times in the last quarter, one of the billing cycles excluded franchise taxes. This was a result of the CRT input being entered incorrectly.
>
> To strengthen internal controls and reduce the chance of future errors, the following steps will be implemented effective January 1, 2012.
>
> 1. Input—The control screen entry form will be completed by the accounting clerk.
> 2. Entry—The CRT entry clerk will enter the control screen input and use local print to produce a hard copy of the actual input.
> 3. Verification—The junior accountant will compare the hard copy to the control screen and initial if correct.
> 4. Notification—The junior accountant will tell data processing to start the billing process.
>
> The new procedure will provide more timely and accurate billing to our customers. Customer service will be happier with accounting, and we will have more pride in our contributions.

Use Positive Words.   In the preceding examples achieving reader benefit, the motivation or value is revealed through positive words. One way to involve your audience is to use positive words throughout your correspondence.

POSITIVE WORDS

accomplish	effective	organize	thank you
achieve	efficient	plan	value
advantage	enjoy	please	
appreciate	ensure	produce	
asset	establish	profit	
assist	favorable	promote	
benefit	guide	raise	
build	happy	recommend	
confident	help	reduce	
coordinate	improve	satisfy	
create	increase	serve	
develop	lead	succeed	
educate	maintain	support	

To clarify how important such words are, let's look at some examples of negative writing, followed by positive revisions.

BEFORE	AFTER
Negative	**Positive**
• We cannot process your request. You failed to follow the printed instructions.	• So that we may process your request rapidly, please fill in line 6 on the printed form.
• The error is your fault. You keep your books incorrectly and cannot complain about our deliveries. If you would cooperate with us, we could solve your problem.	• To ensure prompt deliveries, let's get together to review our bookkeeping practices. Would next Tuesday be convenient?
• We have received your letter complaining about our services.	• Thank you for writing to us about our services.
• Your bill is now three weeks overdue. Failure to pay immediately will result in lower credit ratings.	• If you're as busy as we are, you've probably misplaced our recent bill (mailed three weeks ago). Please send it in soon to maintain your high credit ratings.
• The invoice you sent was useless by the time it arrived. You wasted my time and money.	• We're proud of our ability to maintain schedules. But we need your help. When you return invoices by the fifteenth, we save time and you save money.

Making something positive out of something negative is a challenge, but the rewards for doing so are great. If you attack your readers with negatives, you lose. When you involve your readers through positive words, you motivate them to work with you.

The following example e-mail uses positive words to motivate the reader.

This e-mail achieves audience involvement by speaking directly to the reader ("Renata"), by using pronouns, and by employing positive words to motivate the reader to action.

To: Renata Shuklaper

Subject: Assignment of Contract Administration Project

Renata, thank you for accepting the Contract Administration Project. To accomplish our goal of competitive bidding, I need your help. Please follow these guidelines with your team:

1. Analyze other operating companies to see how they ensure neutrality when administering contracts.
2. Coordinate our consultants to guarantee that they achieve impartiality.
3. Develop standards for all operating procedures across all departments.
4. Prepare a report for our advisory team. They'll need it by the end of the month.

Your successful completion of this project, Renata, will help us improve future contract administrations. I'm confident that you'll do a great job, and I know our company and employees will benefit.

AUDIENCE CHECKLIST

_____ 1. What is my audience's level of understanding regarding the subject matter?
  - High tech
  - Low tech
  - Lay
  - Multiple

_____ 2. Given my audience's level of understanding, have I written accordingly?
  - Have I defined my acronyms, abbreviations, and jargon?
  - Have I supplied enough background data?
  - Have I used the appropriate types of graphics?

_____ 3. Who else might read my correspondence?
  - How many people?
  - What are their levels of understanding?
    - High tech
    - Low tech
    - Lay
  - Will my audience be multicultural?

_____ 4. What is my role in relation to my audience?
  - Do I work for the reader?
  - Does the reader work for me?
  - Is the reader a peer?
  - Is the reader a client?

_____ 5. What response do I want from my audience? Do I want my audience to act, respond, confirm, consider, decide, or file the information for future reference?

_____ 6. Will my audience act according to my wishes? What is my audience's attitude toward the subject (and me)?

  - Negative
  - Positive
  - Noncommittal
  - Uninformed

_____ 7. Is my audience in a position of authority to act according to my wishes? If not, who will make the decision?

_____ 8. What are my audience's personality traits?
  - Slow to act
  - Eager
  - Receptive
  - Questioning
  - Organized
  - Disorganized
  - Oppositional

_____ 9. Have I motivated my audience to act?
  - Have I involved my reader in the correspondence by using pronouns and his or her name?
  - Have I shown my reader the benefit of the proposal by using positive words?
  - Have I considered any questions or objections my audience might have?

_____ 10. Have I considered my audience's preferences regarding style?
  - Is use of a first name appropriate?
  - Will the reader accept contractions?

_____ 11. Have I avoided sexist language?
  - Have I used _their_ or _his or her_ to avoid sexist singular pronouns such as _his_?

- Have I used generic words such as *police officer* versus *policeman*?
- Have I avoided excluding women, writing sentences such as "All present voted to accept the proposal" versus "All the men voted to accept the proposal"?
- Have I avoided patronizing, writing sentences such as "Mr. Smith and Mrs. Brown wrote the proposal" versus "Mr. Smith and Judy wrote the proposal"?

_____12. Have I considered diversity?
- Have I avoided any language that could offend various age groups, people of different sexual orientations, people with disabilities, or people of different cultures and religions?
- Have I considered that people from different countries and people for whom English is a second language will be involved in the communication? This might mean that I should do the following:
  - Clarify time and measurements
  - Define abbreviations and acronyms
  - Avoid figurative language and idiomatic phrases, unique to one culture
  - Avoid humor and puns
  - Consider each country's cultural norms

CHAPTER HIGHLIGHTS

1. High-tech readers understand acronyms, abbreviations, and jargon. However, not all of your readers will be high tech.
2. Low-tech readers need glossaries or parenthetical definitions of technical terms.
3. Lay readers usually will not understand technical terms. Consider providing these readers with extended definitions.
4. In addition to considering high-tech, low-tech, and lay audience levels, you also should consider issues of diversity, such as gender, race, religion, age, and sexual orientation.
5. Multiple audiences have different levels of technical knowledge. This fact affects the amount of technical content and terms you should include in your document.
6. The definition of a technical term usually includes its type and distinguishing characteristics but can also include examples, descriptions, and procedures.
7. For multicultural audiences, define terms, avoid jargon and idioms, and consider the context of the words you use. Avoid cultural biases, be careful with humor, and allow space for translation.
8. Half your audience will be female, so you should avoid sexist language.
9. To get your audience involved in the text, personalize it by using pronouns and the reader's name.
10. Show your reader how he or she will benefit.

CASE STUDY

Home and Business Mortgage

Home and Business Mortgage (HBM) has suffered several lawsuits recently. A former employee sued the company, contending that it practiced "ageism" by promoting a younger employee over him. In an unrelated case, another employee contended that she was denied a raise due to her ethnicity. To combat these concerns, HBM has instituted new human resources practices. The goal is to ensure that the company meets all governmental personnel regulations, including those required by the Equal Employment Opportunity Commission (EEOC), the Family Medical Leave Act (FMLA), and the Americans with Disabilities Act (ADA). HBM is committed to achieving diversity in its workplace.

HBM now needs to hire a new office manager for one of its branch operations. It has three outstanding candidates. Following are their credentials:

- **Carlos Gutierrez.** Carlos is a 27-year-old recent recipient of an MBA (master's in business administration) from an acclaimed business school. At this university, he learned many modern business applications, including capital budgeting, human resources management, organizational behavior, diversity management, accounting and marketing management, and team management strategies. He has been out of graduate school for only two years, but his work during that time has been outstanding. As an employee at one of HBM's branches, he has already impressed his bosses by increasing the branch's market share by 28 percent through innovative marketing strategies he learned in college. In addition, his colleagues enjoy working with him and praise his team-building skills. Carlos has never managed a staff, but he is filled with promise.

- **Cheryl Huff.** Cheryl is a 37-year-old employee of a rival mortgage company, Farm-Ranch Equity. She has a BGS (bachelor of general studies) from a local university, which she acquired while working full time in the mortgage/real estate business. At Farm-Ranch, where she has worked for over 15 years, she has moved up the ladder, acquiring many new levels of responsibility. After working as an executive assistant for five years, she then became a mortgage payoff clerk for two years, a loan processor for two years, a mortgage sales manager for four years, and, most recently, a residential mortgage closing coordinator for the last two years. In the last two positions, she managed a staff of five employees.

  According to her references, she has been dependable on the job. Her references also suggest that she is a forceful taskmaster. Though she has accomplished much on the job, her subordinates have complained about her demanding expectations. Her references suggest, perhaps, that she could profit from some improved people skills.

- **Rose Massin.** Rose is 47 years old and has been out of the workforce for 12 years. During that absence, she raised a family of three children. Now, the youngest child is in school and Rose wants to reenter the job market. She has a bachelor's degree in business, and was the former office manager of this HBM branch. Thus, she has management experience, as well as mortgage experience.

  While she was branch manager, she did an outstanding job. This included increasing business, working well with employees and clients, and maintaining excellent relationships with lending banks and realtors. She was a highly respected employee and was involved in civic activities and community volunteerism. In fact, she is immediate past-president of the local Rotary club. Though she has been out of the workforce for years, she has kept active in the city and has maintained excellent business contacts. Still, she's a bit rusty on modern business practices.

Assignment

Whom will you hire from these three candidates? Based on the information provided, make your hiring decision. This is a judgment call. Be sure to substantiate your decision with as much proof as possible. Then share this finding as follows:

- **Oral Presentations.** Give a 3- to 5-minute presentation to share with your colleagues the results of your decision.
- **Written.** Write an e-mail, memo, or report about your findings.

INDIVIDUAL AND TEAM PROJECTS

Avoiding Sexist Language

1. Revise the following sentences to avoid sexist language.
   a. All the software development specialists and their wives attended the conference.
   b. The foremen met to discuss techniques for handling union grievances.
   c. Every technician must keep accurate records for his monthly activity reports.
   d. The president of the corporation, a woman, met with her sales staff.
   e. Throughout the history of mankind, each scientist has tried to make his mark with a discovery of significant intellectual worth.

Achieving Audience Involvement

1. Revise the following sentences to achieve audience involvement through personalization, adding pronouns and names.
   a. The company will require further information before processing this request.
   b. It has been decided that a new procedure must be implemented to avoid further mechanical failures.
   c. The department supervisor wants to extend heartfelt thanks for the fine efforts expended.
   d. I think you have done a great job. I want you to know that you have surpassed this month's quota by 12 percent. I believe I can speak for the entire department by saying thank you.
   e. If the computer overloads, simultaneously press Reset and Control. Wait for the screen command. If it reads "Data Recovered," continue operations. If it reads "I/O Error," call the computer resource center.
2. Revise the following sentences to achieve reader benefit, using positive words.
   a. We cannot lay your cable until you sign the attached waiver.
   b. John, don't purchase the wrong program. If we continue to keep inefficient records, our customers will continue to complain.
   c. You have not paid your bill yet. Failure to do so might result in termination of services.
   d. If you incorrectly quote and paraphrase, you will receive an *F* on the assignment.
   e. Your team has lost 6 of their last 12 games.
3. Rewrite the following flawed correspondence. Its tone is too negative and commanding, regardless of the audience. Soften the tone to achieve better audience involvement and motivation.

To: Staff

Subject: Fourth Quarter Goals

Due to a severe lack of discipline, the company failed to meet third quarter goals. To avoid repeating this disaster for the fourth quarter, this is what I think you all must do—ASAP.

1. Demand that the sales department increase cold calls by 15 percent.
2. Require weekly progress reports by all sales staff.
3. Penalize employees when reports are not provided on time.
4. Tell managers to keep on top of their staff, pushing them to meet these goals.

Remember, when one link is weak in the chain, the entire company suffers. DON'T BE THE WEAK LINK!

Defining Terms for Different Audience Levels

1. Find examples of definitions provided in computer word processing programs, e-mail packages, Internet dictionaries, and online help screens. Determine whether these examples provide the *term*, its *type*, and its *distinguishing characteristics*. Are the definitions effective? If so, explain why. If not, rewrite the definitions for clarity.

2. Find examples of definitions in manuals such as your car's owner's manual, a computer user manual, or manuals packaged with your MP3 player, Wii, coffeemaker, Bluetooth, iPod, or lawn mower. Note where these definitions are located. Are they placed parenthetically within the text or in a glossary? Next, determine whether the definitions provide the *term*, its *type*, and its *distinguishing characteristics*. Are the definitions effective? If so, explain why. If not, rewrite the definitions for clarity.

3. Find examples of definitions in your textbooks. Where are these definitions placed—parenthetically within the text or in a glossary? Next, determine whether the definitions provide the *term*, its *type*, and its *distinguishing characteristics*. Are the definitions effective? If so, explain why. If not, rewrite the definitions for clarity.

4. Select five to ten terms, abbreviations, or acronyms from your area of expertise or interest (network communications, accounting, electronic engineering, automotive technology, biomedical records, data processing, etc.). First, define each term using words and phrases. Next, define each term in a sentence, conveying the *term*, its *type*, and its *distinguishing characteristics*. Finally, define each term using a longer paragraph. In these longer definitions, include examples, descriptions, or processes.

5. In small groups composed of individuals from like majors, list ten high-tech terms (jargon, acronyms, or abbreviations) unique to your degree programs. Then, envisioning a lay audience, parenthetically define and briefly explain these terms. To test the success of your communication abilities, orally share these high-tech terms with other students who have different majors.
   - State the high-tech term to see if they understand it.
   - If they don't, provide the parenthetical definition. How much does this help? Do they understand now?
   - If they still do not understand the definition, add the third step—the brief explanation. How much information do the readers need before they understand your high-tech terms?

Recognizing Issues of Diversity

1. Rewrite the following sentences for multicultural, cross-cultural audiences.
   a. Let's meet at 8:30 P.M.
   b. The best size for this new component is $16 \times 23 \times 41$.

c. To keep us out of the red, we need to round up employees who can put their pedal to the metal and get us out of this hole.

d. We need to produce fliers/brochures to increase business.

e. The meeting is planned for 07/09/12.

2. Rewrite the following flawed correspondence. Be sure to consider your multicultural or cross-cultural audience's needs.

From: Nachman Sumani

Subject: Brainstorming

I need to pick your brain, guys. We've got a big one coming up, a killer deal with a major European player. Before I can make the pitch, however, let's brainstorm solutions. The client needs a proposal by 12/05/12, so I need your input before that date. Give me your ideas about the following:

- What should we charge for our product, if the client buys in bulk?
- What's our turnaround time for production? I know that your people tend to work slowly, so can you hurry the team up if I promise delivery in six weeks?

E-mail me your feedback by tomorrow, 1:00 p.m. my time, at the latest. Trust me, guys. If we blow this one, everyone's bonuses will be out the window.

PROBLEM-SOLVING THINK PIECES

1. Dove Hill, Georgia, experienced severe thunderstorms on March 15, 2012. GAI (Goodwin & Associates Insurance Co.) insured much of the territory affected by this storm. Over 1,200 houses insured by GAI had water damage. As written, the homeowner's policy provided full-replacement coverage for water damage. In the homeowner's policies, water damage was limited to situations such as the following:

- A hailstorm smashes a window, permitting hail and rain to enter a home.
- A heavy rain soaks through a roof, allowing water to drip through a ceiling.
- A broken water pipe spews water into a home.

    In contrast, homes which suffered water damage due to flooding were not covered, unless the homeowners also had taken out a separate flood damage policy from the National Flood Insurance Program. For insurance purposes, "flood" is defined as "the rising of a body of water onto normally dry land." For example, flood damage can include a river overflowing its banks or a heavy rain seeping into a basement. Of the 1,200 homes suffering water damage, only 680 homes qualified for full-replacement coverage. The remaining homes suffered flood damage and were thus not covered.

Assignment

Write a letter or e-mail message from GAI to a homeowner denying coverage for their water damage claim. You are writing as a high-tech insurance professional to a lay reader. You want to maintain a good business relationship, despite the bad news. Build rapport while specifying in lay terms the denial.

2. Financial Trust & Annuity, a bank with over 100 sites throughout the Southwest, is growing at approximately 15 percent per year. To accommodate this growth, the bank needs to hire more tellers. Good business demands that FTA hire diversity: people with disabilities, employees from many races and cultures, men and women, and a variety of age ranges.

Doing so will help ensure that the bank's tellers represent people from all walks of life and that the bank will design its services to appeal to a diverse customer base. In addition, the bank also wants to hire a percentage of employees who can speak Spanish, Arabic, and Chinese. This will help FTA communicate with a key constituency of its emerging customer base.

Assignment

You are the human resources manager. To meet FTA's hiring needs, write a job advertisement for the teller positions and include a statement addressing the bank's commitment to diversity.

3. Rewrite the following flawed correspondence. Be sure to achieve effective audience understanding and involvement. To do so, avoid sexist language and define high-tech terms. Search online to find any abbreviations or acronyms that need defining.

West Central Auditors
"Your Technical Engineering and Energy Resource Experts"
1890 River
Pocato, Idaho 89022

March 12, 2012

Marks-McGraw, Inc.
2145 Oceanview
Clackamas, Oregon 31002

Gentlemen:

We will visit your plant next week for the TEA. To ensure that our visit goes smoothly, we plan this procedure:

- Your plant foreman will have his engineer assemble his CAD drawings for the power plant at Brush Prairie.
- Our men will review these drawings against TE specs, such as specific cost estimates for energy usage and energy data.
- Our ER group will then help your men plan and prepare your facilities for technology changes mandated by the DOE's office of the EERE.
- We also want to review your engineer's plans for any building envelopes. Make sure he brings all relevant correspondence he has had with his clients.

This audit should take approximately seven man-hours. It's a man-sized job. If we have to go off site for lunch or breaks, the job will take even longer. Therefore, please ask your secretary to provide drinks and food for our six representatives. Tell her that we have no dietary concerns, so anything she provides will be greatly appreciated.

Sincerely,

Jim Wynn, Team Manager

---

4. Find examples of writing directed to high-tech, low-tech, and lay audiences. To do so, read professional journals, find instructions, look at marketing brochures, read trade magazines, or ask your colleagues and co-workers for documents.

Assignment

Once you have found these examples, bring them to class. In groups, discuss your findings to determine whether the documents are written for high-tech, low-tech, or lay readers. Use the Audience Evaluation Form that follows to share your decisions.

AUDIENCE EVALUATION FORM					
Example Number	1	2	3	4	5
Type (Circle One)	High Low Lay	High Low Lay	High Low Lay	High Low Lay	High Low Lay

Criteria
Language
• Abbreviations
• Acronyms
• Jargon
Content
• General
• Specific
• Background
Tone
• Formal
• Informal
Format
• Highlighting
• Graphics
  —type
  —complexity
  —color
  —labeling					

5. In groups, read the following two examples (A and B) to determine whether they are written to a high-tech or low-tech audience. To justify your answer, consider the use of acronyms, abbreviations, or jargon; the presentation of background data; the length of sentences and paragraphs; and the way in which the text has been formatted (how it looks on the page).

EXAMPLE A

This documentation summarizes the items to be tested and the methodology utilized in the testing routine. The test plan tested four new 8536 Switcher features: OIM channel commands, OIM disconnect commands, OIM dump commands, and OIM log-off messages.

The patch files associated with the base must disable the features; consequently, only testing to ensure that this is operational will be addressed.

Software development personnel, after testing the features, will deliver copies of this report and product acceptance releases to program management, who are then responsible for writing functional specifications and submitting them to administration for manufacturing and marketing.

Purpose. This document identifies the time to be tested and the methods used for testing. The tests center on the 8536 Switcher (a network consisting of two nodes) and CN (connect node) bases. The 8536 Switcher and CN bases perform the same function as our currently marketed G9 and G12 systems.

OIM (Operator Interface Module) Features Tested:

- OIM channel commands
- OIM disconnect commands
- OIM dump commands
- OIM log-off messages

Testing Focus. To work appropriately, these features must be disabled by our patch files if an error occurs. Therefore, our testing focus will be the disabling capabilities.

Responsibilities. After testing the features, our software development group will send copies of this report and product acceptance releases to program management. They then will write the functional specifications and send them to administration. Administration is responsible for manufacturing and marketing. The following flowchart graphically depicts this sequence.

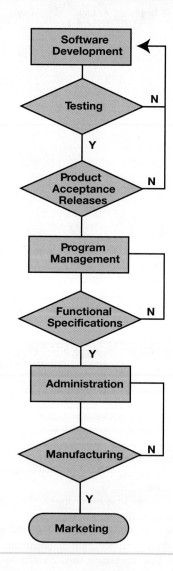

WEB WORKSHOP

1. You are the Director of Human Resources for your company. Your job is to write a nondiscrimination policy. To do so, access an online search engine to find other companies' nondiscrimination policies. Compare and contrast what you find. Then, based on your research, write your company's policy.

2. You are head of International Relations at your corporation. Your company is preparing to go global. To ensure that your company is sensitive to multicultural concerns, research the cultural traits and business practices at five countries of your choice. To do so, access an online search engine and type in "multicultural business practices in _____" (specify the country's name). Report your findings either orally or in an e-mail.

3. Unfortunately, companies occasionally are sued for unfair hiring practices, wrongful termination of employees, or other discriminatory practices on the job. These lawsuits can depend on the company's perceived biases related to age, gender, physical disabilities, or ethnicity. Access an online search engine to learn about corporate challenges with lawsuits based on biases. What was the cause of the lawsuit, and how was the legal issue resolved? Report your findings either orally or in an e-mail.

QUIZ QUESTIONS

1. What are the four main types of audiences based on their level of knowledge?
2. What are the differences among the types of audiences based on their level of knowledge?
3. What are three ways to define terms in technical documents?
4. What are the main parts of a definition?
5. Why should you consider a multicultural audience when you write a technical document?
6. What guidelines should you follow to write to a multicultural audience?
7. What is sexist language?
8. What are three ways to avoid sexist writing?
9. How can you achieve audience involvement in your technical communication?
10. What are two things you can do to express reader benefit in technical communication?

CHAPTER FIVE

Research and Documentation

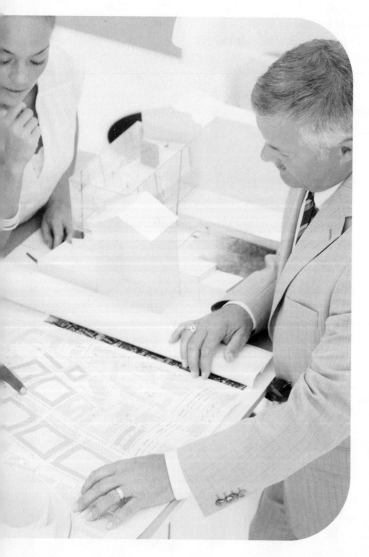

COMMUNICATION AT WORK

The Oak Springs, Iowa, scenario illustrates the use of primary and secondary research before writing construction plans for a new road.

The City of Oak Springs, Iowa, needs to improve a ten-mile stretch of road, Ironhorse Avenue, that runs east and west through the town. The winding, two-lane road was built in 1976. Since then, the city has grown from a population of 45,000 to 86,000. In 1976, the road primarily served farmers whose houses were separated by miles of privately owned land. Now, over 30 years later, many of the landowners have sold their land for development. Ironhorse Avenue is occupied by numerous housing subdivisions, apartment complexes, and strip malls. Due to this development, the road no longer is sufficient for the increased traffic load.

Ironhorse Avenue must be expanded to four lanes and straightened. An expanded and straightened street will allow the city to provide better snow removal; quicker response time to emergencies by fire, police, and medical personnel; easier travel time for commuters; and safer travel for pedestrians. To accomplish these goals, the city must add new sewer lines, sidewalks, easements, esplanades, bicycle lanes, and lighting. However, this construction will impact current homeowners (whose land, through eminent domain, can be expropriated without the owner's consent). In addition, part of the envisioned road construction will impact a wildlife refuge for

Objectives

When you complete this chapter, you will be able to

1. Understand why to conduct research in technical communication

2. Use both primary and secondary research in technical communication

3. Locate information in the library and online

4. Read and analyze sources of information for a report

5. Form an idea for a research report

6. Write a research report

7. Consider ethics and document your sources of information

8. Avoid plagiarism

9. Evaluate researched material using a checklist

waterfowl. This wildlife refuge is important to the ecosystem and is also used for field research by the Oak Springs schools' science classes.

Before construction plans can be made, city engineers must conduct research to determine the public's needs and concerns. First, the engineers will perform primary research by interviewing city residents, as well as employees from the city's parks and recreation, police, fire, and transportation departments. Then, the engineers will conduct secondary research. They will read statutes regarding eminent domain, real estate, environmental considerations, state wildlife refuges, zoning, planning, and land use.

Only after conducting primary and secondary research will the engineers be able to produce a construction plan for the Oak Springs city management for consideration and approval.

Check out our quarterly newsletters at www.pearsonhighered.com/gerson for dot.com updates, new case studies, insights from business professionals, grammar exercises, and facts about technical communication.

Why Conduct Research?

Research skills are important in your school or work environment. You may want to perform research to better understand a technical term or concept; locate a magazine, journal, or newspaper article for your supervisor; or find data on a subject to prepare an oral or written report. Technology is changing so rapidly that you must know how to do research to stay up to date.

You can research information using online catalogs, electronic indexes and data-bases, reference books in print and online, and Internet search engines and directories. Reference sources vary and are numerous, so this chapter discusses only research techniques.

Research, a major component of long, formal business reports, helps you develop your report's content. Often, your own comments, drawn from personal experience, will lack sufficient detail, development, and authority to be persuasive. You need research for the following reasons:

- Create content
- Support commentary and content with details
- Prove points
- Emphasize the importance of an idea
- Enhance the reliability of an opinion
- Show the importance of a subject to the larger business community
- Address the audience's need for documentation and substantiation

Research Including Primary and Secondary Sources

You can use researched material to support and develop content in your formal reports. Researched material generally breaks down into two categories: primary and secondary research.

Primary Research/Field Research

Primary research is research performed or generated by you. You do not rely on books or periodicals for this type of research. With primary research, you will be generating the information based on data or information from a variety of sources that might include observations, tests of equipment, and laboratory experiments.

Another type of primary research is called field research. This can entail the following:

- Preparing a survey or a questionnaire targeting a group of respondents
- Conducting face-to-face interviews
- Interviewing individuals through telephone calls
- Using social media, such as Twitter, Facebook, and LinkedIn, to collect information
- Visiting job sites

You may perform this field research to determine for your company the direction a new marketing campaign should take, how to use social media in the workplace, the importance of diversity in the workplace, the economic impact of relocating the company to a new office site, the usefulness of a new product, or the status of a project.

You also may need to interview people for their input about a particular topic. For example, your company might be considering a new approach to increase security on employee computers. You could ask employees for a record of their logs which would highlight the problems they have encountered with their computer security.

Secondary Research

When you conduct *secondary research*, you rely on already printed and published information taken from sources including books, periodicals, newspapers, encyclopedias, reports, proposals, or other business documents. You also might rely on information taken from a Web site, blog, or YouTube instructional video. All of this secondary research requires parenthetical source citations (discussed in Appendix B).

SPOTLIGHT

Conducting Research to Find Solutions

Tom Woltkamp is Senior Manager of Information Solutions for Teva Neuroscience, a global leader in the "development, production and marketing of generic and proprietary branded pharmaceuticals" (http://www.tevapharm.com/). Teva Pharmaceutical Industries is based in Israel. Teva Neuroscience, a subsidiary of Teva Pharmaceuticals, is based in Kansas City, Missouri, with branches in Florida, Pennsylvania, Texas, North Carolina, California, Minnesota, and Canada.

Approximately 20 employees and consultants report to Tom regarding computer infrastructure (computer networks, help desks, call centers, hardware/software purchases) and application development for programming, software integration, installation, configuration, Web programming, and customized documentation.

On an ongoing basis, Teva employees in finance, human resources, sales, and the company's call centers need new software applications. They ask Tom's group to find solutions to their needs. The research and documentation process for Teva's software development is as follows:

- **Project Charter:** These brief reports "drive the research." In a page or two, the Teva departments with whom Tom works "write down what they want." They clarify their needs, as Tom says, by stating "here's my problem," "here's what I think I want to do about it," and "here's what I think the benefits will be."

That's when Tom begins his research into finding solutions for their software needs. Tom and his team conduct both primary and secondary research:

- **Questionnaires:** Tom and his group create FAQ checklists where they "play 20 questions to find what our end users want." They meet with their clients and discuss the project parameters.

- **Internet searches:** After the initial questionnaire, which helps zero in on the client's needs, Tom and his team search the Internet for "knowledge and solutions." They want to find out more about the topic of research and see if any products already exist that can meet their needs or software they can "tweak" for customization.

- **Consultation:** Another avenue of research involves meeting with consultants—experts in the field—for their take on the topic.

- **Interviews:** Tom works with "key partners," including banks, vendors, professional organizations, and patients using the company's pharmaceuticals. By interviewing these constituents, Tom can learn more about how other entities might have solved similar problems or how the customers using a drug can be better served.

- **Online and hard-copy journals:** An excellent secondary source of research for Tom is professional journals, such as *PC Magazine*, *Call Center Magazine*, and *CRM Magazine* (which focuses on customer relations management).

Tom and his team spend approximately 20 to 30 percent of their work time on research. Why? Research allows Tom to make sure that the software solutions he and his team design and develop are correct. Research helps Tom ensure that his products are on target to meet the client's needs; will be time and cost efficient; and meet national, international, and industry standards. Research helps Tom make certain that he can achieve customer satisfaction and provide value to his organization.

Criteria for Writing Research Reports

As with all types of technical communication, using researched material in reports requires that you

- Recognize your audience
- Use an effective style, appropriate to your reader and purpose
- Use effective formatting techniques for reader-friendly ease of access

Audience

When writing a research report, you first must recognize the level of your audience. Are your readers high tech, low tech, or lay? If you are writing to your boss, consider his or her level of technical expertise. Your boss probably is at least a low-tech reader, but you must determine this based on your own situation. Doing so will help you determine the amount of technical definition necessary for effective communication.

You must decide next whether your reader will understand the purpose of your research report. This will help you determine the amount of detail needed and the tone to take (persuasive or informative). For instance, if your reader has requested the information, you will not need to provide massive amounts of background data explaining the purpose of your research. Your reader has probably helped you determine the scope and purpose. On the other hand, if your research report is unsolicited, your first several paragraphs must clarify your rationale. You will need to explain why you are writing and what you hope to achieve.

If your report is solicited, you probably know that your reader will use your research for a briefing, an article to be written for publication, a technical update, and so forth. Thus, your presentation will be informative. If your research report is unsolicited, however, your goal is to persuade your reader to accept your hypotheses substantiated through your research.

Effective Style

Research reports should be more formal than many other types of technical communication. In a research report, you are compiling information, organizing it, and presenting your findings to your audience using documentation. Because the rules of documentation are structured rigidly (to avoid plagiarism, to create uniformity, and to meet industry standards), a research report is also carefully structured.

The tone doesn't need to be stuffy, but you should maintain an objective distance and let the results of your research support your contentions. Again, considering your audience and purpose will help you decide what style is appropriate. For example, if your boss has requested your opinion, then you are correct in providing it subjectively. However, if your audience has asked for the facts—and nothing but the facts—then your writing style should be more objective.

Formatting

Reading a research report is not always an easy task. As the writer, you must ensure that your readers encounter no difficulties. One way to achieve this is by using effective formatting. Reader-friendly ease of access is accomplished when you use highlighting techniques such as bullets, numbers, headings, subheadings, and graphics (tables and figures).

In addition, effective formatting includes the following:

- Overall organization (including an introduction, discussion, and conclusion)
- Internal organization (various organizational patterns, such as problem/solution, comparison/contrast, analysis, and cause/effect)
- Parenthetical source citations
- Works cited or references page—documentation of sources

Each of these areas is discussed in detail throughout this chapter and in Appendix B.

The Writing Process

Prewriting	Writing	Rewriting

- Decide whether you are writing to inform, instruct, persuade, analyze, or recommend.
- Determine whether your audience is internal, external, or both; whether your audience is high tech, low tech, or lay.
- Gather information from primary and secondary sources of research.

- Use an outline to organize your content.
- Organize your content using modes such as problem/solution, cause/effect, comparison/contrast, argument/persuasion, analysis, chronology, etc.
- Document your sources correctly.

- Revise your draft by
  - Adding details
  - Deleting wordiness
  - Simplifying words
  - Enhancing the tone
  - Reformatting your text
  - Proofreading and correcting errors
- Avoid plagiarism by checking the validity and accuracy of your quotes, paraphrases, summaries, and works cited (references).

The Writing Process at Work

Writing a research report, as with other types of technical correspondence, is easiest when you follow a process. Rather than just wandering into a library and hoping that the correct book or periodical will leap off a shelf and into your hands or expecting your online search to reveal useful information immediately, approach your research systematically. Prewrite (determine objectives and conduct research to gather your data), write a draft, and rewrite to ensure that you meet your goals successfully.

Communication Process

See Chapter 2 for more discussion of prewriting, writing, and rewriting.

Prewriting Research Techniques

1. **Select a general topic (or the topic you have been asked to study).** Your topic may be a technical term, phrase, innovation, or dilemma. If you're in biomedical technology, for example, you might want to focus on biohazards in the workplace. If you work in electronics, select a topic related to nanotechnologies. If your field is computer science, you could focus on developments in online gaming software.

2. **Spot-check sources of information.** Check a library or online sources to find material that relates to your subject. A quick review of your library's online periodical databases, such as Research Library from ProQuest or other equivalent sources, will help you locate periodical articles. Your library may also have a print or electronic edition of the *Reader's Guide to Periodical Literature* and other similar specialized periodical indexes. A keyword search of Internet metasearch engines will

give you an idea of how much information might be readily available through the Internet.

3. **Establish a focus.** After you have chosen a topic for which you can find available source material, decide what you want to learn about your topic. A focus statement can guide you. In other words, if you are interested in biomedical technology, you might write a focus statement such as the following:

EXAMPLE ▶

> I want to research current problems with biohazards in the workplace, including crisis management, disposal of hazardous materials, and training for employees.

For electronics, you could write the following:

EXAMPLE ▶

> I want to research the uses, expenses, and maintenance of nanotechnology.

If you are in computer sciences, you might write the following:

EXAMPLE ▶

> I want to discover the uses, impact on employment, and expenses of software for online gaming.

With focus statements such as these, you can begin researching your topic, concentrating on articles pertinent to your topic.

4. **Research your topic.** You may feel overwhelmed by the prospect of research. But there are many sources that, once you know how to use them, will make the act of research less overwhelming.

Books. All books owned by a library are listed in online catalogs. Books can be searched in online catalogs in a variety of ways: by title, author, subject, keyword, or some combination of these.

Periodicals. Use electronic or print periodical indexes to find articles on your topic. Online indexes can be searched in a variety of ways: by title, author, subject, keyword, or some combination of these.

Indexes to General, Popular Periodicals. Three leaders in providing periodical material in a general database are Gale/Cengage's Academic OneFile, Ebsco's Academic Search Premier, and ProQuest's Research Library. All are "general" resources in that they present material on a wide range of subjects or disciplines. However, each of the three presents this content from scholarly/professional sources, popular magazines, and news sources. The content can be segmented to represent these categories of periodicals. There are electronic and print counterparts for some of these. The electronic indexes provide the full text of many of the articles.

- *Reader's Guide to Periodical Literature*
- *Research Library from ProQuest*
- *Academic Search Premier*
- *Newsbank*

Indexes to Specialized, Scholarly, or Technical Periodicals. Many libraries provide access to one or more specialized indexes covering the literature of a particular subject area. There are electronic and print counterparts for most of these. The electronic indexes provide the full text of many of the articles.

- **Applied Science & Technology Index.** Covers engineering, aeronautics and space sciences, atmospheric sciences, chemistry, computer technology and applications, construction industry, energy resources and research, fire prevention, food and the food industry, geology, machinery, mathematics, metallurgy, mineralogy, oceanography, petroleum and gas, physics, plastics, the textile industry and fabrics, transportation, and other industrial and mechanical arts.
- **Business Periodicals Index.** Covers major U.S. publications in marketing, banking and finance, personnel, communications, computer technology, and so on.
- **ABI/Inform.** Covers business and management.
- **General Science Index.** Covers the pure sciences, such as biology and chemistry.
- **Social Sciences Index.** Covers psychology, sociology, political science, economics, and other social sciences topics.
- **ERIC (Education Resources Information Center).** Provides bibliography and abstracts about educational research and resources. Available for free through the Internet.
- **PubMed.** Covers medical journals and allied health publications.
- **PsycINFO.** Covers psychology and behavioral sciences.
- **LexisNexis.** Includes the full text of newspaper articles, reports, transcripts, law journals, and legal reporters and other reference sources in addition to general periodical articles.

The Internet. Millions of documents from countless sources are found on the Internet. You can find material on the Internet published by government agencies, organizations, schools, businesses, or individuals (see Table 5.1). The list of options grows daily. For example, nearly all newspapers and news organizations have online Web sites.

Accessing Online Newspapers and Magazines. Many periodicals provide a current look at articles but charge for past articles or have otherwise limited offerings available on their Web sites. For example, the *Kansas City Star* stories generally reside on kansascity.com for seven to fourteen days from the date of posting/publication. After that time, they move to the library archives. There is a fee to read stories in the archives. Although many articles are available online, you will have to access other newspaper or magazine articles in hard-copy format.

Finding Information Online

To find information online, use directories, search engines, or metasearch engines.

TABLE 5.1 A Sample of Internet Research Sources

Search Engines, Directories, Metasearch Engines	Online References	Online Libraries	Online Newspapers	Online Magazines	Online Government Sites
Yahoo Excite Lycos AltaVista MetaSearch MetaCrawler Google Britannica.com Ask.com Dogpile	*Webster's Dictionary* *Roget's Thesaurus* *Britannica Online* *Encyclopedia* *Smithsonian* *The Old Farmer's Almanac*	Library of Congress New York Public Library Cleveland Public Library Gutenberg Project Most city and university libraries	*New York Times* CNN *USA Today* *Washington Post* *Kansas City Star* Most city newspapers	*National Geographic* *Atlantic* *The New Republic* *U.S. News Online* *Time Magazine* *Ebony Online* *Slate*	United Nations The White House The IRS U.S. Postal Service USA.gov Most states' supreme courts, legislatures, executive offices, and local governments

Directories. Directories, such as Yahoo, Librarians' Internet Index, Infomine, About.com, and Google Directory, let you search for information from a long list of predetermined categories, including the following:

Arts	Government	Politics and Law
Business	Health and Medicine	Recreation
Computers	Hobbies	Science
Education	Money and Investing	Society and Culture
Entertainment	News	Sports

FAQs: Ethical Considerations when Evaluating Online Sources

Q: Can I quote or paraphrase from any source? Aren't all sources on the Internet equally valid?

A: Not all information, even if it's published online, is valid. Before you quote or paraphrase information, ask yourself these questions:

- What are the author's credentials? Not everyone who publishes a blog entry, a Twitter comment, a Web site, or an editorial in an online newspaper necessarily is an expert in the field. Is the author a college professor, a noted scientist, a government official, or some unknown individual? If you are uncertain about the author's background, you can perform research about the author by going to a search engine, such as Google, to find out the author's level of education, affiliations, and other publications.

- Is the source of information reputable? Not all online publications are equal. Many blog entries, Twitter comments, Facebook pages, YouTube videos, or Web sites do not go through editorial review. Just because someone publishes a comment on their blog doesn't mean it is a valid source of information. In contrast, online journals, magazines, and newspapers generally hire professional writers, conduct research, cite sources, and verify information before it is published.

- Is the information up to date? Some information on the Internet is dated. If content is old, it might no longer be valid. Check the dates of publication to determine whether the content is still relevant and valid.

- Is the information biased? Much of what a person writes has a bias. However, some publications are more biased than others. To determine whether your source of information is biased, check to see who might have funded the research, who publishes the journal or magazine, or what affiliations the author might have.

To access any of these areas, click on the appropriate category and then "drill down," clicking on each subcategory until you get to a useful site.

Search Engines. Search engines, such as Google, Ask.com, Yahoo! search, and Exalead, let you search millions of Web pages by keywords. Type a word, phrase, or name in the appropriate blank space and press the enter key. The search engine will search through documents on the Internet for "hits," documents that match your criteria. One of two things will happen: Either the search engine will report "no findings," or it will report that it has found thousands of sites that might contain information on your topic.

In the first instance, "no findings," you'll need to rethink your search strategy. You may need to check your spelling of the keywords or find synonyms. For example, if you want to research information about online writing, you could try typing "writing online," "online writing," "electronic writing," "writing electronically," and other similar terms. In the second instance, finding too many hits, you'll need to narrow your search. For example, if you are researching illegalities in baseball, you cannot type in "baseball." That's too broad. Instead, try "Shoeless Joe Jackson," "Pete Rose," "crime in baseball," "baseball scandals," "performance enhancing drugs," and so on.

Metasearch Engines. A metasearch engine, such as Ixquick and Clusty, lets you search for a keyword or phrase in a group of search engines at once, saving you the time of searching separately through each search engine.

Researching the Internet presents at least two problems other than finding information. First, is the information you have found trustworthy? Paperbound newspapers, journals, magazines, and books go through a lengthy publication process involving editing and review by authorities. Not all that's published on the Internet is so professional. Be wary. What you read online needs to be filtered through common sense. Second, remember that although a book, magazine, newspaper, or journal can exist unchanged in print form for years, Web sites change constantly. A site you find today online will not necessarily be the same tomorrow. That's the nature of electronic communication.

5. **Read your researched material and take notes.** Once you have researched and located a source (whether it is online, a book, magazine, journal, or newspaper article), study the material. For a book, use the index and the table of contents to locate your topic. When you have found it, refer to the pages indicated and skim, reading selectively. For shorter documents, such as magazine or journal articles or online materials, you can read closely. Reading and rereading the source material is an essential step in understanding your researched information. After you have studied the document thoroughly, go through it page by page and briefly summarize the content.

For a short magazine article, you can make marginal notes on a photocopy. For instance, read a paragraph and then briefly summarize its main point(s) in several words. Such notations are valuable because they are easy to make and provide a clear and concise overview of the article's focus.

You can also take notes on 3 × 5 index cards. If you do this, be sure to write only one fact, quotation, or paraphrase per card, along with the author's name. This will let you organize the information later according to whatever organizational sequence (chronological narrative, analysis by importance, comparison/contrast, problem/solution, cause/effect, etc.) you prefer. Prepare a bibliography of your sources on index cards as well, one source per card.

The following are examples of a **bibliography card** and a **note card** with quotation, respectively.

EXAMPLE ▶

> Swartz, Jon. "Businesses Get Cheap Help from a Little Birdie." *USA Today* 26 June 2009: B1–2.

EXAMPLE ▶

> Uses of Twitter in Business
> "Corporate giants such as Comcast, JetBlue Airways, Whole Foods Market and others are beefing up direct communications with customers through social-media tools such as Twitter, Facebook, and YouTube."
> Swartz, "Businesses Get Cheap Help," B1.

On a *summary* note card, you condense original material by presenting the basic idea in your own words. You can include quotations if you place them in quotation marks, but do not alter the organizational pattern of the original. A summary is shorter than the original.

EXAMPLE ▶

> Uses of Twitter in Business
> Many large corporations use social media to provide customer service and keep in touch with customers.
> Swartz, "Businesses Get Cheap Help," B1.

On a *paraphrase* note card, you restate the original material in your own words without condensing. The paraphrase is essentially the complete version rewritten. A paraphrase is the same length as the original.

EXAMPLE ▶

> Uses of Twitter in Business
> Many large corporations including Comcast, PepsiCo, JetBlue Airways, and Whole Foods Market use social media such as Twitter, Facebook, and YouTube to provide rapid customer service.
> Swartz, "Businesses Get Cheap Help," B1.

6. **Isolate the main points.** After you complete the analysis of the document (whether you use note cards and/or marginal comments), isolate the main points discussed in the books or periodicals. You will find that, of the major points in an article or book, sometimes only three or four ideas will be relevant for your topic. Choose the ones discussed repeatedly or those that most effectively develop the ideas you want to pursue.

7. **Write a statement of purpose.** Once you have chosen two to four main ideas from your research, write a purpose statement that expresses the direction of your research. For example, one student wrote the following statement of purpose after performing research on social media.

The purpose of this report is to reveal the future of social media in business including the application of YouTube instructional videos, "tweets," and Facebook pages.

◀ **EXAMPLE**

8. **Create an outline.** After you have written a purpose statement, formulate an outline. An outline will help you organize your paragraphs and ensure that you stay on track as you develop your ideas through quotes and paraphrases. Figures 5.1 and 5.2 are examples of topic and sentence outlines.

Writing

You now are ready to write your research report.

1. **Review your research.** Prior to writing your report, review your research sources to make sure that you're satisfied with what you've found. Do you have enough information to develop your points thoroughly? Is the information you've found what you want? If not, it's time to do more research. If you're content with what you've discovered in your research, you can start drafting your text.

2. **Organize your report effectively.** When you are ready to write, provide an introductory paragraph, discussion (body) paragraphs, a conclusion or recommendation, and your works cited or references page.

FIGURE 5.1   Topic Outline

I. Sensors used to help robots move
  A. Light
  B. Sound
  C. Touch

II. Touch sensor technology (microswitches)
  A. For gripping
  B. For maintaining contact with the floor

III. Optical sensors (LED/phototransistors)—Like bowling alley foul-line sensors
  A. Less bulky/connected to computer interface
  B. Not just for gripper, but for locating objects by following this sequence:
    1. Scan gripper to locate object
    2. Move gripper arm left and right to center object
    3. Move gripper forward to grasp
    4. Close gripper
  C. Problem—What force to use for gripping?

IV. Force sensors
  A. Spring and microswitch
  B. Optical encoder discs—Microprocessors determine speed of discs to determine force necessary
  C. Integrated circuits with strain gauge and pressure-sensitive paint
  D. Pressure sensors built with conductive foam

V. Conclusion

FIGURE 5.2 Sentence Outline

I. Because robots must move, they need sensors. These sensors could include light sensors, sound sensors, and touch sensors.

II. Touch sensors can have the following technology:

   A. Microswitches can be used for gripping.

   B. Microswitches can also be used for maintaining contact with the floor. This would keep the robot from falling down stairs, for example.

III. Optical sensors might be better than microswitches.

   A. LED/phototransistors are less bulky than switches.

   B. When connected to a computer interface, optical sensors also can help a robot locate an object as well as grip it.

   C. Here is the sequence followed when using optical sensors:

      1. The robot's grippers scan the object to locate it.

      2. The robot moves its gripper arms left and right to center the object.

      3. The robot moves forward to grip the object.

      4. The robot closes the gripper.

   D. The only problem faced is what force should be used when gripping.

IV. Force sensors can solve this problem.

   A. A combination spring and microswitch can be used to determine the amount of force required.

   B. Optical encoder discs can be used also. A microprocessor determines the speed of the disc to determine the required force.

   C. Integrated circuits with strain gauges and pressure-sensitive paint can be used to determine force.

   D. Another pressure sensor can be built from conductive foam.

V. To conclude, all these methods of tactile sensing comprise a field of inquiry important to robotics.

Introduction. In one to three sentences or a short paragraph, tell your readers the purpose of your long report. This purpose statement informs your readers *why* you are writing or *what* you hope to achieve.

Discussion. The number of discussion paragraphs will depend on the number of divisions and the amount of detail necessary to develop your ideas. Use quotes and paraphrases to develop your content. Students often ask how much of a research report should be *their* writing, as opposed to researched information. A general rule is to lead into and out of every quotation or paraphrase with your own writing. In other words,

- Make a statement (your sentences).
- Support this generalization with a quotation or paraphrase (referenced material from another source).
- Provide a follow-up explanation of the referenced material's significance (your sentences).

Q: What's the difference between a quote and a paraphrase?

A: A quotation must be exactly the same as the original word, sentence, or paragraph. You cannot haphazardly change a word, a punctuation mark, or the ideas conveyed. In contrast, when you paraphrase, you rewrite a quote using your own words. Whether you use a quote or a paraphrase, however, you must correctly cite the source (see Appendix B).

Q: Why is correct citation of sources so important?

A: It's all about avoiding plagiarism. Readers need to know where you found your information and from which sources you are quoting or paraphrasing. Therefore, you must document this information. Correct documentation is essential for several reasons:

- You must direct your readers to the books, periodical articles, and online reference sources that you have used in your research report or presentation. If your audience wants to find these same sources, they depend only on your documentation. If your documentation is incorrect, the audience will be confused. You want your audience to be able to rely on the correctness of your research.

- Plagiarism is the appropriation (theft) of some other person's words and ideas without giving proper credit. Communicators are often guilty of unintentional plagiarism. This occurs when you incorrectly alter part of a quotation but still give credit to the writer. Even if you have cited your source, an incorrectly altered quotation constitutes plagiarism.

- On the other hand, if you intentionally use another person's words and claim them as your own, omitting quotation marks and source citations, you have committed theft. This is dishonest and could raise questions about your credibility or the credibility of your research. Teachers, bosses, and colleagues will have little, if any, respect for a person who purposely takes another person's words, ideas, or visuals. It is essential, therefore, for you to cite your sources correctly.

- However, if you use "boilerplate" content (text that resides in your company's archive of research reports), you do not necessarily have to cite the source of the content.

Ethical technical communicators carefully cite the source of their material.

Conclusion/Recommendations. In a final paragraph, summarize your findings, draw a conclusion about the significance of these discoveries, and recommend future action.

Ethics and Citing Sources

On a final page or in an appendix, provide an alphabetized list of your research sources. (We discuss this documentation later in Appendix B.)

3. **Document your sources correctly.** Your readers need to know where you found your information and from which sources you are quoting or paraphrasing. Therefore, you must document this information. Correct documentation is essential to avoid plagiarism.

However, avoiding plagiarism in the workplace is not always clear cut. According to Jessica Reyman, "Communicators commonly perform a variety . . . of composing activities that could be considered plagiarism in the context of the classroom" but are not necessarily plagiarism in the workplace (61). For example, workplace communicators often use boilerplate information and templates—text that has been used before. They rely on "existing designs and layouts"; write collaboratively; and cut, paste, and repurpose corporate content (Reyman 61). Do you need to cite content that has already been used? The answer is yes . . . and no.

If you are writing a proposal to a new client, and you draw from your company's library of existing proposals for content, that is acceptable use of boilerplate content and templates. You do not need to cite the source of your information. However, if you find a proposal from some other company on the Internet, and use their content, that is plagiarism and unethical behavior. You must cite the source of this original information.

To document your research correctly, you must provide parenthetical source citations following the quote or paraphrase within the text. Then, at the end of your document, supply a Works Cited page if you are using the Modern Language Association (MLA) style sheet. If you are using either the American Psychological Association (APA) or Council of Science Editors (CSE) style sheets, you will include a References page. See Appendix B for examples of how to use parenthetical source citations and how to document sources.

Alternative Style Manuals. Although APA, CSE, and MLA are popular style manuals, others are favored in certain disciplines. Refer to these if you are interested or required to do so.

- *U.S. Government Printing Office Style Manual,* 30th edition. Washington, DC: Government Printing Office, 2008.
- *The Chicago Manual of Style,* 15th edition. Chicago: University of Chicago Press, 2007.
- Turabian, Kate L. *A Manual for Writers of Term Papers, Theses, and Dissertations,* 7th edition. Chicago: University of Chicago Press, 2007.

4. **Develop your ideas.** You have learned how to organize your report (through an introduction, discussion, and conclusion/recommendation) and how to document your sources of research (through parenthetical source citations and a works cited or references page). Writing your research report also requires that you use your research effectively to develop your ideas. Successful use of research demands that you correctly quote, paraphrase, or summarize.

TECHNOLOGY TIP

Using Microsoft Word 2010 for Documentation

Microsoft Word 2010 provides students and business employees many new tools related to research and documentation. When you click on the **References** tab, you will find ways to create tables of contents, insert footnotes, insert citations, and create either bibliographies or works cited notations.

For example, to create parenthetical and bibliographic citations, using MLA, follow these steps:

1. Click on the **Insert Citation** down arrow and **Add New Source**.

The following **Create Source** screen will pop up (the screen fields will be blank; we have added the appropriate information—author's name, title, publishing company, city, and date).

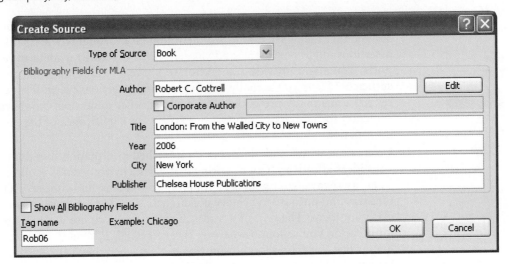

2. Click **OK** and the parenthetical source citation will be inserted.

3. To automatically insert the bibliographical information, click on the **Bibliography** down arrow and then on **Bibliography**.

The following will appear:

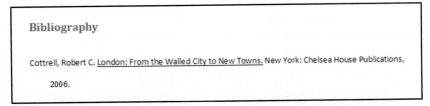

Rewriting

As with all types of writing, drafting the text of your research report is only the second stage of the writing process. To ensure that your report is effective, revise your draft as follows:

1. **Add new detail for clarity and persuasiveness.** Too often, students and employees assume that they have developed their content thoroughly when, in fact, their assertions might be general and vague. This is especially evident in research reports. You might provide a quotation to prove a point, but is this documentation sufficient? Have you actually developed your assertions? If an idea within your report seems thinly presented, either add another quotation, paraphrase, or summary for additional support or explain the significance of the researched information.

2. **Delete dead words and phrases and researched information that does not support your ideas effectively.** Good writing in a work environment is economical writing. Thus, as always, your goal is to communicate clearly and concisely. Delete words that serve no purpose to achieve conciseness. In addition, review your draft for clarity of focus. The goal of a research report is not to use whatever researched information you've found wherever it seems valid. Instead, you want to use quotations, paraphrases, and summaries only when they help develop your statement of purpose. If your research does not support your thesis, it is counterproductive and should be eliminated. In the rewriting stage, delete any documented research that is tangential or irrelevant.

3. **Simplify your words for easy understanding.** The goal of technical writing is to communicate, not to confuse. Write to be understood. Don't say *grain-consuming animal units* if you mean *chickens*. Don't call the July 2008 stock market crash a *fourth-quarter equity retreat*.

4. **Move information within your report to ensure effective organization.** How have you organized your report? Did you use a problem/solution format? Did you use comparison/contrast or cause/effect? Is your report organized as a chronological narrative or by importance? Whichever method you've used, you want to be consistent. To ensure consistency, rewrite by moving any information that is misplaced.

5. **Reformat your text for reader-friendly ease of access.** Look at any technical journal. You will notice that the writers have guided their readers through the text by using headings and subheadings. You will also notice that many journals use graphics (pie charts, bar graphs, line drawings, flowcharts, etc.) to clarify the writer's assertions. You should do the same. To help your readers follow your train of thought, reformat any blocks of wall-to-wall words. Add headings, subheadings, itemized lists, white space, and graphics.

6. **Correct any errors.** This represents your greatest challenge in writing a research report. You not only must be concerned with grammar and mechanics, as you are when writing an e-mail message, report, or proposal, but also with accurate quoting, paraphrasing, summarizing, parenthetical source citations, and works cited.

When revising, pay special attention to these concerns. If you quote, paraphrase, or summarize incorrectly, you run the risk of plagiarizing. If you fail to provide correct parenthetical source citations or works cited, your readers will not be able to find these same sources of information in their research or verify your data. Research demands accuracy and reliability.

_____ 1. Have you considered your purpose and audience in choosing a research topic?

_____ 2. Have you limited the focus of your research topic?

_____ 3. Have you used primary and secondary research sources?

_____ 4. Have you researched using secondary sources, such as books, periodicals, or the Internet; or primary sources, such as interviews and questionnaires?

_____ 5. Do you have enough material to sufficiently develop your topic?

_____ 6. Did you use note cards to gather your information?

_____ 7. Have you used an outline to organize your content?

_____ 8. Have you used both direct quotations and paraphrases?

_____ 9. Have you quoted and paraphrased correctly to avoid plagiarism?

_____ 10. Have you documented your sources according to either the APA, CSE, or MLA formats to avoid plagiarism?

CHAPTER HIGHLIGHTS

1. You can research a topic either using secondary sources, such as books, periodicals, or the Internet; or primary sources, such as interviews or questionnaires.

2. You need to consider the audience's level of technical knowledge when you write a research paper.

3. Primary and secondary research can be conducted to assist you in developing content for reports.

4. Narrowing a topic can help you find sources of information.

5. A focus statement lets you determine the direction of your document.

6. Use discrimination and consider the source when you research on the Internet.

7. Correct source citations help you avoid plagiarism.

8. The Council of Science Editors (CSE), Modern Language Association (MLA), and the American Psychological Association (APA) are three widely used style manuals for citing sources.

9. On a summary note card, condense the original material by using your own words.

10. On a paraphrase note card, restate the original material in your own words without condensing.

CASE STUDIES

1. The City of Oak Springs, Iowa, needs to improve a ten-mile stretch of Ironhorse Avenue that runs east and west through the town. The winding, two-lane road was built in 1976. Since then, the city has grown with several businesses and new housing along this road, and the road no longer is sufficient for the increased traffic load. The road must be expanded to four lanes and straightened. To do so, new sewer lines, sidewalks, easements, esplanades, and lighting must be added. However, this construction will impact current homeowners (whose land, through eminent domain, can be expropriated without the owner's consent). In addition, part of the envisioned road construction will impact a wildlife refuge for waterfowl. Before construction plans can be made, city engineers must conduct research in order to determine needs and considerations of the public. Only after conducting primary and secondary research will the engineers be able to present a construction plan to the Oak Springs city management for their consideration and approval.

Assignment

    a. Perform primary research. Create a questionnaire for interviewing city residents, as well as employees from the city's parks and recreation, police, fire, and transportation departments.

    b. Conduct secondary research. Read statutes online for your city and state regarding eminent domain, real estate, environmental considerations, state wildlife refuges, zoning, planning, and land use. Summarize your findings for your professor either in an e-mail message or oral presentation.

2. You plan to create a new business, which you want to be designated as Minority and Women Owned (MWBD). Before you can incorporate your company, you must research the following:

- How you can become certified as a MWBD
- The standards you must uphold
- The percentage of women and/or minorities that you must employ
- The benefits of operating such a business

Assignment

Research these topics (and any more that interest you) either online for secondary research or interview a MWBD business owner for primary research. Then, write an e-mail message or memo to your professor about your findings.

INDIVIDUAL AND TEAM PROJECTS

1. Summarize in one sentence any paragraph from this textbook. Provide a parenthetical source citation and documentation information (see Appendix B).

2. Read a one- or two-page article from a magazine, journal, or online source. Then practice note taking on a photocopy of the source. Writing in the margins or between paragraphs,

briefly note the key point(s) made in each paragraph. (These marginal notes can be limited to one or two words.)

3. Using a one- or two-page article from a magazine or journal, practice note taking on index cards. To do so, first write the correct documentation information (see Appendix B) on one card. Then, on separate cards, take notes about approximately four key ideas discussed in the article. Write only one note per card; give the card a title for future reference; provide either quotations or paraphrases; and then write the author's name, the article's title, and the page number on the bottom of the card.

4. Select a technical topic from your major field or your job and write a research report. You might want to consider an area of interest such as global warming, hazardous waste management, computer viruses, or nanotechnologies.

PROBLEM-SOLVING THINK PIECES

1. Many communities have recycling projects that allow residents to recycle paper products, cans, and plastic. Not all businesses recycle, however. Research the benefits of recycling, determine how a business or businesses could implement a corporate recycling plan, and write a report recommending action based on your research.

2. In today's global economy, understanding and accommodating multiculturalism and cross-culturalism in business is important. Research the following:

   - The unique challenges that cultural diversity presents to businesses
   - How companies have responded to these challenges

   Write a report recommending why and how a business or businesses can help employees develop cultural awareness.

3. Many companies track the time their employees spend looking at Web sites, sending and receiving Twitter messages, posting blog entries, or viewing Facebook and YouTube. Research the following:

   - Software that companies use to track employee electronic communication usage
   - Corporate guidelines for employee use of company-owned electronic communication hardware and software
   - The legal and ethical ramifications of an employee's private use of corporate-owned e-mail and Internet access
   - The legal and ethical ramifications of an employer eavesdropping on an employee's use of electronic communication

   Write a research report on your findings and provide a corporate guideline for both employee and employer electronic communication responsibilities.

4. Corporate training is big business. Many companies hire outside consultants or staff company training departments to teach employees new skills. These could include training workshops on diet and exercise to improve work efficiency, techniques for avoiding e-mail viruses or screening e-mail spam, resume writing for transitional employees, time management, leadership skills, improved oral presentations, improved customer service skills, PowerPoint usage, or business accounting for non-accountants.

   What training class does your company need? Research possible topics and training approaches. How have other companies offered this training? What benefits do employees derive from this training? How does the company benefit? What are the costs for this training (personnel, time, equipment, and so forth)? Then, write a proposal or an instructional training module based on your research.

5. Entrepreneurialism is one of the fastest growing sectors in business. Many people are opening their own businesses. What does it take to open your own business? Before you can write an effective business plan and seek financing from a bank, you must research the project.

Choose a new business venture, selling a product or service of your choice. What would it cost to open this business? What would be your best location, or should your business be online? What certifications or licensing is needed? How many personnel would you need? What equipment is necessary? Who would be your clientele?

Based on research, write a proposal, appropriate for presentation to a bank. In this proposal, present your business plan for a new entrepreneurial opportunity.

WEB WORKSHOP

1. USA.gov allows you to research a wide variety of topics, such as education and jobs, benefits and grants, consumer protection, environment and energy, science and technology, and public safety and health. This Web site also provides information on breaking news. Access USA.gov and research a topic relevant to your career goals. Write an e-mail message or report to your instructor summarizing your findings.

2. Go to an online news magazine, such as *Slate*, *Time*, or *U.S. News Online*. Type a topic of interest in the magazine's search engine. Research this topic and make a brief oral presentation to your class about the information you have gathered.

3. Access a search engine such as Google, Ask.com, or Yahoo! search. Type in a topic relating to your major field. For example, for computer information technology, medical records and health information, or accounting, look for job openings in your field to learn about salary ranges, benefits, application requirements, and so forth. Report your findings either in an oral presentation to the class or in an e-mail message to your instructor.

4. Using an online newspaper, such as *The New York Times*, *CNN*, or *USA Today*, research business and technology news. Find out the major news stories of the day which relate to your career path. Report your findings in an oral presentation to the class or in an e-mail message to your instructor.

QUIZ QUESTIONS

1. In what places can you locate research information?

2. Explain why you need to consider audience when you write a research report.

3. What is the difference between primary and secondary research?

4. What is the appropriate writing style of a research report?

5. What is the difference between a quote and a paraphrase?

6. How does a focus statement direct you when you write a research report?

7. Explain plagiarism. How can you avoid plagiarizing?

8. Why must you document your sources?

9. What is the difference between a summary note card and a paraphrase note card?

10. In addition to CSE, APA, and MLA, what are additional style manuals used for documentation?

CHAPTER SIX

Routine Correspondence— Memos, Letters, E-Mail Messages, Instant Messages, and Text Messages

COMMUNICATION AT WORK

In this scenario, a biotechnology company frequently corresponds through letters, memos, e-mail, instant messages, and text messages.

CompuMed, a wholesale provider of biotechnology equipment, is based in Reno, Nevada. CompuMed's CEO, Jim Goodwin, plans to capitalize on emerging nanotechnology to manufacture and sell the following:

- Extremely lightweight and portable heart monitors and ventilators
- Pacemakers and hearing aids, 1/10 the size of current products on the market
- Microscopic biorobotics, which can be injected in the body to manage, monitor, and/or destroy blood clots, metastatic activities, arterial blockages, alveoli damage due to carcinogens or pollutants, and scar tissue creating muscular or skeletal immobility

CompuMed is a growing company with over 5,000 employees located in two dozen cities and three states. To manage this business, supervisors and employees write on

Objectives

When you complete this chapter, you will be able to

1. Understand the differences among memos, letters, e-mail messages, instant messages, and text messages

2. Follow an all-purpose template to write memos, letters, and e-mail

3. Use memo samples as guidelines for memo components, organization, writing style, and tone

4. Evaluate your memos, letters, and e-mail messages with checklists

5. Correctly use the eight essential letter components: the writer's address, the date, an inside address for the recipient, a salutation, the body of the letter, a complimentary close, and the writer's signed and typed names

6. Write different types of letters, including the following:
   - Inquiry
   - Cover (transmittal)
   - Complaint
   - Adjustment
   - Bad news
   - Confirmation
   - Recommendation

7. Understand the components of successful e-mail messages

8. Use e-mail samples as guidelines for effective e-mail components, organization, writing style, and tone

9. Recognize techniques for successfully using instant messages and text messages in the workplace

10. Follow the writing process—prewriting, writing, and rewriting—to create memos, letters, and e-mail

---

average over 5 letters, 10 memos, 50 e-mail messages, and numerous instant messages and text messages a day.

The letters are written to many different audiences and serve various purposes. CompuMed must write letters for employee files, to customers, job applicants, outside auditors, governmental agencies involved in biotechnology regulation, insurance companies, and more. They write

- Letters of inquiry to retailers seeking product information (technical specifications, pricing, warranties, guarantees, credentials of service staff, and so forth)
- Cover letters prefacing CompuMed's proposals
- Complaint letters written to parts-manufacturers if and when faulty equipment and materials are received in shipping and adjustment letters to compensate retailers when problems occur

CompuMed's managers and employees also write memos to accomplish a variety of goals:

- Document work accomplished
- Call meetings and establish meeting agendas
- Request equipment from purchasing
- Preface internal proposals

To accomplish the majority of their routine correspondence, CompuMed's employees write many e-mail messages each day. These messages serve different purposes. Some e-mail messages are conversational. Other e-mail messages, however, must be professional in their style, organization, and content. This is especially true for e-mail messages sent to clients, vendors, and customers outside the company. These e-mail messages focus on timelines, deadlines, prices for service, meeting arrangements, cost breakdowns, procedural steps, and a host of other topics.

Finally, when CompuMed employees are working at distant locations, on the road, in hotels, or at the airport, they will use instant messages or text messages to ask each other quick questions or to casually check on the status of a project.

Routinely, CompuMed employees spend a great deal of their time writing memos, letters, e-mail messages, instant messages, and text messages.

Check out our quarterly newsletters at www.pearsonhighered.com/gerson for dot.com updates, new case studies, insights from business professionals, grammar exercises, and facts about technical communication.

153

The Importance of Memos, Letters, E-Mail, Instant Messages, and Text Messages

On a day-to-day basis, employees routinely write memos, letters, e-mail messages, instant messages, reports, and text messages. The National Commission on Writing, in their *Writing: A Ticket to Work . . . Or a Ticket Out, A Survey of Business Leaders*, states that e-mail is "ubiquitous in the American economy," that "more than half of all responding companies also report the following forms of communications as required 'frequently' or 'almost always': technical reports (59 percent), formal reports (62 percent), and memos and correspondence (70 percent)" (11). Figure 6.1 shows the significance of e-mail, memos, reports, and letters in the workplace.

This National Commission on Writing survey of "120 major American corporations employing nearly 8 million people" (3) clearly tells us that you routinely can expect to write many e-mail messages, letters, memos, and reports on the job.

With the growing importance of mobile communication, instant messages and text messages also are written routinely on the job. In 2009, Nielsen Media Ratings stated that 240 million mobile phone users in the United States have text-enabled handsets. Every six months since 2003, text messaging volume has increased by 37 percent (Goforth D15).

Which Communication Channel Should You Use?

Memos, letters, and e-mail messages are three common types of communication channels. Other communication channels include reports, Web sites, blogs, PowerPoint presentations, oral communication, instant messages, text messages, and more. When should you write an e-mail message instead of a memo? When should you write a memo instead of a letter? Is an instant message or a text message appropriate to the situation? You will make these decisions based on your audience (internal or external), the complexity of your topic, the speed with which your message can be delivered, and security concerns.

Communication Channels

See Chapter 1 for more discussion of communication channels.

For example, e-mail is a convenient communication channel. It is easy to write a short e-mail message, which can be sent almost instantaneously to your audience at the click of a button. However, e-mail might not be the best communication channel to use. If you are discussing a highly sensitive topic such as a pending merger, corporate takeover, or layoffs, an e-mail message would be less secure than a letter sent in a sealed envelope. You might need to communicate with employees working in a manufacturing warehouse. Not all of these employees will necessarily have an office or access to a computer. If you sent an e-mail message, how would they access this correspondence? A memo posted in the break room or a text message sent to their handhelds would be better choices of communication channels.

FIGURE 6.1   Percentage of Employees Who Consider E-mail, Memos, Letters, and Reports "Extremely Important"

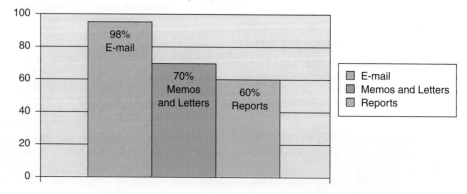

The Differences Among Routine Communication Channels

To clarify the distinctions among memos, letters, e-mail, instant messages (IM), and text messages (TM), review Table 6.1.

TABLE 6.1 Memos vs. Letters vs. E-mail vs. Instant Messages and Text Messages

Characteristics	Memos	Letters	E-mail	IM/TM
Destination	Internal: correspondence written to colleagues within a company.	External: correspondence written outside the business.	Internal *and* external: correspondence written to friends and acquaintances, co-workers within a company, and clients and vendors.	Internal *and* external: correspondence written to friends and acquaintances, co-workers within a company, and clients and vendors.
Format	Identification lines include "Date," "To," "From," and "Subject." The message follows.	Includes letterhead address, date, reader's address, salutation, text, complimentary close, and signatures.	Identification lines: To and Subject. The Date and From are computer generated. Options include cc (complimentary copy), forwarding the e-mail message to others, and replying to numerous readers.	No format.
Audience	Generally high tech or low tech, mostly business colleagues.	Generally low-tech and lay readers, such as vendors, clients, stakeholders, and stockholders.	Multiple readers due to the internal and external nature of e-mail.	Multiple readers due to the internal and external nature of IM and TM.
Topic	Generally topics related to internal corporate information; abbreviations and acronyms often allowed.	Generally topics related to vendor, client, stakeholder, and stockholder interests; abbreviations and acronyms usually defined.	A wide range of diverse topics determined by the audience.	A wide range of diverse topics determined by the audience.
Complexity and length of communication	Memos usually are limited to a page of text. If you need to write longer correspondence and develop a topic in more detail, you might consider using a different communication channel, such as a short report.	Letters usually are limited to a page of text, though you might write a two- or three-page report using a letter format. If you need to develop a topic in greater detail than can be conveyed in one to three pages, you might want to use a different communication channel, such as a longer, formal report.	An effective e-mail message usually is limited to one viewable screen (requiring no scrolling) or two screens. E-mail, generally, is not the best communication channel to use for complex information or long correspondence. If your topic demands more depth than can be conveyed in a screen or two, you might want to write a report instead.	IM and TM are very brief communication channels. Text messages must be limited to 160 characters. The typical size of an instant message is around 200 characters. Generally, the maximum size for mobile messages is approximately 400 characters.
Tone	Informal due to peer audience.	More formal due to audience of vendors, clients, stakeholders, and stockholders.	A wide range of tones due to diverse audiences. Usually informal when written to friends, informal to co-workers, more formal to management or external readers.	Very informal due to the limits placed on size.

(Continued)

Characteristics	Memos	Letters	E-mail	IM/TM
Attachments or enclosures	Hard-copy attachments can be stapled to the memo. Complimentary copies (cc) can be sent to other readers.	Additional information can be enclosed within the envelope. Complimentary copies (cc) can be sent to other readers.	Computer word processing files, HTML files and Web links, PDF files, RTF files, or downloadable graphics can be attached to e-mail. Complimentary copies can be sent to other readers.	Attachments can be sent using IM. You can send photo attachments with TM. However, sending file attachments with TM is limited.
Delivery time	Determined by a company's in-house mail procedure.	Determined by the destination (within the city, state, or country). Letters could be delivered within three days but may take more than a week.	Often instantaneous, usually within minutes. Delays can be caused by system malfunctions or excessively large attachments.	Instantaneous.
Security	If a company's mail delivery system is reliable, the memo will be placed in the reader's mailbox. Then, what the reader sees on the hard-copy page will be exactly what the writer wrote. Security depends on the ethics of co-workers and whether the memo was sent in an envelope.	The U.S. Postal Service is very reliable. Once the reader opens the envelope, he or she sees exactly what was written. Privacy laws protect the letter's content.	E-mail systems are not secure. E-mail can be tampered with, read by others, and sent to many people. E-mail stays within a company's computer backup system and is the property of the company. Therefore, e-mail is not private.	The same problems with e-mail security exist with TM and IM.

FAQs: Memos vs. E-mail

Q: Why write a memo? Haven't memos been replaced by e-mail?

A: E-mail is rapidly overtaking memos in the workplace, but employees still write memos for the following reasons:

1. Not all employees work in offices or have access to computers. Many employees who work in warehouses or in the field cannot easily access an e-mail account. They must depend on hard-copy documentation like memos.

2. Not all companies have e-mail. This may be hard to believe in the twenty-first century, but still it's a fact. These companies depend on hard-copy documentation like memos.

3. Many unions demand that hard-copy memos be posted on walls, in break rooms, in offices, and elsewhere, to ensure that all employees have access to important information. Sometimes, unions even demand that employees initial the posted memos, thus acknowledging that the memos have been read.

4. Some information cannot be transmitted electronically via e-mail. A bank we've worked with, for example, sends hard-copy cancelled checks as attachments to memos. They cannot send the actual cancelled check via e-mail.

5. E-mail messages are very easy to disregard. We get so many e-mail messages (many of them spam) that we tend to quickly delete them. Memos, in contrast, make more of an official statement. People might take hard-copy memos more seriously than e-mail messages.

Memos

Reasons for Writing Memos

Memos are an important means by which employees communicate with each other. Memos, hard-copy correspondence written within your company, are important for several reasons.

First, you will write memos to a wide range of readers. This includes your supervisors, co-workers, subordinates, and multiple combinations of these audiences. Memos usually are copied (cc: complimentary copies) to many readers, so a memo sent to your boss could be read by an entire department, the boss's boss, and colleagues in other departments.

Because of their frequency and widespread audiences, memos could represent a major component of your interpersonal communication skills within your work environment. Furthermore, memos are very flexible and can be written for many different purposes:

- **Documentation**—expenses, incidents, accidents, problems encountered, projected costs, study findings, hiring, firings, and reallocations of staff or equipment
- **Confirmation**—a meeting agenda, date, time, and location; decisions to purchase or sell; topics for discussion at upcoming teleconferences; conclusions arrived at; and fees, costs, or expenditures
- **Procedures**—how to set up accounts, research on the company intranet, operate new machinery, use new software, apply online for job opportunities through the company intranet, create a new company Web site, or solve a problem
- **Recommendations**—reasons to purchase new equipment, fire or hire personnel, contract with new providers, merge with other companies, revise current practices, or renew contracts
- **Feasibility**—studying the possibility of changes in the workplace (practices, procedures, locations, staffing, equipment, missions/visions)
- **Status**—daily, weekly, monthly, quarterly, biannual, yearly statements about where you, the department, or the company is regarding many topics (sales, staffing, travel, practices, procedures, or finances)
- **Directive (delegation of responsibilities)**—informing subordinates of their designated tasks
- **Inquiry**—asking questions about upcoming processes or procedures
- **Cover**—prefacing an internal proposal, long report, or other attachments

Criteria for Writing Memos

Memos contain the following key components:

- Memo identification lines (Date, To, From, and Subject)
- Introduction
- Discussion
- Conclusion
- Audience recognition
- Appropriate memo style and tone

Figure 6.2 shows an ideal, all-purpose organizational template that works well for memos, letters, and e-mail.

Subject Line.   The subject line summarizes the memo's content. One-word subject lines do not communicate effectively, as in the following flawed subject line. The "Before" sample has a *topic* (a what) but is missing a *focus* (a what about the what).

www.pearsonhighered.com/gerson For more information about routine correspondence, visit our companion website.

FIGURE 6.2 All-Purpose Template for Memos, Letters, and E-mail

Introduction: A lead-in or overview stating *why* you are writing and *what* you are writing about.

Discussion: Detailed development, made accessible through highlighting techniques, explaining *exactly what* you want to say.

- •
- •
- •

Conclusion: State *what* is next, *when* this will occur, and *why* the date is important.

BEFORE	AFTER
Subject: Sales	Subject: Report on Quarterly Sales

Introduction. Once you have communicated your intent in the subject line, get to the point in the introductory sentence(s). Write one or two clear introductory sentences which tell your readers *what* topic you are writing about and *why* you are writing. The following example invites the reader to a meeting, thereby communicating *what* the writer's intentions are. It also tells the reader that the meeting is one of a series of meetings, thus communicating *why* the meeting is being called.

EXAMPLE ▶

> In the third of our series of sales quota meetings this quarter, I'd like to review our productivity.

Discussion. The discussion section allows you to develop your content specifically. Readers might not read every line of your memo (tending instead to skip and skim). Thus, traditional blocks of data (paragraphing) are not necessarily effective. The longer the paragraph, the more likely your audience is to avoid reading. Make your text more reader friendly by itemizing, using white space, boldfacing, creating headings, or inserting graphics.

BEFORE

Example—Unfriendly Text

This year began with an increase, as we sold 4.5 million units in January compared to 3.7 for January 2012. In February we continued to improve with 4.6, compared with 3.6 for the same time in 2012. March was not quite so good, as we sold 4.3 against the March 2012 figure of 3.9. April was about the same with 4.2, compared to 3.8 for April 2012.

AFTER

Example—Reader-Friendly Text

Comparative Quarterly Sales (in Millions)

	2012	2013	Increase/Decrease
Jan.	3.7	4.5	0.8+
Feb.	3.6	4.6	1.0+
Mar.	3.9	4.3	0.4+
Apr.	3.8	4.2	0.4+

Conclusion. Conclude your memo with "thanks" and/or directive action. A pleasant conclusion could motivate your readers, as in the following example. A directive close tells your readers exactly what you want them to do next or what your plans are (and provides dated action).

EXAMPLE ▶

> If our quarterly sales continue to improve at the current rate, we will double our sales expectations by 2012. Next Wednesday (12/22/12), please provide next quarter's sales projections and a summary of your sales team's accomplishments.

Audience Recognition.   Since letters go outside your company, your audience is usually a low-tech or lay reader, demanding that you define your terms specifically. In memos your in-house audience is easier to address (usually high tech or low tech). You often can use more acronyms and internal abbreviations in memos than you can in letters.

Usually, you will write the memo to one reader, but you might "cc" (send a "complimentary copy") to other readers. Thus, you might be writing simultaneously to your immediate supervisor (high tech), to his or her boss (low tech), to your colleagues (high tech), and to a CEO (low tech). To accommodate multiple audiences, use parenthetical definitions, such as cash in advance (CIA) or continuing property records (CPR).

Style and Tone.   Because memos are usually only one page long, use simple words, short sentences, specific detail, and highlighting techniques. In addition, strive for an informal, friendly tone. Memos are part of your interpersonal communication abilities, so a friendly tone will help build rapport with colleagues.

In memos, audience determines tone. For example, you cannot write directive correspondence to supervisors mandating action on their part. It might seem obvious that you can write directives to subordinates, but you should not use a dictatorial tone. Though the subordinates are under your authority, they must still be treated with respect. You will determine the tone of your memo by deciding if you are writing vertically (up to management or down to subordinates) or laterally (to co-workers), as shown in Figure 6.3.

> **Audience**
>
> See Chapter 4 for more discussion of audience recognition and involvement.

> **Style, Development, Tone, and Highlighting Techniques**
>
> See Chapters 3 and 9 for more discussion.

BEFORE	AFTER
Example 1—Unfriendly, Demanding Style	**Example 2—Friendly, Personal Style**
We will have a meeting next Tuesday, Jan. 11, 2012. Exert every effort to attend this meeting. Plan to make intelligent comments regarding the new quarter projections.	Let's meet next Tuesday (Jan. 11, 2012). Even if you're late, I'd appreciate your attending. By doing so you can have an opportunity to make an impact on the second quarter projections. I'm looking forward to hearing your comments.

FIGURE 6.3   Vertical and Lateral Communication Within a Company

Sample Memos

See Figures 6.4 and 6.5 for sample memos.

FIGURE 6.4   Comparison/Contrast Feasibility Memo

MEMORANDUM

Date: December 12, 2012
To: George Singh, Manager
From: Luann Brunson, Purchasing Department
Subject: Replacement of Mobile Phones

On December 5, the manufacturing department supervisor informed the purchasing department that our company's mobile phones were malfunctioning. Purchasing was asked to evaluate three mobile communication options (the RPAD, XPO 1690, and MX16). Based on my findings, I have issued a purchase order for 12 RPAD handhelds.

The following points summarize my findings.

1. Performance

During a one-week test period, I found that the RPAD outperformed our current XPO's features. The RPAD has built-in WiFi, multiple messaging capabilities, QWERTY keyboard, and 3.6 mega-pixel camera. Our current XPO does not have photo capabilities, nor does it allow for multiple messaging.

2. Specifications

The RPAD is easier to carry, because of its reduced size. More importantly, the RPAD is easier to read due to its larger screen size and it improved resolution.

	RPAD	XPO 1690	MX16
Screen Size and Resolution	3.5" (400 × 640 pixel)	3" (340 × 240 pixels)	2.8" (300 × 200 pixels)
Phone Size	4.1" × 2.2" × 0.3"	4.21" × 2.17" × 0.55"	4.4" × 2.4" × 0.4"

3. Cost of Equipment

The RPAD is our most cost-effective option because of quantity cost breaks and maintenance guarantees.

	RPAD	XPO 1690	MX16
Cost per unit	$300.00	$500.00	$295.00
Cost per doz.	$3,600.00	$6,000.00	$3,540.00
Guarantees	1 year	6 months	6 months

Purchase of the RPAD will give us improved performance and comfort. In addition, we can buy 12 RPAD handhelds for approximately the cost of 6 XPOs. If I can provide you with additional information, please call. I'd be happy to meet with you at your convenience.

FIGURE 6.5   Cover Memo Prefacing Attachments

MEMO

CompuMed

Date: November 11, 2012

To: CompuMed Management

From: Bill Baker, Human Resources Director

Subject: Information about Proposed Changes to Employee Benefits Package

As of January 1, 2013, CompuMed will change insurance carriers. This will affect all 5,000 employees' benefits packages. I have attached a proposal, including the following:

1. Reasons for changing from our current carrier   page 2

2. Criteria for our selection of a new insurance company   pages 3–4

3. Monthly cost for each employee   pages 5–6

4. Overall cost to CompuMed   page 7

5. Benefits derived from the new healthcare plan   page 8

Please review the proposal, survey your employees' responses to our suggestions, and provide your feedback. We need your input by December 1, 2012. This will give the Human Resources Department time to consider your suggestions and work with insurance companies to meet employee needs.

Enclosure: Proposal

Introduce your memo by stating the topic being discussed and leading into the itemized information.

Provide dated action and the reasons for this request to conclude your memo. This also tells the audience what to do next.

MEMO CHECKLIST

_____ 1. Does the memo contain identification lines (Date, To, From, and Subject)?

_____ 2. Does the subject line contain a topic and a focus?

_____ 3. Does the introduction clearly state
   • Why this memo has been written?
   • What topic the memo is discussing?

_____ 4. Does the body explain exactly what you want to say?

_____ 5. Does the conclusion
   • Tell when you plan a follow-up or when you want a response?
   • Explain why this dated action is important?

_____ 6. Are highlighting techniques used effectively for document design?

_____ 7. Is the memo concise?

_____ 8. Is the memo clear,
   • Achieving specificity of detail?
   • Answering reporter's questions?

_____ 9. Does the memo recognize audience, defining acronyms or abbreviations where necessary for various levels of readers (high tech, low tech, and lay)?

_____ 10. Did you avoid grammatical errors? Errors will hurt your professionalism. See Appendix A for grammar rules and exercises.

Using Memo and Letter Templates in Microsoft Word 2010

1. Click on the **File** button located on the top left of your toolbar and scroll to **New**.

The following window will pop up.

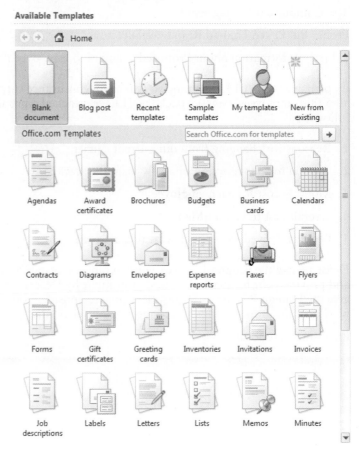

2. Click on the type of document you want to write, such as Letters or Memos. When you choose the communication channel, either of the following windows will pop up.

You can choose from one of twenty-three memo templates or one of hundreds of letter templates. Each of these templates gives you an already-designed letter format, complete with spacing, font selection, and layout. In addition, these templates provide fields in which you type the appropriate information (address, company name, date, salutation, complimentary close, your name and title, etc.).

These templates and wizards are both good and bad. They remind you which components can be included in a memo or letter, they make it easy for you to include these components, and they let you choose ready-made formats.

However, the templates also can create some problems. First, they are somewhat limiting in that they dictate what information you should include and where you should put this information. The content and placement of this information might contradict your teacher's or boss's requirements. Second, the templates are prescriptive, limiting your choice of font sizes and types. Our advice would be to use these templates and wizards with caution.

FAQs: Letters vs. E-mail

Q: Why write a letter? Haven't letters been replaced by e-mail?

A: Though e-mail is quick, it might not be the best communication channel, for the following reasons:

1. E-mail might be too quick. In the workplace, you will write about topics that require a lot of thought. Because e-mail messages can be written and sent quickly, people too often write hurriedly and neglect to consider the impact of the message.

2. E-mail messages tend to be casual, conversational, and informal. Not all correspondence, however, lends itself to this level of informality. Formal correspondence related to contracts, for example, requires the more formal communication channel of a letter. The same applies to audience. You might want to write a casual e-mail to a co-worker, but if you were writing to the president of a company, the mayor of a city, or a foreign dignitary, a letter would be a better, more formal choice of communication channel.

3. E-mail messages tend to be short. For content requiring more detail, a longer letter would be a better choice.

4. We get so many e-mail messages a day that they are easy to disregard—even easy to delete. Letters carry more significance. If you want to ensure that your correspondence is read and perceived as important, you might want to write a letter instead of an e-mail.

5. Letters allow for a "greater paper trail" than e-mail. Most employees' e-mail inboxes fill up quickly. To clean up these inboxes, people tend to delete messages that they don't consider important. In contrast, hard-copy letters are wonderful documentation.

Letters

Reasons for Writing Letters

Letters are external correspondence that you send from your company to a colleague working at another company, a vendor, a customer, a prospective employee, and stakeholders and stockholders. Letters leave your worksite (as opposed to memos, which stay within the company).

Because letters are sent to readers in other locations, your letters not only reflect your communication abilities but also are a reflection of your company. This chapter provides letter components, formats, criteria, and examples to help you write the following kinds of letters:

> **Sales Letters, Letters of Application, and Job Acceptance Letters**
>
> See Chapters 8 and 11 for more discussion of letters related to persuasion and the job search.

- Inquiry
- Cover (transmittal)
- Complaint
- Adjustment
- Bad news
- Confirmation
- Recommendation

Essential Components of Letters

Web Resources

www.pearsonhighered
.com/gerson For more
information about letters,
visit our companion
website.

Your letter should be printed on 8½" × 11" paper. Leave 1" to 1½" margins at the top and on both sides. Choose an appropriately business-like font (size and style), such as Times New Roman or Arial (12 point). Though "designer fonts," such as Comic Sans and Shelley Volante, are interesting, they tend to be harder to read and less professional.

Your letter should contain the essential components shown in Figure 6.6.

FIGURE 6.6 Essential Letter Components

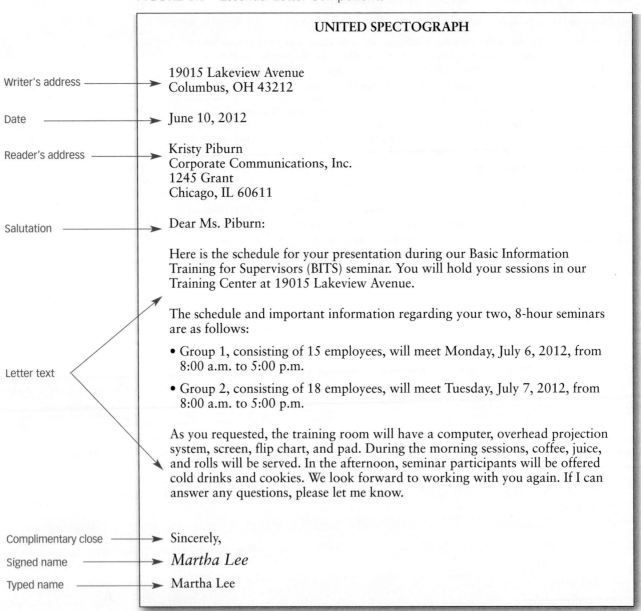

UNITED SPECTOGRAPH

Writer's address →
19015 Lakeview Avenue
Columbus, OH 43212

Date →
June 10, 2012

Reader's address →
Kristy Piburn
Corporate Communications, Inc.
1245 Grant
Chicago, IL 60611

Salutation →
Dear Ms. Piburn:

Here is the schedule for your presentation during our Basic Information Training for Supervisors (BITS) seminar. You will hold your sessions in our Training Center at 19015 Lakeview Avenue.

The schedule and important information regarding your two, 8-hour seminars are as follows:

Letter text

- Group 1, consisting of 15 employees, will meet Monday, July 6, 2012, from 8:00 a.m. to 5:00 p.m.

- Group 2, consisting of 18 employees, will meet Tuesday, July 7, 2012, from 8:00 a.m. to 5:00 p.m.

As you requested, the training room will have a computer, overhead projection system, screen, flip chart, and pad. During the morning sessions, coffee, juice, and rolls will be served. In the afternoon, seminar participants will be offered cold drinks and cookies. We look forward to working with you again. If I can answer any questions, please let me know.

Complimentary close →
Sincerely,

Signed name →
Martha Lee

Typed name →
Martha Lee

Writer's Address. This section contains either your personal address or your company's address. If the heading consists of your address, you will include your street address, the city, state, and zip code. The state may be abbreviated with the appropriate two-letter abbreviation.

If the heading consists of your company's address, you will include the company's name, street address, city, state, and zip code.

Date. Document the month, day, and year when you write your letter. You can write your date in one of two ways: May 31, 2012 or 31 May 2012. Place the date one or two spaces below the writer's address.

Reader's Address. Place the reader's address two lines below the date. This would include the following:

- Reader's name (If you do not know the name of this person, begin the reader's address with a job title or the name of the department.)
- Reader's title (optional)
- Company name
- Street address
- City, state, and zip code

Salutation. The traditional salutation, placed two spaces beneath the reader's address, is *Dear* and your reader's last name, followed by a colon (Dear Mr. Smith:).

You can also address your reader by his or her first name if you are on a first-name basis with this person (Dear John:). If you are writing to a woman and are unfamiliar with her marital status, address the letter *Dear Ms. Jones*. However, if you know the woman's marital status, you can address the letter accordingly: Dear Miss Jones *or* Dear Mrs. Jones.

Letter Body. Begin the body of the letter two spaces below the salutation. The body includes your introductory paragraph, discussion paragraph(s), and concluding paragraph. The body should be single spaced with double spacing between paragraphs. Whether you indent the beginning of paragraphs or leave them flush with the left margin is determined by the letter format you employ.

Complimentary Close. Place the complimentary close, followed by a comma, two spaces below the concluding paragraph. Typical complimentary closes include "Sincerely," "Yours truly," and "Sincerely yours."

Signed Name. Sign your name legibly or use an electronic (digital) signature beneath the complimentary close.

Typed Name. Type your name four spaces below the complimentary close. You can type your title one space beneath your typed name. You also can include your title on the same line as your typed name, with a comma after your name.

Optional Components of Letters

In addition to the letter essentials, you can include the following optional components.

Subject Line. Place a subject line two spaces below the reader's address and two spaces above the salutation.

EXAMPLE ▶

Dr. Ron Schaefer
Linguistics Department
Southern Illinois University
Edwardsville, IL 66205

Subject: Linguistics Conference Registration Payment

Dear Dr. Schaefer:

You also could use a subject line instead of a salutation.

EXAMPLE ▶

Linguistics Department
Southern Illinois University
Edwardsville, IL 662051

Subject: Linguistics Conference Registration Payment

A subject line not only helps readers understand the letter's intent but also (if you are uncertain of your reader's name) avoids such awkward salutations as "To Whom It May Concern," "Dear Sirs," and "Ladies and Gentlemen." In the simplified format, both the salutation and the complimentary close are omitted, and a subject line is included.

New-Page Notations.   If your letter is longer than one page, cite your name, the page number, and the date on all pages after page 1. Place this notation either flush with the left margin at the top of subsequent pages or across the top of subsequent pages. (You must have at least two lines of text on the next page to justify another page.)

Left margin, subsequent page notation	Across top of subsequent pages
Mabel Tinjaca Page 2 May 31, 2012	Mabel Tinjaca 2 May 31, 2012

Writer's and Typist's Initials.   If the letter was typed by someone other than the writer, include both the writer's and the typist's initials two spaces below the typed signature. The writer's initials are capitalized, the typist's initials are typed in lowercase, and the two sets of initials are separated by a colon. If the typist and the writer are the same person, this notation is not necessary.

EXAMPLE ▶

Sincerely,

W. T. Winnery
WTW:mm

Enclosure Notation.   If your letter prefaces enclosed information, such as an invoice or report, mention this enclosure in the letter and then type an enclosure notation two spaces below the typed signature (or two spaces below the writer and typist initials). The enclosure notation can be abbreviated "Enc."; written out as "Enclosure"; show the number of enclosures, such as "Enclosures (2)"; or specify what has been enclosed—"Enclosure: January Invoice."

Copy Notation.   If you have sent a copy of your letter to other readers, show this in a copy notation. A complimentary copy is designated by a lowercase "cc." List the other readers' names following the copy notation. Type the copy notation two spaces below the typed signature or two spaces below either the writer's and typist's initials or the enclosure notation.

◀ EXAMPLE

Sincerely,

Brian Altman

Enclosure: August Status Report

cc: Marcia Rittmaster and Larry Rochelle

Formatting Letters

Three common types of letter formats include **full block** (Figure 6.7), **full block with subject line** (Figure 6.8), and **simplified** (Figure 6.9). Two popular and professional formats used in business are full block and full block with subject line. With both formats, you type all information at the left margin without indenting paragraphs, the date, the complimentary close, or signature. The full block with subject line differs only with the inclusion of a subject line.

Another option is the simplified format. This type of letter layout is similar to the full block format in that all text is typed margin left. The two significant omissions include no

SPOTLIGHT

Why are letters important in a governmental organization?

Dr. Georgia Nesselrode is Director of Government Training for Mid-America Regional Council's Government Training Institute (GTI). The GTI provides training workshops for local government officials (elected and appointed) and their employees. Dr. Nesselrode writes approximately 30 letters each month.

In fact, as Georgia says, she writes letters when other companies might write e-mail messages. Why not write e-mail for routine communication? "Letters are an integral part of our communication for several reasons:

- Local government is more traditional than many corporations.
- Letters provide a better paper trail of documentation.
- Letters are more formal and have greater impact than e-mail.
- Many of my letters preface large, hard-copy attachments.
- The letters convey information that does not always require immediate action (a benefit of e-mail). Instead, my letters often communicate content for future consideration."

Dr. Nesselrode writes cover letters prefacing contracts, letters of recommendation for colleagues, inquiries, responses to inquiries, sales letters promoting her programs, updates, letters of confirmation to potential clients and vendors, and on occasion, responses to seminar participants who were dissatisfied with a training workshop or facilitator.

Her audiences include her local government training liaisons and training workshop instructors. She writes to elected and appointed officials such as mayors, police chiefs, fire captains, city managers, and municipal employees who work in their city's health, accounting, tax, water, road, and parks and recreation departments.

Because she writes to governmental agencies, officials, and employees, Dr. Nesselrode ensures the correspondence is concise, informative, and presents a professional image. In fact, her letters must be perfect. She represents an entire metropolitan area's governmental training, and that's a reflection on city management.

FIGURE 6.7 Full Block Format

State Health Department

1890 Clark Road
Jefferson City, MO 67220

June 6, 2012

Dale McGraw, Manager
Elmwood Mobile Home Park
Elmwood, MO 64003

Dear Mr. McGraw:

On April 19, 2012, Ryan Duran and I, environmental specialists from the Health Department, conducted an inspection of the Elmwood Mobile Home Park Wastewater Treatment Facility. The purpose was to assess compliance with the following: the state's Clean Water Law, Clean Water Commission regulations, and your facility's plan for pollution control. The inspection also would allow the state to promote proper operation of Wastewater Facilities and to provide technical assistance where needed to the Elmwood Mobile Homes management.

Though the Elmwood Mobile Home pollution control plan had expired in 2011, a consent judgment was issued by the state's Attorney Generals Office. The county court stipulated a timeline for correction by connection to an available sewer system. Your mobile home park's wastewater system has continually discharged to the Little Osage River. A copy of the abatement order, which requires that monthly discharge monitoring reports (DMRs) be submitted by the 28th of the month following the reporting periods, is attached. All DMRs for the previous twelve months have been received, and reported pollution parameters are not within limits. Due to the plant's performance, the stream was placed on the 2010, 303 (d) stream for impairment by the Elmwood Mobile Home.

As part of the inspection, a review of the facility's DMR was conducted. Twenty-four-hour composite samples were collected using a composite sampler. Attached are the results of the 24-hour composite samples collected on April 20, 2012. Every one of the problems documented is an infraction that must be addressed.

Within 30 days of receipt of this letter, please submit to the Health Department written documentation describing steps taken to correct each of the concerns identified in the attachments. Also include engineering reports, and submit a timeframe to eliminate the problems. Thank you for your cooperation.

Sincerely,

Harvey Haddix
Environmental Manager

Attachment

FIGURE 6.8   Full Block Format with Subject Line

State Health Department

1890 Clark Road   Jefferson City, MO 67220

June 6, 2012

Dale McGraw, Manager
Elmwood Mobile Home Park
Elmwood, MO 64003

Subject: Pollution Control Inspection

Dear Mr. McGraw:

On April 19, 2012, Ryan Duran and I, environmental specialists from the Health Department, conducted an inspection of the Elmwood Mobile Home Park Wastewater Treatment Facility. The purpose was to assess compliance with the following: the state's Clean Water Law, Clean Water Commission regulations, and your facility's plan for pollution control. The inspection also would allow the state to promote proper operation of Wastewater Facilities and to provide technical assistance where needed to the Elmwood Mobile Homes management.

Though the Elmwood Mobile Home pollution control plan had expired in 2011, a consent judgment was issued by the state's Attorney Generals Office. The county court stipulated a timeline for correction by connection to an available sewer system. Your mobile home park's wastewater system has continually discharged to the Little Osage River. A copy of the abatement order, which requires that monthly discharge monitoring reports (DMRs) be submitted by the 28th of the month following the reporting periods, is attached. All DMRs for the previous twelve months have been received, and reported pollution parameters are not within limits. Due to the plant's performance, the stream was placed on the 2010, 303 (d) stream for impairment by the Elmwood Mobile Home.

As part of the inspection, a review of the facility's DMR was conducted. Twenty-four-hour composite samples were collected using a composite sampler. Attached are the results of the 24-hour composite samples collected on April 20, 2012. Every one of the problems documented is an infraction that must be addressed.

Within 30 days of receipt of this letter, please submit to the Health Department written documentation describing steps taken to correct each of the concerns identified in the attachments. Also include engineering reports, and submit a timeframe to eliminate the problems. Thank you for your cooperation.

Sincerely,

Harvey Haddix
Environmental Manager

Attachment

FIGURE 6.9 Simplified Format Omitting "Dear . . ." and "Sincerely"

State Health Department
1890 Clark Road
Jefferson City, MO 67220

June 6, 2012

Dale McGraw, Manager
Elmwood Mobile Home Park
Elmwood, MO 64003

Subject: Pollution Control Inspection

On April 19, 2012, Ryan Duran and I, environmental specialists from the Health Department, conducted an inspection of the Elmwood Mobile Home Park Wastewater Treatment Facility. The purpose was to assess compliance with the following: the state's Clean Water Law, Clean Water Commission regulations, and your facility's plan for pollution control. The inspection also would allow the state to promote proper operation of Wastewater Facilities and to provide technical assistance where needed to the Elmwood Mobile Homes management.

Though the Elmwood Mobile Home pollution control plan had expired in 2011, a consent judgment was issued by the state's Attorney Generals Office. The county court stipulated a timeline for correction by connection to an available sewer system. Your mobile home park's wastewater system has continually discharged to the Little Osage River. A copy of the abatement order, which requires that monthly discharge monitoring reports (DMRs) be submitted by the 28th of the month following the reporting periods, is attached. All DMRs for the previous twelve months have been received, and reported pollution parameters are not within limits. Due to the plant's performance, the stream was placed on the 2010, 303 (d) stream for impairment by the Elmwood Mobile Home.

As part of the inspection, a review of the facility's DMR was conducted. Twenty-four-hour composite samples were collected using a composite sampler. Attached are the results of the 24-hour composite samples collected on April 20, 2012. Every one of the problems documented is an infraction that must be addressed.

Within 30 days of receipt of this letter, please submit to the Health Department written documentation describing steps taken to correct each of the concerns identified in the attachments. Also include engineering reports, and submit a timeframe to eliminate the problems. Thank you for your cooperation.

Harvey Haddix
Environmental Manager

Attachment

salutation ("Dear _____:") and no complimentary close ("Sincerely,"). Omitting a salutation is useful in the following instances:

- You do not know your reader's name (NOTE: avoid the trite salutation, "To Whom It May Concern:")
- You are writing to someone with a non-gender-specific name (Jesse, Terry, Stacy, Chris, etc.) and you do not know whether to use "Mr.," "Mrs.," or "Ms."

The Administrative Management Society (AMS) suggests that if you omit the salutation, you also should omit the complimentary close. Some people feel that omitting the salutation and the complimentary close will make the letter cold and unfriendly. However, the AMS says that if your letter is warm and friendly, these omissions will not be missed. More importantly, if your letter's content is negative, beginning with "Dear" and ending with "Sincerely" will not improve the letter's tone or your reader's attitude toward your comments.

The simplified format includes a subject line to aid the letter's clarity.

Criteria for Different Types of Letters

Though you might write different types of letters, including inquiry, cover (transmittal), complaint, adjustment, bad news, confirmation, and recommendation, consider using the all-purpose memo, letter, and e-mail template from Figure 6.2 (on page 158) to format your correspondence.

Letter of Inquiry

If you want information about degree requirements, equipment costs, performance records, turnaround time, employee credentials, or any other matter of interest to you or your company, then write a letter requesting that data. Letters of inquiry require that you be specific. For example, if you write, "Please send me any information you have on your computer systems," you are in trouble. You will either receive any information the reader chooses to give you or none at all. Look at the following flawed letter of inquiry from a biochemical waste disposal company.

BEFORE

Dear Mr. Jernigan:

Please send us information about the following filter pools:

1. East Lime Pool
2. West Sulphate Pool
3. East Aggregate Pool

Thank you.

The reader replied as follows:

Dear Mr. Scholl:

I would be happy to provide you with any information you would like. However, you need to tell me what information you require about the pools.

I look forward to your response.

The first writer, recognizing the error, rewrote the letter as follows:

AFTER

Dear Mr. Jernigan:

My company, Jackson County Hazardous Waste Disposal, Inc., needs to purchase new waste receptacles. One of our clients used your products in the past and recommended you. Please send us information about the following:

1. Lime Pool—costs, warranties, time of installation, and dimensions
2. Sulphate Pool—costs, material, and levels of acidity
3. Aggregate Pool—costs, flammability, maintenance, and discoloration

We plan to install our pools by March 12. We would appreciate your response by February 20. Thank you.

To compose your letter of inquiry, include the following:

Introduction. Clarify your intent in the introduction. Until you tell your readers why you are writing, they do not know. It is your responsibility to clarify your intent and explain your rationale for writing. Also tell your reader immediately what you are writing about (the subject matter of your inquiry). You can state your intent and subject matter in one to three sentences.

Discussion. Specify your needs in the discussion. To ensure that you get the response you want, ask precise questions or list specific topics of inquiry. You must quantify. For example, rather than vaguely asking about machinery specifications, ask more precisely about "specifications for the 12R403B Copier." Rather than asking, "Will the roofing material cover a large surface?" you need to quantify—"Will the roofing material cover 150' × 180'?"

Conclusion. Conclude precisely. First, explain when you need a response. Do not write, "Please respond as soon as possible." Provide dated action and tell the reader exactly when you need your answers. Second, to sell your readers on the importance of this date, explain why you need answers by the date given.

Figure 6.10 will help you understand the requirements for effective letters of inquiry.

Cover (Transmittal) Letters

In business, you are often required to send information to a client, vendor, or colleague. You might send multipage copies of reports, invoices, drawings, maps, letters, memos, specifications, instructions, questionnaires, or proposals.

A cover letter accomplishes two goals. First, it lets you tell readers up front what they are receiving. Second, it helps you focus your readers' attention on key points within the enclosures. Thus, the cover letter is a reader-friendly gesture to help your audience. To compose your cover letter, include the following:

Introduction. In the introductory paragraph, tell your reader why you are writing and what you are writing about. What if the reader has asked you to send the documentation? Do you still need to explain why you are writing? The answer is yes. Although the reader requested the information, time has passed, other correspondence has been written, and your reader might have forgotten the initial request.

Discussion. In the body of the letter, you can accomplish two things. You tell your reader either exactly what you have enclosed or exactly what of value is within the enclosures. In both instances, provide an itemized list or easily accessible, short paragraphs.

FIGURE 6.10   Letter of Inquiry Using the Simplified Format

CompuMed

8713 Hillview   Reno, NV 32901   1-800-551-9000   Fax: 1-816-555-0000

September 12, 2012

Sales Manager
OfficeToGo
7622 Raintree
St. Louis, MO 66772

Subject: Request for Product Pricing and Shipping Schedules

My medical technology company has worked well with OfficeToGo (OTG) for the past five years. However, in August I received a letter informing me that OTG had been purchased by a larger corporation. I need to determine if OTG remains competitive with other major office equipment suppliers in the Reno area.

In the introduction, explain why you are writing and introduce yourself to establish the letter's context.

Please provide the following information:

1. What discounts will be offered for bulk purchases?
2. Which freight company will OTG now be using?
3. Who will pay to insure the items ordered?
4. What is the turnaround time from order placement to delivery?
5. Will OTG be able to deliver to all my satellite sites?
6. Will OTG technicians set up the equipment delivered, including desks, file cabinets, bookshelves, and chairs?
7. Will OTG be able to personalize office stationery onsite, or will it have to be outsourced?

In the discussion, specify your needs. To ensure accuracy of response, ask precise questions.

Please respond to these questions by September 30 so I can prepare my quarterly orders in a timely manner. I continue to expand my business and want assurances that you can fill my growing office supply needs. You can contact me at the phone number provided above or by e-mail (jgood@CompuMed.com). Thank you for your help.

Jim Goodwin
Owner and CEO

In the conclusion, state when you need a response and explain why this date is important. Providing contact information will help the reader respond.

Conclusion. Your conclusion should tell your readers what you want to happen next, when you want this to happen, and why the date is important.

See Figure 6.11 for an example of a cover letter from a healthcare provider.

Complaint Letters

You are purchasing director at an electronics firm. Although you ordinarily receive excellent products and support from a local manufacturing firm, two of your recent orders have been filled incorrectly and included defective merchandise. You don't want to have to look for a new supplier. You should express your complaint as pleasantly as possible.

To compose your complaint letter, include the following:

Introduction. In the introduction, politely state the problem. Although you might be angry over the service you have received, you want to suppress that anger. Blatantly negative comments do not lead to communication; they lead to

FIGURE 6.11   Cover Letter in Block Format

AMERICAN HEALTHCARE
1401 Laurel Drive
Denton, TX 76201
November 11, 2012

Jan Pascal
Director of Outpatient Care
St. Michael's Hospital
Westlake Village, CA 91362

Dear Ms. Pascal:

Thank you for your recent request for information about our specialized outpatient care equipment. American Healthcare's stair lifts, bath lifts, and vertical wheelchair lifts can help your patients. To show how we can serve you, we have enclosed a brochure including the following information:

	Page
• Maintenance, warranty, and guarantee information	1–3
• Technical specifications for our products, including sizes, weight limitations, colors, and installation instructions	4–6
• Visuals and price lists for our products	7–8
• An American Healthcare order form	9
• Our 24-hour hotline for immediate service	10

Early next month, I will call to make an appointment at your convenience. Then we can discuss any questions you might have. Thank you for considering American Healthcare, a company that has provided exceptional outpatient care for over 30 years.

Sincerely,

Toby Sommers

Enclosure

A positive tone in the introduction builds rapport and informs the reader why this letter is being written: in response to a request.

An itemized body clarifies what is in the enclosure. Adding page numbers in the list helps readers find the information in the enclosed material.

The conclusion provides dated action, noting when a follow-up call will be made.

combat. Because angry readers will not necessarily go out of their way to help you, your best approach is diplomacy.

To strengthen your assertions, in the introduction, include supporting details such as the following: serial numbers, dates of purchase, invoice numbers, check numbers, names of salespeople involved in the purchase, and/or receipts. When possible, include copies documenting your claims.

Discussion. In the discussion paragraph(s), explain in detail the problems experienced. This could include dates, contact names, information about shipping, breakage information, an itemized listing of defects, and poor service.

Be specific. Generalized information will not sway your readers to accept your point of view. In a complaint letter, you suffer the burden of proof. Help your audience understand the extent of the problem. After documenting your claims, state what you want done and why.

Conclusion. End your letter positively. Remember, you want to ensure cooperation with the vendor or customer. You also want to be courteous, reflecting your company's professionalism. Your goal should be to achieve a continued rapport with your reader. In this concluding paragraph, include your contact information and the times you can best be reached.

See Figure 6.12 for a sample complaint letter to an automotive supplies company.

Creating a Positive Tone. Audiences respond favorably to positive words. If you use negative words, you could offend your reader. In contrast, positive words will help you control your readers' reactions, build goodwill, and persuade your audience to accept your point of view. Choose your words carefully. Even when an audience expects bad news, they still need a polite and positive response.

FIGURE 6.12   Complaint Letter in Block Format

1234 18th Street
Galveston, TX 77001
May 10, 2012

Mr. Holbert Lang
Customer Service Manager
Gulfstream Auto
1101 21st Street
Galveston, TX 77001

Dear Mr. Lang:

On February 12, 2012, I purchased two shock absorbers in your automotive department. Enclosed are copies of the receipt and the warranty for that purchase. One of those shocks has since proved defective.

> The introduction includes the date of purchase (to substantiate the claim) and the problems encountered.

I attempted to exchange the defective shock at your store on May 2, 2012. The mechanic on duty, Vernon Blanton, informed me that the warranty was invalid because your service staff did not install the part. I believe that your company should honor the warranty agreement and replace the part for the following reasons:

1. The warranty states that the shock is covered for 48 months and 48,000 miles.
2. The warranty does not state that installation by someone other than the dealership will result in warranty invalidation.
3. The defective shock absorber is causing potentially expensive damage to the tire and suspension system.

> The body explains what happened, states what the writer wants done, and justifies the request. This letter develops its claim with warranty information.

I can be reached between 1 p.m. and 6 p.m. on weekdays at 763-9280 or at 763-9821 anytime on weekends. You also can email me at cdelatorre12@hotmail.com. I look forward to hearing from you. Thank you for helping me with this misunderstanding.

> The letter concludes by providing contact information and an upbeat, pleasant tone.

Sincerely,

Carlos De La Torre

Enclosures (2)

See Chapter 4 for more discussion of how to achieve audience involvement through positive word usage.

Tone

TABLE 6.2 Positive Words

advantage	efficient	meaningful
asset	enjoyable	please
benefit	favorable	positive
certain	good	profit
confident	grateful	quality
constructive	happy	successful
contribution	helpful	thank you
effective	improvement	value

TABLE 6.3 Positive Verbs

accomplish	improve
achieve	increase
assist	initiate
assure	insure (ensure)
build	maintain
coordinate	organize
create	plan
develop	produce
encourage	promote
establish	satisfy
help	train

TABLE 6.4 Negative vs. Positive Sentences

Before	After
1. The error is your fault. You scheduled incorrectly and cannot complain about our deliveries. If you would cooperate with us, we would work with you to solve this problem.	1. To improve deliveries, let's work together on our companies' scheduling practices.
2. I regret to inform you that we will not replace the motor in your dryer unless we have proof of purchase.	2. When you provide us proof of purchase, we will be happy to replace the motor in your dryer.
3. The accounting records your company submitted are incorrect. You have obviously miscalculated the figures.	3. After reviewing your company's accounting records, please recalculate the numbers to ensure that they correspond to the new X44 tax laws (enclosed).
4. Your letter suggesting an improvement for the system has been rejected. The reconfigurations you suggest are too large for the area specifications. We need you to resubmit if you can solve your problem.	4. Thank you for your suggestions. Though you offer excellent ideas, the configurations you suggest are too large for our area specifications. Please resubmit your proposal based on the figures provided online.
5. You have not paid your bill yet. Failure to do so will result in termination of services.	5. Prompt payment of bills ensures continued service.

Web Resources

www.pearsonhighered .com/gerson See our companion website for sample complaint letters

The positive words in Table 6.2 and positive verbs in Table 6.3 will help you create a pleasant tone and build audience rapport.

Table 6.4 gives you a *before* and *after* view of negative sentences rewritten using a positive tone.

Adjustment Letters

Responses to letters of complaint, also called adjustment letters, can take three different forms.

- **100% Yes.** You could agree 100 percent with the writer of the complaint letter.
- **100% No.** You could disagree 100 percent with the writer of the complaint letter.
- **Partial adjustment.** You could agree with some of the writer's complaints but disagree with other aspects of the complaint.

Table 6.5 shows you the differences among these three types of adjustment letters.

TABLE 6.5 Differences Among Adjustment Letters

	100% Yes	100% No	Partial Adjustment
Introduction	State the good news.	Begin with a buffer, a comment agreeable to both reader and writer.	State the good news.
Discussion	Explain what happened and what the reader should do and/or what the company plans to do next.	Explain what happened, state the bad news, and provide possible alternatives.	Explain what happened, state the bad news, and provide possible alternatives—what the reader and/or company should do next.
Conclusion	End upbeat and positive.	Resell (provide discounts, coupons, follow-up contact names and numbers, etc.) to maintain goodwill.	Resell (provide discounts, coupons, etc.) to maintain goodwill.

Writing a 100% Yes response to a complaint is easy. You are telling your audience what they want to hear. The challenge, in contrast, is writing a 100% No response or a partial adjustment. In these letters, you must convey bad news, but you do not want to convey bad news too abruptly. Doing so might offend, anger, or cause hurt feelings. Using a **buffer statement** delays bad news in written communication and gives you an opportunity to explain your position.

Buffers to Cushion the Blow. Use the following techniques to buffer the bad news:

- **Establish rapport with the audience through positive words to create a pleasant tone.** Instead of writing, "We received your complaint," be positive and say, "We always appreciate hearing from customers."
- **Use persuasive facts to help your reader accept the bad news to come.** "In the last quarter, our productivity has decreased by 16%, necessitating cost-cutting measures."
- **Provide information that both you and your audience can agree upon.** "With the recession, many jobs in business and industry have been lost."
- **Compliment your reader or show appreciation.** "Thank you for your June 9 letter commenting on fiscal year 2012."
- **Make your buffer concise, one to two sentences.** "Thank you for writing. Customer comments give us an opportunity to improve service."
- **Be sure your buffer leads logically to the explanation that follows. Consider mentioning the topic, as in the following example about billing practices.** "Several of our clients have noted changes in our corporate billing policies. Your letter was one that addressed this issue."
- **Avoid placing blame or offending the reader.** Rather than stating, "Your bookkeeping error cost us $9,890.00," write, "Mistakes happen in business. We are refining our bookkeeping policies to ensure accuracy."

See Figures 6.13, 6.14, and 6.15 for sample adjustment letters.

Web Resources

www.pearsonhighered.com/gerson See our companion website for sample adjustment letters

Bad-News Letters

Unfortunately, you occasionally will be required to write bad-news letters. These letters might reject a job applicant, tell a vendor that his or her company's proposal has not been accepted, or reject a customer's request for a refund. Maybe you will have to write to a corporation to report that its manufacturing is not meeting your company's specifications. Maybe you will need to write a letter to a union documenting a grievance. You might even have to write a bad-news letter to fire an employee.

DOT.COM UPDATES

For more information about delivering bad news, check out the following link:

- http://www.articles911.com/Communication/Delivering_Bad_News/

FIGURE 6.13   100% Yes Adjustment Letter, Complete with Letter Essentials

Gulfstream Auto

1101 21st Street
Galveston, TX 77001
(712) 451–1010
May 31, 2012

Mr. Carlos De La Torre
1234 18th Street
Galveston, TX 77001

Dear Mr. De La Torre:

Thank you for your recent letter. Gulfstream will replace your defective shock absorber according to the warranty agreement.

The Trailhandler Performance XT shock absorber that you purchased was discontinued in April 2012. Mr. Blanton, the mechanic to whom you spoke, incorrectly assumed that Gulfstream was no longer honoring the warranty on that product. Because we no longer carry that product, we either will replace it with a comparable model or refund the purchase price. Ask for Mrs. Cottrell at the automotive desk on your next visit to our store. She is expecting you and will handle the exchange.

We appreciate your business, Mr. De La Torre. I'm glad you brought this problem to my attention. If I can help you in the future, please contact me at hlang@gulfstream.com.

Sincerely,

Holbert Lang
Sales Manager

cc: Jordan Cottrell, Supervisor
     Jim Gaspar, CEO

Positive word usage ("Thank you") achieves audience rapport.

The introduction immediately states the good news.

The discussion explains what created the problem and provides an instruction telling the customer what to do next.

The conclusion ("We appreciate your business") resells to maintain customer satisfaction.

In any of these instances, tact is required. You cannot berate a customer or client. You should not reject a job applicant offensively. Even your grievance must be worded carefully to avoid future problems. Because the point of a bad-news letter is bad news, you will need to structure your correspondence carefully to avoid offending your reader.

Introduction. We have suggested throughout this chapter that your introduction should explain why you're writing and what you're writing about. However, such conciseness and clarity in a bad-news letter would be harsh and abrupt. Therefore, to avoid these lapses in diplomacy, begin your bad-news letter with a buffer. Start your letter with information that your reader can accept as valid and will not disagree with but will sway your reader to accept the bad news to come.

Discussion. Once you have provided the buffer, swaying your audience to your point of view, you can no longer delay the inevitable. The discussion paragraph states the bad news. However, to ensure that the reader accepts the bad news, preface your assertions with quantifiable proof.

FIGURE 6.14   100% No Adjustment Letter, Beginning with a Buffer Statement

1101 21st Street
Galveston, TX 77001
(712) 451–1010
May 31, 2012

Gulfstream Auto

Mr. Carlos De La Torre
1234 18th Street
Galveston, TX 77001

Dear Mr. De La Torre:

Thank you for your May 10 letter. Gulfstream Auto always appreciates hearing from its customers.

The Trailhandler Performance XT shock absorber that you purchased was discontinued in April 2012. Mr. Blanton, the mechanic to whom you spoke, correctly stated that Gulfstream was no longer honoring the warranty on that product. Because we no longer carry that product, we cannot replace it with a comparable model or refund the purchase price. Although we can not replace the shock absorber for free, we want to offer you a 10% discount off of a replacement.

We appreciate your business, Mr. De La Torre. I'm glad you brought this problem to my attention. If I can help you in the future, please contact me.

Sincerely,

Holbert Lang
Sales Manager

cc: Jordan Cottrell, Supervisor
    Jim Gaspar, CEO

The introduction begins with a buffer. The writer establishes rapport with the audience through positive words to create a pleasant tone.

The discussion explains the company's position, states the bad news, and offers an alternative.

Conclusion. If you end your bad-news letter with the bad news, then you leave your reader feeling defeated and without hope. You want to maintain a good customer-client, supervisor-subordinate, or employer-employee relationship. Therefore, you need to conclude your letter by giving your readers an opportunity for future success. Provide your readers options which will allow them to get back in your good graces, seek employment in the future, or reapply for the refund you have denied. Then, to leave your readers feeling as happy as possible, given the circumstances, end upbeat and positively.

The example in Figure 6.16, from an insurance company canceling an insurance policy, shows how a bad-news letter should be constructed.

Confirmation Letter

In business, letters represent an official contract. Often, when clients and vendors make arrangements for the purchase of services or products, a letter of confirmation must be written. This confirmation letter verifies the details of the agreement. For example, as head of corporate training, you met with a consulting firm to discuss services they could provide

1101 21st Street
Galveston, TX 77001
(712) 451–1010
May 31, 2012

Mr. Carlos De La Torre
1234 18th Street
Galveston, TX 77001

Dear Mr. De La Torre:

Thank you for your recent letter. Gulfstream will replace your defective shock absorber according to the warranty agreement.

The Trailhandler Performance XT shock absorber that you purchased was discontinued in April 2012. Mr. Blanton, the mechanic to whom you spoke, incorrectly assumed that Gulfstream was no longer honoring the warranty on that product. However, we no longer carry that product. We will replace the shock absorber with a comparable model, but you will have to pay for installation.

We appreciate your business, Mr. De La Torre. I'm glad you brought this problem to my attention. If I can help you in the future, please contact me at hlang@gulfstream.com.

Sincerely,

Holbert Lang
Sales Manager

cc: Jordan Cottrell, Supervisor
     Jim Gaspar, CEO

Begin your letter with the good news.

Explain what happened, state the bad news, and provide a possible alternative.

your company. After the meeting, you must summarize the results of the discussion, confirming payment, dates, and training content.

Maybe you are an accountant. A client has contracted with you to prepare documents for a complex divorce case. This will include reports of joint property, taxes paid, and cash and investments on hand. When you return to the office, you will write a letter confirming your responsibilities, fees, and turnaround time.

Introduction. In a confirmation letter's introduction, state the context of the letter so that the reader can relate to the discussion. This could include when you met, spoke on the telephone, or communicated by e-mail; where the meeting took place; what the topic of the letter is; and why this letter is being written.

Discussion. The letter's body clarifies the details of the agreement. Since this constitutes a legally binding document, you must specify anything agreed upon. Using highlighting techniques to make your content accessible, consider including any of the following:

FIGURE 6.16   Bad-News Letter in Block Format

State of Mind Insurance
11031 Bellbrook Drive
Stamford, CT 23091
213-333-8989

September 7, 2012

John Chavez
4249 Uvalde
Stamford, CT 23091

Dear Mr. Chavez:

Thank you for letting us provide you and your family with insurance for the last 10 years. We have appreciated your business.

However, according to our records, you have filed three claims in the past three years:

1. Damage to your basement due to a failed sump pump
2. Flooring damage due to a broken water seal in your second-floor bathroom
3. Fender repair coverage for a no-fault automobile accident

SMI's policy stipulates that no more than two claims may be filed by a client within a three-year period. Therefore, we must cancel your policy as of October 15, 2012.

If you have any questions about this cancellation, please call our 24-hour assistance line (913-482-0000). Our transition team can help you find new coverage. Thank you for your patronage and your understanding.

Sincerely

Darryl Kennedy
Darryl Kennedy

> Begin with a buffer statement in a bad-news letter.

> In the body, explain what happened and then state the bad news.

> In the conclusion, give your reader follow-up options.

Date	Makes/models/ serial number	Locations	Retainer fess
Times	Personnel and certification requirements	Technology requirements	Length of agreement
Costs	Menus and decorations	Parking	Room setup

Conclusion. The conclusion of a confirmation letter tells the reader what to do next. You might include a request for signature, payment due dates, or method of payment.

See Figure 6.17 for a sample confirmation letter.

FIGURE 6.17 Confirmation Letter in Block Format

Metro Consulting

600 Broadway   Albuquerque, NM 23006
510-234-1818   www.metrocon.com

January 23, 2012

Mr. Justin Meyers
ProfCom
1999 Saguaro Dr.
Santa Fe, NM 23012

Dear Mr. Meyers:

In response to your request, this letter confirms our discussion from last week. Below I summarize the agreement we reached regarding the training services your company will provide and the fees contracted.

ProfCom is scheduled to offer the following workshops in 2012:

- Fifteen days of "Customer Service" training to Albuquerque municipal employees. The workshops will be held throughout the city at locations to be determined later. Each location will provide required technology, per your specifications. ProfCom will be paid $500 for each half-day workshop and $700 for each full-day workshop.
- Thirty days of "Managing Diversity" training to Albuquerque municipal employees. These workshops will be held in the Albuquerque Civic Center, 1800 Mountainview Dr. ProfCom will be paid $900 a day for each of these workshops. Please contact Mr. Silvio Hernandez, 1-800-ALCIVIC, to request your technical support needs.
- Ten days of "Supervisor/Management" training, leading to a "Supervisor/Management Certificate." These workshops will be held in the Albuquerque City Hall, Conference Room A. ProfCom will be paid $1,000 a day for each workshop. Please contact Mary O'Sullivan, 1-510-222-5150, to request your technical support needs.
- Undecided additional training workshops, including "Ethical Decision Making," "Accounting for Non-Accountants," and "Dynamic Presentation Skills." We will work with you to schedule these as enrollment figures are calculated. MetroConsulting will pay ProfCom a monthly retainer fee of $3,000 to ensure your availability for these workshops. The $3,000 retainer will be adjusted against complete payment for services rendered.

Thank you for agreeing to provide us these services, Mr. Meyers. MetroConsulting is excited about the prospect of working with your firm. By signing and dating this letter of confirmation below, you indicate your agreement with the assigned services and compensation.

Sincerely,

Rob Harken, Director
MetroConsulting

_____, Date: _____   _____, Date: _____

The introduction explains why the letter is being written ("in response to your request"), specifies the type of letter (confirmation), and clarifies the topic ("training services" and "fees contracted").

The letter's discussion provides specific details regarding the agreement: costs, locations, and contact.

The conclusion ends positively and instructs the reader, stating what must be done next to confirm the agreement.

Recommendation Letter

You might write a letter of recommendation for many reasons:

- An employee deserves a promotion
- An ex-employee asks for a reference for a new job
- A colleague is nominated for an award
- An acquaintance is applying for an education scholarship
- A governmental agency is checking references
- A consultant requests a reference for a new client

When someone asks you to write a letter of recommendation, consider the *Do's* and *Don'ts* in Table 6.6.

Introduction. In the introduction, include any of the following:
- Your position (or title)
- Your relationship with the person
- The length of your relationship
- The applicant's name
- The position, scholarship, or award

Discussion. In the body, include the specific details of the applicant's skills:
- Examples of the applicant's job performance
- Illustrations proving how and why the person will benefit the company
- Evaluations of the person's chances of success in the company or program
- Differences from other people
- Examples of projects the applicant worked on
- Projects supervised
- Team skills
- Communication abilities
- Names of classes attended or certifications acquired to enhance the applicant's skills
- Honors earned at work, school, or military

Conclusion. In the conclusion, sum up why this person is deserving of consideration for the job, award, or scholarship. Be sure to include contact information, such as your telephone number or e-mail address. This will help the recipient of the recommendation letter reach you for a follow-up discussion.

See Figure 6.18 for a sample recommendation letter.

TABLE 6.6 Do's and Don'ts of Recommendation Letters

Do's	Don'ts
- Agree to write the letter only if you can be supportive - Request a current resume - Obtain information about the position, scholarship, or award the person is applying for - Write the letter with that specific position in mind - Study any information about the person to avoid omissions - Keep your letter to a reasonable length of one to two pages	- Avoid writing if you feel the candidate is weak - Avoid writing if you cannot be positive - Avoid writing letters longer than two pages - Avoid writing these letters if you have only vague memories of the person's work - Avoid writing a reference letter if you cannot also talk with confidence about the person to the interviewer

FIGURE 6.18   Recommendation Letter in Simplified Format

MIDWEST TECHNOLOGICAL COLLEGE

15431 College Blvd.   Milwaukee, WI 32556   451-987-0101

March 15, 2012

Dr. Anne Cohen
University of Wisconsin Medical Center
1900 E. 39th Street
Madison, WI 35567

Subject: Letter of Recommendation for the University of Wisconsin Medical Center, Department of Health Information Management

I am pleased to write this letter recommending Pekkahm Shoumavong for admission to the WU Department of Health Information Management. Pekkahm was a student in my Technical Communication class last semester (Fall 2011). She impressed me not only with her scholarship, but also with her team skills and her conscientious desire to excel.

Pekkahm earned an A in my class due to her excellent written and oral communication skills (letters, memos, reports, e-mail, instructions, and speeches accompanied by PowerPoint presentations). More importantly, I was impressed with her ability to work well with others and to take on leadership roles. My technical communication class emphasizes collaboration. Pekkahm's team was one of my more successful groups due to their ability to work in concert toward a team goal and to communicate professionally with each other—valuable skills in any environment. This was an especially impressive achievement since Pekkahm's team was multicultural and comprised of students with vastly different abilities.

In addition, I remember that her team encountered a problem with lost computer files. Rather than become distressed, Pekkahm remained calm and worked through the problem professionally. Her even-keeled and congenial personality, interpersonal communication talents, and ability to get along with others while demanding high quality work helped her team succeed. Finally, Pekkahm attended all classes and turned in all assignments on time, evincing her discipline, conscientiousness, and dedication to excellence.

Self-motivation, professionalism, problem-solving skills, and outstanding communication abilities suggest a high level of potential. I believe these traits distinguish Pekkahm and indicate her ability to succeed in the WU Department of Health Information Management. Pekkahm is deserving of your consideration for admission to your program. Thus, I am happy to recommend her and serve as a reference. If you would like to discuss her attributes further, please call me at 451-987-0101, ext. 59, or e-mail me at cescobar@mtc.edu.

Dr. Carmen Escobar
Technical Communication Professor

The introduction gives the applicant's name, how the writer knows her, and the topic for this letter.

The discussion specifies ways in which the applicant excels. These include grades, team projects, interpersonal communication abilities, and personality traits.

The conclusion sums up the candidate's assets, focusing on ways in which this person meets the needs of a specific program. The conclusion also provides contact information for follow-up.

LETTERS CHECKLIST

_____ 1. **Letter Essentials:** Does your letter include the eight essential components (writer's address, date, reader's address, salutation, text, complimentary close, writer's signed name, and writer's typed name)?

_____ 2. **Introduction:** Does the introduction state what you are writing about and why you are writing?

_____ 3. **Discussion:** Does your discussion clearly state the details of your topic depending on the type of letter?

_____ 4. **Highlighting/Page Layout:** Is your text accessible? To achieve reader-friendly ease of access, use headings, boldface, italics, bullets, numbers, underlining, or graphics (tables and figures). These add interest and help your readers navigate your letter.

_____ 5. **Organization:** Have you helped your readers follow your train of thought by using appropriate modes of organization? These include chronology, importance, problem/solution, or comparison/contrast.

_____ 6. **Conclusion:** Does your conclusion give directive action (tell what you want the reader to do next and when) and end positively?

_____ 7. **Clarity:** Is your letter clear, answering reporter's questions and providing specific details that inform, instruct, or persuade?

_____ 8. **Conciseness:** Have you limited the length of your words, sentences, and paragraphs?

_____ 9. **Audience Recognition:** Have you written appropriately to your audience?

This includes avoiding biased language, considering the multicultural/cross-cultural nature of your readers, and your audience's role (supervisors, subordinates, co-workers, customers, or vendors). Have you created a positive tone to build rapport?

_____ 10. **Correctness:** Is your text grammatically correct? Errors will hurt your professionalism. See Appendix A for grammar rules and exercises.

FAQs: Professionalism in E-mail Messages

Web Resources

www.pearsonhighered.com/gerson See our companion website for sample e-mail messages

Q: Aren't e-mail messages just casual communication? Isn't it easy to write an e-mail message since you don't have to worry about grammar or correct style?

A: Nothing could be further from the truth. E-mail might be your major means of communication in the workplace. Therefore, you must pay special attention to correctness.

Listen to what managers at an engineering company say about e-mail messages:

- "Most workplace communication is now via e-mail. Business e-mail needs to be almost as formal and as carefully written as a letter because it is a formal and legal document. Never send an e-mail that you would not be comfortable seeing on the front page of a newspaper, because some day you may."

- "I see more and more new hires wanting to rely on e-mail. It is a totally ineffective way to resolve many issues on an engineering project. But they seem to feel it is OK for almost any communication. I suspect the general acceptance by their peers for this form of communication has led them to mistakenly assume the same is true for a business setting."

- "Many people tend to be very 'social' in e-mails. Your employer owns your e-mails written on your work computers. They are NOT private. They can be used not only against you, but against your firm in court. For example, if I send an e-mail to a co-worker that states in it somewhere what a lousy job Frank is doing on the such-and-such project and that project goes bad, it is possible that e-mail could end up in court and be used against my employer. In my mind all I was doing was venting my frustrations to an understanding friend and co-worker. But, in reality, I am creating a permanent record of anything I say."

(Gerson et al. "Core Competencies")

E-Mail

Why Is E-mail Important?

E-mail has become a predominant means of routine correspondence. Many companies are "geared to operate with e-mail," creating what the Harvard Business School calls "e-mail cultures" for the following reasons ("The Transition to General Management").

Time. "Everything is driven by time. You have to use what is most efficient" (Miller et al. 10). The primary driving force behind e-mail's prominence is time. E-mail is quick. Whereas a posted letter might take several days to deliver, e-mail messages can be delivered within seconds.

Convenience. With wireless communication, you can send e-mail from notebooks to handhelds. Handheld communication systems combine a voice phone, a personal digital assistant, and e-mail into a package that you can slip into a pocket or purse. Then, you can access your e-mail messages anywhere, anytime.

Internal/External. E-mail allows you to communicate internally to co-workers and externally to customers and vendors. Traditional communication channels, like letters and memos, have more limited uses. Generally, letters are external correspondence written from one company to another company; memos are internal correspondence transmitted within a company.

Cost. E-mail is cost effective because it is paper-free. With an ability to attach files, you can send many kinds of documentation without paying shipping fees. This is especially valuable when considering international business.

Documentation. E-mail provides an additional value when it comes to documentation. Because so many writers merely respond to earlier e-mail messages, what you end up with is a "virtual paper trail" (Miller et al. 15). When e-mail is printed out, often the printout will contain dozens of e-mail messages, representing an entire string of dialogue. This provides a company an extensive record for future reference. In addition, most companies archive e-mail messages in backup files.

Reasons for Writing E-Mail Messages

E-mail is used to convey many types of information in business and industry. You can write an e-mail message to accomplish any of the following purposes:

- **Directive.** Tell a subordinate or a team of employees to complete a task.
- **Cover/transmittal.** Inform a reader or readers that you have attached a document, and list the key points that are included in the attachment.
- **Documentation.** Report on expenses, incidents, accidents, problems encountered, projected costs, study findings, and reallocations of staff or equipment.
- **Confirmation.** Inform a reader about a meeting agenda, date, time, and location; decisions to purchase or sell; topics for discussion at upcoming teleconferences; conclusions arrived at; fees, costs, or expenditures.
- **Procedures.** Explain how to set up accounts, research on the company intranet, operate new machinery, use new software, apply online for job opportunities through the company intranet, or solve a problem.
- **Recommendations.** Provide reasons to purchase new equipment, hire personnel, contract with new providers, revise current practices, or renew contracts.
- **Feasibility.** Study the possibility of changes in the workplace (practices, procedures, locations, staffing, or equipment).
- **Status.** Provide a daily, weekly, monthly, quarterly, biannual, or yearly report about where you, the department, or the company is regarding any topic (project, sales, staffing, travel, practices, procedures, or finances).
- **Inquiry.** Ask questions about upcoming processes, procedures, or tasks.

Techniques for Writing Effective E-mail Messages

To convey your messages effectively and to ensure that your e-mail messages reflect professionalism, follow these tips for writing e-mail.

How is e-mail used and misused in business today?

Michael Smith, PE, President of George Butler Associates, Inc., an architectural-engineering consulting firm, supervises 270 employees, located in Kansas, Missouri, and Illinois. These include architects, engineers, scientists, and administrative support staff in human resources, marketing, information systems, and accounting.

Mike receives about 40 "meaningful e-mail" a day—and about 60 spam e-mail, attempting to tell him how to lose weight, buy gifts for his loved ones, make cost-effective travel arrangements, and so on. Mike deletes the spam, which is obvious at a glance in his e-mail inbox, and responds to the business-related correspondence.

He says that e-mail has been essential at GBA for at least two reasons:

The Good News:

1. **E-mail is fast and efficient.** For example, Mike finds that e-mail is ideal for setting up meetings with groups of individuals, including co-workers and clients. Before e-mail, he had to make multiple calls to his staff or customers, usually missing them while they were off-site. Or, he'd have to walk down the halls and up the stairs to leave messages on desks. Instead, with e-mail, Mike knows that the message will arrive at its intended location and that the recipients will respond.

2. **E-mail provides an "electronic record."** When GBA used to rely more heavily on the telephone, people had trouble remembering to document conversations. With e-mail, you have a communication trail: multiple e-mail messages reporting all communication that has occurred regarding a topic.

Unfortunately, Mike tells us that e-mail also has a few downsides.

The Bad News:

1. **E-mail lacks privacy.** Mike says that though phone calls made record keeping hard, they ensured privacy. All you needed to do was shut your office door or speak quietly. E-mail, in contrast, inadvertently can be sent to many people.

2. **E-mail can lead to inaccurate communication.** "Didn't I read somewhere that 65 percent of communication is body language?" Mike asks. This leads to a second problem. With e-mail, the people involved can't see each other's eyes, hand motions, shrugs, smiles, or frowns, nor can they hear the grunts, groans, or laughs. Much of this body language is lost with electronic communication.

3. **E-mail can be depersonalized.** Body language not only helps communication, but also it personalizes. E-mail, Mike notes, can be impersonal. In some instances, especially those regarding "contentious situations," Mike knows that a face-to-face discussion is the best way to solve a problem. In fact, Mike concludes that some people who are "conflict resistant" use e-mail as a way of avoiding person-to-person communication. To maintain good business relationships (with personnel as well as with customers), face-to-face talks often are needed.

Still, for speed and efficiency, e-mail is hard to beat.

Recognize Your Audience.   E-mail messages can be sent to managers, co-workers, subordinates, vendors, and customers, among other audiences. Your e-mail readers will be high tech, low tech, and lay. Thus, you must factor in levels of knowledge.

If an e-mail message is sent internationally, you also might have to consider your readers' language. Remember that abbreviations and acronyms are not universal. Dates, times, measurements, and monetary figures differ from country to country. In addition, your reader's e-mail system might not have the same features or capabilities that you have. Hard-copy text will look the same to all readers. E-mail platforms, such as in AOL, Outlook, Gmail, HotMail, and Yahoo, display text differently. To communicate effectively, recognize your audience's level of knowledge, language, and technology needs.

> **Multicultural Audiences**
>
> See Chapter 4 for more discussion of multiculturalism.

Identify Yourself.   Identify yourself by name, affiliation, or title. You can accomplish this either in the "From" line of your e-mail or by creating a signature file or .sig file. This .sig file acts like an online business card. Once this identification is complete, readers will be able to open your e-mail without fear of corrupting their computer systems.

Provide an Effective Subject Line.   Readers are unwilling to open unsolicited or unknown e-mail, due to fear of spam and viruses. In addition, corporate employees receive approximately 50 e-mail messages each day. They might not want to read every message sent to them. To ensure that your e-mail messages are read, avoid uninformative subject lines, such as "Hi," "What's New," or "Important Message." Instead, include an effective subject line, such as "Subject: Meeting Dates for Tech Prep Conference."

Keep Your E-mail Message Brief.   E-readers skim and scan. To help them access information quickly, "Apply the 'top of the screen' test. Assume that your readers will look at the first screen of your message only" (Munter et al. 31). Limit your message to one screen (if possible).

Organize Your E-mail Message.   Successful writing usually contains an introductory paragraph, a discussion paragraph or paragraphs, and a conclusion. Although many e-mail messages are brief, only a few sentences, you can use the introductory sentences to tell the reader why you are writing and what you are writing about. In the discussion, clarify your points thoroughly. Use the concluding sentences to tell the reader what is next, possibly explaining when a follow-up is required and why that date is important.

Use Highlighting Techniques Sparingly.   Many e-mail packages will let you use highlighting techniques, such as boldface, italics, underlining, computer-generated bullets and numbers, centering, font color highlighting, and font color changes. Many other e-mail platforms will not display such visual enhancements. To avoid having parts of the message distorted, limit your highlighting to asterisks, numbers, double spacing, and headings.

Proofread Your E-mail Message.   Errors will undermine your professionalism and your company's credibility. Recheck your facts, dates, addresses, and numerical information before you send the message. Try these tips to help you proofread an e-mail message:

- Type your text first in a word processing package, like Microsoft Word.
- Print it out. Sometimes it is easier to read hard-copy text than text online. Also, your word processing package, with its spell check and/or grammar check, will help you proofread your writing.

Once you have completed these two steps, copy and paste the text from your word processing file into your e-mail.

Make Hard Copies for Future Reference.   Making hard copies of all e-mail messages is not necessary because most companies archive e-mail. However, in some instances, you might want to keep a hard copy for future reference. These instances could include transmissions of good news. For example, you have received compliments about your work and want to save this record for your annual job review. You also might save a hard copy of an e-mail message regarding flight, hotel, car rental, or conference arrangements for business-related travel.

Be Careful When Sending Attachments.   When you send attachments, tell your reader within the body of the e-mail message that you have attached a file. Specify the file name of your attachment and the software application that you have used (HTML, PowerPoint, PDF, RTF [rich text format], Word, or Works); and use compression (zipped) files to limit your attachment size. Zipped files are necessary only if an attachment is quite large.

Practice Netiquette.   When you write your e-mail messages, observe the rules of "netiquette":

- **Be courteous.** Do not let the instantaneous quality of e-mail negate your need to be calm, cool, deliberate, and professional.
- **Be professional.** Occasionally, e-mail writers compose excessively casual e-mail messages. They will lowercase a pronoun like "i," use ellipses (. . .) or dashes instead of more traditional punctuation, use instant messaging shorthand

Online E-readers

See Chapter 14 for more discussion about the characteristics of e-readers.

DOT.COM UPDATES

For more information about e-mail, check out the following links:

- "Top 26 Most Important Rules of Email Etiquette" http://email.about.com/cs/netiquettetips/tp/core_netiquette.htm
- ABC News Radio "E-mail Rules" http://www.abc.net.au/newsradio/txt/s1620247.htm
- Australian Department of Education and Training "Rules for Email Users" http://www.det.wa.edu.au/education/cmis/eval/downloads/curriculum/emailrules.pdf
- The University of Washington's E-mail "Things to Consider" http://www.washington.edu/itconnect/email/index.html

FIGURE 6.19 Flawed E-mail Message

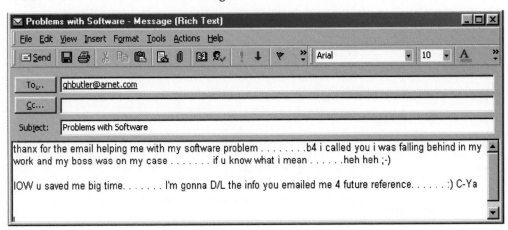

This e-mail is flawed due to its use of instant messaging abbreviations, lowercase text, ellipses (...), slang, and emoticons. Though these techniques might be appropriate for personal communication, avoid overly casual e-mail messages for job-related communication.

language or "textese" such as "LOL" or "BRB," and depend on emoticons. These e-mail techniques might not be appropriate in all instances. Don't forget that your e-mail messages represent your company's professionalism. Write according to the audience and communication goal.

- **Avoid abusive, angry e-mail messages.** Because of its quick turnaround abilities, e-mail can lead to negative correspondence called flaming. Flaming is sending angry e-mail, often TYPED IN ALL CAPS.

Many e-mail messages only require a sentence or two. If you need to convey more information than can be accomplished in only a few sentences, use the all-purpose template for memos, letters, and e-mail provided earlier in this chapter, in Figure 6.2 (page 158).

Figure 6.19 is a flawed e-mail message. See Figure 6.20 for an example of a well-written e-mail message.

E-MAIL CHECKLIST

_____ 1. Does the e-mail use the correct address?

_____ 2. Have you identified yourself? Provide a sig (signature) line.

_____ 3. Did you provide an effective subject line? Include a *topic* and a *focus*.

_____ 4. Have you effectively organized your e-mail? Consider including the following:

- Opening sentence(s) telling *why* you are writing and *what* you are writing about.
- Discussion unit with itemized points telling *what exactly* the e-mail is discussing.
- Concluding sentence(s) *summing up* your e-mail message or telling your audience what to do next.

_____ 5. Have you used highlighting techniques sparingly?

- Avoid boldface, italics, color, or underlining.
- Use asterisks for bullets, numbers, and double spacing for access.

_____ 6. Did you practice netiquette?

- Be polite, courteous, and professional.
- Don't flame.
- Don't overuse textese.

_____ 7. Is the e-mail concise?

_____ 8. Did you identify and limit the size of attachments?

- Tell your reader(s) if you have attached files and what types of files are attached (PPt, PDF, RTF, Word, etc.).
- Zip large files.

_____ 9. Does the memo recognize audience?

- Define acronyms or abbreviations where necessary.
- Consider a diverse audience (factoring in issues, such as multiculturalism or gender).

_____ 10. Did you avoid grammatical errors?

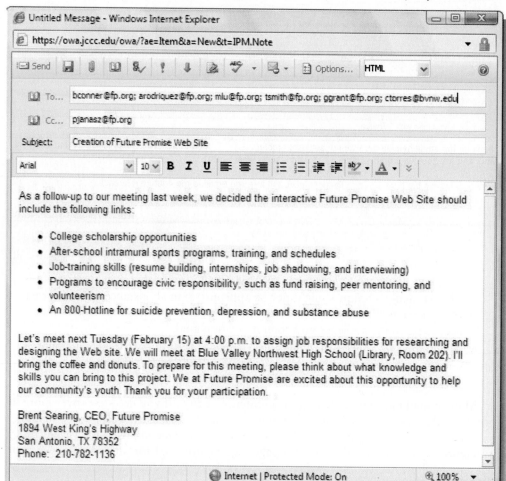

The e-mail's subject line clearly states the *topic* and the *focus*.

This e-mail message answers reporter's questions (who, what, when, where, why, and how) to develop its content.

The e-mail provides a sig. line which acts as an online business card to identify the writer.

In Figure 6.21 an information technology supervisor writes an e-mail to be distributed companywide to many different employees with varied levels of technical expertise. Note how the writer uses a simple style of writing, carefully omitting any technical terms. The e-mail only includes the information necessary to assure audience comfort with the new identification system.

In contrast to Figure 6.21 designed for an audience with multiple levels of expertise, Figure 6.22 is written for a high-tech audience. This high-tech audience is familiar with the technical terms included in the e-mail, so the writer does not have to define such terminology. In addition, the writer includes sophisticated details and content which the technically proficient audience can understand easily.

Instant Messages

E-mail could be too slow for today's fast-paced workplace. Instant message (IM) pop-ups are already providing businesses many benefits.

FIGURE 6.21 E-mail for Multiple Audience Levels

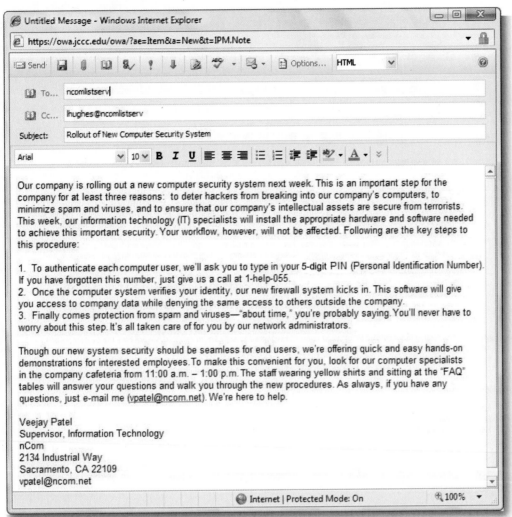

The e-mail's subject line clearly states the *topic* and the *focus*.

To communicate effectively to multiple audiences, this writer defines high-tech terms like "IT" and "PIN." The e-mail uses a conversational tone and simple instructions to make this challenging topic more friendly and informative. The emphasis is on helpfulness vs. depth of detail.

This e-mail message answers reporter's questions (who, what, when, where, why, and how) to develop its content.

Benefits of Instant Messages

Following are benefits of instant messages:

- Increased speed of communication
- Improved efficiency for geographically dispersed workgroups
- Collaboration by multiple users in different locations
- Communication with colleagues and customers at a distance in real time
- Avoidance of costly long distance telephone rates (Note: Voice over IP [VoIP] services, which allow companies to use the Internet for telephone calls, could be more cost-efficient than IM.)
- More personal link than e-mail
- Communication channel that is less intrusive than telephone calls

FIGURE 6.22   E-mail for High-Tech Audience

For a more high-tech audience, this e-mail does not define terms, such as "stateful inspection," "Trojan horses," "honeypots," "IPS" and "PIN." The e-mail contains technical information about system security that the audience both needs and will understand. The tone of this e-mail is businesslike, since the writer is a subordinate writing to management.

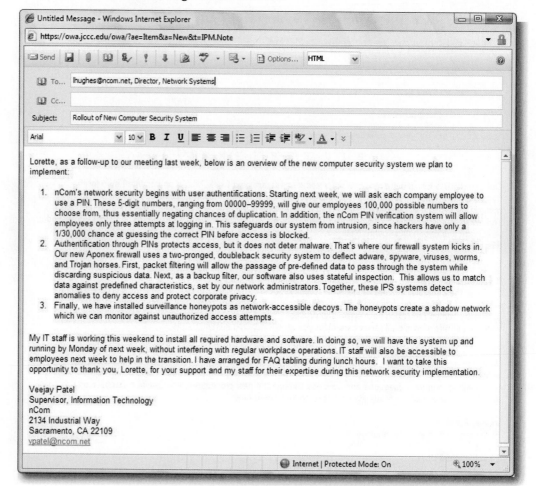

Web Resources

www.pearsonhighered
.com/gerson For more
information about IM,
visit our companion
website.

- Communication channel that allows for multitasking (With IM, you can speak to a customer on the telephone or via an e-mail message and simultaneously receive product updates from a colleague via IM.)
- Quick way to find out who is in the office, out of the office, available for conversation, or unavailable due to other activities (Hoffman; Shinder)

Challenges of Instant Messages

IM has potential problems in the workplace, including security, archiving, monitoring, and employee misuse (Hoffman; Shinder):

- **Security issues.** This is the biggest concern. IM users are vulnerable to hackers, electronic identity theft, and uncontrolled transfer of documents. With unsecured IM, a company could lose confidential documents, internal users could download copyrighted software, or external users could send virus-infected files.
- **Lost productivity.** Use of IM on the job can lead to job downtime. First, we tend to type more slowly than we talk. Next, the conversational nature of IM leads to "chattiness." If employees are not careful, or monitored, a brief IM conversation can lead to hours of lost productivity.

- **Employee abuse.** IM can lead to personal messages rather than job-related communication with co-workers or customers.
- **Distraction.** With IM, a bored colleague easily can distract you with personal messages, online chats, and unimportant updates.
- **Netiquette.** As with e-mail, due to the casual nature of IM, people tend to relax their professionalism and forget about the rules of polite communication. IM can lead to rudeness or just pointless conversations.
- **Spim.** IM lends itself to "spim," instant message spam—unwanted advertisements, pornography, pop-ups, and viruses.

Techniques for Successful Instant Messages

To solve potential problems, consider these ten suggestions:

1. **Choose the correct communication channel.** Use IM for speed and convenience. If you need length and detail, other options—e-mail messages, memos, reports, letters—are better choices. In addition, sensitive topics or bad news should never be handled through IM. These deserve the personal attention provided by telephone calls or face-to-face meetings.
2. **Document important information.** For future reference, you must archive key text, in compliance with SEC Rule 17a-4 and the Sarbanes-Oxley Act.
3. **Summarize decisions.** IM is great for collaboration. However, all team members might not be online when decisions are made. Once conclusions have been reached that affect the entire team, the designated team leader should e-mail everyone involved. In this e-mail, the team leader can summarize the key points, editorial decisions, timetables, and responsibilities.
4. **Tune in, or turn off.** The moment you log on, IM software tells everyone who is active online. Immediately, your IM buddies can start sending messages. IM pop-ups can be distracting. Sometimes, in order to get your work done, you might need to turn off your IM system. Your IM product might give you status options, such as "on the phone," "away from my desk," or "busy." Turning on IM could infringe upon your privacy and time. Turning off might be the answer.
5. **Limit personal use.** Your company owns the instant messages you write in the workplace. IM should be used for business purposes only.
6. **Create "buddy" lists.** Create limited lists of IM users, including legitimate business contacts (colleagues, customers, and vendors).
7. **Avoid public directories.** This will help ensure that your IM contacts are secure and business related.
8. **Disallow corporate IM users from installing their own IM software.** A company should require standardized IM software for safety and control.
9. **Never use IM for confidential communication.** Use another communication channel if your content requires security.
10. **Remember that IM in the workplace is not confidential.** As with e-mail, IM programs can let systems administrators log and review IM conversations (Hoffman; Shinder).

See Figure 6.23 for an example of an instant message.

IM/TM Corporate Usage Policy

To clearly explain the role of IM and TM (text messages) in the workplace, a company should establish a corporate IM and TM usage policy. Many industries already have

DOT.COM UPDATES

For more information about instant messaging, check out the following links:

- Instant Messaging: The next wave in business communications http://www.expressitpeople.com/20040524/cover.shtml
- 10 Tips for Using Instant Messages for Business http://www.microsoft.com/smallbusiness/resources/technology/communications/10-tips-for-using-instant-messaging-for-business.aspx
- IM for Business: Commercial Uses for the IM Community http://im.about.com/od/imforbusiness/IM_for_Business_Commercial_Uses_for_the_IM_Community.htm

Teamwork

See Chapter 1 for more discussion of collaboration.

FIGURE 6.23 Instant Message

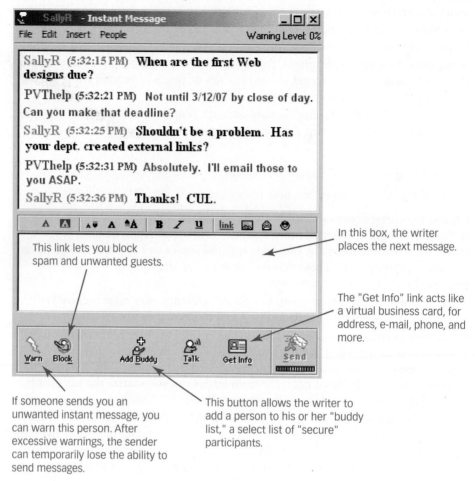

In this box, the writer places the next message.

This link lets you block spam and unwanted guests.

The "Get Info" link acts like a virtual business card, for address, e-mail, phone, and more.

If someone sends you an unwanted instant message, you can warn this person. After excessive warnings, the sender can temporarily lose the ability to send messages.

This button allows the writer to add a person to his or her "buddy list," a select list of "secure" participants.

such policies in place, in relation to existing restrictions established by Sarbanes-Oxley or HIPAA (Health Insurance Portability and Accountability Act). Consider a policy as follows:

- Train employees to use IM and TM effectively for business.
- Teach employees IM and TM security concerns.
- Explain which services are allowable in the workplace. For example, conversation between co-workers is acceptable, but chat between employees and outside individuals might not be advisable. File transfers of proprietary information need to be restricted and monitored.
- Do not allow employees to store IM or TM passwords on computer desktops or other sites easily accessed.
- Install appropriate security measures, including automated encryption, to protect against external threats, such as spam, spim, and viruses.
- Log and archive instant messages and text messages for compliance with company policy (Ollman; "How To"; Bradley).

Text Messages

Text messages, also known as SMS (short message service), are a growing presence in workplace communication. Billions of text messages are sent within the United States every month. "American cell phone users are sending more text messages than they are making phone calls, according to a Nielsen Mobile survey . . . . U.S. mobile subscribers sent and received on average 357 text messages per month, compared with making and receiving 204 phone calls a month, according to Nielsen. The new statistic is a clear indication that Americans have jumped onto the SMS text bandwagon" (Reardon).

A unique characteristic of TM is its demand for conciseness. Text messages over 160 characters are delivered in multiple segments. Each segment is billed as a separate message. That's why users try to limit messages to 160 characters—to save expenses.

"Pretty soon, you're going to have to teach text messaging in your classes," Robert Clark said. "Why's that?" we asked. He responded, "30% of my interaction with staff is through text messaging" (Clark). Robert, a facilities manager for a real estate company, drives between eight apartment complexes to supervise his staff at each location. Robert says that he depends on text messages instead of e-mail for the following reasons:

Reasons for Using TM

- **Cost.** Though Robert's cell phone is supplied by his employer, the people he supervises pay for their own cell phones. It's cheaper for them to use text messaging if their cell phone plans allow for unlimited TM.
- **Technological access.** Staff members who work under Robert do not have computers at their worksites, nor does Robert have a computer in his car as he drives around town. However, all of Robert's employees have access to a cell phone.
- **Speed.** TM is a quick and easy way to communicate short messages like, "John will be late for work today. Is it OK if I stay overtime?" That's 63 characters (counting the letters, spaces, and punctuation marks). Robert and his staff use other means of communication for longer correspondence.
- **Multitasking.** TM is a great way for Robert to multitask. Robert might be in a meeting, for example. He says, "If three of my staff need to attend training on a certain day, I can text all three employees right then and there and not disrupt the meeting. They'll respond quickly, allowing me to tell the meeting facilitator which of my employees can or can't attend. Then we can reschedule accordingly."
- **Decrease the "intimidation factor."** Many people don't like to write letters, memos, reports, or even e-mail messages. Many people don't like face-to-face communication, either. They're intimidated by writing or by bosses. Robert says that TM decreases this "intimidation factor" for his employees. With TM, they can avoid person-to-person situations.
- **Documentation.** Finally, TM allows Robert to document his conversations, something that's not always possible with phone calls. TM is an instant record of a dialogue. A TM account saves incoming and outgoing calls for a few days until it's full. This allows an employee to clarify any later misconceptions.

The Writing Process at Work

Effective writing follows a process of prewriting, writing, and rewriting. Each of these steps is sequential and yet continuous. The writing process is dynamic, with the three steps frequently overlapping. To clarify the importance of the writing process, look at how Jim Goodwin, CEO of CompuMed, used prewriting, writing, and rewriting to write a memo to his employees.

The Writing Process

Prewriting	Writing	Rewriting

• Decide whether you are writing to inform, instruct, persuade, or build rapport.
• Determine whether your audience is internal, external, high tech, low tech, lay, or multiple.
• Choose the correct formal or informal tone and content for your audience and purpose.
• Choose the correct communication channel (e-mail, memo, letter, IM, or TM) for your audience and purpose.
	• Organize your content using modes such as problem/solution, cause/effect, comparison/contrast, argument/persuasion, analysis, chronology, etc.
	• Use headings, bullets, and numbered lists for access.
	• Let a peer or colleague read your draft and provide input.
		• Revise your draft by
		  • Adding details
		  • Deleting wordiness
		  • Simplifying words
		  • Enhancing the tone
		  • Reformatting your text
		  • Proofreading and correcting errors

Prewriting

Prewriting

See Chapter 2 for more discussion of the writing process and prewriting techniques.

No single method of prewriting is more effective than another. Throughout this textbook, you will learn many different types of prewriting techniques, geared uniquely for different types of communication. The goal of all prewriting is to help you overcome the blank page syndrome (writer's block). Prewriting will allow you to spend time before writing your document, gathering as much information as you can about your subject matter. In addition, prewriting lets you determine your objectives. Jim used mind mapping/clustering to gather data and determine objectives (Figure 6.24).

Clarify Your Audience.   Before writing the draft, consider your audience. Are you writing laterally to colleagues or vertically up to management or down to subordinates? Is the audience multiple?

Gather Data.   Prewriting allows you to discover information about your topic.

Determine Objectives.   Use prewriting to decide why you are writing. Are you writing to inform, persuade, build rapport, instruct, recommend, etc.?

Writing

Once you have considered audience, gathered data, and determined objectives in prewriting, your next step is to draft your document. In doing so, you need to organize content.

Organize Your Ideas.   If your supporting details are presented randomly, your audience will be confused. As a writer, develop your content logically. When you draft your document,

FIGURE 6.24 Mind Mapping/Clustering to Gather Data

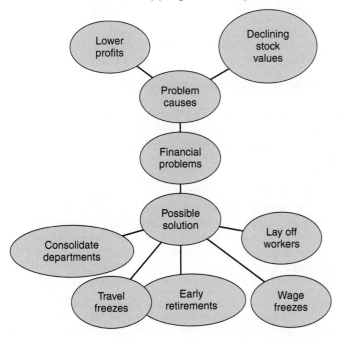

Clustering helps you see the different parts of any subject. Clustering helps you develop ideas and see the relationship between these concepts. Because clustering is less restrictive and less structured than outlining, this prewriting technique might allow you to gather data more creatively.

choose a method of organization that will help your readers understand your objectives. This could include comparison/contrast, problem/solution, chronology, cause/effect, and more.

Jim drafted a memo, focusing on the information he discovered in prewriting and then received revision suggestions from his administrative assistant (Figure 6.25).

Rewriting

Jim considered his administrative assistant's suggestions and rewrote the memo. See Figure 6.26 for the finished product.

FIGURE 6.25 Rough Draft with Revision Suggestions

Date: October 14, 2012
To: CompuMed Employees
From: Jim Goodwin
Subject: Problems

As you know, we are experiencing some problems at CompuMed. These include lower profits and stock value declines. We have alot of unhappy stockholders. Its up to me to help everyone figure out how to solve our problems.

I have some ideas I want to share with you. I'm happy to have you share your ideas with me too. Here are my ideas: we need to consider consolidating departments and laying off some employees. We also might need to freeze wages and certainly its time to freeze travel.

The best idea I have is for some of you to take early retirement. If all of you who have over twenty years vested in the company would retire, that would save us around 2.1 million dollars over the next fiscal year. And, you know, saving money is good for all of us in the long run.

Add a focus to the subject line, such as "Problems with . . . "

I'd consider removing words like "as you know," "some," and "alot." Replace them with stronger words. Also, "alot" and "its" are spelled wrong.

List these problems and solutions to make them more accessible. Also, could you add more details?

Jim, I think you need to alter the tone of this memo. Is there some way to avoid talking about saving money by firing people?

Date: October 14, 2012
To: CompuMed Employees
From: Jim Goodwin
Subject: Suggestions for Improving Company Finances

CompuMed is experiencing lower profits and declining stock value. Consequently, stockholders are displeased with company performance. I have been meeting with the Board of Directors and division managers to determine the best course of action. Here are ideas to improve our company's financial situation.

1. Consolidating departments—By merging our marketing and advertising departments, for example, we can reduce redundancies. This could save CompuMed approximately $275,000 over a six-month period.
2. Reducing staff—We need to reduce employees by 15 percent. This does not necessarily mean that layoffs are inevitable. One way, for instance, to reduce staff is through voluntary retirements. We will be encouraging employees with over 20 years vested in the company to take our generous early-retirement package.
3. Freezing wages—For the next fiscal quarter, no raise increases will go into effect. Internal auditors will review the possibility of reestablishing raises after the first quarter.
4. Freezing travel—Conference attendance will be stopped for six months.

I encourage you to visit with me and your division managers with questions or suggestions. CompuMed is a strong company and will bounce back with your help. Thank you for your patience and understanding.

CHAPTER HIGHLIGHTS

1. Memos, letters, e-mail, instant messages, and text messages are an important part of your interpersonal communication on the job.
2. Memos, letters, e-mail, instant messages, and text messages differ in destination, purpose, format, audience, tone, delivery time, and security.
3. Use an effective subject line including a topic and a focus.
4. Follow the all-purpose template for memos, letters, and e-mail including an introduction, a body or discussion section, and a conclusion.
5. In the template, the introduction states what you want and why you are writing.
6. In the discussion section, you state the details.
7. In the template, conclude by telling the reader what you plan to do next or what you expect him or her to do next. You also can date this action and thank the readers for their time.
8. Consider the audience whether you are writing a memo, letter, e-mail, instant message, or text message.
9. Wizards allow you to format your memos and letters but can be somewhat restrictive.
10. Follow techniques from this chapter to create effective instant messages and text messages.

CASE STUDIES

1. As director of human resources at CompuMed biotechnology company, Andrew McWard helps employees create and implement their individual development plans (IDPs). Employees attend 360-Degree Assessment workshops, where they learn how to get feedback on their job performance from their supervisors, co-workers, and subordinates. They also provide self-evaluations.

    Once the 360-Degree Assessments are complete, employees submit them to Andrew, who, with the help of his staff, develops IDPs. Andrew sends the IDPs to the employees, prefaced by a cover letter. In this cover letter, he tells them why he is writing and what he is writing about. In the letter's body, he focuses their attention on the attachment's contents: supervisor's development profile, schedule of activities which helps employees implement their plans, courses designed to increase their productivity, costs of each program, and guides to long-term professional development.

    In the cover letter's conclusion, Andrew ends upbeat by emphasizing how the employees' IDPs help them resolve conflicts and make better decisions.

Assignment

Based on the information provided, write this cover letter for Andrew McWard. He is sending the letter to Sharon Baker, Account Executive, 1092 Turtle Hill Road, Evening Star, GA 20091.

2. ITCom is committed to increasing the diversity of its workforce and its clientele. ITCom realizes that a diverse population of employees and customers (in terms of gender, ages, races, and religions) makes good business sense.

    To ensure that ITCom achieves diversity awareness, the company plans the following:

    - Develop a diversity committee
    - Focus on ways the company can be a responsible member of the community's diverse constituency
    - Hire a diverse workforce
    - Train employees to respect diversity
    - Write corporate communication (Web sites, e-mail, letters, corporate reports) that accommodate the unique needs of a diverse audience

    As CEO of ITCom, you want buy-in throughout the company. You want the company to realize that diversity is good for society and good for business.

Assignment

Write an e-mail message to your employees. In this e-mail, explain the company's diversity goals, highlight what the company plans to do to accomplish these goals, and ask the audience to participate in upcoming diversity workshops. Follow the criteria for effective e-mail provided in the textbook.

3. Mark Shabbot works for Apex, Inc., at 1919 W. 23rd Street, Denver, CO 80204. Apex, a retailer of computer hardware, wants to purchase 125 new flat-screen monitors from a vendor, Omnico, located at 30467 Sheraton, Phoenix, AZ 85023. The monitors will be sold to Northwest Hills Educational Cooperative. However, before Apex purchases these monitors, Mark needs information regarding bulk rates, shipping schedules, maintenance agreements, equipment specifications, and technician certifications. Northwest Hills needs this equipment before the new term (August 15).

Assignment

Write a letter of inquiry for Mr. Shabbot based on the preceding information.

4. Gregory Peña (121 Mockingbird Lane, San Marcos, TX 77037) has written a letter of complaint to Donya Kahlili, the manager of TechnoRad (4236 Silicon Dr., San Marcos, TX 77044). Mr. Peña purchased a computer from a TechnoRad outlet in San Marcos. The *San Marcos Tattler* advertised that the computer "came loaded with all the software you'll need to write effective letters and perform basic accounting functions." (Mr. Peña has a copy of this advertisement.) When Mr. Peña booted up his computer, he expected to access word processing software, multiple fonts, a graphics package, a grammar check, and a spreadsheet. All he got was a word processing package and a spreadsheet. Mr. Peña wants Ms. Kahlili to upgrade his software to include fonts, graphics, and a grammar check; a computer technician from TechnoRad to load the software on his computer; and TechnoRad to reimburse him $400 (the full price of the software) for his trouble. Ms. Kahlili agrees that the advertisement is misleading and will provide Mr. Peña software including the fonts, graphics, and grammar check (complete with instructions for loading the software).

Assignment

Write Ms. Kahlili's 100% Yes adjustment letter to Mr. Peña based on the information provided.

5. Bob Ward, an account manager at HomeCare Health Equipment, has not received the raise that he thinks he deserves. When Bob met with his boss, Tasha Williams, last Thursday for his annual evaluation, she told him that he had missed too many days of work (eight days during the year), was unwilling to work beyond his 40-hour work week to complete rush jobs, and had not attended two mandatory training sessions on the company's new computerized inventory system.

   Bob agrees that he missed the training sessions, but he was out of town on a job-related assignment for one of those sessions. He missed eight days of work, but he was allowed five days of sick leave as part of his contract. The other three days missed were due to his having to stay home to take care of his children when they were sick. He believed that these absences were covered by the company's parental leave policy. Finally, he does not agree that employees should be required to work beyond their contractual 40 hours.

Assignment

Write a memo or e-mail message to Tasha Williams, stating Bob's case.

6. You are the manager of WhiteOut, a store that sells snowboarding equipment and clothing. You have seven employees. Since you believed that today would be a light day for sales, you asked only one employee to work. However, suddenly, more customers than you had expected showed up, and you and your employee are overwhelmed. You need help—fast. E-mailing your other employees might not work, since they aren't necessarily near a computer. In contrast, you know they all have cell phones. The fastest and most trustworthy way to communicate with them is through instant messaging or text messaging.

Assignment

Write an instant message or a text message to the employees. Tell them the circumstances at your store and ask them for assistance. Be sure to limit your text to 160 characters.

INDIVIDUAL AND TEAM PROJECTS

1. Write a letter of inquiry. You might want to write to a college or university requesting information about a degree program, to a company requesting information about a summer internship, or to a manufacturer for information about a product or service. Whatever the subject matter, be specific in your request.

2. Write a cover letter. Perhaps your cover letter will preface a report you are working on in school, a report you are writing at work, or documentation you will need to send to a client.
3. Write a letter of complaint. You might want to write to a retail store, manufacturing company, restaurant, or governmental organization. Whatever the subject matter, be specific in your complaint.
4. Write an adjustment letter. Envision that a client has written a complaint letter about a problem he or she has encountered with your product or service. Write a 100% Yes letter in response to the complaint.
5. Write a memo or e-mail message to your boss requesting updated office equipment. Your memo or e-mail will explain your office's needs. Specifically state what equipment or furniture you want and why these purchases are important.
6. Visit five to ten local companies. When meeting with these companies' employees, ask them how much time they spend writing e-mail, instant messages, and/or text messages. Ask them what they generally write about, who their audiences are, and whether their e-mail, instant messages, and/or text messages are important parts of their jobs. Then, report your findings in an e-mail message to your instructor.

PROBLEM-SOLVING THINK PIECES

1. Northwest Regional Governmental Training Consortium (GTC) provides educational workshops for elected and appointed officials, as well as employees of city and state government offices.

   One seminar participant, Mary Bloom, supervisor of the North Platte County Planning and Zoning Department, attended a GTC seminar entitled "Developing Leadership Skills," on February 12, 2012. Unfortunately, she was disappointed in the workshop and the facilitator. On February 16, Mary wrote a complaint letter to GTC's director, Leonard Liss, stating her dissatisfaction. She said that the training facilitator's presentation skills were poor. According to Mary, Doug Aaron, the trainer, exhibited the following problems:

   - Late arrival at the workshop
   - Too few handouts for the participants
   - Incorrect cables for his computer, so he could not use his planned PowerPoint presentation
   - Old, smudged transparencies as a backup to the PowerPoint slides

   Mary also noted that the seminar did not meet the majority of the seminar participants' expectations. She and the other government employees had expected a hands-on workshop with breakout sessions. Instead, Mr. Aaron lectured the entire time. In addition, his information seemed dated and ignored the cross-cultural challenges facing today's supervisors.

   Neither Leonard nor his employees had ever attended this workshop. They offered the seminar based on the seemingly reliable recommendation of another state agency, the State Data Collection Department. From Doug Aaron's course objectives and resume, he appeared to be qualified and current in his field.

   However, Mary deserves consideration. Not only are her complaints justified by others' comments, but also she is a valued constituent. The GTC wants to ensure her continued involvement in their training program.

Assignments

- Leonard needs to write a 100% Yes adjustment letter. In this letter, Leonard wants to accommodate the dissatisfied client. How should he recognize Mary's concern, explain what might have gone wrong, and offer satisfaction?

- Leonard needs to write an internal memo or an e-mail message to his staff. In this memo or e-mail message, he will provide standards for hiring future trainers. What should the standards include?

2. You are the manager of a human resources department. You are planning a quarterly meeting with your staff (training facilitators, benefits employees, and personnel directors). The staff works in three different cities and 12 different offices. You need to accomplish four goals: get their input regarding agenda items, find out what progress they have made on various projects, invite them to the meeting, and provide the final agenda.

   How should you communicate with them? Should you write an e-mail, instant message, text message, letter, or memo? Write an e-mail message to your instructor explaining your answer. Be sure to give reasons for and against each option.

WEB WORKSHOP

1. Many state governments provide guidelines for writing letters either to government officials or for government employees. In addition, you can go online, type an entry such as "how to write state government letters" in a search engine, and find sites with instructions and sample letters for writing to state officials or state agencies.

   Research the Internet and find letter samples and guidelines. Based on your research, do the following:

   - Review sample letters to or from governmental agencies.
   - Determine whether these letters are successful, based on the criteria provided in this chapter.
   - Write either a letter, memo, or e-mail message to your instructor explaining how and why the letters succeeded.
   - Rewrite letters that can be improved, using the guidelines provided in this chapter.

2. Most city, county, and state governments have written guidelines for their employees' e-mail. For example, see "Electronic and Voice Mail: Connecticut's Management and Retention Guide for State and Municipal Government Agencies" (http://www.cslib.org/email.htm). Similarly, most companies, colleges, and universities have written guidelines for their employees' and students' e-mail use. For example, see what Florida Gulf Coast University says about e-mail use (http://admin.fgcu.edu/compservices/policies3.htm).

   Research your city, county, and/or state e-mail policies. Research your city, college, or university's e-mail policies. Determine what they say, what they have in common, and how they differ.

   Write an e-mail to your instructor reporting your findings.

3. Research information about text messages by checking out the following sites. Then, write an e-mail message to your professor, summarizing your findings:

   - Calvey, Mark. "RU Ready? Wells Initiates Banking by Text Message." *Kansas City Business Journal.* http://kansascity.bizjournals.com/kansascity/othercities/sanfrancisco/stories/2007/10/22/daily9.html?b=1193025600%5E1539015
   - Kedrosky, Paul. "Why We Don't Get the (Text) Message." *CNN.Money.* http://money.cnn.com/magazines/business2/business2_archive/2006/08/01/8382255/index.htm
   - Noguchi, Yuki. "Life and Romance in 160 Characters or Less: Brevity Gains New Meaning as Popularity of Cell Phone Text Messaging Soars." *Washington Post.* http://www.washingtonpost.com/wp-dyn/content/article/2005/12/28/AR2005122801430.html

- Tedeschi, Bob. "Reaching More Customers with a Simple Text Message." *The New York Times*. http://www.nytimes.com/2007/07/16/business/media/16ecom.html
- "Text Messaging: Frequently Asked Questions." *AT&T*. http://www.wireless.att.com/learn/messaging-internet/messaging/faq-text-messaging.jsp

QUIZ QUESTIONS

1. What are five distinctions among memos, letters, e-mail, instant messages, and text messages?
2. Why is a subject line in a letter, memo, or e-mail important, and what are the components of an effective subject line?
3. What do you hope to accomplish in the introduction to a memo, letter, or e-mail?
4. Why should you be concerned with document design in a memo, letter, or e-mail?
5. Why should you be careful when using highlighting techniques in e-mail?
6. What should you include in a conclusion to a memo, letter, or e-mail?
7. How do the audiences differ among memos, letters, and e-mail?
8. How can clustering or mind mapping be valuable to you as a form of prewriting a memo or e-mail?
9. What are three benefits of using instant messages or text messages in the workplace?
10. What are three challenges of using instant messages or text messages in the workplace?

CHAPTER SEVEN

Building Online Communities Through Blogging, YouTube, Twitter, Facebook, and Virtual World

COMMUNICATION AT WORK

In this scenario, Spiral16 studies social media and its relation to business.

Spiral16 is an informatic software company that builds products related to social media monitoring and analysis. One of the company's products, Spark ©, monitors what people say online to help companies

- Manage and protect their brands
- Determine customer opinions
- Spot potential public relations crises
- Understand "language trends" through semantic analysis

Spiral16 helps companies make sense of the immense social media space by mapping the terrain of the Internet. This is done by visualizing near real-time results pulled from all categories of Web sites, including blogs, forums, news organizations, and videos. Spark assesses social media trends by accessing, studying, and reporting on public data on the Web. Some examples of Web sites Spark gathers include

- Blogs
- Online news organizations

Objectives

When you complete this chapter, you will be able to

1. Understand the purpose of social media in the workplace

2. Communicate effectively using social media

3. Apply ethical considerations when using social media in the business world

4. Follow guidelines for effective use of social media

5. Use the checklist to determine if your text is appropriate

6. Apply your knowledge

- Social network communities, including Facebook, MySpace, Flickr, Ning, and LinkedIn
- YouTube and Vimeo
- Reference Web sites, such as MSNMoney, YahooFinance, and Wikipedia
- Twitter

Here are three examples of Spiral16's clients who have used social media to enhance their business:

- **NFL/NBA clients.** During the NFL draft, Spiral16 provided one NFL team feedback from online sentiment regarding the team's draft picks. The team was able to see what people were saying and where they were saying it, giving the team the information needed to better market their players online.

- **Major clothing manufacturers.** One blue jeans manufacturer decided to promote bands and venues overseas, asking that the band members wear the company's brand of clothing. The company used Spark to monitor the campaign. One of Spark's features revealed a language trend used by their online audience of which the company was unaware. The company then used that language in their marketing to relate to their audience.

- **Movie studios.** Movie makers advertise their releases and tend to focus promotions on whichever actor they consider to be the "headliner." Usually, it takes months to determine if their promotional strategy is sound. However, Spiral16 uses Spark to access audience feedback in just a few hours by listening to conversations on blogs and social networks. Then, if fans are talking about a different actor in the movie than the one the studio had initially promoted, the studio can alter its promotional approach within days instead of months.

Spiral16 effectively uses social media to grow its business as it helps other companies and organizations connect with their audience's needs and interests.

Check out our quarterly newsletters at www.pearsonhighered.com/gerson for dot.com updates, new case studies, insights from business professionals, grammar exercises, and facts about technical communication.

Why Is Social Media Important in the Workplace?

A recent survey published by Forrester Research shows that "the number of social-media 'spectators'—people who read or watch social media—has increased . . . to 69 percent of people who venture online" (Ranii). Companies need to listen to what all of these people are saying, understand what it means, and participate in the conversations. As Whitney Mathews, social media manager at Spiral16, says, social media is the "new customer help line" (Mathews). Many companies are discovering "the benefits of blogs, online video and social-networking sites" (Ranii). According to Marius Ciortea, senior manager of project management and office Web architecture at Oracle, using social media in the workplace makes employees approachable, allowing for transparent conversations with customers and quick response to customer concerns. Michael Brito, social media strategist for Intel, says that with social media, companies can "humanize their brand" and achieve a sense of community and participation with customers.

Sun Microsystems is one example of a company that is using social media channels to connect its customers and employees. According to Sumaya Kazi, senior social media manager at Sun Microsystems, employees at Sun are expected to participate in social networking (wikis, blogs, tweets, and Facebook). "It's part of their jobs" to build community, listen, share, learn, and communicate with each other and with their customers. Social media is such a large part of life at Sun that the company has created "social media ecosystems"—multiple avenues of connectivity. Recently, Sun employees participated in social media as follows:

- 5,426 Sun bloggers
- 4,437 Sun blogs (including group blogs)
- 71,000 contributors to wikis.sun.com
- 946,000 contributors to forums.sun.com
- 200 communities on Sun's Facebook site
- 5,240 followers on Sun's Twitter site

According to a study performed by the University of Massachusetts Dartmouth Center for Marketing Research, "500 of the fastest growing companies in the US [have] found that familiarity with and use of blogs, podcasting, wikis, online video and social networking has skyrocketed." Seventy-seven percent of the businesses that responded used social media in their business (Kirkpatrick).

What Is Social Media?

Teamwork

See Chapter 1 for more discussion on using wikis for teamwork.

Social media can take many different forms, including Internet discussion forums, blogs, vlogs, podcasts, online pictures, videos, and wikis. Social media is unique from traditional media (newspapers, radio, television, print and broadcast journalism) primarily because of the following:

1. Anyone with Internet access can participate in social media. Users are part of a global community.
2. Social media takes no training.
3. Social media is affordable.
4. Social media can be instantaneously accessed and altered. The community of users can change online text and affect the message being communicated.
5. Social media is both archivable and searchable. Companies can store information in their Facebook or Twitter sites. Corporate videos can be archived on YouTube. Then, users can search for and retrieve the stored information.

In contrast to traditional media, social media allows for and encourages person-to-person interaction among people and corporations. For example, by participating in a blog, a customer can let a company know preferences about products and service. In turn, a company can use a blog to transmit information about the company, its personnel, and its products and services. This communication can be delivered quickly and efficiently to an individual's cell phone, iPod, or laptop computer—anytime, anywhere. Social media, in combination with mobile communication, allows for real-time video instructions to be delivered through YouTube and Q/A feedback to be delivered through Twitter.

By having a Facebook presence, a company can include a link to videos about events, personnel activities, or online training options. Moreover, through a Facebook "wall," customers can post directly to the company and communicate with many different levels of management. Bill Brelsford, owner of Rebar Business Builders in Lenexa, Kansas, says, "I believe the 'cool' factor lies not in the technology but in the fact that now I can have a conversation with someone like the president of a large corporation, something I would probably not be able to do without social media tools" (Goforth). See Table 7.1 for examples of social media software applications. Figure 7.1 provides commonly used social media terms.

TABLE 7.1 Examples of Social Media Software Applications

Blogger—a blog publishing system	**Meetup**—an online social networking site that facilitates group meetings and allows members to find and join groups unified by common interests	**TypePad**—a blog publishing service
Delicious—a social bookmarking or social tagging Web service for storing and sharing Web bookmarks	**MySpace**—a social networking Web site	**Ustream.tv**—a network of channels for live video streaming of online events
Digg—a social news Web site allowing users to submit Internet links and stories, vote, and comment on submitted stories	**Open Diary**—a blog publishing system	**vBulletin**—an Internet forum, or message board, software for online discussions
Facebook—a social networking Web site	**Photobucket**—an image and video hosting Web site	**Vimeo**—a social networking site that supports video sharing, video storage, and user comments
Flickr—an image and video hosting Web site	**Plaxo**—an online address book and social networking service	**Vox**—an Internet blogging service
Friendster—a social networking Web site	**Plurk**—a social networking, microblogging service	**Wikis**— community Web sites, such as Wetpaint, PBWiki, and Wikispaces, that allow for collaborative creation and editing of Web pages
Google Groups—a service from Google allowing discussions about common interests	**Pownce**—a social networking, microblogging Web site	**WordPress**—an open source blog publishing application
Jaiku—a social networking, microblogging service	**Revver**—a video sharing Web site	**Yahoo! Groups**—an electronic mailing list and Internet forum
Justin.tv—a network of channels for online live video streaming	**Second Life**—a virtual world	**Yelp**—a social networking online opinion site for user reviews
LinkedIn—a business-oriented social networking site	**Skype**—a software application that lets users make telephone calls over the Internet	**YouTube**—a video sharing Web site allowing users to upload and share video clips
LiveJournal—a blog publishing system	**Twitter**—a social networking, microblogging service	**Zooomr**—a Web site for sharing digital photos

FIGURE 7.1 Commonly Used Social Media Terms

- **Cloud services**—a variety of social media resources, such as audio, video, music, chat, search, and postings, combined in one site.
- **Crowd sourcing**—using the social media audience as the source of new ideas.
- **Data portability**—get it here; get it everywhere (podcasts, instant messages, text messages, Facebook, YouTube videoclips, music downloads, Twitter, cell phone capabilities, and more).
- **Evangelists**—social media followers and users who promote a company or product through their tweets or Facebook postings.
- **IdeaStorm**— Dell's technology for crowd sourcing.
- **Intracorporate social networking**—password-protected social media sites within a company where employees connect with other employees.
- **Monetization**—a company's use of social media to make money.
- **Social ecosystems**—multiple social networking communities, all connected to each other.
- **Stickiness**—the amount of time an individual spends on a social media site over a given time period. Stickiness is anything about a Web site that encourages a visitor to stay longer and to revisit the site.
- **Viral marketing**—marketing techniques that use social networks to increase brand awareness.

Crisis Communication and Social Media

Another reason social media is important in the workplace is crisis communication. Today, workplace communicators must be prepared to communicate information about crises that occur in the workplace. A crisis could occur for any number of reasons including an armed intruder on the premises, severe weather, bomb threats, receipt of potentially hazardous material through the mail system, the threat of terrorism, and more.

According to Gerard Braud, award-winning journalist and specialist in crisis management, one of the most important aspects of crisis management is to have a "crisis communication plan" ("Crafting a Crisis Communication Plan"). A crisis communication plan establishes the procedures to follow in a crisis situation, including who does what and which communication channels will be used to warn employees of a crisis situation. These channels could include telephone calls, e-mail messages, text messages, or loudspeaker announcements. Additional means of crisis communication include social media, such as Web postings, blogs, podcasts, and Twitter messages (Griffin). By communicating through social media channels, you can warn an audience anywhere and anytime of potential threats in the workplace. For example, you could send a tweet and reach employees instantaneously. Because of the near real-time nature of communicating through social media, many police and fire organizations, as well as health agencies, are making social media a part of their crisis communication plans.

Ethical Considerations when Using Social Media for Business

Though many companies are using social media channels, this means of communication has risks. What you write on a Facebook wall or depict visually in a YouTube video can have significantly negative results. For example, your initial reaction to the mention of YouTube and business might be a memory of the two employees at Domino's Pizza who posted a video of themselves doing unsanitary things with food that was then to be served to customers. As a result of this video, the two employees were fired and faced felony charges (Clifford). Neither of these employees considered the ethical ramifications of their actions when they created and posted the video on YouTube. Similar problems can occur when you write a blog entry or tweet. An effective communicator must assess "the risk of digital writing" (Rife). Tweets, blogs, and Facebook text rarely are read by only one person because of the nature of these "globally circulating digital writing spaces" (Rife).

To avoid problems with social media in the workplace, consider these ethical guidelines:

- Follow your employer's rules regarding Internet conduct.
- Never publish negative, confidential, or proprietary comments about your employer.
- Link to your sources so you can avoid plagiarizing content.

DOT.COM UPDATES

To learn more about one company's rules for appropriate use of social media, check out "IBM's Social Computing Guidelines."

- http://www.ibm.com/blogs/zz/en/guidelines.html

Guidelines of Social Media Usage on the Job

When using social media, consider the following:

- Be careful about what you say. It will be there for anyone to see, for everyone to see, and forever.
- Don't share your personal or professional drama; separate your personal life from your work. For example, if you are tweeting, avoid "twitter fights." If you are on Facebook, don't post personal videos that might portray you or your company negatively.
- Censure yourself—avoid swear words.
- Leave the phone in your purse or pocket when you're in a meeting or speaking to clients face-to-face, thereby avoiding the urge to get on the Internet and check tweets, texts, e-mail messages, and status updates. You are being discourteous when you constantly look at your PDA or cell phone.
- Use spell check. No matter what the situation, poor spelling and grammar errors make you look unprofessional, and you'll lose credibility.
- Don't overuse "text speak." Too much "text speak" lingo, such as "ur," "b4," "omg," or "lol" can negate your professionalism.

Types of Social Media

Although social media involves many different channels of delivery, this chapter focuses on the following:

- Blogging
- YouTube
- Twitter
- Facebook
- Virtual World (Second Life)

Understanding these types of social media will allow you to participate in online communities and incorporate social media into your workplace. Your company could benefit from a social media presence, for according to Brad Keown, marketing representative for Facebook in China, "A Facebook marketing program can help spread a brand message virally from one user to another" (Goforth). In fact, "social media can help level the playing field with larger businesses that may have greater marketing clout" (Goforth). Both large and small businesses can benefit from understanding how to use social media effectively.

Blogging for Business

Jonathan Schwartz, president and chief operating officer of Sun Microsystems, says that blogging is a "must-have tool for every executive." "It'll be no more mandatory that they have blogs than that they have a phone and an e-mail account" (Kharif). Bill Gates, Microsoft's CEO, says that blogs could be a better way for firms to communicate with customers, staff, and partners than e-mail and Web sites. "More than 700 Microsoft employees are already using blogs to keep people up to date with their projects" ("Gates Backs Blogs for Business"). Marc Cuban, owner of the Dallas Mavericks basketball team, has a blog (Ray). IBM is planning "the largest corporate blogging initiative" by encouraging its 320,000 employees to become active bloggers with the goal of achieving "thought leadership" in the global information technology market (Foremski). Nike launched an "adverblog" to market its products. Technorati, a blog search engine, currently tracks "112.8 million blogs and over 250 million pieces of tagged social media" ("Welcome to Technorati").

What is "blogging," and why are so many influential companies and business leaders becoming involved in this social media channel?

Blogging—a Definition.   A blog, the shortened version of the words "Web log," is created online and can include text, graphics, links to other Web sites, and video. Blogs not only allow but also encourage input from many readers. Daily, millions of people use blogs to publish their thoughts. Though many blogs feature text, blogs take different forms, including the following:

- **Vlogs.** The posting of videos through RSS feeds (Really Simple Syndication or Rich Site Summary) to create miniprograms, such as seen on YouTube.
- **Podcasting.** News, market information, product announcements, public service announcements, how-to videos, photographs, audio files, and more can be conveyed through podcast blogs. The term "podcast" is a combination of the words "iPod" and "broadcast."
- **Microblogging.** Brief text, usually less than 200 characters. Facebook and MySpace, two popular blog sites, include microblogging features.
- **Twitter.** A free microblogging service. Users can send short tweets or updates limited to 140 characters to cell phones or other blog sites.

Reasons for Blogging.   Following are reasons why companies blog:

1. **Communicate with colleagues.** Chris Winfield, president of an online marketing company, says that he uses blogging to improve "communication flow to his employees" (Ray). Many companies encourage their employees to use blogging for project updates, issue resolutions, and company announcements. The engineering department at Disney ABC Cable Networks Group uses blogs "to log help desk inquiries" (Li).
2. **Communicate with customers.** In contrast to private, intranet-based blogs used for internal corporate communication, companies also have public blogs. Through public blogs, a company can initiate question/answer forums, respond to customer concerns, allow customers to communicate with each other, create interactive newsletters, introduce new products and services, and build rapport with customers, vendors, and stakeholders.
3. **Improve search engine rankings.** Marketing is a key attraction of corporate blogs. A blog post using keywords, allowing for comments and responses, and providing references and links to other sites tends to rank in the top 10 to 20 listings in Internet search engines (Ray).
4. **Network through "syndication."** To access a Web site, a reader must know the URL or use a search engine, such as Google. In contrast, blogs can be distributed directly to the end users through a "Web feed." The writers of a blog can syndicate their content. Then, through the Web feed, readers can subscribe to the

blog. Syndication makes Web feeds available for people to access. By using feed programs, such as RSS, Atom, or Current Awareness, bloggers can syndicate their blogs or be notified when topics of interest are published. Thus, blogs can be very personalized, essentially delivered to your door on a moment's notice.

5. **Track public opinion.** "Trackback" features, available from many blog services, let companies track blog usage. Tracking lets companies monitor their brand impact and learn what customers are saying (good news or bad news) about their products or services.

Guidelines for Effective Corporate Blogging. If you and your company decide to enter the "blogosphere," the world of blogging, follow these guidelines:

1. **Identify your audience.** As with all technical communication, audience recognition and involvement are crucial. Before blogging, decide what topics you want to focus on, what your unique spin will be, what your goals are in using a blog, and who your blog might appeal to.

2. **Achieve customer contact.** Blogs are innately personal. Take advantage of this feature. Make your blogs fun and informal. You can give your blog personality and encourage customer outreach by including "interesting news of the day, jokes," personnel biographies, question/answer forms, updates, an opportunity to add comments, as well as information about products and services (Ray). In addition, make sure the blog is interactive, allowing for readers to comment, check out new links, or add links.

3. **Start "blogrolling."** Once you have determined audience and blog location, it's time to start blogrolling. Start talking. Not only do you need to start the dialogue by adding content to your blog, but also you want to link your blog to other sites (Wuorio "Blogging").

4. **Keep it fresh.** Avoid a stale blog. By posting frequently (daily or weekly) and responding quickly to comments and criticism, you encourage bloggers to access your site and return to it often.

5. **Develop guidelines for corporate blog usage.** Because blogs encourage openness from customers as well as employees, a company must establish guidelines for corporate blog usage. Figure 7.2 offers a sample code of ethics created by Forrester Research (Li).

Figure 7.3 is a sample blog from Spiral16 showing responses.

FIGURE 7.2 Blogging Code of Ethics

1. I will tell the truth.
2. I will write deliberately and with accuracy.
3. I will acknowledge and correct mistakes promptly.
4. I will preserve the original post, using notations to show where I have made changes so as to maintain the integrity of my publishing.
5. I will never delete a post.
6. I will not delete comments unless they are spam or off-topic.
7. I will reply to e-mails and comments when appropriate and will do so promptly.
8. I will strive for high quality with every post—including basic spell-checking.
9. I will stay on topic.
10. I will disagree with other opinions respectfully.
11. I will link to online references and original source materials directly.
12. I will disclose conflicts of interest.
13. I will keep private issues and topics private because discussing private issues would jeopardize my personal and work relationships.

FIGURE 7.3 Sample Blog with Responses

spirall6 informatic software

Home | Spark | Blog | Engage | Client Login

<u>**BIG Omaha: @Jeffrey on transparency**</u>

Do you classify your online strategy as **"transparent?"** How so? What elements go into your version of transparency? I think of a few elements when it comes to being transparent on the Web. **Honesty. Openness. Humility.**

<u>Jeffrey Kalmikoff</u> from <u>Threadless</u> gave a great presentation at BIG Omaha on what it means to be transparent online. Listen to the audio of his speech and follow along with the slides from <u>SlideShare</u>. They are posted on <u>Flickr</u>, too.

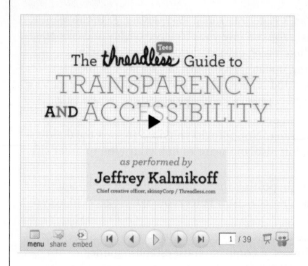

As a bonus, I included the panel that preceded Jeffrey. It was a Q&A with Jeffrey, Jason Fried, Gary Vaynerchuk and Micah Baldwin.

Warning: These podcasts contain graphic language that some may find offensive.

Published by Whitney on 14 May 2009 in <u>Content Of Interest</u>, <u>Industry Buzz</u>, <u>Resources</u>, <u>Shared Content</u>

Tagged:

2 Responses to "BIG Omaha: @Jeffrey on transparency"

1.

   <u>Jeffrey</u>

   - 3:48 pm on 05/14/2009

FIGURE 7.3 (Continued)

Thanks for putting this up! One weird thing is that some of the slides in that presentation are flip-flopped, so if it doesn't make a ton of sense - that's why. Also, you can view my slides up on slideshare… http://www.slideshare.net/fancyjeffrey/big-omaha

2.

Whitney

- 4:12 pm on 05/14/2009

Awesome, Jeffrey! I'll switch out the slideshare w/ the Flickr so it makes more sense.

Leave a Reply of Your Own

Name (required)

Mail (will not be published) (required)

Website

Submit Comment

Source: Courtesy of Spiral16

YouTube and Business

Why Should Businesses Have a YouTube Presence? Another social media opportunity for a business is YouTube. YouTube is a way for a company to connect with the customer. Because television, radio, and print advertising have become very expensive, companies need a more affordable means of advertising and marketing products and services. YouTube videos can be posted for free and allow businesses to reach millions of potential customers. "YouTube is a video sharing website allowing members to upload, share, and comment on a large database of videos. In one month alone, YouTube had a staggering 2.4 billion page views. Online video is hot" (Zahorsky). Research performed by the Kelsey Group "revealed that 59 percent of survey respondents watch online video. Emarketer estimates online video ad spending to reach over $700 million this year" (Zahorsky).

What Should Businesses Post on YouTube? A business can benefit from posting the following on YouTube:

- Presentations from executives to discuss finances, corporate activities, company performance, and so forth
- Announcements of important corporate news
- News releases of company activities
- Training for purchasers of products
- Videotapes of consumer questions and help-desk answers with procedural steps
- Marketing of new products with pictures
- Events such as corporate celebrations, meetings, and other activities
- Instructional video for use of products (Ramel)

Instructions

See Chapter 13 for more discussion of instructions.

Guidelines for Shooting Video to Post on YouTube. You have to shoot a video that people actually want to watch. A picture might be worth a thousand words, but if no one sees the picture, then you've wasted time. To shoot a successful video and ensure "proper production values" for viewing on a person's computer screen, which is much smaller and has poorer resolution than a television equipped with high definition, do the following (Miller):

- Title your video and use appropriate subtitles periodically in the video.
- Include the company name and contact information such as a telephone number and the Web site URL.
- Use onscreen graphics to enliven the video.
- Edit a longer video into "several shorter segments" (Miller). YouTube allows videos 10 minutes long.
- Keep focused "on one key message with the strongest lead benefit," and avoid attempting to cover all your products or all your services (Zahorsky).
- Avoid plagiarism by creating your own content. "Before using music clips, television clips, or movie clips, you need to get permission to do so. If you fail to get permission, you are, essentially, stealing someone's work" (Brewer).

See Figure 7.4 for a sample YouTube screen from Spiral16.

Tweeting for Business

What's Twitter? What's a Tweet? To use Twitter, you tweet. Each tweet is a text-based post of up to 140 characters in length. Twitter is a free service that lets you subscribe to, share, or follow as many twitter messages as you like. Twitter.com says that businesses can use tweets to "quickly share information with people interested in your company, gather real-time market intelligence and feedback, and build relationships with customers, partners and other people who care about your business."

Jessica Vascellaro of the *Wall Street Journal* writes, "Email has had a good run as king of communications. But its reign is over. In its place, a new generation of services is starting to take hold—services like Twitter . . . [promise] to profoundly rewrite the way we communicate" (R1). The primary reason for this change is access. Twitter is a distributed communication system, and you don't need to open a Web browser to access this service. Instead, you and your readers (called "followers") can receive tweets in a variety of ways. "People download . . . programs like Twitterific, Feedalizr or Twinkle, they get the updates on their cell phones as text messages, or they use something like PocketTweets, Tweetie or iTweet for the iPhone" (Pogue). If you want to access tweets through your desktop or laptop, Twitterific opens up as a "tall, narrow window at the side of the screen. Incoming tweets scroll up" (Pogue).

FIGURE 7.4 YouTube Screen

Source: Courtesy of Spiral16

Tweets don't have to be intrusive, either. Not everyone needs to read what you tweet. You can be more discerning by sending a Direct Message, which only one "follower" sees versus responding with a public tweet. Tweets can be sent directly to your RSS reader, and you can republish (retweet) a stream of content on your Web site or blog (Jantsch). Figure 7.5 defines commonly used Twitter terms.

Why Do Companies Use Twitter?   Restaurants can tweet their daily specials, realtors can tweet new homes on the market, and a chamber of commerce could tweet about local city events and promotions (Brooks). Companies can accomplish the following with Twitter:

- Direct people's attention to corporate news, products, services, and events.
- Give customers and co-workers a forum, allowing a company to study concerns, manage issues, initiate discussions, answer questions, and respond to complaints.
- Assist with business development—promotional and marketing collateral.
- Add to and improve customer service. Through Twitter, you can send service updates and respond to customer concerns. Best Buy, for example, has created "Twelpforce," a customer service team to answer customer questions about products and services.
- Help you show the "human" side of your company (tell what employees are doing, show how your company is involved in charitable activities, display photos of corporate events, etc. After all, Twitter is "social media").

Which Companies Have Twitter Sites?   Table 7.2 shows companies with Twitter sites, their uses of these sites, and a few examples of tweets.

FIGURE 7.5   Definitions of Commonly Used Twitter Terms

Follower	People who read your tweets on Twitter
Retweet	When you copy someone else's tweet or republish a stream of tweets
Tweet	A single Twitter message
Tweetdeck	Software that runs on your desktop. You can post tweets from it, respond to tweets, set up searches, and get updates in real time.
Tweetie	The Twitter application for iPhone and iPod Touch
Tweet scan	Twitter's search tool
Tweeple	People who tweet
Twellow	A Twitter directory (something like the Yellow pages)
Twhirl	A desktop application that updates incoming tweets
Twitturly	A service for tracking URLs that people are talking about

TABLE 7.2   Companies Using Twitter

Companies Using Twitter	Purposes	Tweet Examples
"Hey there! JetBlue is using Twitter." (http://twitter.com/jetblue).	Responses to customer questions	Q: "Which destination would you pick for a short vacation?: Las Vegas (1) or New Orleans (2)?" A: "2 to 1 in favor of Vegas for a short trip after the initial volley of replies."
"Hey there! TheHomeDepot is using Twitter." HomeDepot (http://twitter.com/HomeDepot)	Responses to customer concerns	"Sorry you didn't get the attention you deserved. We'd still like to help with your project. Can I help?"
"Hey there! Dell is using Twitter." Dell (www.dell.com/twitter)	Highlights of individual Dell employee activities	"RT@LionelatDell Dell donates $5 to Susan G. Komen for the Cure for each Promise Pink laptop we sell. Lots of options (http://twitter.com/richardatdell)."
"Hey there! Delta Airlines is using Twitter." (http://twitter.com/deltaairlines)	Helpful hints for improved customer service	"Traveling light not only helps you get thru the airport quicker-it helps us get in & out of the gates quicker!"
"Hey there! SouthwestAir is using Twitter." (http://twitter.com/SouthwestAir)	Links to online articles for more information	"Did you know Southwest Pilots mentor thousands of students in science, math, geography, and writing? Read more: http://tinyurl.com/bpf6wb"
"Hey there! Apple is using Twitter." (http://twitter.com/appleinc)	Advertisements	"You can also download music over WiFi and 3G on the iPhone."
"Hey there! GM is using Twitter." (http://twitter.com/GMblogs)	Sharing customer comments	"Retweeting @WarLordwrites: I've been using the GM OnStar Navigation system. I'm pleased. GM got this very right!"
"Hey there! M&Ms is using Twitter." (http://twitter.com/msgreen)	Keeping customers up to date about company promotional activities	"Need a little va-va-voom in your love life? Tell me why over at www.mms.com/green and you could win a romantic Paris getaway!"
"Hey there! HRBlock is using Twitter." (http://twitter.com/hrblock)	Pleasant comments to build rapport	"Happy to hear it went well! It's a relief to have your taxes filed isn't it? You're ahead of me;)"

Rules for Twittetiquette. The bottom line to etiquette of any sort is treating people with respect. After that essential consideration, following are rules for etiquette when using Twitter—Twittetiquette:

1. Use services like Twitter Search to find who might be writing about you or your company.
2. Don't tweet all day while at work. That's not an effective use of your time or corporate time.
3. Don't use Twitter just as another channel for advertising. It should have a more personal, social media component—people talking to people.
4. Separate business from fun. Have two Twitter accounts: one that is strictly business and one that is for friends, family, and fun.
5. Don't tweet in anger. A tweet sent in anger can be read by thousands of people. They might not appreciate your haste and negativity. In contrast, positive interaction with a company representative could have a more positive impact (Brogan, Mathews).

See Figure 7.6 for a sample Twitter page from Spiral16.

Facebook

Business Uses for Facebook. Isn't Facebook just an online Web site where teens can connect for purely social purposes? Why would a business want to have a Facebook presence? Perhaps the greatest reason for companies to use Facebook is the enormous growth of this social media site. Facebook has surpassed 175,000,000 active users, growing by over 600,000 users per day ("Facebook Surpasses").

DOT.COM UPDATES

Check out these URLs to learn more about Twitter:

- Brogan, Chris. "50 Ideas on Using Twitter for Business." 4 Feb. 2009. http://www.chrisbrogan.com/50-ideas-on-using-twitter-for-business/
- Brooks, Rich. "How to Use Twitter For Business." 4 Feb. 2009. *Flyte.* http://www.flyte.biz/resources/newsletters/08/06-twitter-for-business.php
- King, Rachael. "How Companies Use Twitter to Bolster Their Brands." 6 Sep. 2008. *BusinessWeek.* http://www.businessweek.com/technology/content/sep2008/tc2008095_320491.htm
- Pogue, David. "Twittering Tips for Beginners." *New York Times.* 15 Jan. 2009. http://www.nytimes.com/2009/01/15/technology/personaltech/15pogue-email.html?_r=2&sq=Pogue%20twitter&st=Search&scp=4&pagewanted=all

FIGURE 7.6 Sample Twitter Page

Source: Courtesy of Spiral16

FIGURE 7.7   Total Facebook Users in the United States

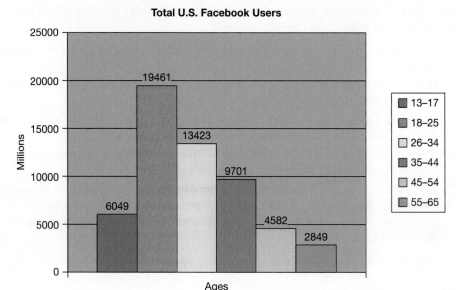

"Facebook's user base has expanded beyond college students" (Goforth D15). As shown in Figure 7.7, though the largest increase in total Facebook users is the 18–25 year cohort, the second largest rise is in the 26–34 age range. In fact, "45% of Facebook's US audience is now 26 years old or older," and the "fastest growing segment" of Facebook users is "women over 55" ("Fastest Growing Demographic on Facebook: Women Over 55"). See Figure 7.7 for an overview of Facebook users in the United States.

With such large numbers of potential customers, a company can use Facebook as a one-stop location to market products or services, create brand recognition, recruit new employees, and connect with the customer on a personal basis. On their Facebook site, companies can host their corporate blog, provide access to their Twitter entries, post company YouTube videos, show photographs of corporate employees, list upcoming events and product releases, and provide access to discussion forums. Businesses use Facebook in a variety of ways, including the following:

- Collaborating in teams—sharing documents and tasks and collaborating on projects
- Uploading PowerPoint slides to share presentations, tutorials, and demonstrations
- Conducting polls to find out end-user needs and preferences
- Managing a crisis by providing updates and responding to customer concerns

Guidelines for Businesses Participating on Facebook.   For you and your company to participate on Facebook, follow these guidelines:

1. **Create a clear business goal.** What do you hope to accomplish in your company's Facebook site? Do you want to sell, promote your charitable activities, share information about your employees, show videos of how-to tutorials, allow for give and take with customers through blog entries, and so forth? What's your plan? Determine this before going online in Facebook ("Bosses").

FIGURE 7.8   Spiral16 Facebook Page

Source: Courtesy of Spiral16

2. **Keep it fresh.** You should update your Facebook profile consistently. Add new videos, new blog postings, new tweets, new information about company products and services, and news about company activities.

3. **Add links to social media tools.** Your company's Facebook site will be enhanced by other social media tools such as Twitter, YouTube videos, your company's blog, LinkedIn, etc.

4. **Create reasonable conduct policies with employees.** Write down and disseminate corporate policies for Facebook usage regarding security, confidentiality, and personal conduct online. "This way staff will be less likely to get an unpleasant surprise when they find out their employer doesn't approve of something they said or did" ("Facing Up").

See Figure 7.8 for a sample of Spiral16's Facebook site.

Virtual World

What's Second Life?   Created by Linden Research, Inc., Second Life® is a 3-D virtual world, a digital continent populated by avatars. To access this virtual world, all you need to do is download a free Second Life (SL) viewer application which runs on Macs, PCs, and Linux servers. You create your avatar (your virtual self) and "teleport" into alternative worlds.

What Goes on in These Virtual Worlds?   Isn't this just an MMORPG (massive multiplayer online role-playing game) for geeks? No. SL isn't a game. It doesn't have goals, prizes, or ulterior motives. Instead, SL provides a virtual site where businesses, educational institutions, and governmental organizations create "public and private spaces for communication, collaboration, and training" (*Secondlife Grid*). Figure 7.9 is IBM's SL Courtyard.

FIGURE 7.9   IBM's SL Courtyard

Source: Courtesy of Marziah Karch

Which Companies Have SL Presences?   Here's a quick look at a diverse list of organizations that have SL sites:

Adidas	American Apparel	AOL	BMW	Calvin Klein	Cisco Systems	Coca-Cola
Coldwell Banker	Comcast	Dell	H&R Block	IBM	Kraft Foods	Major League Baseball
Mazda	Mercedes-Benz	Microsoft	National Basketball Association	National Public Radio	Nissan	Pontiac
Reebok	Sears	Sony Ericsson	Sprint	Starwood Hotels	Sun Microsystems	Toyota

Why Would You Want to Enter This Virtual World?   SL provides its end users and its host organizations a multitude of technical/business communication values. Here's a list of reasons why organizations use SL (Webber).

Marketing.   Companies use SL to tell customers about new products. SL allows companies to develop "working models of new and existing products" and hold virtual events with guest speakers (Webber). In SL, a company can build brand recognition, "test-market future product lines or host events to foster brand loyalty and generate buzz" (Jana).

Training.   First there were hard-copy manuals, then instructional videos. Now, with SL, you and your avatar can actually be involved in virtual hands-on tutorials. SL is an especially valuable tool for training when employees and clients are geographically dispersed, actual equipment is expensive, and situations are dangerous. In a virtual world, machines can't break and no one can be injured following a procedure (Webber).

Modeling, Monitoring, and Product Testing.   Whether it's a new car, software product, machine, or a hotel resort, it can be cheaper to create an SL version and get feedback from potential end users than it would be to build an actual prototype in RL (real life) (Webber).

Providing Customer Service.   IBM has opened an SL business center staffed by an international core of IBM sales representatives. In this virtual business center, customers can buy hardware and software. Plus, these avatars of actual clients can get answers to their computing questions and solutions to their business problems. IBM's virtual sales representatives can "speak the following languages: English, Portuguese, German, Spanish, Dutch, Italian and French, and Asian languages" (Brodkin).

Virtual Conferencing.   You can travel to off-site conferences, you can engage in webinars, you can participate in teleconferences, or you could attend a virtual conference in SL. For example, National Public Radio's Science Friday has an SL link. In this site, you can "chat with other listeners during the show, ask questions and meet host Ira Flatow's avatar" (sciencefriday.com).

Guidelines for Participating in SL.   To participate in the virtual world of Second Life, follow these guidelines:

- For SL use, you will need the correct hardware, software, and Internet connectivity. Also ensure security with a password-protected SL site.
- You can sign up for free. Currently, SL is publicly available for business and education. You lease "land" in SL from the owning company, Linden Labs, but almost 90 percent of everything else in SL is created by SL users. If you choose to buy land in SL, you pay $9.95 a month for a subscription fee. Additional optional fees involve clothing and home or office decorations. These fees are paid in Linden dollars (see Figure 7.10 for definitions of commonly used Second Life terms).
- Treat SL professionally. It's an extension of your company. Therefore, do not make fun of another avatar's appearance, interrupt online conversations, or create an unprofessional name or look for your avatar.

DOT.COM UPDATES

Check out these URLs to learn more about Second Life:

- Benner, Katie. "I Got My Job Through Second Life." *Fortune.* 23 Jan. 2007. http://money.cnn.com/2007/01/22/magazines/fortune/secondlife_recruit.fortune/index.htm.
- "Business Communicators of Second Life." Thomson NetG. 5 Feb. 2009. http://freshtakes.typepad.com/sl_communicators/2006/09/thomson_netg_se.html.
- "Companies in Second Life." *Second Life Research Blog.* 4 July 2007. http://secondliferesearch.blogspot.com/2007/07/companies-in-second-life.html.
- Pichardo, Mallory and Shiang Kwei Wang. "How Companies Are Using the Second Life Virtual." *Slideshare.* 3 Feb. 2009. http://www.slideshare.net/skwang/how-companies-are-using-the-second-life-virtual-presentation.

FIGURE 7.10   Definitions of Commonly Used SL Terms

Avatar	The computer user's 3-D representation of himself or herself
CRPG	Computer Role-playing Game
Linden dollar (L$)	The internal currency of SL. Used to buy, sell, rent, or trade land, goods, buildings, vehicles, clothing, hair, jewelry, art, and more
MMORPG	Massively Multiplayer Online Role-playing Game
RL	Real Life
Resident	The user of an SL avatar
SL	Second Life
SLurl	Second Life URL (the Web address for SL sites), lets you create a teleport link to a location in Second Life

Social Media Connects Spiral16 to the World

Whitney Mathews, social media manager at Spiral16, has an interesting and varied job with numerous responsibilities. She often works 55 hours or more each week and rarely turns off her iPhone. "I even watch television with the remote in one hand and my iPhone in the other!" For Whitney, a typical workday consists of

- Researching Spiral16 every morning when she gets to work using "our software Spark." Whitney performs this research to find out what is being said about Spiral16 and to find out who is talking about the company. She then leaves blog posts in response to the commentary and thanks the people who have communicated about Spiral16. She also likes to let her co-workers know about the comments and "flags" entries for them.
- Updating case study research about clients.
- Writing entries on the company's blog or working on current projects. One recent project was a PowerPoint presentation for Spiral16's president who was making the presentation at a meeting in Chicago.
- Attending social networking events. These events include "blog meetups" or "Tweetups" with business professionals she has met through her social media connections.
- Participating in sponsorships with other companies or individuals through bartering. Whitney will monitor other events for free "as long as they put our logo in their literature." In this way, Whitney advertises her company.
- Attending social media conferences to network and to acquire content for the company blog.

According to Whitney, organization is a critical part of performing her job. She finds that the following tools are critical to her success:

- iPhone allows 24/7 access for Whitney.
- Spark gives Whitney the software capability to study the information found on the Web.
- My laptop "has my whole work life on it."
- TweetDeck is an application designed around Twitter that filters Twitter information.

Whitney usually knows what she wants to say before she writes her blog postings, so she focuses on organization of content. She starts with a storyboard of the content so she can see how it will lay out. The storyboard merges what she plans to say with its design. She says, "I know people read online text in small sections. If I put the key words in boldface and then use bullets to itemize the information, they'll read it. People react to placement, so I focus on that when I format. I know that everyone who blogs is also a designer because blogging is such a visual process."

After Whitney begins writing, she then breaks content into sections containing two or three sentences. She adds links and format such as color and videos. "I let the words interact with the image."

Whitney says that the challenge in her job is keeping up with the constantly changing world of social media. "I have so much to learn. When I write something, I worry if I'll be taken seriously. When I write on the company blog, I am representing Spiral16. I have to stop and consider if I want to publish that idea. My primary goal is to generate discussion so I can get Spiral16's name out there. I want people to talk about our Web site or our blog. By generating more discussion, I can help market the company and its products."

SOCIAL MEDIA CHECKLIST

_____ 1. Have you considered your audience's needs in your social media communication?

_____ 2. If you are using social media, how will you engage your audience and encourage participation in the blog, tweet, or Facebook site?

_____ 3. If you are tweeting, have you limited your tweet to 140 characters?

_____ 4. If you are using Facebook, have you observed the guidelines for participating?

_____ 5. If you have created a video to post to YouTube, have you followed the guidelines, such as limited length, appropriate subject matter, titles and subtitles, contact information, and so forth?

_____ 6. Have you considered ethical standards when using social media?

_____ 7. Did you review your communication for grammar, punctuation, or spelling errors?

_____ 8. If you have used material such as pictures, video, or words created by someone other than yourself, have you linked to the source or provided copyright information?

_____ 9. If you are participating in Second Life (the virtual world), have you been appropriate in your choice of avatar, and do you treat other avatars respectfully?

_____ 10. Do you avoid spending excessive amounts of work time communicating via social media?

The Writing Process at Work

Effective communication follows a process of prewriting, writing, and rewriting. To illustrate the importance of the writing process in social media, look at how Whitney Mathews used prewriting, writing, and rewriting to create an effective blog entry for Spiral16.

The Writing Process

Prewriting	Writing	Rewriting

• Determine your goals in using social media.
• Try brainstorming, listing, storyboarding, and/or answering reporter's questions to clarify what content you will need to include.
• Decide whether your audience is high tech, low tech, or lay. This will help you decide which abbreviations and acronyms need to be defined.
• Choose the correct communication channel (blog, Facebook, YouTube, Twitter, Second Life, and others) for your audience and purpose.
	• Organize your content using modes such as problem/solution, cause/effect, comparison, argument/persuasion, analysis, chronological narratives, etc.
	• Use figures and tables to clarify content.
	• Use headings and talking headings for access.
	• Create hypertext links to your other sources of information.
		• Revise your draft by
		  • Adding details
		  • Deleting wordiness
		  • Simplifying words
		  • Enhancing the tone
		  • Reformatting your text
		  • Proofreading and correcting errors
		  • Enhancing content with visual aids

Prewriting

Figure 7.11 illustrates Whitney Mathews's listing to gather and organize data for her blog entry for Spiral16.

Writing

Figure 7.12 is Whitney's draft of the blog entry that she wrote. She then thought about redesign issues to revise the text.

Rewriting

See Figure 7.13 for Whitney's successful blog entry from Spiral16.

Web Resources

www.pearsonhighered.com/gerson For more information about the writing process, visit our companion website.

FIGURE 7.11   Listing for a Blog Entry

1. What do I want the readers to learn about social media from my blog?
2. Where should people begin when they use social media?
3. What can the different departments in a company contribute to the use of social media?
4. What questions should be asked of the different departments to learn more about their use of social media with customers?
5. What do we hope to accomplish with social media?

FIGURE 7.12   Whitney's Draft of the Blog Entry with Her Thoughts on Revision

I'll add headings and subheadings to help our readers navigate the text more easily.

I'll add hypertext links to help our readers find the original sources.

I'll break the text into smaller chunks, smaller paragraphs for easier access.

This paragraph would be much more reader friendly if I bulletized the questions.

A graphic would be a great way to show how we at Spiral16 help our readers visualize data. I'll try one of our bubble charts that show connections between data points.

One of the toughest things for companies to grasp when building a social media strategy is that they need to write the definition for their own success. When it comes to social media, objectives are different for everyone. That's the point. You can take your message to your audience in a way that fits what you want to achieve. We've talked about this before on the blog, but it's becoming more important as more companies create legit strategies.

So, if we come to a demo and ask what you are looking for when it comes to social media monitoring, take a second to think it through. Define your own goals and objectives. If you have no idea where to start, head to the whiteboard and have an old-fashioned brainstorming session. Bring a representative from every department. They will all have different ideas and voices on the value of your results. If they (or you) need a better understanding of why social media and monitoring matters. Also, if each department with a valuable interest has access to your data, it can keep your company on the same page while satisfying individual needs. Sales can identify potential clients, marketing can form strategy around results, communications can respond quickly to crises, and HR can police potential problems with employees. All these different departments can get these insights from the same data set.

Ask the group questions like the following: Do you respond to your audience online? If yes, are you satisfied with your response time? Do you want to improve customer service? What about sales? Do you want to increase your brand loyalty? Should you monitor specific campaigns or just your overall presence? What are some of your goals? What are suggestions you give to clients when it comes to developing strategy?

FIGURE 7.13   Whitney's Blog Entry for Spiral16

New rule: You must define your objectives

In my experience, one of the toughest things for companies to grasp when building a social media strategy is that they need to write the definition for their own success.

When it comes to social media, **objectives are different for everyone.** That's the point. You can take your message to your audience in a way that fits what you want to achieve.

We've talked about this before on the blog (see Three Steps for SocMed Solution Shopping), but it's becoming more important as more companies create legit strategies.

Define your objectives...

There are no standard social media metrics.

What is important to you?
• **Improved** customer service
• **Increased** brand loyalty
• **Growth** in sales

Go after it.

This will help you **define ROI.**

So, if we come to a demo and ask what you are looking for when it comes to social media monitoring, take a second to think it through. **Define your own goals and objectives.**

If you have no idea where to start, head to the whiteboard and have an old-fashioned brainstorming session. **Bring a representative from every department** (sales, marketing, HR, etc.). They will all have different ideas and voices on the value of your results. If they (or you) need a better understanding of why social media and monitoring matters, check out our recent presentation given at Magnet Interactive.

Also, if each department with a valuable interest has access to your data, it can keep your company on the same page while satisfying individual needs. Sales can identify potential clients, marketing can form strategy around results, communications can respond quickly to crises, and HR can police potential problems with employees. All these different departments can get these insights from the same data set.

Ask the group questions like...
• Do you respond to your audience online?
• If yes, are you satisfied with your response time?
• Do you want to improve customer service?
• What about sales?
• Do you want to increase your brand loyalty?
• Should you monitor specific campaigns or just your overall presence?
• ... Or both?

The list goes on...

Make it clear what you want, and your monitoring software should be flexible enough to give you what you need.

What are some of your goals? What are suggestions you give to clients when it comes to developing strategy?

Source: Published by Whitney on 18 May 2009 in Content of Interest, Industry Buzz

CHAPTER HIGHLIGHTS

1. Social media can take many different forms, including Internet discussion forums, blogs, vlogs, podcasts, online pictures, videos, and wikis.
2. Social media is unique from traditional media (newspapers, radio, television, print and broadcast journalism).
3. Social media takes no training. Anyone can participate in social media.
4. Social media is affordable. Anyone with Internet access can use social media.
5. Social media can be instantaneously accessed and altered. The community of users can change online text and affect the message being communicated.
6. Thousands of companies have Facebook sites.
7. Link to your sources so you can avoid plagiarizing content.
8. On their Facebook site, companies can host their corporate blog, provide access to their Twitter entries, post company YouTube videos, show photographs of corporate employees, list upcoming events and/or product releases, and provide access to discussion forums.
9. Organizations use Second Life for marketing; training; modeling, monitoring, and product testing; and providing customer service.
10. Companies can use social media to communicate during a crisis situation.

APPLY YOUR KNOWLEDGE

CASE STUDY

S&L Engineering has been working on the construction of a multipurpose site that would inhabit five blocks in downtown Los Angeles. The proposed site includes a boutique hotel, office space, condos, and retail shops. However, the project has met financial and legal problems that have negatively impacted the construction schedule. The project is two years behind schedule. Although the site is 60 percent complete, the construction company that S&L hired has filed for bankruptcy. S&L has hired a new construction company that promises to complete the project in 18 months.

S&L Engineering needs to bring the community up to date on the status of the project. This includes damage control, because S&L wants to be a good neighbor to other businesses and residents who work and live near the construction site.

To accomplish their communication goals, S&L plans the following:

a. Post a YouTube video of S&L's CEO speaking to the public about the project's challenges.
b. Write a tweet announcing the status of the project.
c. Post a blog entry from the construction manager of the building site, detailing plans for the project's completion.
d. Post a Facebook wall entry about the new construction company (its local reputation and its plans for hiring from the local labor force).

Assignment

What content should be included in each of the above communication channels? Write the scripts for a, b, c, and d.

INDIVIDUAL AND TEAM PROJECTS

Second Life

1. Your skateboard company's profits declined 12 percent worldwide. Management wants to enlarge the company's market share. Currently, the company has a Web site and brick and mortar facilities in numerous shopping malls around the world. To reach an untapped market, the company now wants to create a larger presence on the Internet. To do so, it plans on creating an SL, interactive, online shop.

   With this goal, accomplish the following:

   - Research the SL trend to learn how other companies are using this virtual world for shopping.
   - Write a report proposing your market plan.
   - Create a SWOT analysis (strengths, weaknesses, opportunities, and threats) of an SL shopping site.
   - Design and create the site, using Linden dollars as the SL currency.

   Give an oral presentation about your SL shopping site, including a demonstration in SL.

2. You work for an accounting company with a large, national staff of tax preparation specialists. With the tax season nearing, you are responsible for training new hires, some of whom are part time. Your company has hard-copy manuals providing instructions on tax preparation and face-to-face training. However, to accommodate a dispersed workforce, including part-time employees who might work from home, you need to modernize your training options. To make training more accessible to your dispersed staff, create a tutorial in SL that will allow staff to experience hands-on, virtual instruction. In this tutorial, focus on the following:

   - Understanding basic accounting terminology
   - Familiarizing staff with withholding guidelines
   - Reviewing existing tax laws
   - Implementing strategies for updating employees on new tax laws

   Give an oral presentation about your SL tutorial, including a demonstration in SL.

Facebook

1. Research the use of Facebook for posting online resumes by checking out these articles:

   - Doyle, Allison. "Treat Your Facebook Page Like a Resume." *About.Com.* http://jobsearch.about.com/b/2007/08/31/treat-your-facebook-page-like-a-resume.htm
   - Hernandez, Jenn. "From Facebook to Resume: Marketing Yourself Via Social Networks." *Interface.* http://bcis.pacificu.edu/journal/2009/05/article.php?id=73
   - Russell, C. M. "Secrets of the Job Hunt." http://secretsofthejobhunt.blogspot.com/2008/01/your-resume-on-facebook.html

   Report your findings to your class and/or professor in an oral presentation or e-mail report.

2. Create an online resume presence by designing a Facebook page. In this site, convey your career objectives, summary of qualifications, work experience, education, and professional skills.

3. Create a Facebook site for your college or technical writing/business communication/workplace communication class.

4. Review the Facebook sites of various businesses, governmental organizations, or industries. What do these sites have in common? What is unique to each site you visit? What is each company trying to accomplish in its Facebook site? How would you improve a site if you

were the social media manager of this organization or company? For a few samples, check out the following:

- The Missouri Department of Transportation. http://www.modot.mo.gov/facebook/index.htm
- Sun Microsystems. http://www.facebook.com/sunmicrosystems
- Sprint. http://www.facebook.com/sprint
- Harley-Davidson. http://www.facebook.com/harley-davidson
- Nike. http://www.facebook.com/pages/nike/37522773001
- Gap. http://www.facebook.com/gap
- H&R Block. http://www.facebook.com/hrblock

Twitter

After reading the following scenarios, respond to each company's communication need by writing a 140-character tweet.

1. Your bank, Hearth and Home Savings and Loan, has new, low rates for mortgages (4.5 percent for 30-year loans—this is the lowest rate for home loans in over 30 years) and high rates for certificates of deposit (5.5 percent for three year CDs—CDs have been higher in the past, but recent rates have been much lower). Market these benefits to your bank's customers by sending a tweet.

2. At a recent X-Games motorcycle jumps and tricks competition, hosted by the EnVi Hotel in Las Vegas, Nevada, a rider, Jason Whitworth, performed a first-ever trick, which he called the "double loop superman kick out." Jason rides your company's top-of-the line motorcycle, the Turbo ZZ210, and wears your company's gear (helmet, boots, gloves, knee pads, and jumpsuit). To inform your company's audience of this new trick and to market your brand, send a tweet.

3. Four employees of your internationally marketed fast-food restaurant, Hot Wings & Cool Things, created a video of themselves acting inappropriately in one of your restaurants. Posted on YouTube, they can be seen eating food from customers' trays and using cooking utensils in an unhealthy manner. Yes, you fired the employees and created new rules of appropriate behavior, incorporated into your training manuals, but still, you must manage this publicity crisis. Write a tweet to your international audience in which you confront bad publicity quickly and show how you manage poor employee behavior.

4. Your software company is rolling out a new product, a software program that will help potential college students apply for college loans, write their letters of application, and gather all application materials in a single portfolio file. To help customers use this new software, your company is planning a free-to-the-public informational webinar on March 11. The webinar will provide hands-on tutorials for using the new software and provide a "virtual" help-desk for questions and answers. Share this update with potential customers by writing a tweet.

5. Today at 11:00 a.m., your city of Greenwood Springs experienced a major explosion at a chemical plant downtown. This led to over 50 injuries, closed streets (Elm Street going north and south from 45th Street to 91st Street), a portion of a closed highway (from Highway 62 to the I48 southbound exit), and pollution problems due to heavy smoke. As city manager, you want to quickly inform your city residents of this crisis situation. Write a tweet to share the news, to warn residents, and to provide updates on traffic detours.

YouTube

1. Following are YouTube sites providing step-by-step instructional videos:

   - Installing Mac OS X onto a RunCore and Dell Mini
   - Battery Operated Holiday LED Lights
   - How to Replace RAM in Your Dell Inspiron Laptop

- Sun Microsystems t2000 Server
- How to Adjust Honda 50/70 cc Valves

Access any of the above YouTube instructional videos and analyze ways in which they succeed or could be improved. Base your comments on the guidelines provided in this chapter. Write an e-mail message to your professor with a summary of your findings.

2. Create a company, product, or service. Write a script for a YouTube video that introduces or markets the new company, product, or service.

PROBLEM-SOLVING THINK PIECES

1. Check out Stonyfield Farm's blog site at http://stonyfield.typepad.com/bovine/. This site says, "Welcome to the bovine bugle."

Assignment

Read entries from the blog site. Then answer these questions and share your responses with your class or professor either in an oral presentation or e-mail report:

- What are Stonyfield's goals in creating this blog?
- How does this blog relate to corporate business?

2. Sanya Guptel is attending a social media conference in San Francisco. She is presenting a speech about her company's use of YouTube for customer service training. In addition, she has attended workshops on ethics regarding the use of social media in the workplace, how to target audience demographics through social media, and the development of new social media channels of communication. To bring her colleagues up to date on her experiences at the conference, Sanya can write on her company's Facebook wall, send a tweet, post a blog entry, or make a YouTube video.

Assignment

Explain the benefits and drawbacks for each of Sanya's options. Then, write the text for Sanya, using the social media option that you consider best. Create any additional text you would like to complete this task.

WEB WORKSHOP

1. Bloggers provide up-to-date information on newsbreaking events and ideas.

- For e-commerce news, visit http://blog.clickz.com/ (ClickZ Network—Solution for Marketers).
- For business blogging, visit http://www.businessweek.com/the_thread/blogspotting/ (Business Week Online) to learn where the "worlds of business, media and blogs collide."
- For technology news (information technology, computer information systems, biomedical informatics, and more), visit http://blogs.zdnet.com/. Once in this site, use their search engine to find a technology topic that interests you.
- For information about criminal justice, visit http://lawprofessors.typepad.com/ crimprof_blog/criminal_justice_policy/ (Law Professor Blog Network).
- For information about science, visit http://www.scienceblog.com/cms/index.php (ScienceBlog), which has links to blogs about aerospace, medicine, anthropology, geoscience, and computers.
- The Car Blog (http://thecarblog.com/) has news items about automotives.
- Every news agency, such as ABC, NBC, CBS, CNBC, has a news blog.

To see what's new in your field of interest, check out a blog. Then, report key findings to your instructor in an e-mail message.

2. There are many published guidelines for accepted use of social media. Review the following links to learn what these guidelines entail. Then write an e-mail message to your professor reporting your findings.
   - "Social Media Guide's Top 10 Twitter Tips." http://mashable.com/2009/06/25/twitter-seo-tips/
   - "IBM Social Computing Guidelines: Blogs, wikis, social networks, virtual worlds and social media." http://www.ibm.com/blogs/zz/en/guidelines.html
   - "Sun Microsystems's Guidelines on Public Discourse." http://www.sun.com/communities/guidelines.jsp
   - "Intel Social Media Guidelines." www.intel.com/sites/sitewide/en_us/social-media.htm

3. Forrester Research has studied the impact of social media on business. Read their article entitled, "The Future of the Social Web: In Five Eras," written by Jeremiah Owyang (http://www.web-strategist.com/blog/2009/04/27/future-of-the-social-web/). After reading the article, write an e-mail message to your professor reporting on your findings.

4. Not everyone believes that social media is appropriate for the workplace. Read the following articles about potential problems using social media, Twitter, and Facebook in the workplace.
   - Chris Crum's "Is Social Media Good or Bad For Business?" http://www.webpronews.com/topnews/2008/10/27/is-social-media-good-or-bad-for-business
   - Edmund Lee's "Why Facebook Can't Succeed." http://www.thebigmoney.com/articles/0s-1s-and-s/2009/04/01/why-facebook-cant-succeed

   Based on your readings, write an e-mail message to your instructor or give an oral presentation about the pros and cons of social media in the workplace.

QUIZ QUESTIONS

1. How does social media differ from traditional media?
2. What are three ethical considerations you must have when you use social media?
3. What is a blog?
4. What are three guidelines to follow when blogging on the job?
5. What are four reasons why companies blog?
6. How do you ensure good production values of the YouTube videos you post?
7. What can a company achieve on their Facebook site?
8. List three guidelines for participating on Facebook.
9. What are five reasons why a company has a Second Life presence?
10. What are five reasons why companies or organizations use Twitter?

Communicating Effectively to Get a Job

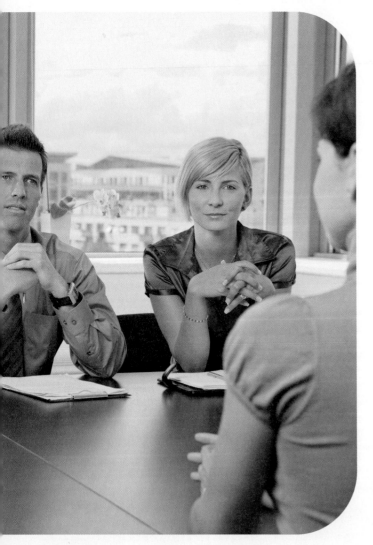

COMMUNICATION AT WORK

In this scenario, a business owner interviews potential job applicants.

The job search involves at least two people—the applicant and the individual making the hiring decision. Usually more than two people are involved, however, because companies typically hire based on a committee's decision. That is the case at DiskServe. This St. Louis–based company is hoping to hire a customer service representative for its computer technology department. DiskServe is eager to hire a new employee because one of its best workers has just advanced to a new position in the company. DiskServe asked applicants to apply using e-mail. Thus, the applicants submitted an application by way of a cover e-mail message and an attached scannable resume.

DiskServe advertised this opening in the career placement centers at local colleges, in the city newspaper, and online at its Web site: http://www.DiskServe.com.

In addition to DiskServe's chief executive officer (CEO), Sarah Beske, the hiring committee will consist of two managers from other DiskServe departments, the former employee whose job is being filled, and two co-workers in the computer technology department.

Ten candidates were considered for the position. All candidates first had teleconference interviews. While Sarah talked with the candidates, the other hiring committee members

Objectives

When you complete this chapter, you will be able to

1. Search for job openings applicable to your interests, education, and experience

2. Compose effective letters of application that gain attention and are persuasive

3. Choose either to write a functional or reverse chronological resume

4. Write effective resumes consisting of your objectives, summary of qualifications, work experience, education, and professional skills

5. Decide on the correct method of delivery for your resume, either through the mail, as an e-mail attachment, or scannable

6. Understand effective interview techniques that demonstrate your professionalism

7. Write appropriate follow-up correspondence to restate how you can benefit the company

8. Write an acceptance letter

9. Use the job search checklists to evaluate your resume, letter of application, interview, follow-up letter, and job acceptance letter

listened on a speaker phone. After the telephone interviews, five candidates were invited to DiskServe's worksite for personal interviews—Macy Heart, Aaron Brown, Rosemary Lopez, Quisha Southerland, and Robin Scott.

Sarah Beske, who has worked hard to create a family-oriented environment at DiskServe, values three traits in her employees: technology know-how, an ability to work in teams, and a positive attitude toward customers and co-workers. When the candidates arrived at DiskServe, Sarah gave them a tour of the facilities, introducing them to many employees. Then the interviews began.

Each job candidate was asked a series of questions that included the following:

- What is your greatest strength? Give an example of how this reveals itself on the job.

- What did you like most about your previous jobs?

- How have you handled customer complaints in the past?

- Where do you see yourself in five years?

Then, each candidate was taken to the computer repair lab and confronted with an actual hardware or software problem. The candidates were asked to solve the problem, and their work was timed. Finally, the applicants were allowed to ask questions about DiskServe and their job responsibilities.

Sarah is a stickler for good manners and business protocol. She waited 48 hours after the final interview to make her hiring decision. The wait time allowed her to check references. More important, she wanted to see which of the candidates wrote follow-up thank-you notes, and she planned to assess the quality of their communication.

Sarah takes the hiring process seriously. She wants to hire the best people because she hopes those employees will stay with the company a long time. Hiring well is a good corporate investment.

Check out our quarterly newsletters at www.pearsonhighered.com/gerson for dot.com updates, new case studies, insights from business professionals, grammar exercises, and facts about technical communication.

The Three *R*s of Searching for a Job

When beginning your job search, consider the three *R*s: research yourself, research the company, and research the position. By performing this research, you will improve the focus of your job search. Learning more about your goals, the company's culture, and the specifics of the position will help you determine your goals, decide if the company and job meet your needs, and prepare you for an interview.

Research Yourself

You are not right for every job, nor is every job right for you. To ensure that you are applying for the correct position, ask yourself the following questions:

- What are your skills, attributes, and accomplishments?
- What do you have to offer this company?
- What can you bring to the company that is unique?
- What are your selling points?
- How do your qualifications meet the position's requirements?
- What interests you about this company and the position?

Research the Company

By researching the company, you educate yourself about the company's culture, values, products, and services. To research the company, view the employer's Web site, read annual reports, or speak with an employee. Consider these questions:

- How does the employer's future look?
- What is the employer's product or service?
- How do your values agree with the company's values?
- What is the company's vision or mission statement?
- What is the company's culture or philosophy?
- What are the needs or problems of the employer?
- What role does the employer play in the community?
- Is the employer expanding?

Research the Position

When you consider the job opening, review its requirements and duties. Doing so will help you to determine if this job is right for you. Ask yourself these questions:

- What are the job responsibilities?
- What are the pay and benefits for this job?
- What skills and talents are needed for the position?
- What freedoms does the position allow?
- What is the structure of the department?
- Who does the position report to?

("Resume Guidelines")

How to Find Job Openings

When it's time for you to look for a job, how will you begin your search? You know you can't just wander up and down the street, knocking on doors randomly. That would be time consuming, exhausting, and counterproductive. Instead, you must approach the job search more systematically.

Visit Your College or University Job Placement Center

A college's job placement center is an excellent place to begin a job search, for several reasons.

- Your school's job placement service has counselors who will counsel you regarding your skills and job options.
- Your job placement center can give you helpful hints on preparing resumes, letters of application, and follow-up letters.
- The center will post job openings within your community and possibly in other cities.
- The center will be able to tell you when companies will visit campus for job recruiting.
- The service might keep on file your letters of recommendation or portfolio. The job placement center can send these out to interested companies upon your request.

Attend a Job Fair

Many colleges and universities host job fairs. A job fair will allow you to research job openings, make contacts for internships, or submit resumes for job openings. If you attend a job fair, treat it like an interview. Dress professionally and take copies of your resume and letters of recommendation.

Talk to Your Instructors

Whether in your major field or not, instructors can be excellent job sources. They will have contacts in business, industry, and education. They may know of job openings or people who might be helpful in your job search. Furthermore, because your instructors know a great deal about you (having spent a semester or more working closely with you), they will know which types of jobs or work environments might best suit you.

Network with Friends and Past Employers

A *Smart Money* magazine article reported that 62 percent of job searchers find employment through "face-to-face networking" (Bloch 12). A study performed by Drake Beam Morin, an internationally renowned career services organization, confirms this, stating that "64 percent of . . . almost 7,500 people surveyed said they found their jobs through socializing and meeting people" (Drakeley 5). Tell friends and acquaintances that you are looking for a job. They might know someone for you to call. Why is networking so important? It is simple math. Your friends and past employers know people. Those people know people. By visiting acquaintances, you can network with numerous individuals. The more people you talk to about your job quest, the more job opportunities you will discover.

Use Social Media for Your Job Search

In the past, networking was primarily achieved face to face. Now, you can network virtually through social media. Twitter, for example, allows you "to connect directly with recruiters and employees at companies you want to work for" (Schawbel).

Use blogs to search for job openings. Technorati provides you access to "over 130 million blogs" (Schawbel) which you can search to find information about job openings. In addition, search engines like Pipl, PeekYou, and Wink can help you find people to contact at companies you're interested in. Finally, you can join LinkedIn, a social media site for professionals. LinkedIn not only lets you post your own profile, but also through this social media site, you can find information about other professionals in your work environment. LinkedIn provides both the job seeker and prospective employer the following benefits:

- LinkedIn Jobs lets members of LinkedIn search for or post jobs by key word, country, and postal code.
- Human resource managers can access a network of over 65 million professionals to find candidates with specific skills and required experience.
- Job seekers can use their connections at potential employers to help them land a job within the company.
- Job seekers can review a company's profile to learn more about its products, services, projects, clients, employees, ratings, community involvement, and corporate contacts.

Social Media

See Chapter 7 for more discussion of social media.

Get Involved in Your Community

There are many ways to network. In addition to using social media and talking to your family or previous employers, you can network by getting involved in your community. Consider volunteering for a community committee, pursuing religious affiliations, joining community clubs, or participating in fund-raising events. Join Toastmasters. Take classes in accounting, HTML, RoboHelp, or Flash at your local community college. Each instance not only teaches you a new skill but also provides an opportunity to meet new people. Then, "when a job comes open, you'll be on their radar screen" (Drakeley 6).

Check Your Professional Affiliations and Publications

If you belong to a professional organization, this could be a source of employment in three ways. First, your organization's sponsors or board members might be aware of job openings. Second, your organization might publish a listing of job openings. Finally, you can gain professional experience by serving on boards or being an officer in an organization. These experiences can be included in your resume under professional skills, work experience, and affiliations.

Read the Want Ads

Check the classified sections of your local newspapers or in newspapers in cities where you might like to live and work. These want ads list job openings, requirements, and salary ranges. In addition, by reading the classifieds, you can learn keywords in your industry and incorporate them in your resume.

Read Newspapers or Business Journals to Find Growing Businesses and Business Sectors

Which companies are receiving grants, building new sites, winning awards, or creating new service or product lines? Which companies have just gained new clients or received expanded contracts? Newspapers and journals report this kind of news, and a growing company or business sector might be good news for you. If a company is expanding, this means more job opportunities for you to pursue.

Take a "Temp" Job

Temporary ("temp") jobs, accessed through staffing agencies, pay you while you look for a job, help you acquire new skills, allow you to network, and can lead to full-time employment.

Get an Internship

Internships provide you outstanding job preparedness skills, help you meet new people for networking, and improve your resume. An unpaid internship in your preferred work area might lead to full-time employment. An internship "gives you the opportunity to show your skills, work ethic, positive attitude, and passion for your work." By interning, you can prove that you should be "the next employee the company hires" (Drakeley 7). In fact, the National Association of Colleges and Employers lists internships as one of the top ten skills employers want ("Top Ten Qualities").

Job Shadow

Job shadowing allows you to visit a worksite and follow employees through work activities for a few hours or days. This allows you to learn about job responsibilities in a certain field or work environment. In addition, job shadowing also helps you find out if a company or industry is hiring, allows you to make new contacts, and places you in a favorable position for future employment at that company.

Set Up an Informational Interview

In an informational interview, you talk with people currently working in a field, asking them questions about career opportunities and contacts. Informational interviews allow you to

- Explore careers and clarify your career goal
- Expand your professional network
- Build confidence for your job interviews
- Access the most up-to-date career information
- Identify your professional strengths and weaknesses

Research the Internet

You should make the Internet part of your job search strategy. The Internet provides numerous benefits. Internet job search engines provide excellent job search resources (see Table 8.1). These sites and many more provide the following career information resources:

- **Resumes.** Explaining the difference between resumes and curriculum vitae (CV), addressing gaps in your career history, avoiding typical resume mistakes, and writing winning resumes
- **Interviews.** Interviewing to get the job and handling illegal questions
- **Cover letters and thank-you letters.** Providing sample cover letter techniques and ways to write a better thank-you letter
- **Job search tips.** Employing the correct netiquette and job search techniques
- **Job locations.** Listing jobs both in the United States and worldwide

Criteria for Effective Resumes

Once you have found a job that interests you, it is time to apply. Your job application will start when you send the prospective employer your resume. Resumes are usually the first impression you make on a prospective employer. If your resume is effective, you have opened the door to possible employment—you have given yourself the opportunity to sell your skills during an interview. If, in contrast, you write an ineffective resume, you have closed the door to opportunity.

Your resume should present an objective, easily accessible, detailed biographical sketch. However, do not try to include your entire history. Because the primary goal of your resume, together with your letter of application, is to get an interview, you can use your interview to explain in more detail any pertinent information that does not appear on your resume. When writing a resume, you have two optional approaches. You can write either a reverse chronological resume or a functional resume.

TABLE 8.1 Online Job Search Links

Search Engines for Jobs in the United States

- http://online.onetcenter.org—"the nation's primary source of occupational information, providing comprehensive information on key attributes and characteristics of workers and occupations"
- http://www.monster.com—lets you post resumes and search for jobs, and provides career advice
- http://www.careerbuilder.com—lets you search for jobs by company, industry, and job type
- http://www.collegerecruiter.com—lists the latest job postings, "coolest career resources, and most helpful employment information"
- http://www.careerjournal.com—the *Wall Street Journal's* career search site; provides salary and hiring information, a resume database, and job hunting advice
- http://www.flipdog.com—provides national job listings in diverse fields by job title, location, and date of listing
- http://hotjobs.yahoo.com/—lets you search for jobs by keyword, city, and state
- http://www.wantedtech.com/—lists jobs in the United States
- http://www.wetfeet.com—gives company and industry profiles, resume help, city profiles, and international job sites
- http://www.careerlab.com—provides a cover letter library
- http://www.truecareers.com/—lets you search jobs, post resumes, research industry news, and read online articles about the job search
- http://www.careermag.com—allows for online job searches by keyword, industry, job title, and zip code
- http://www.dice.com/—focuses on technology careers
- http://www.jobweb.com/—helps new college graduates with career development and the job search
- http://www.job-hunt.org/—called by *PC Magazine* and *Forbes* the Internet's Best Web site for job hunting and resources
- http://usajobs.opm.gov/—a site dedicated to government job searches and advice
- http://www.vault.com—links to jobs, resume samples, salary information, and tips for posting video resumes

Search Engines for International Jobs

- http://www.monster.com/geo/siteselection.aspx—allows you to select the country and search for jobs worldwide
- http://www.rileyguide.com/internat.html—lets you look for jobs in Canada, Western Europe, the British Isles, Central and Eastern Europe, Scandinavia, Africa, the Middle East, Asia, and the Pacific Rim
- http://www.careerbuilder.com/JobSeeker/Jobs/jobfindil.aspx—allows you to select the country and search for jobs
- http://www.4icj.com/—lets you search for jobs by country and category
- http://www.job-hunt.org/international.shtml—provides a listing of Web sites with jobs located outside of the United States

Reverse Chronological Resume

Write a reverse chronological resume if you

- Are a traditional job applicant (a recent high school or college graduate, aged 18 to 25)
- Hope to enter the profession in which you have received college training or certification
- Have made steady progress in one profession (promotions or salary increases)
- Plan to stay in your present profession

Functional Resume

Write a functional resume if you

- Are a nontraditional job applicant (are returning to the workforce after a lengthy absence, are older, are not a recent high school or college graduate)
- Plan to enter a profession in which you have not received formal college training or certification
- Have changed jobs frequently
- Plan to enter a new profession

Ethical Considerations when Writing a Resume

If your resume contains inaccurate information, your potential employer can find out. Prospective employers can perform follow-up reference checks, contacting your previous bosses to verify your resume's assertions. In addition, it is very easy to perform Internet background checks related to a job candidate's credit issues, prior criminal offenses, driving records, military history, or even presence on terrorism watch lists. Many employers use social networking sites, such as Facebook and MySpace, to learn details about a job candidate.

Due to the ease with which an employer can verify your resume's assertions, you must honestly present your qualifications. Make sure that your resume is ethical and honest. Following are examples of resume fraud that you must avoid:

- Including a degree that you have not earned
- Listing colleges or universities that you have not attended
- Inflating your job title
- Inflating your job responsibilities
- Inflating your job achievements
- Listing jobs that you have not held
- Claiming technical knowledge that you do not have
- Omitting large amounts of time from your school and job histories
- Including military rank you have not earned

Failing to follow ethical standards and being completely honest on your resume could result in your not being offered a job.

Key Resume Components

Whether you write a reverse chronological or a functional resume, include the following key components.

Identification. Begin your resume with the following:

- Your name (full first name, middle initial, and last name). Your name can be in boldface and printed in a larger type size (14 point, 16 point, etc.).
- Contact information. Include your street address, your city, state (use the correct two-letter abbreviation), and zip code. If you are attending college or serving in the armed forces, you might also want to include a permanent address. By including alternative addresses, you will help your prospective employer contact you more easily.
- Your area code and phone numbers. Limit yourself to two phone numbers, and don't provide a work phone. Having prospective employers call you at your present job is not wise. First, your current employer will not appreciate your receiving this sort of personal call. Second, your future employer might believe that you often receive personal calls at work—and will continue to do so if he or she hires you.
- Your e-mail, Web site address, or fax number. Be sure that your e-mail address is professional sounding. An e-mail address such as "ILuvDaBears," "Hotrodder," or "HeavyMetalDude" is not likely to inspire a company to interview you. Also avoid using your e-mail address at your current place of employment.

> **State Abbreviations**
>
> See Appendix A for more discussion of state abbreviations.

Career Objectives. The career objectives line is like a subject line in a memo, e-mail, or report. Your career objective informs the reader of your resume's focus. Be sure your career objective is precise. Too often, career objectives are so generic that their vagueness does more harm than good.

> This poorly written career objective provides no focus. What kind of business? What kind of opportunities for professional growth? Employers don't want to hire people who have only vague notions about their skills and objectives.

BEFORE

Flawed Career Objective

Career Objective: Seeking employment in a business environment offering an opportunity for professional growth.

Improved Career Objective

Career Objectives: To market financial planning programs and provide financial counseling to ensure positive client relations.

This improved career objective not only specifies which job the applicant is seeking but also how he or she will benefit the company.

Summary of Qualifications. After the career objectives, provide your audience a summary of qualifications. According to Monster.com, "resumes normally get less than a 15-second glance at the first screening" (Isaacs). A summary of qualifications allows the employer an immediate opportunity to see how you can add value to the company.

A summary of qualifications should include the following:

- An overview of your skills, abilities, accomplishments, and attributes
- Your strengths in relation to the position for which you are applying
- How you will meet the employer's goals

To write an effective summary of qualifications, list your top three to seven most marketable credentials.

EXAMPLE ▶

Summary of Qualifications
- Over four years combined experience in marketing and business
- Developed winning bid package for promotional brochures
- Promoted to Manager in less than two years
- Maintained database of customers, special ordered merchandise, and tracked inventory
- Managed over ten employees

FAQs: Resume Guidelines

Q: Do I really need to limit my resume to one page?

A: Do not worry about limiting yourself to the traditional one-page resume. Conciseness is important in all technical communication, but if your education, work experience, and professional skills merit more than a page, you must show your accomplishments. In addition, if you submit your resume as part of an e-mail message, readers will scroll. However, don't pad the resume. Limit yourself to jobs within the last ten years and skills that fit the job you're seeking.

Q: In my resume, which should I list first, my work experience or my education?

A: It's all about location, location, location. You should present your most important section first. If education is your strength and will help you get the job, lead with education. If, in contrast, your work experience is stronger, begin your resume with work.

Q: Can I omit jobs that I didn't like?

A: Yes and no. You cannot have any large gaps in your resume, such as a missing year or more. If you have any large gaps, you must either explain the gap or fill it with other activities (education, volunteerism, or childrearing, for instance). However, a missing month or so is not a problem. If you worked at a job for a month, left that job, and then found other employment, you do not need to list the short-term job.

Employment.   The employment section lists the jobs you've held. This information must be presented in reverse chronological order (your current job listed first, previous jobs listed next). This section must include the following:

- Your job title (if you have or had one)
- The name of the company you worked for
- The location of this company (city and state)
- The time period during which you worked at this job
- Your job duties, responsibilities, and accomplishments

This last consideration is important. This is your chance to sell yourself. Merely stating where you worked and when you worked there will not get you a job. Instead, what did you achieve on the job? In this part of the resume, you should detail how you met deadlines, trained employees, cut expenses, exceeded sales expectations, and so forth. Plus, you want to quantify your accomplishments.

Assistant Manager

McConnel Oil Change, Beauxdroit, LA
    2010 to present

- Track and maintain over $25,000 inventory
- Train a minimum of four new employees quarterly
- Achieved a 10 percent growth in service performed for three consecutive years
- Developed a user manual for hazardous waste disposal, earning a "Citizen's Recognition Award" from the Beauxdroit City Council

---

◀ **EXAMPLE**

Listing your job title, company name, location, and dates of employment merely shows where you were in a given period of time. To prove your contributions to the company, provide specific details highlighting achievements.

Education.   In addition to work experience, you must include your education. Document your educational experiences in reverse chronological order (most recent education first; previous schools, colleges, universities, military courses, and training seminars next). When listing your education, provide the following information:

- **Degree.** Tell your audience whether you've earned a BA, BS, MA, etc., and name the field (Business, Engineering, Computer Sciences, etc.). If you have not yet received your degree, you can write "Anticipated date of graduation June 2012" or "Degree expected in 2012."
- **School attended.** Do not abbreviate. Although you might assume that everyone knows what *UT* means, your readers won't understand this abbreviation. Is UT the University of Texas, the University of Tennessee, the University of Tulsa, or the University of Toledo?
- **Location.** Include the city and state.
- **Year of graduation** or years attended.

As you can see, this information is just the facts and nothing else. Many people might have the same educational history as you. For instance, just imagine how many of your current classmates will graduate from your school, in the same year, with the same degree. Why are you more hirable than they are? The only way you can differentiate yourself from other job candidates with similar degrees is by highlighting your unique educational accomplishments. These might include any or all of the following:

Grade point average	Academic honors, scholarships, and awards
Academic club memberships and leadership offices held	Fraternity or sorority leadership offices held
Unique coursework	Number of hours you worked while attending school
Special class projects	Software and hardware certifications, or technical equipment you can operate

Please note a key concern regarding your work experience and education. You should have no chronological gaps when all of your work and education are listed. You can't omit a year without a very good explanation. (A missing month or so is not a problem.)

Professional Skills. If you are changing professions or reentering the workforce after a long absence, you will write a functional resume. Therefore, rather than beginning with education or work experience, which won't necessarily help you get a job, focus your reader's attention on your unique skills. These could include the following:

Proficiency with computer hardware and software	New techniques you have invented or implemented
Procedures you can perform	Numbers of and types of people you have managed
Special accomplishments and awards you have earned	Publications you have created or worked on
	Certifications you have earned
On-the-job training you have received	Languages you speak, read, and write
Training you have provided	

These professional skills are important because they help show how you are different from all other applicants. In addition, they show that although you have not been trained in the job for which you are applying, you can still be a valuable employee.

EXAMPLE ▶

Highlight professional skills that will set you apart from other potential employees.

Professional Skills
- Proficient in Microsoft Word, Excel, Publisher, and PowerPoint
- Knowledge of HTML, Java, Visual Basic, and C++
- Certified OSHA Hazardous Management Safety Trainer
- Fluent in Spanish and English
- Completed Second Shift Administration Certificate

Military Experience. If you served for several years in the military, you might want to describe this service in a separate section. You would state the following:

Rank	Discharge status
Service branch	Special clearances
Location (city, state, country, ship, etc.)	Achievements and professional skills
Years in service	Training seminars attended and education received

Professional Affiliations. If you belong to regional, national, or international clubs or have professional affiliations, you might want to mention these. Such memberships might include the Rotarians, Lions Club, Big Brothers and Big Sisters, or Junior League. Maybe you belong to the Society for Technical Communication, the Institute of Electrical and Electronic Engineers, the National Office Machine Dealers Association, or the American Helicopter Society. Listing such associations emphasizes your social consciousness and your professional sincerity.

Optional Resume Components

Portfolios.   As an optional component for your job search, consider using a portfolio. If you are in fashion merchandising, heating/ventilation/air conditioning (HVAC), engineering, drafting, architecture, technical writing, or graphic design, for example, you might want to provide samples of your work. These samples could include schematics, drawings, photographs, certifications, or samples of documents you have written.

However, avoid sending an unsolicited portfolio electronically. Many companies are reluctant to open unknown attachments. Make hard copies of your work or a disk. When you have an interview with a prospective employer, you can give the employer the hard copy or disk.

References.   Avoid a reference line that reads "Supplied on request," "Available on request," or "Furnished on request." Every employer knows that you will provide references if asked. Instead of wasting valuable space on your resume with unnecessary text, use this space to develop your summary of qualifications, education, work experience, or professional skills more thoroughly. Create a second page for references, and bring this reference page to your interview. On the reference page, list three or four colleagues, supervisors, teachers, or community individuals who will recommend you for employment. (Note: Obtain their permission before listing them as references.) Provide their names, titles, addresses, and phone numbers. By bringing the reference page to your interview, you will show your prospective employer that you are proactive and organized.

Personal Data.   Do not include any of the following information: birth date, race, gender, religion, height, weight, religious affiliations, marital status, or pictures of yourself. Equal opportunity laws disallow employers from making decisions based on these factors.

Effective Resume Style

The preceding information suggests *what* you should include in your resume. Your next consideration is *how* this information should be presented. As mentioned throughout this textbook, format or page layout is essential for effective technical communication. The same holds true for your resume.

Choose Appropriate Font Types and Sizes.   As with most technical communication, the best font types are Times New Roman, Calibri, and Arial. These are readable and professional looking. Avoid designer fonts, such as Comic Sans, and cursive fonts, such as Shelley Vollante. In addition, use a 10- to 12-point font for your text. Smaller font sizes are hard to read; larger font sizes look unprofessional. Headings can be boldface and 14- to 16-point font size. Limit your resume to no more than two font types.

Avoid Sentences.   Sentences create three problems in a resume. First, if you use sentences, the majority of them will begin with the first-person pronoun *I*. You'll write, "I have . . .," "I graduated . . .," or "I worked. . . ." Such sentences are repetitious and egocentric. Second, if you choose to use sentences, you'll run the risk of committing grammatical errors: run-ons, dangling modifiers, agreement errors, and so forth. Third, sentences will take up room in your resume, making it longer than necessary.

Format Your Resume for Reader-Friendly Ease of Access.   Instead of sentences, highlight your resume with easily accessible lists. Set apart your achievements by bulletizing your accomplishments, awards, unique skills, and so on. In addition to bullets, make your resume accessible by capitalizing and boldfacing headings, indenting subheadings to create white space, and italicizing subheadings and major achievements. Avoid underlining headings. Studies show that most people find underlined text hard to read (Vogt).

Begin Your Lists with Verbs.   To convey a positive, assertive tone, use verbs when describing your achievements. Use present tense verbs for current jobs and education; use past tense verbs for past jobs and education. Table 8.2 provides a list of verbs you might use.

TABLE 8.2 Active Verbs to Highlight Achievements

Accomplished	Designed	Initiated	Planned
Achieved	Developed	Installed	Prepared
Analyzed	Diagnosed	Led	Presented
Awarded	Directed	Made	Programmed
Built	Earned	Maintained	Reduced
Completed	Established	Managed	Resolved
Conducted	Expanded	Manufactured	Reviewed
Coordinated	Gained	Negotiated	Sold
Created	Implemented	Ordered	Supervised
Customized	Improved	Organized	Trained

Quantify Your Achievements. Your resume should not tell your readers how great you are; it should prove your worth. To do so, quantify by precisely explaining your achievements.

BEFORE	AFTER
Maintained positive customer relations with numerous clients.	Maintained positive customer relations with 5,000 retail and 90 wholesale clients.
Improved field representative efficiency through effective training.	Improved field representative efficiency by writing corporate manuals for policies and procedures.
Achieved production goals.	Achieved 95% production, surpassing the company's desired goal of 90%.
Trained employees.	Trained 20 employees annually in methods for safely removing hazardous waste from the workplace.

Make It Perfect. You cannot afford to make an error in your resume. Remember, your resume is the first impression you'll make on your prospective employer. Errors in your resume will make a poor first impression.

Methods of Delivery

When writing either a reverse chronological or a functional resume, you can deliver your document in several ways.

Mail Version

The traditional way to deliver a resume is to insert it into an envelope and mail it. This resume can be highly designed, using bullets, boldface, horizontal rules, indentations, and different font sizes. Because this document will be a hard copy, what the reader sees will be exactly what you mail.

Do not be tempted to over design your resume, however. For example, avoid decorative fonts, clip art, borders, or photos. Do not print your resume on unusual colors, like salmon, baby blue, tangerine, or yellow. Instead, stick to heavy white paper and standard fonts, like Times New Roman, Calibri, and Arial.

The Technology Tips resume templates provide benefits but can create problems. On the positive side, the templates are great reminders of what to include in your resume, such as objectives, work experience, education, and skills. In contrast, the templates also limit you and perhaps suggest that you include information that isn't needed. First, the templates mandate font sizes and page layout. Second, a few of the templates suggest

Using Resume Templates in Microsoft Word 2010

Microsoft Word 2010 provides you a resume template if you want help getting started.

1. Click on the **File** tab located on the top left of your toolbar and scroll to **New**.

The following **Available Templates** window will pop up.

2. Click on the **New resume samples** or **Resumes and CVs** tab to find 44 optional resume layouts, including "Nursing assistant resume," "High school student resume," "Database Administer resume," "Customer Service resume," and more.

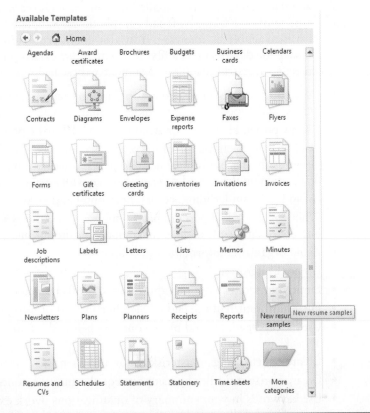

that you include information about "Interests" and "References." Rarely should you include "Interests," such as hobbies. Furthermore, most experts suggest that you omit the "References" line, saving valuable resume space for more important information. You can include references on a separate page, especially for interviews. More important, if you use the same templates that everyone else does, then how will your resume stand out as unique? A good compromise is to review the templates for ideas and then create your own resume with your unique layout.

Figures 8.1 and 8.2 are excellent examples of traditional resumes, ready to be mailed.

E-mail Resume

Delivering your resume by mail can take several days. The quickest way to get your resume to the prospective employer is as an e-mail attachment. Speed isn't the only issue. "Hiring managers and recruiters have become as addicted to e-mail as everyone else. More than one-third of human-resource professionals reported a preference for e-mailed resumes, according to . . . the Society for Human Resource Management" (Dixson).

If you choose to submit the resume as an attachment, be sure to write a brief e-mail cover message (we discuss this later in the chapter). In addition, you must clarify what software you have used for this attachment—Microsoft Word, Works, WordPerfect, or Rich Text Format (rtf), for example.

Scannable Resume

Many companies use computers to screen resumes with a technique called electronic applicant tracking. The company's computer program scans resumes as raster (or bitmap) images. Next, the software uses artificial intelligence to read the text, scanning for keywords. If your resume contains a sufficient number of these keywords, the resume will then be given to someone in the human resources department for follow-up.

A scannable resume can be e-mailed or sent through the mail. To create a scannable resume, type your text using Notepad for Windows, Simpletext for Macintosh, or Note Tab, which is available as freeware. You also could type your resume using Microsoft Word and save the document as a text file, with a *.txt* extension (Dikel 3).

To create a successful scannable resume, try these techniques:

- Use high-quality paper (send an original), optimum contrast (black print on white paper), and paper without wrinkles or folds (if you are mailing the resume).
- Use a Courier, Helvetica, or Arial typeface (10- to14-point type).
- Place your name at the top of the page. "Scanners assume that whatever is at the top is your name. If your resume has two pages, place your name and a 'page two' designation on the second page, and attach with a paper clip—no staples" (Kendall).
- Avoid italics, underlining, colors, horizontal and vertical bars, and iconic bullets.
- White space is still important, but do not use your Tab key for spacing. Tabs will be interpreted differently in different computer environments. Use your space bar instead.
- Avoid organizing information in columns.
- Do not center text.
- Limit your screen length to approximately 60 characters per line.
- Hit a hard return at the end of each line instead of using word wrap.
- Use all-cap headings and place your text below the headings, spacing for visual appeal.
- Create bullets using an asterisk (*) or a hyphen (-).
- Create horizontal rules by using the equal sign (=) or underline (_).
- Use keywords in your summary of qualifications, work experience, and professional skills.

FIGURE 8.1 Chronological Resume

Sharon J. Barenblatt

1901 Rosebud Avenue
Boston, MA 12987
Cell phone: 202-555-2121
E-mail: sharonbb@juno.com

Objective

Employment as an Account Manager in public relations, using my education, work experience, and interpersonal communication skills to generate business.

Summary of Qualifications

- Over five years customer service experience
- Experience in public relations, writing public service announcements
- Proven record of written and interpersonal communication abilities
- Outstanding leadership skills, shown by my management background
- Fluent in Spanish

Education

BS, Business. Boston College. Boston, MA 2010

- 3.2 GPA
- Social Justice Chair, Sigma Delta Tau, 2010
- Study Abroad Program, Madrid, Spain 2009
- Internship, Ace Public Relations, Boston, MA 2008

Frederick Douglas High School. Newcastle, MA 2007

- 3.5 GPA
- Member, Honor Society
- Captain, Frederick Douglas High School tennis team

List your education and work experience in reverse chronological order.

Work Experience

Salesperson/assistant department manager. Jessica McClintock Clothing Store. Boston, MA 2009 to present.

- Prepare nightly deposits, input daily receipts
- Open and close the store
- Provide customer service
- Trained six new employees

Salesperson. GAP Clothing. Newcastle, MA 2007 to 2008

- Assisted customers
- Stocked shelves

List current jobs using present tense verbs and previous jobs using past tense verbs.

Do not only list where you worked and when you worked there. Also include your job responsibilities.

Professional Skills

- Made oral presentations to the Pan-Hellenic Council to advertise sorority philanthropic activities
- Helped plan community-wide "Paul Revere's Ride Day"
- Created advertising brochures and fliers

FIGURE 8.2 Functional Resume

JODY R. SEACREST
1944 W. 112th Street
Salem, OR 64925
(513) 451-4978   jseacrest12@hotmail.com

Objectives
Management position using my expertise in merchandising, administration, interpersonal communication, and computer skills to help a company meet its revenue goals

Professional Skills
- Operated sporting goods/sportswear mail-order house. Business began as home-based but experienced 125% growth and was purchased by a national retail sporting goods chain.
- Managed a retail design studio producing over $500,000 annually.
- Hired, trained, and supervised an administrative staff of 15 employees for a financial planning institution.
- Sold copiers through on-site demonstrations. Exceeded corporate sales goals by 10% annually.
- Provided purchaser training for office equipment, reducing labor costs by 25%.
- Acquired modern management skills through continuing education courses.

Work Experience
Office manager, Simcoe Designs, Salem, OR 2010 to present.
Sales representative, Hi-Tech Office Systems, Salem, OR 2006 to 2010.
Office manager, Lueck Finances, Portland, OR 2004 to 2006.
President, Good Sports, Inc., Portland, OR 2002 to 2004.

Computer Proficiency
Microsoft Office XP, Visual Basic 6, C++, Oracle, Microsoft SQL Server, Network Administration

Military Experience
Corporal, U.S. Army, Fort Lewis, WA 1998–2002. Honorably discharged.
- Served as Company network administrator.
- Planned and budgeted all IT purchases.

Education
BA, Communication Studies, Portland State University, Portland, OR 1998.

In a functional resume, emphasize skills you have acquired which relate to the advertised position. Also quantify your accomplishments.

In a functional resume, you list education and work experience in reverse chronological order.

A functional resume is organized by importance. Begin with the skills or accomplishments that will get you the job. Place less important information lower in the resume.

Keywords are the most important feature of scannable resumes. OCR (optical character recognition) searches focus on keywords and phrases specifically related to the job opening. The keywords include job titles, skills and responsibilities, corporate buzzwords, acronyms and abbreviations related to hardware and software, academic degrees, and certifications.

You can find which keywords to focus on by carefully reading the following:

- Job advertisements
- Your prospective employer's Web site

- Government job descriptions
- Industry-specific Web sites
- The *Occupational Outlook Handbook* (found online at http://www.bls.gov/oco/)
- Career-related discussion groups or blogs
- Sample resumes found online

When using keywords, be specific; avoid vague words and phrases.

BEFORE	AFTER
Knowledge of various software products	Can create online help using Author-iT and have expertise with PageMaker and Quark

BEFORE	AFTER
Familiar with computer technology	Proficient in multimedia, HTML, and Windows and Macintosh platforms

Figure 8.3 shows an excellent example of a scannable resume.

YouTube Video Resume

Another option for delivering your resume is social media. In addition to creating a written resume, consider a "video resume." YouTube video resumes are growing in popularity. A search for "resume" on YouTube gives you over 163,000 results, as of August 2010. A YouTube video resume can benefit you in several ways:

- YouTube entries place high on search engine results.
- Video resumes differentiate you from other job applicants.
- Short video resumes show you in action. If you have a pleasant and dynamic personality, you can reveal this visually to a prospective employer.
- Through a YouTube video, you can explain the value you will bring to an organization by discussing your background, achievements, and talents "in a story-like format" (Schawbel).

To create an effective video resume (or what some call a "vesume"), consider these five tips:

1. **Video quality.** The video must have clear resolution, excellent sound, and no background distractions (noises or images).
2. **Appearance and diction.** Just as you would strive to dress and behave professionally in an interview, you must also present a professional appearance in your video resume. See our tips for professional appearance discussed later in this chapter. In addition to appearance, prospective employers will judge you on the content of your comments, what you say, and how you say it. Speak with confidence, clarity, and professionalism. Avoid mumbling, looking down, shuffling notes, or checking cell phones.
3. **Content.** Your video resume, just like your hard-copy or electronic resume, must focus on your objectives, education, work history, and special skills. Begin your video with an introductory overview introducing yourself and your goals; end the video resume with a conclusion, summing up your value to the organization.

FIGURE 8.3 Scannable Resume

Place your name at the top of a scannable resume and avoid centering text.

Rochelle J. Kroft
1101 Ave. L
Tuscaloosa, AL 89403
Home: (313) 690-4530
Cell: (313) 900-6767
E-mail: rkroft90@aol.com

OBJECTIVES
==

To use HAZARDOUS WASTE MANAGEMENT experience and knowledge to ensure company compliance and employee safety.

SUMMARY OF QUALIFICATIONS
==

* HAZARDOUS WASTE MANAGEMENT with skills in teamwork, end-user support, quality assurance, problem solving, and written documentation.
* Five years experience working with international and national businesses and regulatory agencies.
* Skilled in assessing environmental needs and implementing hazardous waste improvement projects.
* Able to communicate effectively with multinational, cross-cultural teams, consisting of clients, vendors, co-workers, and local and regional stakeholders.
* Excellent customer service, using strong problem-solving techniques.
* Effective project management skills, able to multitask.

Use key words to summarize your accomplishments.

COMPUTER PROFICIENCY
==

Microsoft Windows XP, PowerPoint, C++, Visual Basic, Java, CAD/CAM

EXPERIENCE
==

Hazardous Waste Manager
Shallenberger Industries, Tuscaloosa, AL (2008 to present)
* Assess client needs for root cause analysis and recommend strategic actions.
* Oversee waste management improvements, using project management skills.
* Conduct and document follow-up quality assurance testing on all newly developed applications to ensure compliance.
* Develop training manuals to ensure team and stakeholder safety. Shallenberger has had NO injuries throughout my management.
* Manage a staff of 25 employees.
* Achieved "Citizen's Recognition" Award from Tuscaloosa City Council for safety compliance record.

Use dashes or equal signs for design elements such as line dividers and asterisks (*) for bullets.

Hazardous Waste Technician
CleanAir, Montgomery, AL (2003–2008)
* Developed innovative solutions to improve community safety.
* Created new procedure manuals to ensure regulatory compliance.

Type your scannable resume in Courier, Arial, Verdana, or Helvetica. Avoid designer fonts like Comic Sans, Lucida, or Corsiva.

EDUCATION
==

B.S. Biological Sciences, University of Alabama, Tuscaloosa, AL (2003)
* Biotechnology Honor Society, President (2002)
* Golden Key National Honor Society

AFFILIATIONS
==

Member, Hazardous Waste Society International

4. **Length.** A video resume isn't a movie. Limit your video resume to three to five minutes.

5. **Editing.** To achieve the desired length and to ensure that your video has a high quality, edit for errors, distractions, noises, and content. Review your video numerous times to make sure you haven't missed any problems. Then, if you need to, shoot the video again so you are an appealing job candidate (Nale).

Criteria for Effective Letters of Application

Your resume, whether hard copy or electronic, will be prefaced by a letter of application (or cover e-mail if you submit the resume as an e-mail attachment). These two components of your job package serve different purposes.

The resume is generic. You'll write one resume and use it over and over again when applying for numerous jobs. In contrast, the letter of application is specific. Each letter of application will be different, customized specifically for each job. Criteria for an effective letter of application include the following elements.

Letter Essentials

Letters contain certain mandatory components: your address, the date, your reader's address, a salutation, the letter's body, a complimentary close, your signed name, your typed name, and an enclosure notation if applicable. If you are submitting an electronic resume along with an e-mail cover message, you will not need these letter essentials. We discuss the e-mail cover message later in this chapter.

Letters

See Chapter 6 for more discussion of letter formats and cover letters.

Introduction.   In your introductory paragraph, include the following:

- Tell where you discovered the job opening. You might write, "In response to your advertisement in the May 31, 2012, *Lubbock Avalanche Journal* . . ." or "Bob Ward, manager of human resources, informed me that. . . ."

- State which specific job you are applying for. Often, the classified section of your newspaper or a company's online resource for job seekers will advertise several jobs at one company. You must clarify which of those jobs you're interested in. For example, you could write, "Your advertisement for a computer maintenance technician is just what I have been looking for."

- Sum up your best credentials. "My BS in chemistry and five years of experience working in a hazardous materials lab qualify me for the position."

Discussion.   In the discussion paragraph(s), sell your skills. To do so, describe your work experience, your education, and your professional skills. This section of your letter of application, however, is not meant to merely repeat your resume. The resume is generic; the letter of application is specifically geared toward your reader's needs. Therefore, in the discussion, follow these guidelines:

- Focus on your assets uniquely applicable to the advertised position. Select only those skills from your resume that relate to the advertisement and which will benefit the prospective employer.

- Don't explain how the job will make you happy: "I will benefit from this job because it will teach me valuable skills." Instead, using the pronouns *you* and *your*, show reader benefit: "My work with governmental agencies has provided me a wide variety of skills from which your company will benefit."

- Quantify your abilities. Don't just say you're great ("I am always looking for ways to improve my job performance"). Instead, prove your assertions with quantifiable facts: "I won the 2012 award for new ideas saving the company money."

Conclusion. Your final paragraph should be a call to action. You could say, "I hope to hear from you soon" or "I am looking forward to discussing my application with you in greater detail." You could tell your reader how to get in touch with you: "I can be reached at 913-469-8500." If you are more daring, you could write, "I will contact you within the next two weeks to make an appointment at your convenience. At that time I would be happy to discuss my credentials more thoroughly."

In addition to these suggestions, you should mention that you have enclosed a resume. You can do this either in the introduction, discussion, or conclusion. Select the place that best lends itself to doing so.

E-mail Cover Message

E-mail

See Chapter 6 for more discussion of e-mail messages.

If you submit the resume as an attachment to e-mail, you should write a brief e-mail cover message. This e-mail message will serve the same purpose as a hard-copy letter of application. Therefore, you want to include an introductory paragraph, a body, and a conclusion.

Introduction. Tell the reader which job you are applying for and where you learned of this position.

Body. State that you have attached a resume. Tell the reader which software you have used to write your resume. Briefly explain why you are the right person for the job. You can do this in a short paragraph or by listing three to five key assets.

EXAMPLE ▶

> I have attached a resume for your review. To open this document, you will need Microsoft Word.
>
> *or*
>
> I have saved the resume as an RTF (rich text file).

Web Resources

www.pearsonhighered.com/gerson See our companion website for sample e-mail cover messages

Conclusion. Sum up your e-mail message pleasantly. Tell your reader that you would enjoy meeting him or her and that you look forward to an interview.

Online Application Etiquette

If you send your resume as an attachment to an e-mail message, be sure to follow online etiquette:

- **Do not use your current employer's e-mail system.** That clearly will tell your prospective employer that you misuse company equipment and company time.
- **Avoid unprofessional e-mail addresses.** Addresses such as Mustang65@aol.com, Hangglider@yahoo.com, or HotWheels@juno.com are inappropriate for business use. When you use e-mail to apply for a job, it is time to change your old e-mail address and become more professional. Use your initials or your name instead.
- **Send one e-mail at a time to one prospective employer.** Do not mass mail resumes. No employer wants to believe that he or she is just one of hundreds to whom you are writing.
- **Include a clear subject line.** Announce your intentions or the contents of the e-mail: "Resume—Vanessa Diaz" or "Response to Accountant Job Opening."

Figure 8.4 shows an effective cover e-mail message prefacing an attached resume.

Figures 8.5 is an example of an effective letter of application.

Techniques for Interviewing Effectively

The goal of writing an effective resume and letter of application is to get an interview. The resume and letter of application open the door; only a successful interview will win you the job. In fact, some sources suggest that the interview is the most important stage of your job search. The Society for Human Resource Management states that 95 percent of respondents ranked "interview performance" as a "very influential" factor when deciding to hire an employee.

FIGURE 8.4 E-mail Cover Message

Subject line clarifying the purpose of the e-mail

Introduction stating where you found the job listing and what job you are applying for.

Lead-in to body stating what software has been used for the attachment.

A brief, itemized list of your assets, specifically related to the job opening.

A conclusion ending upbeat and positive.

NOTE: Standard keystrokes, such as asterisks for bullets and paragraph spacing, are used for highlighting. This ensures that any e-mail package can read the text.

FIGURE 8.5   Effective Letter of Application

11944 West 112th Street
St. Louis, MO 66221

December 11, 2012

Ms. Sarah Beske
CEO
DiskServe
9659 W. 157th St.
St. Louis, MO 78580

Dear Ms. Beske:

I am responding to your advertisement in the November 24, 2012, issue of *The St. Louis Courier* for a job in your computer technology department. Because of my two years experience in computer information systems, I believe I have the skills you require.

Although I have enclosed a resume, let me elaborate on my achievements. While working as a computer technician for Radio Shack, I troubleshot motherboards, worked the help desk, and made service calls to businesses and residential customers. Your advertisement listed the importance of customer service and technical skills. I have expertise in both areas.

Your job description also mentioned the importance of working in a team environment. At Radio Shack, I frequently made service calls with a team of technicians. Together, we provided quality service.

I would like to meet with you and discuss employment possibilities at your company. Please call me at (913) 469-8500 so that we can set up an interview at your convenience. I appreciate your consideration.

Sincerely,

Macy G. Heart

Enclosure (Resume)

The introduction mentions where the writer found the job, which job he is applying for, and key qualifications for the position.

The letter's body quantifies the writer's abilities. The body also shows how the writer meets the job requirement.

The conclusion ends positively, using words and phrases like "convenience," "Please," and "appreciate." It also provides follow-up information.

"Interview performance [is] more influential than 17 other criteria, including years of relevant work experience, resume quality, education levels, test scores or references" (Stafford L1).

Dress Professionally

Professionalism starts with your appearance. The key to successful dressing is to wear clean, conservative clothing. No one expects you to spend money on high-fashion, stylish clothes, but everyone expects you to look neat and acceptable. Business suits are still best for both men and women.

Be on Time

Plan to arrive at your interview at least 20 to 30 minutes ahead of schedule. That way, you won't have to worry about unexpected traffic, crowded parking lots, or finding the correct building.

Watch Your Body Language

To make the best impression, don't slouch, chew your fingernails, play with your hair or jewelry, or check your watch. These actions will make you look edgy and impatient. Sit straight in your chair, even leaning forward a little to show your enthusiasm and energy. Look your interviewer in the eye. Smile and shake your interviewer's hand firmly.

Don't Chew Gum, Smoke, or Drink Beverages during the Interview

The gum might distort your speech; the cigarette will probably offend the interviewer, particularly if he or she is a nonsmoker; and you might spill the beverage.

Turn Off Your Cell Phone, Smart Phone, or PDA

Today, cell phones, smart phones, and PDAs are commonplace. However, the interview room is one place where these devices must be avoided. Taking a call while you are being interviewed is rude and will ensure that you will not be hired. Reading a text message or sending one during an interview is not appropriate.

Watch What You Say and How You Say It

Speak slowly, focus on the conversation, and don't ramble. Once you have answered the questions satisfactorily, stop.

Bring Supporting Documents to the Interview

Supporting documents can include extra copies of your resume, a list of references, letters of recommendation, employer performance appraisals, a portfolio (hard copies and electronic version) of technical documents you have written, and transcripts.

Research the Company

Show the interviewer that you are sincerely interested in and knowledgeable about the company. Dr. Judith Evans, vice president of Right Management Consultants of New York, says that the most successful job candidates show interviewers that they "know the company inside and out" (Kallick D1).

Be Familiar with Typical Interview Questions

You want to anticipate questions you will be asked and be ready with answers. Some typical questions include the following:

What are your strengths and weaknesses?	Can you travel for work-related activities?
Why do you want to work for this company?	Will you relocate?
Why are you leaving your present employment?	What do you want to be doing in five years, ten years?
What did you like least about your last job?	How would you handle this (hypothetical) situation?
What computer hardware are you familiar with, and what computer languages do you know?	What was your biggest accomplishment in your last job or while in college?
What machines can you use?	What about this job appealed to you?
What special techniques do you know, or what special skills do you have?	What starting salary would you expect?
What did you like most about your last job?	How do you get along with colleagues and with management?

When Answering Questions, Focus on the Company's Specific Need

For example, if the interviewer asks if you have experience using RoboHelp, explain your expertise in that area, focusing on recent experiences or achievements. Be specific. In fact, you might want to tell a brief story to explain your knowledge. This is called "behavioral description interviewing" (Ralston et al. 9). It allows an interviewer to learn about your speaking abilities, organization, and relevant job skills. To respond to a behavioral description interview question, answer questions as follows (Ralston et al. 11):

- Organize your story chronologically.
- Tell who did what, when, why, and how.
- Explain what came of your actions (the result of the activity).
- Depict scenes, people, and actions.
- Make sure your story relates exactly to the interviewer's needs.
- Stop when you are through—do not ramble on. Get to the point, develop it, and conclude. If, however, you do not have the knowledge required, then "explain how you can apply the experience you *do* have" (Hartman 24). You could say, "Although I've never used RoboHelp, I have created online help using HTML coding. Plus, I'm a quick learner. I was able to learn FlashHelp and XML well enough to create online help screens in only a week. Our customer was very happy with the results." This will show that you understand the job and can adapt to any task you might be given.

Criteria for Effective Follow-Up Correspondence

Once you have interviewed, don't just sit back and wait, hoping that you will be offered the job. Write a follow-up letter or e-mail message. This follow-up accomplishes three primary goals: It thanks your interviewers for their time, keeps your name fresh in their memories, and gives you an opportunity to introduce new reasons for hiring you.

SPOTLIGHT

How do human resource professionals conduct a job search?

Maria Levit, Human Resources Director, says that for each open staff position at DeVry University, Kansas City, over 100 people apply. How do you find the best applicant out of so many possibilities?

Maria follows a step-by-step approach, which includes

- Reviewing the resumes
- Following with a prescreen telephone call to the top 10 to 15 applicants
- Inviting between 6 and 8 individuals to face-to-face interviews
- Calling references

Ms. Levit looks for two strengths above all others in the resumes. First, she wants to see "evidence of skills and credentials applicable to the job." Next, Maria needs proof. "I look for evidence of success. I want concrete examples that prove the applicant's accomplishments. For example, I want a resume to read something like, 'reduced turnover by a specific percent within such and such a time.'"

If an applicant's resume meets the job's criteria, Maria conducts the phone prescreen. In these phone calls, she wants evidence of good communication skills and a positive attitude. Maria wants to hear that the applicant "cares about people, wants to make a positive impact on students and co-workers, and has a passion" for the job. Most importantly, Maria uses the phone prescreen to detect "red flags." She is wary of applicants who make negative comments about current co-workers, bosses, or work environment.

In the face-to-face interview, Maria looks for applicants who come prepared with extra copies of resumes, names and telephone numbers of references, or transcripts. In contrast, if they "have poor posture, speak in a monotone, don't make eye contact, and dress inappropriately," then their general day-to-day job preparedness might be lacking as well.

Maria's most challenging activity is the follow-up reference call. Current and past employers are cautious about discussing employees. To overcome this reticence, Maria begins with a nonthreatening question: "What was John's basic job description?" Next, she moves on to more challenging probes, like "Identify two to three of John's strengths and weaknesses."

Whatever Maria does, it must be working. Many faculty and staff have worked at the Kansas City DeVry for over ten years. Hiring the right person for the right job leads to continuity in the workplace.

A follow-up letter or e-mail message contains an introduction, discussion, and conclusion:

- **Introduction.** Tell the readers how much you appreciated meeting them. Be sure to state the date on which you met and the job for which you applied.
- **Discussion.** In this paragraph, emphasize or add important information concerning your suitability for the job. Add details that you forgot to mention during the interview, clarify details that you covered insufficiently, and highlight your skills that match the job requirements. In any case, sell yourself one last time.
- **Conclusion.** Thank the readers for their consideration, or remind them how they can get in touch with you for further information. Don't, however, give them any deadlines for making a decision.

See Figure 8.6 for a sample follow-up correspondence.

This letter succeeds for several reasons. First, it is short, merely reminding the reader of the writer's interest, instead of overwhelming him or her with too much new information. Second, the letter is positive, using words such as *enjoyed, ability, welcome, opportunity, exciting,* and *thank you.* Finally, the letter provides the reader an e-mail address for easy follow-up.

Job Acceptance Letter

Great news! After working hard to find a job, your efforts have paid off. You've just received a job offer. Now what? Sometimes, accepting the offer over the phone isn't enough. Your new employer might want you to write and sign an official acceptance letter. In this brief letter, you will want to accomplish the following goals:

- Thank the company for the job opportunity
- Officially accept the job offer
- Restate the terms of employment (salary, benefits, location, position, job responsibilities, and/or start date)

Address the letter to the individual who offered you the position for a more personalized touch. Be sure to include your phone number, e-mail address, or mailing address, just in case the company needs to contact you.

This acceptance letter actually could be seen as your first day on the job. Therefore, take as much care in writing this letter as you did in applying for the job. Make sure your letter is grammatically correct, well organized, and conveys a positive tone. Show your new boss or colleagues that you are a professional asset to the company. Figure 8.7 provides a sample job acceptance letter.

FIGURE 8.6 Follow-Up Correspondence

Thank you for allowing me to interview with Acme Corporation on July 8. I enjoyed meeting you and the other members of the team to discuss the position of Account Representative.

You stated in the interview that Acme is planning to expand into international marketing. With my Spanish speaking ability and my study-abroad experience, I would welcome the opportunity to become involved in this exciting expansion.

Again, thank you for your time and consideration. I look forward to hearing from you. Please e-mail me at gfiefer21@aol.com.

For an e-mail follow-up, you would include your reader's e-mail address and a subject line, such as "Thank You for the July 8 Interview" or "Follow-up to July 8 Interview."

For a hard-copy follow-up letter, you would include all letter components: writer's address, date, reader's address, salutation, complimentary close, and signature.

Job Openings

_____ 1. Did you visit your college or university job placement center?

_____ 2. Did you talk to your professors about job openings?

_____ 3. Have you networked with friends or previous employers, and have you considered using social media for networking?

_____ 4. Have you checked with your professional affiliations or looked for job openings in trade journals?

_____ 5. Did you search the Internet for job openings?

Resume

_____ 1. Are your name, address, and phone number correct?

_____ 2. Is your job objective specific?

_____ 3. Have you included a summary of qualifications?

_____ 4. Is all information within your education, work experience, and military experience sections accurate?

_____ 5. Have you used lists beginning with verbs?

_____ 6. Have you quantified each of your achievements?

_____ 7. Have you avoided using sentences and the word _I_?

_____ 8. Does your resume use highlighting techniques to make it reader friendly?

_____ 9. Have you proofread your resume to find grammatical and mechanical errors?

_____ 10. Have you decided whether you should write a reverse chronological resume or a functional resume?

Letter of Application

_____ 1. Have you included all of the letter essentials?

_____ 2. Does your introductory paragraph state where you learned of the job, which job you are applying for, and your interest in the position?

_____ 3. Does your letter's discussion unit pinpoint the ways in which you will benefit the company?

_____ 4. Does your letter's concluding paragraph end cordially and explain what you will do next or what you hope your reader will do next?

_____ 5. Is your letter free of errors?

Interview

_____ 1. Will you dress appropriately?

_____ 2. Will you arrive ahead of time?

_____ 3. Have you practiced answering potential questions?

_____ 4. Have you researched the company so you can ask informed questions?

_____ 5. Will you bring to the interview additional examples of your work or copies of your resume?

Follow-Up Letter

_____ 1. Have you included all the letter essentials?

_____ 2. Does your introductory paragraph remind the readers when you interviewed and what position you interviewed for?

_____ 3. Does the discussion unit highlight additional ways in which you might benefit the company?

_____ 4. Does the concluding paragraph thank the readers for their time and consideration?

_____ 5. Does your letter avoid errors?

Job Acceptance Letter

_____ 1. Have you included the letter essentials?

_____ 2. Does your introductory paragraph explain _why_ you are writing (in response to a job offer) and _what_ you are writing about (accepting the job)?

_____ 3. Does the discussion unit confirm the particulars of the offer (salary, benefits, job duties, location, start date, and so on)?

_____ 4. Does the concluding paragraph thank the reader for the job opportunity?

_____ 5. Does your letter avoid errors and show your professionalism?

Checklists

Seeking a job is not easy. However, it can be a manageable activity if you approach it as a series of separate but equal tasks. To get the job you want, you must search for an appropriate position, write an effective resume and letter of application, interview successfully, write a follow-up letter, and send a job acceptance letter.

The Writing Process at Work

Effective communication in the job search follows a process of prewriting, writing, and rewriting. To illustrate the importance of the writing process in the job search, look at how LaShanda E. Brown used prewriting, writing, and rewriting to create an effective resume so she could apply for a job in the field of human resources.

FIGURE 8.7   Job Acceptance Letter

Amy Zhang
9103 Stonefield Rd.
Georgetown, TX 77829

May 15, 2012

George Smithson
Dell Computers
2134 Silicon Way
Round Rock, TX 77112

Dear Mr. Smithson:

In response to your phone call yesterday (5/14/12), I am very happy to accept the
position as a Technical Sales Representative II at Dell Computers. I know that with
my education, experience, and energy, I will be an asset to your workforce.

As we discussed, my salary will be $35,000 with health and life insurance
benefits provided after 90 days of employment. I will work in your Round
Rock, Texas, facility, where my job will entail working with a team to meet the
schedules, budgets, product costs, and production ramp rates for projects
assigned to my team and me. In addition, I will manage small to medium
platform/peripheral development programs.

I look forward to joining the Dell team on June 10, 2012. If you have any
questions or need additional information, please let me know. Thank you,
Mr. Smithson, for this outstanding opportunity.

Sincerely,

Amy Zhang
713-555-0112
azhang@gt.edu

Prewriting

Before writing a resume, use prewriting to help you begin a job search and determine your
objectives. Beginning with a list of goals, audience, channels of communication, and material to include will get you started. LaShanda used listing to prewrite and plan her job
search (Figure 8.8).

A second method of prewriting will help you to gather information to include in your
resume. Answering the reporter's questions is a very effective way to prewrite. LaShanda
asked herself reporter's questions before she wrote her resume (Figure 8.9).

Writing

After determining objectives by prewriting, you can begin drafting your resume with the
information learned by answering the reporter's questions.

LaShanda went to her husband's military career center for help writing her resume.
They told her to use a functional resume because that's what they tended to advocate for
people leaving the military. LaShanda drafted her resume, creating a functional resume
that used information she discovered in prewriting (Figure 8.10).

Web Resources

www.pearsonhighered
.com/gerson For more
information about the
writing process, visit our
companion website.

The Writing Process

Prewriting	Writing	Rewriting

• Determine your goals in the job search.
• Try listing to determine your objectives and answering reporter's questions to gather data about the job search.
• Decide who will read your resume and letter of application (human resources employee, potential supervisor, interview committee?).
• Choose the correct method of delivery for your audience and purpose.
	• Organize your content using reverse chronology.
	• Use headings to organize your resume.
	• Avoid full sentences in the resume.
	• Start with present tense verbs for current education and work achievements.
	• Start with past tense verbs for past education and work achievements.
	• Use bullets to aid ease of access.
	• Quantify your activities.
		• Revise your draft by
		  • Adding details
		  • Deleting wordiness
		  • Simplifying words
		  • Enhancing the tone
		  • Reformatting your text
		  • Proofreading and correcting errors

FIGURE 8.8 LaShanda's Listing

- Goal—to find a job in the field of human resources
- Audience—directors of human resources
- Channels—resume and letter of application, submitted electronically
- Data—information drawn from my experiences and education

FIGURE 8.9 LaShanda's Reporter's Questions

- **Who** should I send my resume to?
- **What** should I include about my experience in human resources? What should I include about education?
- **When** is the resume due?
- **Where** should I send the resume (what's the city/state address, e-mail address, or URL)?
- **Why** am I interested in this job posting?
- **How** should I send the resume—by e-mail, online form, or hard copy?
- **How** can I show benefit to the company hiring me?

FIGURE 8.10   Draft of a Functional Resume

LaShanda E. Brown
833 Hampton Rd., Apt. 1
Virginia Beach, VA 23460
757-555-5555
labrown10@yahoo.com

Objective: Retail Management

HIGHLIGHTS OF QUALIFICATIONS

- 2 years retail management and recruiting experience.
- Expertise in oral and written communications. Able to present ideas and goals clearly.
- Motivated team player that thrives in a fast-paced, multi-faceted environment.
- Proficient in Microsoft Office (Word, Excel, PowerPoint, Outlook, Access), as well as basic office hardware (fax, copiers, cash registers)
- Familiar with most video game consoles (PS1, PS2, Xbox, NES, etc.)

EXPERIENCE

Management

- Supervised 5 management trainees and taught them all aspects of the rental car industry to include accounts receivable, marketing accounts as well as competition, managing a rental fleet, etc.
- Oversee 4 telemarketers and 2 sales representatives. Managed training and quality customer service when representing the company via telephone.
- Improved loss prevention programs within a retail setting by breaking the store up into teams and zones so that all areas were covered at all times. Loss prevention decreased 6% in 3 months.
- Earned All-stars sales award 4 months in a row. Average of 91% optional coverage sales on rentals within that four-month period.

Customer Service

- Implemented new strategies for customer satisfaction, such as interactive games to give customers an opportunity to win coupons on current and future rentals. Customer satisfaction increased 15%.
- Greeted customers enthusiastically and efficiently to provide a positive company image and impression.
- Researched and provided solutions to customer service issues, ensuring customer needs were met and they were satisfied.
- Coordinated customer follow-ups ensuring they had a good rental experience with us and would return if they needed our rental services again.

EMPLOYMENT

2010–Present	Office Manager	Best Value Remodeling
2008–2010	Insurance Sales Agent	Geico Direct
2008	Management Assistant	Enterprise Rent-A-Car

EDUCATION AND TRAINING

- **Bachelors in Communication, emphasis in Public Relations**
  Missouri Western State College, St. Joseph, MO
- Elite Sales Training, Enterprise Rent-a-car
- Effective Management Training, Best Value Remodeling

Rewriting

Editing and revising a resume will help you make it look professional and achieve the desired result of getting an interview. After getting no job interviews, LaShanda realized that her functional resume wasn't working. She had to consider the following aspects of her resume that needed revision:

- Her career objective's line did not match her new search for a job in human resources.
- The functional resume listed jobs in one part of the resume and then responsibilities in another part of the text.
- Directors of human resources could not match dates and duties.
- The functional resume left too many holes.
- The functional resume did not clarify her achievements and education.
- She needed to delete information from past employment that was not relevant for her current job search.
- She needed to include courses and professional development that proved she was up to date in the field of human resources.

She changed strategies and restructured the functional resume into a reverse chronological resume. With the revised resume (Figure 8.11), LaShanda got an interview. Today she works in her desired field of human resources and gets to review job applicants' resumes and interview them.

CHAPTER HIGHLIGHTS

1. Use many different resources to locate possible jobs, such as college placement centers, instructors, friends, professional affiliations, want ads, social media, and the Internet.
2. Use either a reverse chronological resume or a functional resume.
3. Write a traditional hard-copy resume, an e-mail resume, or a scannable resume.
4. Indicate a specific career objective on your resume.
5. Write your letter of application so that it targets a specific job.
6. Prepare before your interview so you can anticipate possible questions.
7. Your follow-up letter or e-mail message will impress the interviewer and remind him or her of your strengths.
8. Over 60 percent of jobs are found through networking.
9. Use the Internet to help you job search.
10. On your resume, place education first if that is your strongest asset, or begin with work experience if this will help you get the job.

FIGURE 8.11   Revised Reverse Chronological Resume

LaShanda E. Brown
833 Hampton Rd., Apt. 1
Virginia Beach, VA 23460
757-555-5555
labrown10@yahoo.com

EDUCATION
Bachelor of Science in Communication, Public Relations emphasis, May 2010
Missouri Western State College,   St. Joseph, MO
- Related courses: small group communication, presentational communication, consumer marketing, persuasive speaking, media in communications, public relations communication analysis, human resources management, nonverbal communication, advertising, desktop publishing.

EMPLOYMENT
Office Manager
Best Value Remodeling   Virginia Beach, VA   2010–Present
- Screen and hire new employees, as well as review employee performance
- Make spreadsheets using MS Excel to track sales by outside sales teams and manage commissions
- Update Microsoft Access database to track results from sales calls
- Create PowerPoint presentations for general manager to use when going on corporate sales leads
- Oversee 4 telemarketers and 2 sales representatives

Insurance Sales Agent
Geico Direct   Virginia Beach, VA   2008–2010
- Provided excellent customer service on the phone
- Performed data entry by setting up insurance policies

Management Assistant
Enterprise Rent-a-Car   Oklahoma City, OK   2008
- Recruited at career fairs
- Won award as "Best New Hire 2008"
- Supervised 5 management trainees
- Created advertisements locally for the branch
- Improved loss prevention programs by 6% in 3 months.
- Earned All-stars sales award 4 months in a row. Average of 91% optional coverage sales on rentals within that four-month period.

Customer Service Supervisor
Office Max   St. Joseph, MO   2006–2008
- Trained new hires
- Worked front desk
- Provided customer service

Data Entry Clerk
Internal Revenue Service   Oklahoma City, OK   2005–2006
- Input data entry of tax forms from scanned images
- Downloaded data entry from scanned microfilm into IRS database

SOFTWARE KNOWLEDGE
- Proficient with Microsoft Office programs, including MS Word, MS Excel, MS Access, MS PowerPoint, MS Outlook, and MS Project
- Proficient with PeopleSoft program (HR module)
- Proficient with Lotus programs, including Lotus Notes, Lotus 1-2-3

By leading with education, LaShanda emphasizes her skills as they relate to a job search in human resources.

The employment section highlights human resource experience, allowing the city's human resources department to see that LaShanda meets the requirements for a position.

The chronological resume better suits her needs in that it matches job responsibilities with specific jobs and dates.

CASE STUDIES

1. Rewrite the flawed resume on page 265. In doing so, revise the errors and create three different types of resumes for Macy G. Heart—a chronological resume, a functional resume, and a scannable resume.

2. DiskServe, a St. Louis–based company, is hoping to hire a customer service representative for their computer technology department. In addition to DiskServe's CEO, Sarah Beske, the hiring committee will consist of two managers from other DiskServe departments and two co-workers in the computer technology department.

   The position requires a bachelor's degree in information technology (or a comparable degree) and/or four years experience working with computer technology. Candidates must have knowledge of C++, Visual Basic (VB), SQL, Oracle, and Microsoft Office applications. In addition, customer service skills are mandatory. Four candidates were invited to DiskServe's worksite for personal interviews—Macy Heart, Aaron Brown, Rosemary Lopez, and Robin Scott.

Macy has a bachelor's degree in computer information systems. He has worked two years part time in his college's technology lab helping faculty and students with computer hardware and software applications, including Microsoft Office and VB. He worked for two years at a computer hardware/software store as a salesperson. His supervisor considers Macy to be an outstanding young man who works hard to please his supervisors and to meet customer needs. According to the supervisor, Macy's greatest strength is customer service, since Macy is patient, knowledgeable, and respectful.

Aaron has an information technology certificate from Microsoft, where he has worked for five years. Aaron began his career at Microsoft as a temporary office support assistant, but progressed to a full-time salesperson. When asked where he saw himself in five years, Aaron stated, "The sky's the limit." References proved Aaron's lofty goals by calling him "a self-starter, very motivated, hardworking, and someone with excellent customer service skills." He is taking programming courses at night from the local community college, focusing on C++, Visual Basic, and SQL.

Rosemary has an associate's degree in information technology. She has five years experience as the supervisor of Oracle application. Prior to that, Rosemary worked with C++, VB, and SQL. She also has extensive knowledge of Microsoft Office. Rosemary was asked, "How have you handled customer complaints in the past?" and responded, "I rarely handle customer complaints. In my last job, I assigned that work to my subordinates."

Robin has a bachelor's degree in information technology. To complete her degree, Robin took courses in C++, VB, Oracle, and SQL. She is very familiar with Microsoft Office. Since Robin just graduated from college, she has no full-time experience in the computer industry. However, she worked in various retail jobs (food services, clothing stores, and book stores) during high school, summers, and in her senior year at college. She excelled in customer service, winning the "Red Dragon Employee of the Month Award" from her last job as a server in a Chinese restaurant.

Assignment

Who would you hire? Give an oral presentation or write an e-mail or memo to Sarah Beske, DiskServe's CEO, explaining which of the candidates she should hire.

INDIVIDUAL AND TEAM PROJECTS

1. Practice a job search. To do so, find examples of job openings in newspapers, professional journals, at your college or university's job placement service, and online. Bring these job possibilities to class for group discussions. From this job search, you and your peers will get a better understanding of what employers want in new hires.

1890 Arrowhead Dr.
Utica, MO 51246
710-235-9999

Resume of Macy G. Heart

Objective: Seeking a position in Computer Technology Customer Service where I can use my many technology and people-person skills. I want to help troubleshoot software and hardware problems and work in a progressive company which will give me an opportunity for advancement and personal growth.

Work Experience

Jan. 2010 to now Aramco.net   St. Louis, Missouri   Tech Support Specialists
Primarily I provide customer support for customer problems with C++, Visual Basic, Java, Networking, and Databases (Access, Oracle, SQL). I provide solutions to software and hardware problems and respond to e-mail queries in a timely manner.

Oct. 2008 to Dec. 22, 2010   DocuHelp   Chesterfield, MO Computer Consultant
Provide technical support for PC's and Mac's. I also trained new PC and Mac users in hardware applications. When business was slow, I repaired computer problems, using my many technology skills.

May 2006 to Oct. 2008   Ram-on-the-Run East St. Louis, Illinois   Computer Salesman
Sold laptops, PCs, printers, and other computer accessories to men and women. Answered customer questions. Won "Salesman of the Month Award" three months in a row due to exceeding sales quotas.

Jan. 2004 to May 2006   Carbondale High School   Carbondale, IL   Lab Tech
Worked in the school's computer lab, helping Mr. Jones with computer-related classwork. This included fixing computer problems and tutoring new students having trouble with assignments.

Education

Aug. 2010   Bachelor's Degree, Information Technology, Carbondale Institute of Technology, Carbondale, IL
Concentration: Database/Programming Applications
Relevant classes: Business Information Systems, Hardware Maintenance, Database Management, Visual Basic, Systems Analysis and Design, C++, Web Design

May 2006   Graduate Carbondale High School
Member of the Computer Technology Club
Member of FFA and DECA
Principal's Honor Roll, senior year

Computer Expertise

Cisco Certified, knowledge of C++, VB, Microsoft Office Suite XP, HTML, Java, SML

References

Mr. Oscar Jones, Computer Applications Teacher, Carbondale High School
Mr. Renaldo Gomez, Manager, Aramco.net
Mr. Ted Harriot, Technical Support Supervisor, DocuHelp

Additional Information

A good team player, who works well with others
Maintained 4.0 in my college major classes
Built my own computer from scratch in high school
Starting football player in high school, Junior Varsity tight end

2. An informational interview can help you learn about the realities of a specific job or work environment. Interview a person currently working in your field of interest. You can find such employees in the following areas:

- **Alumni office.** Ask your college or university for a list of alums who are willing to speak to students about careers.
- **Career center.** Your school's career center might give you names of people to contact.
- **Human resources.** Visit a company in your field of interest and ask the human resources staff for help.
- **Networking.** Do you have friends or family who know of employees in your field of interest?

Once you find an employee willing to help, visit with him or her and find out the following:

- What job opportunities exist in your field?
- What does a job in your field require, in terms of writing, education, interpersonal communication skills, teamwork, and so on, as well as the primary job responsibilities?

After gathering this information, write a thank-you letter to the employee who helped you. Then, write an e-mail to your professor documenting your findings and give an oral presentation to your classmates.

3. Write a resume. To do so, follow the suggestions provided in this chapter. Once you have written this resume, bring it to class for peer review. In small groups, discuss each resume's successes and areas needing improvement.

4. Write a letter of application. To do so, find a job advertisement in your newspaper's classified section, in your school's career planning and placement office, at your worksite's personnel office, in a trade journal, or online. Then write the letter according to the suggestions provided in this chapter. Next, in small groups, review your letter of application for suggested improvements.

5. Practice a job interview in small groups, designating one student as the job applicant and other students as the interview committee. Ask the applicant the sample interview questions provided in this chapter or any others you consider valid. This will give you and your peers a feel for the interviewing process.

PROBLEM-SOLVING THINK PIECES

1. You need to submit a resume for a job opening. However, you have problems with your work history, such as the following:

- You have had no jobs.
- Your work experience is limited to baby sitting or mowing lawns.
- You have been fired from a job.
- You have had five (or more) jobs in one year.
- You have been out of work for an extended period of time.

Consider how you would meet the challenges of your job history.

2. You have found a job that you want to apply for. The job requires a bachelor's degree in a specific field. Though you had been enrolled in that specific degree program for three years, you never completed the degree.

What should you say in your resume and letter of application to apply for this position, even though you do not meet the degree requirement? Which type of resume (reverse chronological or functional) should you write? Explain your answer.

3. You are in the middle of an interview. Though the interview had been scheduled for 1:00 p.m. to 1:45 p.m., it is running late. You had planned to pick up your son from daycare at 2:00 p.m. How do you handle this problem?

4. You have just completed the interview process for a job in your field. You have impeccable credentials, meeting and exceeding all of the requirements. Your interview went very well.

At the close of the interview, one of the interviewers says, "Thank you for interviewing with us. You did a great job answering our questions, and your credentials are truly excellent. However, I think you would find the job unchallenging, maybe even boring. You are overqualified."

How should you handle this situation?

5. During an interview, you are asked to "describe a problem you encountered at work and explain how you handled that challenge." How do you answer this typical question, but avoid giving an answer that paints a negative picture of a boss, co-worker, or your work environment?

WEB WORKSHOP

1. Access any Internet search engine to find information about the job search. You can go online to research resumes, cover letters, and follow-up thank-you notes. For example, check out Monster.com for resume guidelines, samples, resume makeovers, and do's and don'ts. Research CareerJournal.com for up-to-date articles about resume writing (tips, samples, whether you should pay to have someone write a resume for you, and case studies about "red-flag" resumes). CareerPerfect.com provides guidelines for electronic resume submission and techniques for creating scannable versions of your resume. Check out JobSearchTech .com to learn why and how to write follow-up thank-you letters, and to see samples.

   Once you research any of these sites, analyze your findings. What information have you learned from the online articles?
   a. Report your findings, either in an oral presentation or in writing (e-mail, memo, letter, or report).
   b. If you find sample letters or resumes that you dislike, rewrite them according to the criteria provided in this textbook.

2. Using an Internet search engine, find job openings in your area of interest. Which companies are hiring, what skills do they want from prospective employees, and what keywords are used to describe preferred skills in this work field? Report your findings either to your instructor by writing a memo or e-mail, or give an oral presentation to your class about the job market in your field.

3. Social media is a growing resource for the job search. Go to a search engine and type in "social media+job search" or "social networking+job search." Find online articles related to the importance of social media and employment. Summarize your findings in an e-mail message to your instructor.

4. Research "Vault video guide to video resumes." Based on the YouTube video, report your findings either orally or in an e-mail message, or create your own video resume.

QUIZ QUESTIONS

1. What are four places you can search for a job?
2. What is the primary goal of your resume?
3. What are two types of resumes?
4. When would you write a reverse chronological resume?
5. What should you include in a job acceptance letter?
6. In what order should you list jobs you have held?
7. What should you include when you list your education?
8. What are some optional inclusions in your education section?
9. Why would you write a functional resume?
10. What types of personal data should you exclude?
11. Why should you avoid sentences in a resume?
12. Why should you bulletize on a resume?
13. How can you prove "you are great" on a resume?
14. How can you design a resume effectively?
15. What are the benefits of social media in the job search?
16. What is the main difference between a resume and a letter of application?
17. What are two ways to conclude a letter of application?
18. What are five things you can do to interview effectively?
19. What are three things you can accomplish in a follow-up letter?
20. How can networking help you find a job?

Document Design

In this scenario, Roger Traver relies on meeting minutes to record information for his company.

DesignGlobal Incorporated (DGI), an engineering consulting company, holds monthly meetings at the Carriage Club, a privately owned banquet facility. To communicate to its 3,750 employees located internationally, these meetings are simulcast through videoconferences. Roger Traver, CEO of DGI, invites speakers to make presentations related to news of interest to his employees. Last month, Roger invited George Smith, a university chancellor, to speak about the connection between industry and academia. George hopes to attract more professionals such as the engineers at DGI to mentor students.

Figure 9.1 shows you the meeting minutes taken at the monthly meeting where George Smith spoke. These minutes are clear, answering reporter's questions such as who, what, when, where, why, and how. The minutes also are concise in terms of word usage and sentence length. However, the minutes are unsuccessful technical communication. The wall-to-wall words not only are visually unappealing but also disallow easy access of information. Improved document design would help readers understand the meeting minutes. This flawed document design causes readability challenges.

Objectives

FIGURE 9.1   Flawed Document Design

MINUTES

The meeting at the Carriage Club was attended by thirty members and guests. After the dinner, Roger Traver introduced the guest speaker, George Smith, university chancellor, and noted his accomplishments and experiences prior to education—U.S. Navy commander, Oak Ridge Laboratory researcher, and politician. Dr. Smith's talk, "Industry and Education Collaboration," was very interesting and included a history of special projects enjoyed by both academics and corporate heads. Dr. Smith suggested that we engineers could work with education to (1) provide training seminars, (2) help in urban development, and (3) provide intern opportunities. Recent industry–education collaborations include training seminars in computers, fiber optics, and human resource options. The chancellor's primary thrust was a request for $100,000 in financial aid for urban development. He said money had already been donated from three sources: a large realty firm, Capital Homes, had given $20,000; a philanthropic group, We Care, had donated a matching $20,000; Dr. Smith's university gave a matching $20,000. The remaining $40,000, Dr. Smith hoped, would come from industry donations. Finally, the chancellor noted that industry could help itself, as well as the community, by providing internships for university undergraduate majors. These internships could either be semester- or year-long arrangements, whereby students would work for minimum wage to learn more about the day-to-day aspects of their chosen fields. The chancellor said that these internships would not only increase the students' theoretical knowledge of engineering by giving them hands-on experience but also make them better future employees for the host engineering companies. Everyone would benefit. Dr. Smith noted that the students would receive a grade and credit for their work. After the speech, our VP introduced new business, calling for nominations for next year's officers; gave us the agenda for our next meeting; and adjourned the meeting.

Check out our quarterly newsletters at www.pearsonhighered.com/gerson for dot.com updates, new case studies, insights from business professionals, grammar exercises, and facts about technical communication.

Importance of Document Design

Web Resources

www.pearsonhighered
.com/gerson For more
information about
document design, visit
our companion website.

In technical communication, words are not your only concern. What you write is important, but how the text looks on the page is equally important. If you give your readers excessively long paragraphs, or pages full of wall-to-wall words, you have made a mistake. Ugly blocks of unappealing text negatively impact readability. In contrast, effective technical communication allows readers rapid access to the information, highlights important information, and graphically expresses your company's identity.

The Technical Communication Context

Why do people read correspondence? Although individuals read poetry, short stories, and novels for enjoyment, few people read memos, letters, reports, or instructions for fun. They read these types of technical communication for information about a product or service. They read this correspondence while they talk on the telephone, commute to work, or walk to meetings.

Given these contexts, your readers want you to provide them information quickly, information they can understand at a glance. Reading word after word, paragraph after paragraph takes time and effort, which most readers cannot spare. Therefore, if your technical communication is visually unappealing, your audience might not even read your words. Readers will either give up before they have begun or be unable to remember what they have read. You cannot assume they will labor over your text to uncover its worth. Good technical communication allows readers rapid access to information.

Damages and Dangers

If your intended readers fail to read your text because it is visually inaccessible, imagine the possible repercussions. They could damage equipment by not recognizing important information which you have buried in dense blocks of text. The readers could give up on the text and call your company's toll-free hotline or send an e-mail message for assistance. This wastes your readers' and your co-workers' time and energy. Worse, your readers might hurt themselves and sue your company for failing to highlight potential dangers in a user manual.

Corporate Identity

Your document—whether a memo, letter, report, instruction, Web site, or brochure—is a visual representation of your company, graphically expressing your company's identity. It might be the only way you meet your clients. If your text is unappealing, that is the corporate image your company conveys to the customer. If your text is not reader friendly, that is how your company will appear to your client. Visually unappealing and inaccessible correspondence can negatively affect your company's sales and reputation. In today's competitive workplace, any leverage you can provide your company is a plus. Document design is one way to appeal to a client.

Time and Money

Successful technical communication involves the audience. Your goal is to help the readers understand the organization of your text, recognize its order, and access information at a glance. You also want to use varied types of communication, including graphics, for readers who absorb information better visually, rather than through words. What's the payoff? Studies tell us that effective document design saves money and time. One international customs department, after revising the design of its lost-baggage forms, reduced its error rate by over 50 percent. When a utilities company changed the look of its billing statements, customers asked fewer questions, saving the company approximately $250,000 per year. The U.S. Department of Commerce, Office of Consumer Affairs, reported that when several companies improved their documents' visual appeal, the companies increased business and reduced customer complaints (Schriver 250–51). Document design isn't a costly frill. Effective document design is good for your company's business.

To clarify how important document design is, look at the inaccessible meeting minutes in this chapter's opening "Communication at Work" scenario (see Figure 9.1). The meeting minutes are neither clear nor concise. You are given so much data in such an unappealing format that your first response upon seeing the correspondence probably is to say, "I'm not going to read that."

How can you make these minutes more inviting? How can you break up the wall-to-wall words and make key points more accessible? To achieve effective document design, you should provide your readers visual

- Organization
- Order
- Access
- Variety

Organization

The easiest way to organize your document's design is to break text into smaller chunks of information, a technique called *chunking*. When you use chunking to separate blocks of text, you help your readers understand the overall organization of your correspondence. They can see which topics go together and which are distinct.

Chunking to organize your text is accomplished by using any of the following techniques:

- **White space** (horizontal spacing between paragraphs, created by double or triple spacing)
- **Rules** (horizontal lines typed across the page to separate units of information)
- **Section dividers and tabs** (used in longer reports to create smaller units)
- **Headings and talking headings**

Headings and Talking Headings

To improve your page layout and make content accessible, use headings and talking headings. Headings—words or phrases such as "Introduction," "Discussion," "Conclusion," "Problems with Employees," or "Background Information"—highlight the content in a particular section of a document. When you begin a new section, you should use a new heading. In addition, use subheadings if you have a long section under one heading. This will help you break up a topic into smaller, more readable units of text.

Talking headings, in contrast, are more informative than headings. A heading helps your readers navigate the text by guiding them to key parts of a document. However, headings such as "Introduction," "Discussion," and "Conclusion" do not tell the readers what content is included in the section. Talking headings, such as "Human Resources Committee Reviews 2012 Benefits Packages," informatively clarify the content that follows.

One way to create a talking heading is to use a subject (someone or something performing the action), a verb (the action), and an object (something acted upon).

◀ EXAMPLE

TABLE 9.1 Talking Headings

Sentences Used as Talking Heads
Rude Customer Service Leads to Sales Losses
Accounting Department Requests Feedback on Benefits Package
Corporate Profit Sharing Decreases to 27 Percent
Parking Lot Congestion Angers Employees

Phrases Used as Talking Heads
Budget Increases to Be Frozen until 2012
Outsourced Workers Leading to Corporate Layoffs
Harlan Cisneros—New Departmental Supervisor
EEOC: Questions about Company Hiring Practices

Another way to create talking headings is to use informative phrases, such as "Problems Leading to Employee Dissatisfaction," "Uses of Company Cars for Personal Errands," and "Cost Analysis of Technology Options for the Accounting Department."

Table 9.1 provides examples of informative talking headings.

Notice how using some of these techniques improves the document design in Figure 9.2.

Order

Once a wall of unbroken words has been separated through chunking to help the reader understand the text's organization, the next thing a reader wants from your text is a sense of order. What's most important on the page? What's less important? What's least important? You can help your audience prioritize information by ordering—or *queuing*—ideas. The primary way to accomplish this goal is through a hierarchy of headings set apart from each other through various techniques:

- **Typeface.** There, are many different *typefaces* (or *fonts*), including Times New Roman, Courier, **Verdana**, Helvetica, Arial, Bauhaus 93, **Comic Sans MS**, *Lucida Calligraphy,* **Cooper Black,** and **STENCIL.** Whichever typeface you choose, it will either be a *serif* or *sans serif* typeface. Serif type has "feet" or decorative strokes at the edges of each letter. This typeface is commonly used in text because it is easy to read, allowing the reader's eyes to glide across the page.

Sans Serif is a block typeface that omits the feet or decorative lines. This typeface is best used for headings.

Though you have many font typefaces to choose from, all are not appropriate for every technical document. Times New Roman, Arial, and Calibri are best to use for letters, memos, e-mail, reports, resumes, and proposals, because these font types are professional looking and easy to read. Arial and Verdana are considered

FIGURE 9.2 Document Design Using Chunking to Organize the Information

MINUTES

First-level heading

The meeting at the Carriage Club was attended by 30 members and guests. After the dinner, Roger Traver introduced the guest speaker, George Smith, university chancellor, and noted his accomplishments and experiences prior to education—U.S. Navy commander, Oak Ridge Laboratory researcher, and politician. Dr. Smith's talk "Industry and Education Collaboration" was very interesting and included a history of special projects enjoyed by both academics and corporate heads. Dr. Smith suggested that we engineers could work with education to accomplish three goals.

Horizontal rule

Training Seminars

Recent industry–education collaborations include training seminars in computers, fiber optics, and human resource options.

Urban Development

Second-level heading

The chancellor's primary thrust was a request for $100,000 in financial aid for urban development. He said money had already been donated from three sources: a large realty firm, Capital Homes, had given $20,000; a philanthropic group, We Care, had donated a matching $20,000; Dr. Smith's university gave a matching $20,000. The remaining $40,000, Dr. Smith hoped, would come from industry donations.

Internships

The chancellor noted that industry could help itself, as well as the community, by providing internships for university undergraduate majors. These internships could either be semester- or year-long arrangements, whereby students would work for minimum wage to learn more about the day-to-day aspects of their chosen fields. The chancellor said that these internships would not only increase the students' theoretical knowledge of engineering by giving them hands-on experience but also make them better future employees for the host engineering companies. Everyone would benefit. Dr. Smith noted that the students would receive a grade and credit for their work.

Conclusion

After the speech, our VP introduced new business, calling for nominations for next year's officers; gave us the agenda for our next meeting; and adjourned the meeting.

best for Web sites since these fonts are very readable online. If you want to use "designer fonts," such as Comic Sans, Lucida Calligraphy, or Stencil, limit them to brochures and sales letters, for example.

- **Type size.** Another way of queuing for your readers is through the size of your type. A primary, first-level heading should be larger than subsequent, less important headings: second level, third level, and so forth. For example,

DOT.COM UPDATES

To access hundreds of designer fonts, check out the following link: http://flamingtext.com/

FIGURE 9.3   Examples of Typefaces and Type Sizes

Sans Serif Typefaces	Serif Typefaces
Avant Garde 12 point	Courier 12 point
Avant Garde 14 point	Courier 14 point
Avant Garde 18 point	Courier 18 point
Futura 12 point	Bookman 12 point
Futura 14 point	Bookman 14 point
Futura 18 point	Bookman 18 point
Helvetica 12 point	Goudy 12 point
Helvetica 14 point	Goudy 14 point
Helvetica 18 point	Goudy 18 point

a first-level heading could be in 18-point type. The second-level heading would then be set in 16-point type, the third-level heading in 14-point type, and the fourth-level heading in 12-point type.

Figure 9.3 shows examples of different typefaces and type sizes.

- **Density.** The *weight* of the type also prioritizes your text. Type *density* is created by boldfacing words.
- **Spacing.** Another queuing technique to help your readers order their thoughts is the *amount of horizontal space* used after each heading. You can emphasize headings with white space.
- **Position.** Your headings can be centered, aligned with the left margin, indented, or outdented (*hung heads*). No one approach is more valuable or more correct than another. The key is consistency. If you center your first-level heading, for example, and then place subsequent heads at the left margin, this should be your model for all chapters or sections of that report.

Figure 9.4 shows an *outdented* first-level heading with *indented* subsequent headings. Figure 9.5 shows a *centered* heading with subsequent headings aligned with the left margin.

Figure 9.6 reformats the meeting minutes seen in Figure 9.1 and uses queuing to order the hierarchy of ideas. The outdented first-level heading is set in a 12-point bold sans serif typeface, all caps. The second-level heading is set in a 10-point bold serif typeface and is separated from the preceding text by horizontal white space. The third-level heading is set in a 10-point bold serif typeface and is separated from the preceding text by double spacing. It is also set on the same line as the following text. Hierarchical heading levels shown in Figure 9.6 allow the readers to visualize the order of information to see clearly how the writer has prioritized text.

Access

Chunking helps the reader see which ideas go together, and a hierarchy of headings helps the reader understand the relative importance of each unit of information. Nonetheless, the document design in Figure 9.6 needs improvement. The reader still must read every

DOT.COM UPDATES

For more information about typography (links and news about font selection), check out the following link: http://www.microsoft.com/typography/default.mspx

FIGURE 9.4   Outdented and Indented Headings

Outdented
Heading

Text

Indented Heading

Text

Indented Heading

Text

word carefully to see the key points within each chunk of text. Readers are not that generous with their time. As the writer, you should make your reader's task easier.

A third way to assist your audience is by helping them *access* information rapidly—at a glance. You can use any of the following highlighting techniques to help the readers filter out extraneous or tangential information and focus on key ideas:

- **White space.** In addition to horizontal space, created by double or triple spacing, you also can create *vertical space* by indenting. This vertical white space breaks up the monotony of wall-to-wall words and gives your readers breathing room. White space invites your readers into the text and helps the audience focus on the indented points you want to emphasize.
- **Bullets.** Used to emphasize items within an indented list, bullets are created by using asterisks (*), hyphens (-), a lowercase *o*, degree signs (°), typographic symbols (■, ❑, •, or ♦), or iconic webdings and wingdings (☞, ☺, or ✔).
- **Numbering.** *Enumeration* creates itemized lists that can show sequence or importance and allow for easy reference.

FIGURE 9.5   Centered and Left-Margin Aligned Headings

Centered Heading

Left–Margin–Aligned Heading

_____

_____

_____ Text _____

_____

_____

Left–Margin–Aligned Heading

_____

_____

_____

_____ Text _____

_____

Left–Margin–Aligned Heading

_____

_____

_____

_____ Text _____

_____

- **Boldface.** To emphasize a key word or phrase in text, use boldface.
- **All caps.** The technique of capitalizing text is an excellent way to highlight a WARNING, DANGER, CAUTION, or NOTE. However, capitalizing other types of information is not suggested because reading lowercase words is easier for your audience. All caps creates a block of letters in which individual letters aren't easily distinguished from each other. Avoid using all caps in correspondence since typing in all caps can create a negative tone. Readers might think you are yelling at them or "flaming."
- **Underlining.** Underlining should be used cautiously. If you underline too frequently, none of your information will be emphatic. One underlined word or phrase will call attention to itself and achieve reader access. Several underlined words or phrases will overwhelm your readers.
- **Italics.** Both italics and underlining are used similarly as highlighting techniques.

> **MINUTES**   The meeting at the Carriage Club was attended by 30 members and guests. After the dinner, Roger Traver introduced the guest speaker, George Smith, university chancellor, and noted his accomplishments and experiences prior to education—U.S. Navy commander, Oak Ridge Laboratory researcher, and politician. Dr. Smith's talk, "Industry and Education Collaboration," was very interesting and included a history of special projects enjoyed by both academics and corporate heads. Dr. Smith suggested that we engineers could work with education to accomplish three goals.
>
> **Urban Development**
> The chancellor's primary thrust was a request for $100,000 in financial aid for urban development. He said money had already been donated from three sources: a large realty firm, Capital Homes, had given $20,000; a philanthropic group, We Care, had donated a matching $20,000; Dr. Smith's university gave a matching $20,000.
> The remaining $40,000, Dr. Smith hoped, would come from industry donations.
>
> **Internships**
> The chancellor noted that industry could help itself, as well as the community, by providing internships for university undergraduate majors. These internships could either be semester- or year-long arrangements, whereby students would work for minimum wage to learn more about the day-to-day aspects of their chosen fields. The chancellor said that these internships would not only increase the students' theoretical knowledge of engineering by giving them hands-on experience but also make them better future employees for the host engineering companies. Everyone would benefit. Dr. Smith noted that the students would receive a grade and credit for their work.
>
> **Training Seminars.** Recent industry–education collaborations include training seminars in computers, fiber optics, and human resource options.
>
> **Conclusion**
> After the speech, our VP introduced new business, calling for nominations for next year's officers, gave us the agenda for our next meeting, and adjourned the meeting.

This hanging head is set in all caps and a sans serif font.

Create a hierarchy of headings by changing font size and style.

Headings create accessible content.

- **Text boxes.** Place key points in a text box for emphasis. You also can italicize and use all caps within the text box, as we have in the example.

> NOTE: Only place glass containers in the *red recycle bins*.
>
> Use *green bins* for paper and plastic.

◀ EXAMPLE

- **Fills.** You can highlight text boxes through fills (color, gradients, and shadings).
- **Inverse type.** You can help readers access information by using inverse type—printing white on black, versus the usual black on white.
- **Color.** Another way to make key words and phrases leap off the page is to color them. *Danger* would be red, for example, *Warning* orange, and *Caution* yellow. You can use color to help a reader access the first-level heading, a header, or a footer. (*Headers* contain information placed along the top margin of text; *footers* contain information placed along the bottom margin of text.)

  For instance, if your text is typed in a black font, then headings typed in a blue font would stand out more effectively. However, as with all highlighting techniques, a little bit goes a long way. Do not overuse color. Do not type several headings in different colors. Doing so could produce a very unprofessional impression.

EXAMPLE ▼
Effective Use of Colored Headings

Committee Action
The Budget and Personnel Committee will vote to approve the audit report at the July meeting.

Recommendation
The committee will recommend that a proposal be submitted to improve roadway construction.

Staff Contacts
Mel Henderson
Sean Thomson

EXAMPLE ▼
Ineffective Use of Colored Headings

Committee Action
The Budget and Personnel Committee will vote to approve the audit report at the July meeting.

Recommendation
The committee will recommend that a proposal be submitted to improve roadway construction.

Staff Contacts
Mel Henderson
Sean Thomson

All colors are not equal in visual value. Generally, dark-colored fonts provide the most contrast against light-colored backgrounds, or vice versa. For example, a black font on a white background (or a white font on a black background) creates optimum contrast. On the other hand, a light-colored font on a light background does not improve access.

EXAMPLE ▶   **Good Contrast Helps Access**

Committee Action
The Budget and Personnel Committee will vote to approve the audit report at the July meeting.

Recommendation
The committee will recommend that a proposal be submitted to improve roadway construction.

Staff Contacts
Mel Henderson
Sean Thomson

Inverse print (white on black) creates optimum contrast.

Bad Contrast Hurts Access

◀ EXAMPLE

Light-colored text on a
light-colored background
harms readability.

Committee Action
The Budget and Personnel Committee will vote to
approve the audit report at the July meeting.

Recommendation
The committee will recommend that a proposal be
submitted to improve roadway construction.

Staff Contacts
Mel Henderson
Sean Thomson

You also should use colors tastefully. Avoid garish color combinations in graphics or backgrounds (red and orange, pink and green, or purple and yellow, and so forth). Colors that clash will distract the reader more than aid access.

Clashing Color Combinations that Hurt Access

◀ EXAMPLE

Committee Action
The Budget and Personnel Committee will vote to
approve the audit report at the July meeting.

Recommendation
The committee will recommend that a proposal be
submitted to improve roadway construction.

Staff Contacts
Mel Henderson
Sean Thomson

Here is a very important consideration: When it comes to using highlighting techniques, *more is not better*. A few highlighting techniques help your readers filter out background data and focus on key points. Too many highlighting techniques are a distraction and clutter the document design. Be careful not to overdo a good thing. Figure 9.7 gives examples of several highlighting methods. Notice how Figure 9.8 uses highlighting techniques to help the readers access information in the meeting minutes.

Variety

Each of the document designs in Figures 9.2, 9.6, and 9.8 uses one column and is printed vertically on a traditional 8½ × 11-inch page (a type of printing called portrait; see Figure 9.9).

This is not your only option. Your reader might profit from more variety. You might want to use smaller or larger paper; vary the weight of your paper (10-pound, 12-pound, or heavier card stock paper), or even print your text on colored paper.

FIGURE 9.7   Highlighting Techniques for Access

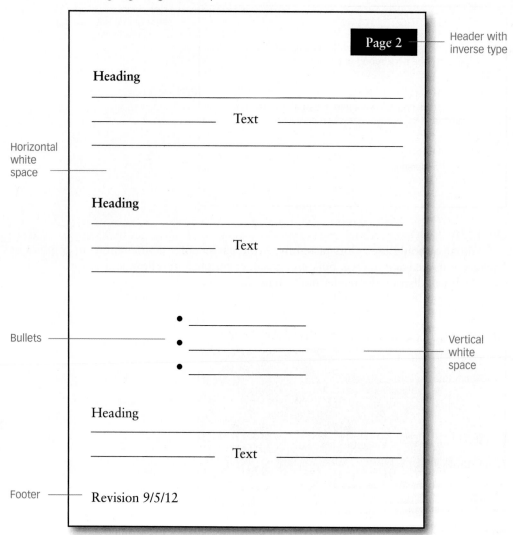

More important, you can vary the document design as follows:

- **Choose a different page orientation.** Rather than use portrait orientation (8½ × 11 inch), you could choose landscape orientation (11 × 8½ inch).
- **Use more columns.** Provide your reader two to five columns of text.
- **Vary gutter width.** Columns of text are separated by vertical white space called the *gutter*.
- **Use ragged-right margins.** Some text is fully justified (both right and left margins are aligned). Once this was considered professional, giving the text a clean look. Now, however, studies confirm that right-margin-justified text is harder for the audience to read. It's too rigid. In contrast, *ragged-right* type (the right margin is not justified) is easier to read and more pleasing to the eye. You can use this method to vary page layout.

Figure 9.10 shows how you can use columns, landscape orientation, and ragged-right margins to vary your document design.

Although you can vary your document design through page orientation and columns, the audience is still confronted by words, words, and more words. The majority of readers do not

FIGURE 9.8 Document Design Using Highlighting Techniques to Help Readers Access Key Ideas

MINUTES The meeting at the Carriage Club was attended by 30 members and guests. After the dinner, Roger Traver introduced the guest speaker, George Smith, university chancellor, and noted his accomplishments and experiences prior to education:

- U.S. Navy commander
- Oak Ridge Laboratory researcher
- Politician

Bulletized list

Dr. Smith's talk, "Industry and Education Collaboration," was very interesting and included a history of special projects enjoyed by both academics and corporate heads. Dr. Smith suggested that we engineers could work with education to accomplish three goals.

Urban Development
The chancellor's primary thrust was a request for $100,000 in financial aid for urban development. He said money had already been donated from three sources:

1. A large realty firm, Capital Homes, had given $20,000.
2. A philanthropic group, We Care, had donated a matching $20,000.
3. Dr. Smith's university also gave $20,000.

Numbered list

The remaining $40,000 would come from industry.

Internships
The chancellor noted that industry could help itself, as well as the community, by providing internships for university undergraduate majors:

1. Semester-long internships
2. Year-long internships

Students would work for minimum wage to learn more about the day-to-day aspects of their chosen fields. The chancellor said that these internships would not only increase the students' theoretical knowledge of engineering by giving them hands-on experience but also make them better future employees for the host engineering companies. <u>Everyone would benefit.</u> Dr. Smith noted that the students would receive a grade and credit for their work.

Underlining

Training Seminars. Recent industry–education collaborations include training seminars in computers, fiber optics, and human resource options.

Conclusion
After the speech, our VP introduced new business, calling for nominations for next year's officers, gave us the agenda for our next meeting, and adjourned the meeting.

---

want to wade through text. Luckily, words are not your only means of communication. You can reach a larger audience with different learning styles by varying your method of communication. Graphics are an excellent alternative. Many people are more comfortable grasping information visually than verbally. Although it's a cliché, a picture is often worth a thousand words.

To clarify our point about the value of variety, see Figure 9.11, which adds a graphic to the meeting minutes.

Visual Aids

See Chapter 10 for more discussion of tables and figures.

FIGURE 9.9 Portrait Orientation with One Column Fully Justified

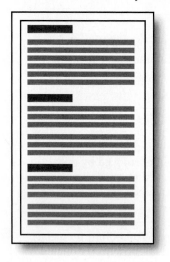

FIGURE 9.10 Landscape Orientation with Two Columns and Ragged Right Margins

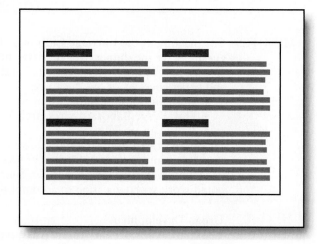

FIGURE 9.11 Document Design Using a Pie Chart to Add Variety

MINUTES   The meeting at the Carriage Club was attended by 30 members and guests. After the dinner, Roger Traver introduced the guest speaker, George Smith, university chancellor, and noted his accomplishments and experiences prior to education:

- U.S. Navy commander
- Oak Ridge Laboratory researcher
- Politician

Dr. Smith's talk, "Industry and Education Collaboration," was very interesting and included a history of special projects enjoyed by both academics and corporate heads. Dr. Smith suggested that we engineers could work with education to accomplish three goals.

Urban Development

The chancellor's primary thrust was a request for $100,000 in financial aid for urban development. He said money had already been donated from three sources, but business and industry can still help significantly. The following pie chart clarifies what money has been encumbered and how industry donations are still needed.

Internships

The chancellor noted that industry could help itself, as well as the community, by providing internships for university undergraduate majors:

1. Semester-long internships
2. Year-long internships

Students would work for minimum wage to learn more about the day-to-day aspects of their chosen fields. The chancellor said that these internships would not only increase the students' theoretical knowledge of engineering by giving them hands-on experience but

Bulletized list for accessibility

FIGURE 9.11 (Continued)

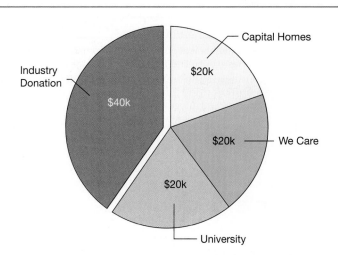

Graphic for visual appeal and accessible data

Figure 1 Donations

also make them better future employees for the host engineering companies. <u>Everyone would benefit.</u> Dr. Smith noted that the students would receive a grade and credit for their work.

Training Seminars. Recent industry–education collaborations include training seminars in computers, fiber optics, and human resource options.

Conclusion
After the speech, our VP introduced new business, calling for nominations for next year's officers, gave us the agenda for our next meeting, and adjourned the meeting.

TECHNOLOGY TIP

Document Design Using Microsoft Word

When it comes to document design, you have a world of options at your fingertips. Not only does your word processing software offer you possibilities to enhance page layout, but also the Internet provides unlimited resources.

Word Processing

In a word processing software program like Microsoft Word, you enhance your document's design in many ways. From the **Home** tab ribbon, you can make changes to **Font**, **Paragraphs**, and **Styles**.

(Continued)

Font category	Paragraph category	Styles category
• Choose your **FONT TYPE** and **SIZE**	• Create bulleted, numbered, or multilevel lists (as with outlines)	• Format text (Quick Styles) and change colors and fonts throughout a document (Change Styles)
• **Bold face**, *italicize*, and <u>underline</u> the text	• Decrease or increase an indentation	
• ~~strikethrough~~, create subscripts and superscripts	• Change the margins (ragged right, centered, block right, or full block)	
• Highlight the color of text, change the font color, or change the case of selected text to uppercase or lowercase	• Change the spacing between lines	
• Increase or decrease the font size	• Color background behind selected text, create borders, alphabetize, and show or hide paragraphing	

From the **Insert** tab ribbon shown below, you can add **Tables, Illustrations, Links, and Headers & Footers.**

The **Illustrations** category lets you insert pictures, clip art, and charts (pie charts, bar charts, line charts, etc.). **SmartArt** provides access to more complex charts such as Venn diagrams and organizational charts. The **Shapes** pull down menu lets you insert lines, arrows, flowchart symbols, callouts, and stars and banners.

The **Text** category lets you insert the following:

• 36 different kinds of text boxes including sidebars Text boxes can be enhanced further by changing the color of the lines or fill. You also can create shadows, as shown in this example.	• **WordArt** *This is an example of WordArt*

_____ 1. Have you broken up the text in your document into smaller chunks (units of information)?

_____ 2. Have you added white space to enhance readability of the text?

_____ 3. Did you use headings or talking headings to improve page layout and make content accessible?

_____ 4. Did you prioritize information by ordering (queuing) ideas?

_____ 5. Did you create a hierarchy of headings by relying on different typefaces and sizes, boldface, spacing, and varied positions?

_____ 6. Have you relied on a variety of highlighting techniques (indented text, bullets, numbers, boldface text, all caps, underline or italicizing, textboxes, or color) to make content accessible?

_____ 7. Have you avoided overuse of highlighting techniques?

_____ 8. Have you added variety to your technical document through landscape versus portrait orientation, columns, margin alignments, or graphics?

_____ 9. Have you considered using a different weight, size, or color paper?

_____ 10. Have you used color sparingly, and avoided overloading the document with excessive variety?

CHAPTER HIGHLIGHTS

1. Breaking your text into smaller chunks of information will help you create a more readable document.

2. When you create an order among the items in a document through use of typeface and type size, your audience can more easily prioritize the information.

3. Your reader should be able to glance at the document and easily pick out the key ideas. Highlighting techniques will help accomplish this goal.

4. Vary the appearance of your document by using columns, varying gutter widths, and printing in portrait or landscape orientation.

5. Your audience will access the content easily if you use white space, bullets, numbering, underlining, and text boxes.

APPLY YOUR KNOWLEDGE

CASE STUDIES

1. Reformat the following memo to improve its document design.

Date: November 30, 2012
To: Jan Hunt
From: Tom Langford
Subject: Cleaning Procedures for Manufacturing Walk-in Ovens #98731, #98732, and #98733

The above mentioned ovens need extensive cleaning. To do so, vacuum and wipe all doors, walls, roofs, and floors. All vents and dampers need to be removed, and a tack cloth must be used to remove all loose dust and dirt. Also, all filters need to be replaced.

   I am requesting this because loose particles of dust and dirt are blown onto wet parts when placed in the air-circulating ovens to dry. This causes extensive rework. Please perform this procedure twice per week to ensure clean production.

2. Reformat the following summary to improve its document design.

SUMMARY

The City of Waluska wants to provide its community with a safe and reliable water treatment facility. The goal is to protect Waluska's environmental resources and to ensure community values.

To achieve these goals, the city has issued a request for proposal to update the Loon Lake Water Treatment Plant (LLWTP). The city recognizes that meeting its community's water treatment needs requires overcoming numerous challenges. These challenges include managing changing regulations and protection standards, developing financially responsible treatment services, planning land use for community expansion, and upholding community values.

For all of the above reasons, DesignGlobal, Inc. (DGI) Engineering is your best choice. We understand the project scope and recognize your community's needs.

We have worked successfully with your community for a decade, creating feasibility studies for Loon Lake toxic control, developing odor-abatement procedures for your streams and creeks, and assessing your water treatment plant's ability to meet regulatory standards.

DGI personnel are not just engineering experts. We are members of your community. Our dynamic project team has a close working relationship with your community's regulatory agencies. Our Partner in Charge, Julie Schopper, has experience with similar projects worldwide, demonstrated leadership, and the ability to communicate effectively with clients.

DGI offers the City of Waluska an integrated program that addresses all your community's needs. We believe that DGI is your best choice to ensure that your community receives a water treatment plant ready to meet the challenges of the twenty-first century.

3. The following text is visually unappealing and inaccessible. Using the techniques discussed in this chapter (organization, order, access, and variety), improve the document's design.

In 1998, the Transportation Equity Act for the 21st Century (TEA-21) was passed and authorized Federal programs for roadway safety.

To limit its focus, the TEA-21 discussed safety as it was affected by facility planning, roadway design, and maintenance.

The TEA-21 was concerned about roadway safety for pedestrians, motorized vehicles, and non-motorized vehicles. Motorized vehicles included cars, vans, trucks, SUVs, and motorcycles. Non-motorized vehicles include bicycles, roller skates, rollerblades, and scooters.

The TEA-21's planning guidelines consisted of many goals. These included communicating revised safety programs to pedestrians, non-motorized vehicle users, and motor vehicle users; organizing safety data for analysis by highway and city road departments; strategies for reducing fatalities for pedestrians and vehicle users; and developing city safety blueprints based on statewide prototypes.

One prototypical model that the TEA-21 recommends following is Operation Green Light. This model uses a traffic signal coordination system to maintain a steady flow of traffic. Steady traffic flow has been proven to reduce fatalities by 15 percent.

4. The following short report is poorly formatted. The text is so dense that readers would have difficulty understanding the content easily. Improve the document's design to aid access. Use highlighting techniques discussed in this chapter to revise the text.

Date: May 18, 2012
To: Martha Collins
From: Richard Davis
Subject: 2012 Switch Port Carriers

Attached are the supplemental 2012 Switch Port Carriers that are required to support this year's growth patterns. As we have discussed in previous phone conversations, the May numbers show a decrease in traffic, but forecasts still suggest increased traffic. Therefore, the ports being placed in the network via these plans will support our future growth except for areas where growth cannot be predicted. Some areas, for example, are too densely populated for forecasting because the company did not hire enough survey personnel to do a thorough job.

Following is an update of our suggestions port additions. For Port 12ABR, add 16 ports in Austin. For Port 13RgX, add 27 ports in Houston. For Port 981D, add 35 ports in San Antonio. For Port 720CT, add 18 ports in Dallas. The total port additions will be 96 and cost $3,590,625.

After working long hours on these suggestions, I learned that port additions should be considered mandatory. However, follow-up forecasts are probably needed due to the short time we were provided to do these studies. If you are going to perform these follow-up forecasts, do so before September 1. The survey teams, if you want a successful forecast, need at least three months. Twenty-five team members should be sufficient.

5. Review the following PowerPoint screen. First, explain how the screen is visually flawed. Second, revise it for improved access using the techniques suggested in this chapter.

Recommendations

Table 1 outlines the investments recommended for *80 percent of the fiscal year 2011 funds.*

The table focuses on 80 percent of the available funds. The remaining *20 percent are being withheld until fiscal year 2012.*

Company	Costs
Harness	$2,300
MarTT3	$4,500
Notary Rex	$5,200
Mobile CRT	$2,750

INDIVIDUAL AND TEAM PROJECTS

1. Bring samples of technical communication to class. These could include letters, brochures, fliers, instructions, proposals, or reports.

    Assess the document design of each sample. Determine which samples have successful document designs and which samples have poor document designs. Base your decisions on the criteria provided in this chapter: organization, order, access, and variety. Either orally or in writing, share your findings with other teams in the class.

2. Take one of the less successful samples from team project 1 and reformat it to improve its document design. Focus on improving the sample's organization, order, access, and variety.

    Once your team has completed reformatting the text, share your revisions with the class. Then review the various team projects, determining which team created the best document design. Select a winner, and explain why this text now has a successful document design.

3. Read the following headings and make them more accessible by creating a hierarchy using different font sizes and font types:

    Meeting Minutes

    Agenda

    Discussion of Ongoing Projects

    Recommendations

    Pricing

    Cost of Equipment

    Cost of Facilities Update

    Cost of Insurance Benefits

    New Hiring Policies

    Job Requirements

    Employee Credentials

    Licensing

4. Revise the following headings into "talking headings" by changing them to phrases or complete sentences. To do so, add any information you choose to clarify your content.

    - Computer Problems
    - WiFi Compatibility
    - Biotechnology Advances
    - Accounting Regulations
    - Facilities Update

PROBLEM-SOLVING THINK PIECES

1. The following two lists need to separate and highlight the information more clearly. But which highlighting technique should you use for each list—bullets or numbers? Explain why you would use bullets vs. numbers, or vice versa for both lists.

List 1	List 2
To access your online course, follow these steps:	To choose the right car for your family and business needs, consider these factors:
Turn on the computer.	Price
Double click the Internet icon on your desktop.	Options
Type in the following URL:	Fuel economy
http://webct.acc.edu.	Cost of repairs
Type your username and password.	Availability of dealerships
Click on the online course of your choice.	Financing
Complete assignment 1.	Capacity

2. Look at the following two lists with headings. Both use Comic Sans and colored headings, but should they? Assessing the two lists below, explain when and why it's okay to use designer fonts and color or when and why you should avoid their use.

EXAMPLE 1   RESUME

Objective	Use my information technology skills to improve a company's network capabilities, computer security, and troubleshooting.
Qualifications	• Four years experience in information technology • BS in Computer Science • Security certified
Work History	Information Technologist Pantheon Corp. St. Louis, MO 2005–present • Networked 24 computer stations • Created computer passwords for all employees • Installed Spam and Virus protection on all systems

EXAMPLE 2   BROCHURE

Prices	• Buy one, get one free • Guaranteed lowest in the market • Last year's prices—today!
Options	• Sizes—4″ × 2″, 3″ × 2″, and 2″ × 2″ • Colors—red, green, black, and silver • WiFi compatible
Service	• 24/7 • On site or online help

3. Which is the best way to show a comparison/contrast to your audience: a list with headings or a table? Assess the following two choices, decide which document design is best, and justify your answer.

OPTION 1 LIST WITH HEADINGS
Renner Road
• Two lanes
• No sidewalk
• Limited lighting
Bannister Road
• Four lanes
• Sidewalks on both sides of the road
• Street lighting on one side of the road
Shawnee Road
• Four lanes
• Sidewalks on both sides of the road
• Street lighting in the center esplanade

OPTION 2 TABLE		
Renner Road	**Bannister Road**	**Shawnee Road**
• Two lanes	• Four lanes	• Four lanes
• No sidewalk	• Sidewalks on both sides of the road	• Sidewalks on both sides of the road
• Limited lighting	• Street lighting on one side of the road	• Street lighting in the center esplanade

WEB WORKSHOP

On the Internet, access ten corporate Web sites. Study them and make a list of the techniques used for visual communication. Which Web sites are successful, and why? Which Web sites are unsuccessful, and why? How would you redesign the less successful Web sites to achieve better visual communication?

QUIZ QUESTIONS

1. What are three reasons for designing a technical document effectively?
2. What is chunking?
3. What do you achieve by chunking?
4. In what four ways can you accomplish chunking?
5. What is the purpose of creating order or queuing your ideas?
6. How can you achieve a hierarchy of headings in your technical document?
7. What's the difference between headings and talking headings?
8. How can you help your readers access your information, filtering out extraneous or tangential information?
9. What can happen if you overuse highlighting techniques?
10. What are four ways in which you can vary your document design?

CHAPTER TEN

Using Visual Aids in Technical Communication

COMMUNICATION AT WORK

To help his clients understand complex figures, Bert Lang includes visual aids in his proposals.

Bert Lang is an investment banker at Country Commercial Bank. He is writing a proposal to a potential client, Sylvia Light, a retired public health nurse. Sylvia is 68 years old and worked for the Texas Public Health Department for 36 years. She has earned her State of Texas retirement and Social Security benefits. She now has $315,500 allocated as follows: $78,000 in an individual retirement account (IRA), $234,000 in a low-earning certificate of deposit (CD), and $3,500 in her checking account.

Sylvia contacted Bert, asking him to help her organize her portfolio for a comfortable retirement. Bert has studied Sylvia's various accounts and considered her lifestyle and expenditures. Now, he is ready to write the proposal.

In this proposal, Bert wants to show Sylvia how to reallocate her funds. She should keep some ready money available and invest a portion of capital for long-term returns. Currently, too

Objectives

much of her money is tied up in a CD earning 1.1%. Bert plans to propose that Sylvia could reallocate her funds as follows:

- $110,000 in an annuity
- $78,000 in an individual retirement account (IRA)
- $45,000 in municipal bonds
- $37,000 in a stock fund
- $42,000 in certificates of deposit (CD)
- $3,500 in a checking account

Like most people, Sylvia is unfamiliar with financial planning. Though she was an expert in her health field, tuberculosis treatment, money matters confuse her. Numbers alone will not explain Bert's vision for her money management.

To make this proposal visually appealing and more readily understandable to Sylvia, Bert will use visual aids. He will provide Sylvia a pie chart to show how he wants to invest her money. Bert will use a line graph to predict how much more money she can earn by reallocating her assets. Finally, he will create a table to clarify the types of investments, the amount in each investment, the interest to be earned, and the fees.

Although Sylvia has always been fiscally conservative, Bert hopes that he can explain the need for growth of capital even in retirement. The graphic aids will visually enhance his written explanation.

The Benefits of Visual Aids

Although your writing may have no grammatical or mechanical errors and you may present valuable information, you won't communicate effectively if your information is inaccessible. Consider the following paragraph:

EXAMPLE ▶

> According to a Nielsen Wire report, Social Media usage from April 2008 to April 2009 grew substantially. For example, in April 2008, Facebook's total minutes used were 1,735,698. This number grew to 13,872,640 in April 2009. This equals a 699 percent growth. During the same time span, Myspace usage decreased from 7,254,645 to 4,973,919. This is a 31 percent decrease in usage. In April 2008, Blogger usage was 448,710 minutes and increased to 582,683 minutes in April 2009. This equals a 30 percent increase in usage. During the same time period, LinkedIn accounted for 119,636 minutes used. LinkedIn usage increased 69 percent in April 2009, or 202,407 minutes. Finally, from April 2008 to April 2009, Twitter usage increased 3,712 percent. This was 7,865 minutes to 299,836 minutes of usage ("Time Spent on Facebook up 700%").

If you read the preceding paragraph in its entirety, you are an unusually dedicated reader. Such wall-to-wall words mixed with statistics do not create easily readable writing. The goal of effective technical communication is accessible information. The example paragraph fails to meet this goal. No reader can digest the data easily or see clearly the comparative changes in social media usage from year to year.

To present large blocks of data or reveal comparisons, you can supplement, if not replace, your text with graphics. In technical communication, visual aids accomplish several goals. Graphics (whether hand drawn, photographed, or computer generated) will help you achieve conciseness, clarity, and cosmetic appeal.

Conciseness

Visual aids allow you to provide large amounts of information in a small space. Words used to convey data (such as in the example paragraph) double, triple, or even quadruple the space needed to report information. By using graphics, you can also delete many unnecessary words and phrases.

Clarity

Visual aids can clarify complex information, such as trends, comparison, percentages, and facts and figures. See Table 10.1 for examples of how you can use graphics to clarify content.

Cosmetic Appeal

Visual aids help you break up the monotony of wall-to-wall words. If you only give unbroken text, your reader might tire, lose interest, and overlook key concerns. Graphics help you sustain your reader's interest. Let's face it; readers like to look at pictures.

Color

All graphics look best in color; don't they? Not necessarily. Without a doubt, a graphic depicted in vivid colors will attract your reader's attention. However, the colors might not aid communication. For example, colored graphics could have these drawbacks (Reynolds and Marchetta 5–7):

1. The colors might be distracting (glaring orange, red, and yellow combinations on a bar chart would do more harm than good).

TABLE 10.1   Use of Graphics to Clarify Content

Use of Visual Aid	Sample Graphic
Trends—Certain trends, such as increasing or decreasing sales figures, enrollment numbers, revenues, stock prices, and so on, are most evident in line graphs.	**FIGURE 1** Online Class Growth 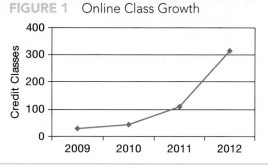
Comparisons between like components—Comparisons of sizes, costs, or amounts can be seen in grouped bar charts.	**FIGURE 2** Online Classes vs. On-Campus Classes 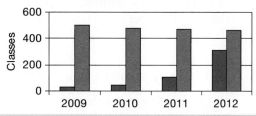
Percentages—Pie charts help readers discern the divisions of parts of a whole.	**FIGURE 3** Ages of Online Students 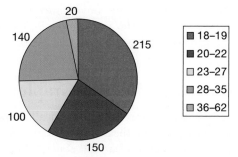

Facts and figures—A table states statistics/numbers more clearly than a wordy paragraph.

TABLE 1 Online Student Facts

Student Age Cohort	Number of Online Students	Number of Online Credit Hours
18–19	215	1,290
20–22	150	900
23–27	100	600
28–35	140	840
36–62	20	120
Total	625	3,750

2. Colors that look good today might go out of style in time.

3. Colored graphics increase production costs.

4. Colored graphics consume more disk space and computer memory than black-and-white graphics.

5. If your reader prints out the text in black and white, your original colors will have no meaning. In fact, colors, when printed in black and white, will not be distinguishable.

6. The colors you use might not look the same to all readers. Just because you see the colors one way does not mean your readers will see them the same. The color on a computer monitor depends on its resolution (the number of pixels displayed) and the monitor's RGB values (how much red, green, and blue light is displayed). Because all monitors do not display these same values, what you see on your monitor will not necessarily be the same as what your reader sees.

   To solve the problems with color and resolution, test your graphics on several monitors. Also, limit your choices to primary colors instead of the infinite array of other color possibilities. Use patterns to distinguish your information so that the color becomes secondary to the design.

Three-Dimensional Graphics

Many people are attracted to three-dimensional (3-D) graphics. After all, they have obvious appeal. Three-dimensional graphics are more interesting and vivid than flat, one-dimensional graphics. However, 3-D graphics have drawbacks. A 3-D graphic is visually appealing, but it does not convey information quantifiably. A word of caution: Use 3-D graphics sparingly. Better yet, use the 3-D graphic to create an impression; then include a data table to quantify your data (Figure 10.1).

Criteria for Effective Graphics

Figure 10.2 is an example of a cosmetically appealing, clear, and concise graphic. At a glance, the reader can pinpoint the comparative prices per barrel of crude oil between 2006 and 2012. Thus, the line graph is clear and concise. In addition, the writer has included an interesting artistic touch. The oil gushing out of the tower shades parts of the graph to emphasize the dollar amounts. Envision this graph without the shading. Only the line would exist. The shading provides the right touch of artistry to enhance the information communicated.

The graph shown in Figure 10.2 includes the traits common to effective visual aids. Successful tables and figures have these characteristics:

1. Integrated with the text (i.e., the graphic complements the text; the text explains the graphic).

2. Appropriately located (preferably immediately following the text referring to the graphic and not a page or pages later).

3. Enhance the material explained in the text (without being redundant).

4. Communicate important information that could not be conveyed easily in a paragraph or longer text.

5. Do not contain details that detract from rather than enhance the information.

6. Sized effectively (large enough to be readable but not so large as to overwhelm the page).

7. Correctly labeled (with numbers, titles, legends, and headings).

8. Follow the style of other figures or tables in the text (same font size, font style, color, size of the graphic, and so on).

FIGURE 10.1 3-D Bar Chart and Data Table

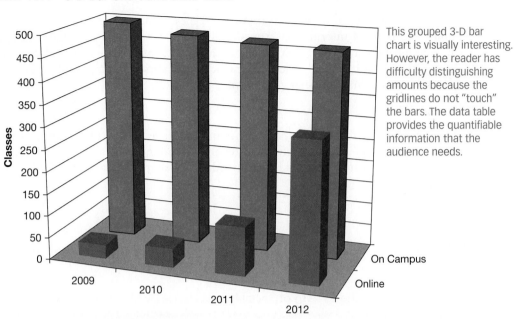

This grouped 3-D bar chart is visually interesting. However, the reader has difficulty distinguishing amounts because the gridlines do not "touch" the bars. The data table provides the quantifiable information that the audience needs.

	2009	2010	2011	2012
■ Online	30	42	110	315
■ On Campus	500	480	470	465

FIGURE 10.2 Line Graph with Shading

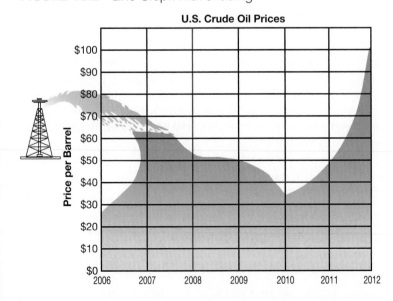

Types of Graphics

Graphics can be broken down into two basic types: tables and figures. Tables provide columns and rows of information. You should use a table to make factual information, such as numbers, percentages, and monetary amounts, easily accessible and understandable. Figures, in contrast, are varied and include bar charts, line graphs, photographs, pie charts, schematics, line drawings, and more.

Tables

Let's tabulate the information about social media usage from 2008 to 2009 presented earlier on page 294. Because effective technical communication integrates text and graphic, you will want to provide an introductory sentence prefacing Table 10.2, as follows:

EXAMPLE ▶

Table 10.2 compares the amount in minutes of social media usage in April 2008 and April 2009.

TABLE 10.2   The Growth of Social Media Usage by Minutes

Site	April 2008 Total Minutes	April 2009 Total Minutes	Growth/Decrease
Facebook	1,735,698	13,872,640	+ 699%
Myspace	7,254,645	4,973,919	− 31%
Blogger	448,710	582,683	+ 30%
LinkedIn	119,636	202,407	+ 69%
Twitter	7,865	299,836	+3,712%

Source: "Time Spent on Facebook up 700%, but MySpace Still Tops for Video"

Table 10.2 has advantages for both the writer and the reader. First, the headings eliminate needless repetition of words, thereby making the text more readable. Second, the audience can see easily the comparison between the minutes of social media usage in April 2008 versus April 2009. Thus, the table highlights the content's significant differences. Third, the table allows for easy future reference. Tables could be created for each year. Then the reader could compare quickly the changes in social media usage. Finally, if this information is included in a report, the writer will reference the table in the list of illustrations. This creates ease of access for the reader.

Criteria for Effective Tables.   To construct tables correctly, do the following:

1. Number tables in order of presentation (i.e., Table 1, Table 2, Table 3).
2. Title every table. In your writing, refer to the table by its number, not its title. Simply say, "Table 1 shows . . . ," "As seen in Table 1," or "The information in Table 1 reveals . . ."
3. Present the table as soon as possible after you have mentioned it in your text. Preferably, place the table on the same page as the appropriate text, not on a subsequent, unrelated page or in an appendix.
4. Don't present the table until you have mentioned it.

5. Use an introductory sentence or two to lead into the table.

6. After you have presented the table, explain its significance. You might write, "Thus, the use of social media grew for all options except for Myspace."

7. Write headings for each column. Choose terms that summarize the information in the columns. For example, you could write "% of Error," "Length in Ft.," or "Amount in $."

8. Because the size of columns is determined by the width of the data or headings, you may want to abbreviate terms (as shown in item 7). If you use abbreviations, however, be sure your audience understands your terminology.

9. Center tables between right and left margins. Don't crowd them on the page.

10. Separate columns with ample white space, vertical lines, or dashes.

11. Show that you have omitted information by printing two or three periods or a hyphen or dash in an empty column.

12. Be consistent when using numbers. Use either decimals or numerators and denominators for fractions. You could write 3 1/4 and 3 3/4 or 3.25 and 3.75. If you use decimal points for some numbers but other numbers are whole, include zeroes. For example, write 9.00 for 9.

13. If you do not conclude a table on one page, on the second page write *Continued* in parentheses after the number of the table and the table's title.

14. Cite the source of your information, if you have taken content from primary or secondary sources.

Table 10.3 is an excellent example of a correctly prepared table.

Figures

Another way to enhance your technical communication is to use figures. Whereas tables eliminate needless repetition of words, figures highlight and supplement important points in your writing. Like tables, figures help you communicate with your reader.

TABLE 10.3 Student Headcount Enrollment by Age Group and Student Status, Fall 2012

Age Group	New Students	Continuing Students	Readmitted	Other	Total
15–17	453	33	2	2	490
18–20	1,404	1,125	132	—	2,661
21–23	339	819	269	—	1,427
24–26	263	596	213	—	1,072
27–29	250	436	134	—	820
30–39	524	1,168	372	—	2,064
40–49	271	510	186	—	967
50–59	76	121	54	—	251
60+	19	48	16	—	83
Unknown	109	92	27	2	230
Total	3,708	4,948	1,405	4	10,065

Source: "City Community College Enrollment Statistics"

(Annotations: Table number and title — Column headings — Indicate if no data is available by using a dash, periods, or hyphens. — Provide a final row for totals when necessary.)

Creating Graphics in Microsoft Word 2010 (Pie Charts, Bar Charts, Line Graphs, etc.)

You can create customized graphics in Microsoft Word as follows:

1. Click on **Insert** on the menu bar.

2. Click on **Chart**. Word will open a datasheet and bar chart template that you can customize by inserting your own text and numbers.

		A	B	C	D	E
		1st Qtr	2nd Qtr	3rd Qtr	4th Qtr	
1	East	20.4	27.4	90	20.4	
2	West	30.6	38.6	34.6	31.6	
3	North	45.9	46.9	45	43.9	
4						

Once you have opened Word's graphic's datasheet and template, you can customize the graphic further as follows:

3. Choose the type of graphic you want by clicking on **Chart** on the menu bar and scrolling to and selecting **Chart Type**. The **Chart Type** dialog box displays, allowing you to select a chart type.

4. To add figure numbers, figure titles, legends, gridlines, data labels, and data tables, click on **Chart** and then select **Chart Options**.

Types of figures include the following:

- Bar charts
  - Grouped bar charts
  - 3-D (tower) bar charts
  - Pictographs
  - Gantt charts
  - 3-D topographical charts
- Pie charts
- Line charts
  - Broken line charts
  - Curved line charts
- Combination charts
- Flowcharts
- Organizational charts
- Schematics
- Geologic maps
- Line drawings
  - Exploded views
  - Cutaway views
  - Renderings
  - Virtual reality drawings
- CAD drawings
- Photographs
- Icons
- Online graphics

All of these types of figures can be computer generated using an assortment of computer programs. The program you use depends on your preference and hardware.

Criteria for Effective Figures. To construct figures correctly, do the following:

1. Number figures in order of presentation (i.e., Figure 1, Figure 2, Figure 3).
2. Title each figure. When you refer to the figure, use its number rather than its title: "Figure 1 shows the relation between the average price for houses and the actual sales prices."
3. Preface each figure with an introductory sentence.
4. Avoid using a figure until you have mentioned it in the text.
5. Present the figure as soon as possible after mentioning it instead of several paragraphs or pages later.
6. After you have presented the figure, explain its significance. Don't let the figure speak for itself. Remind the reader of the important facts you want to highlight.
7. Label the figure's components. For example, if you are using a bar or line chart, label the x- and y-axes clearly. If you're using line drawings, pie charts, or photographs, use clear *callouts* (names or numbers that indicate particular parts) to label each component.
8. When necessary, provide a legend or key at the bottom of the figure to explain information. For example, a key in a bar or line chart will explain what each differently colored line or bar means. In line drawings and photographs, you can use numbered callouts in place of names. If you do so, you will need a legend at the bottom of the figure explaining what each number means.
9. If you abbreviate any labels, define these in a footnote. Place an asterisk (*) or a superscript number (^{1}, ^{2}, ^{3}) after the term and then at the bottom of the figure where you explain your terminology.
10. If you have drawn information from another source, note this at the bottom of the figure.
11. Frame the figure. Center it between the left and right margins or place it in a text box.
12. Size figures appropriately. Don't make them too small or too large.

Bar Charts. Bar charts show either vertical bars (as in Figure 10.3) or horizontal bars (as in Figure 10.4). These bars are scaled to reveal quantities and comparative values. You can shade, color, or crosshatch the bars to emphasize the contrasts. If you do so, include a legend explaining what each bar represents, as in Figure 10.3. *Pictographs* (as in Figure 10.5) use picture symbols instead of bars to show quantities. To create effective pictographs, do the following:

1. The picture should be representative of the topic discussed.
2. Each symbol equals a unit of measurement. The size of the units depends on your value selection as noted in the key or on the x- and y-axes.
3. Use more symbols of the same size to indicate a higher quantity; do not use larger symbols.

Gantt Charts. *Gantt charts*, or *schedule charts* (as in Figure 10.6), use bars to show chronological activities. For example, your goal might be to show a client phases of a project. This could include planned start dates, planned reporting milestones, planned completion dates, actual progress made toward completing the project, and work remaining. Gantt charts are an excellent way to represent these activities visually. They are often included in proposals to project schedules or in reports to show work completed. To create successful Gantt charts, do the following:

1. Label your x- and y-axes. For example, the y-axis represents the various activities scheduled, and then the x-axis represents time (either days, weeks, months, or years).

FIGURE 10.3 Vertical Bar Chart of Social Media Usage by Minutes ◄─────── Figure number and title

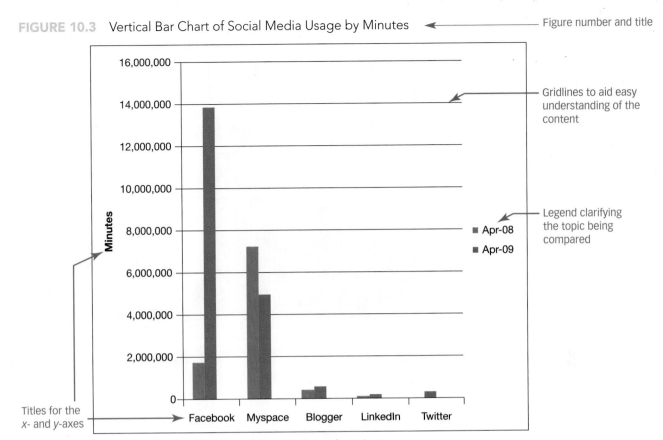

Gridlines to aid easy
understanding of the
content

Legend clarifying
the topic being
compared

■ Apr-08
■ Apr-09

Titles for the
x- and y-axes

Source: "Time Spent on Facebook up 700%, but MySpace Still Tops for Video"

FIGURE 10.4 Horizontal Bar Chart for High-Tech Readers

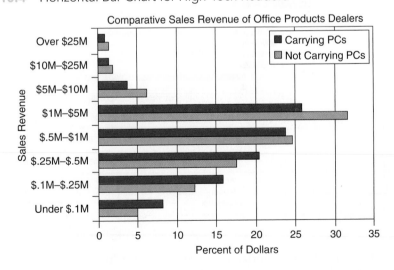

Comparative Sales Revenue of Office Products Dealers

■ Carrying PCs
☐ Not Carrying PCs

This bar chart, geared
toward office supply
sales managers, is as
factual as the pictograph
in Figure 10.5. However,
Figure 10.4, which omits
the drawings of the PCs,
keeps the graphic more
businesslike for the
intended audience of
high-tech peers.

FIGURE 10.5   Pictograph for Lay Audience

The pictograph is as factual as the bar chart in Figure 10.4. However, since the pictograph is designed for lay readers, it uses symbols of PCs to enhance the visual appeal of this topic. In addition, the PCs make the subject matter more interesting for the audience.

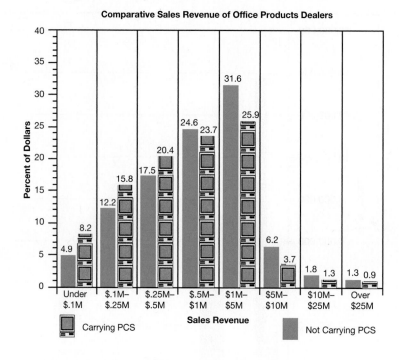

FIGURE 10.6   Gantt Chart

The Gantt chart shows the schedule of activities, including start and stop dates, plus percentage of completion.

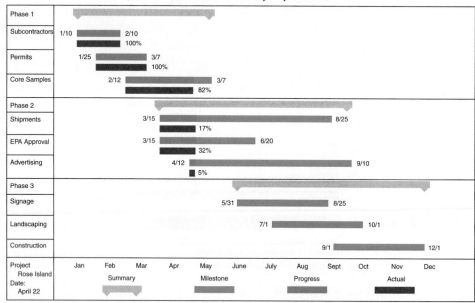

2. Provide gridlines (either horizontal or vertical) to help your readers pinpoint the time accurately.
3. Label your bars with exact dates for start or completion.
4. Quantify the percentages of work accomplished and work remaining.
5. Provide a legend to differentiate between planned activities and actual progress.

FIGURE 10.7 Topographical Chart

FIGURE 10.8 Pie Chart for Social Media Usage in 2009

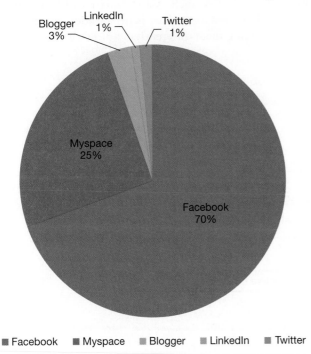

■ Facebook ■ Myspace ■ Blogger ■ LinkedIn ■ Twitter

Source: "Time Spent on Facebook up 700%, but MySpace Still Tops for Video"

3-D Topographical Charts.   Three-dimensional contour representations are not limited to land elevations. A three-dimensional surface chart could be used to represent many different forms of data. These 3-D "topos" (as shown in Figure 10.7) are used in industries as varied as aerospace, defense, education, research, oil, gas, and water. Applications include CAD/CAM, statistical analysis, and architectural design.

Pie Charts.   Use pie charts (as in Figure 10.8) to illustrate portions of a whole. The pie chart represents information as pie-shaped parts of a circle. The entire circle equals 100 percent or 360 degrees. The pie pieces (the wedges) show the various divisions of the whole.

To create effective pie charts, do the following:

1. Be sure that the complete circle equals 100 percent or 360 degrees.
2. Begin spacing wedges at the twelve o'clock position.
3. Use shading, color, or crosshatching to emphasize wedge distributions.
4. Use horizontal writing to label wedges.
5. If you don't have enough room for a label within each wedge, provide a legend defining what each shade, color, or crosshatching symbolizes.
6. Provide percentages for wedges when possible.
7. Do not use too many wedges—this would crowd the chart and confuse readers.
8. Make sure that different sizes of wedges are fairly large, readable, and dramatic.

Line Charts. Line charts reveal relationships between sets of figures. To make a line chart, plot sets of numbers and connect the sets with lines. These lines create a picture showing the upward and downward movement of quantities. Line charts of more than one line (see Figure 10.9) are useful in showing comparisons between two sets of values. However, avoid creating line charts with too many lines, which will confuse your readers.

Combination Charts. A combination chart reveals relationships between two sets of figures. To do so, it uses a combination of figure styles, such as a bar chart and a line chart (as shown in Figure 10.10). The value of a combination chart is that it adds interest and distinguishes the two sets of figures by depicting them differently.

FIGURE 10.9 Line Chart Comparing Departmental Budgets

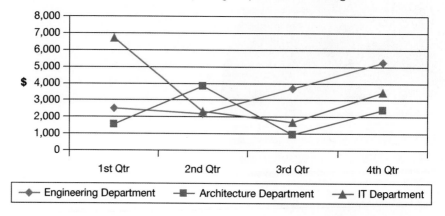

FIGURE 10.10 Combination Chart (Bar and Line)

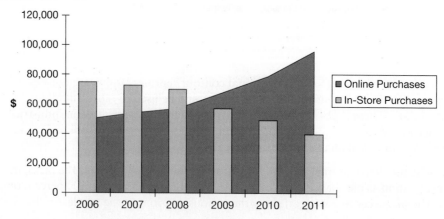

Flowcharts.   You can show a chronological sequence of activities using a flowchart. Flowcharts are especially useful for writing technical instructions. When using a flowchart, remember that ovals represent starts and stops, rectangles represent steps, and diamonds equal decisions (see Figure 10.11).

Instructions and flowcharts

See Chapter 13 for more discussion of flowcharts in instructions.

Organizational Charts.   The chart in Figure 10.12 shows the chain of command in an organization. You can use boxes around the information or use white space to distinguish among levels in the chart. An organizational chart helps your readers see where individuals work within a business and their relation to other workers.

Schematics.   Schematics are useful for presenting abstract information in technical fields such as electronics and engineering. A schematic diagrams the relationships among the parts of something such as an electrical circuit. The diagram uses symbols and abbreviations familiar to highly technical readers.

The schematic in Figure 10.13 shows various electronic parts (resistors, diodes, and condensors) in a radio.

Geologic Maps.   Maps help us understand locations. Usually these show cities, streets, roads, highways, rivers, lakes, mountains, and so forth. Geologic maps do more. They also show terrain, contours, heat ranges, the surface features of a place or object, or an analysis of an area. Often, to help readers orient themselves, these maps are printed on top of a regular map (called a base map). The base map is black and white. The geologic map, in contrast, uses colors, contact and fold lines, and special symbols to reveal the geology of an area. These features are then defined on a map legend (as in Figure 10.14).

FIGURE 10.11   Flowchart

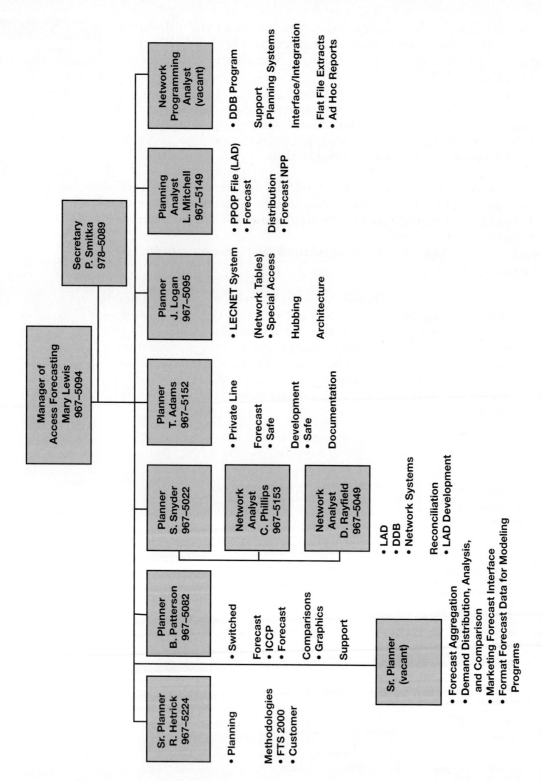

FIGURE 10.12 Organizational Chart

FIGURE 10.13 Schematic of a Radio

FIGURE 10.14 Geologic Map

Source: Courtesy of George Butler Associates, Inc.

Line Drawings.   Use line drawings to show the important parts of a mechanism. To create line drawings, do the following:

1. Maintain correct proportions in relation to each part of the object drawn.
2. If a sequence of drawings illustrates steps in a process, place the drawings in left-to-right or top-to-bottom order.
3. Using callouts to name parts, label the components of the object drawn (see Figure 10.15).
4. If there are numerous components, use a letter or number to refer to each part. Then reference this letter or number in a key or legend (see Figure 10.16).
5. Use exploded views (Figures 10.17 and 10.18) or cutaways (Figures 10.17 and 10.18) to highlight a particular part of the drawing.

FIGURE 10.15   Line Drawing of Ventilator (Exploded View with Callouts)

Source: Courtesy of Nellcor Puritan Bennett Corp.

Renderings, Virtual Reality Drawings, and Section Elevation. Three different types of line drawings are renderings, virtual reality views, and section elevations. They offer 3-D representations of buildings, sites, or objects. Often used in the architectural/engineering industry, these 3-D drawings (as shown in Figures 10.19, 10.20, and 10.21) help clients get a visual idea of what services your company can provide. Renderings, virtual reality drawings, and section elevations add lighting, materials, and shadow reflection mapping to mimic the real world and allow customers to see what a building or site will look like in a photorealistic setting.

FIGURE 10.16 Line Drawing of Exhalation Valve (Exploded View with Legend)

Exhalation Valve Parts List		
Item	Part Number	Description
1	000723	Nut
2	003248	Cap
3	T50924	Diaphragm
4	Reference	Valve Body
5	Reference	Elbow Connector
—	T11372	Exhalation Valve

Source: Courtesy of Nellcor Puritan Bennett Corp.

FIGURE 10.17 Line Drawing of Cable (Cutaway View)

Source: Courtesy of Nellcor Puritan Bennett Corp.

FIGURE 10.18 Cutaway View of a Railcar Braking System

Source: Courtesy of Burlington Northern Santa Fe Railroad

FIGURE 10.19 Architectural Rendering

Source: Courtesy of George Butler Associates, Inc.

CAD Drawings.   Computer-aided design (CAD) drawings, such as the site plan in Figure 10.22, use geometric shapes and symbols to provide a graphic view of a setting drawn to a particular scale. CAD drawings include *notations* to define scale and a *title block*. The title block gives the date of completion, name of the draftsperson, company, client, and project name.

Photographs.   A photograph can illustrate your text effectively. Like a line drawing, a photograph can show the components of a mechanism. If you use a photo for this purpose, you will need to label (name), number, or letter parts and provide a key. Photographs are excellent visual aids because they emphasize all parts equally. Their primary advantage is that they show something as it truly is.

FIGURE 10.20   3-D Drawing

Source: Courtesy of Johnson County Community College

FIGURE 10.21   Section Elevation of Pharmacy for Lay Audience of City Council Members

Source: Courtesy of Landplan

The 3-D section elevation drawing provides an impressionistic view of what the site will potentially look like. It not only factors in elevation and landscaping but also allows the audience to relate to and envision the design.

Photographs have one disadvantage, however. They are difficult to reproduce. Whereas line drawings photocopy well, photographs do not. See Figure 10.23.

Icons.   Approximately 23 percent of America's population is functionally illiterate. In today's global economy, consumers speak diverse languages. Given these two facts, how can technical writers communicate to people who cannot read and to people who speak different languages? Icons offer one solution. Icons (as in Figures 10.24, 10.25, 10.26, and 10.27) are visual representations of a capability, danger, direction, acceptable behavior, or unacceptable behavior. For example, the computer industry uses icons, open manila folders, to represent computer files. In manuals, a jagged lightning stroke iconically represents the danger of electrocution. On streets, an arrow represents the direction we should travel; on computers, the arrow shows us which direction to scroll. Universally depicted stick figures of men and women greet us on restroom doors to show us which rooms we can enter and which rooms we must avoid.

FIGURE 10.22 CAD/CAM Site Plan of Pharmacy for High-Tech Audience of Engineers

In contrast to the section elevation in Figure 10.21, this CAD/CAM site plan for the same construction project provides an objective, high-tech depiction of the project, specifying the limits of the land (streets, right of way, property lines, and boundaries).

Landscape schedule including types of plantings

Title block

Notation showing the scale of the drawing

Source: Courtesy of Landplan

When used correctly, icons can save space, communicate rapidly, and help readers with language problems to understand the writer's intent. To create effective icons, follow these suggestions:

1. **Keep it simple.** You should try to communicate a single idea. Icons are not appropriate for long discourse.

2. **Create a realistic image.** This could be accomplished by representing the idea as a photograph, drawing, caricature, outline, or silhouette.

3. **Make the image recognizable.** A top view of a telephone or computer terminal is confusing. A side view of a playing card is completely unrecognizable. Select the view of the object that best communicates your intent.

4. **Avoid cultural and gender stereotyping.** For example, if you are drawing a hand, you should avoid showing any skin color, and you should stylize the hand so it is neither clearly male nor female.

5. **Strive for universality.** Stick figures of men and women are recognizable worldwide. In contrast, letters—such as *P* for *parking*—will mean very little in China, Africa, or Europe. Even colors can cause trouble. In North America, red represents danger, but red is a joyous color in China. Yellow calls for caution in North America, but this color equals happiness and prosperity in the Arab culture (Horton 682–93).

FIGURE 10.23 Photograph of Mechanical Piping

Source: Courtesy of George Butler Associates, Inc.

FIGURE 10.24 Icon of Explosives

FIGURE 10.25 Icon of Dangerous Machinery

FIGURE 10.26 Icon of Electric Shock

FIGURE 10.27 Icon of Corrosive Material

Ethical Considerations when Downloading Existing Online Graphics or Using Screen Captures

Downloading Existing Online Graphics. The Internet has thousands of Web sites that contain graphics, including photographs, line drawings, cartoons, icons, animated images, arrows, buttons, horizontal lines, balls, letters, bullets, hazard signs, and more. You can download images from any Web site.

To download graphics, place the cursor on the graphic and right-click on the mouse. A pop-up menu will appear. Either click on "Copy" or click on "Save Picture." You can save your image in the file of your choice. The images from the Internet will already be GIF (graphics interchange format) or JPEG (joint photographic experts group) files.

Using Screen Captures. By hitting the "Print Screen" button on your keyboard, you can capture any image shown on your monitor. Then, with a graphics software program, you can alter that image to fit your needs.

However, when downloading images or capturing screens, you must abide by copyright laws and ethical considerations. "Every element of a Web page—text, graphics, and HTML code—is protected by U.S. and International copyright laws, whether or not a formal copyright application has been submitted" (LeVie 20–21). If you and your company "borrow" from an existing Internet site or capture images that were created by someone else, thus infringing upon that site's copyright, you can be assessed actual or statutory damages. To meet ethical considerations,

- Assume that any information on the Internet is covered under copyright protection laws unless proven otherwise.
- Obtain permission for use from the original creator of graphics or text.
- Copyright any information you create.
- Include a copyright notice.
- Modify existing images.
- Create your own graphics and text (Johnson 17).

Can you ethically make changes to a graphic taken from your own company's archives or from boilerplate documents? Do these graphics need a source citation? If your company owns these images, you can modify them and use the changed graphics without citing the source. See Figures 10.28 and 10.29 for examples of original and modified photographs taken from a company's archives.

FIGURE 10.28   Original Photograph with Two Men and Pipe

This photograph includes two men working on a pipe. Note the writing on the pipe.

Source: Courtesy of George Butler Associates, Inc.

FIGURE 10.29   Modified Photograph (with the two men and *PLO5* removed)

In this computer-altered photograph, the two men and the writing on the pipe have been digitally erased.

Source: Courtesy of George Butler Associates, Inc.

Modifying or Creating New Graphics

To modify graphics, you can download them in two ways. First, you can print the screen by pressing the Print Screen key (usually found on the upper right of your keyboard). This captures the entire screen image in a clipboard. Then you can open a graphics program and paste the captured image.

Second, you can save the image in a file and then open the graphic in a graphics package. Most graphics programs will allow you to customize a graphic. Popular programs include Paint, Paint Shop Pro, PhotoShop, Corel Draw, and Adobe Illustrator. In these graphics programs, you can manipulate the images by changing colors, adding text, reversing the images, cropping, resizing, redimensioning, rotating, retouching, deleting or erasing parts of the images, overlaying multiple images, joining multiple images, and so forth.

Another option is to create your own graphic. If you are artistic, draw your graphic in a graphics program. This option might be challenging and time consuming. However, creating your own graphic gives you control over the finished product, provides a graphic precisely suited to your company's needs, and helps avoid infringement of copyright laws.

Now that you know the criteria for creating different types of visual aids, you next need to create the visual aid. How do you begin? As always, *prewrite, write,* and *rewrite.* Remember that the process is dynamic and the steps frequently overlap.

VISUAL AID CHECKLIST

_____ 1. Will a visual aid add to your technical communication and make the document concise, clear, and add cosmetic appeal?

_____ 2. Should you use color or three-dimensional graphics in your visual aids?

_____ 3. Are visuals integrated with the text?

_____ 4. Do the visuals add to the text and enhance it without being redundant?

_____ 5. Do the graphics communicate information visually that could not easily be conveyed in text?

_____ 6. Are the visuals the correct size, labeled, readable, and similar in style?

_____ 7. Did you use a table when you presented factual information, such as numbers, percentages, or monetary amounts?

_____ 8. Have you used a figure to highlight and supplement important points in your writing?

_____ 9. Have you included a figure or table number and title for your graphic?

_____ 10. Have you cited your source of information?

The Writing Process at Work

The Writing Process

Prewriting	→ ←	Writing	→ ←	Rewriting

Prewriting	Writing	Rewriting
• Determine whether your graphic will inform, clarify content, analyze, or persuade. • Determine whether your audience is internal or external, high tech, low tech, lay, or multiple. • Research your topic to gather information.	• Organize your content. • Choose the correct type of visual aid appropriate for your content.	• Revise your visual aid by adjusting the color. • Add a figure number, table number, title, and legend. • Add callouts as appropriate.

Prewriting

You first should prewrite to gather data and determine your goals and audience. One employee, Bert Lang, worked for the trust department of a bank and was going to make a presentation to a client, Sylvia Light, about asset allocations. Bert considered the following:

- Audience—specialists (bank employees) and lay reader (client)
- Channels—financial proposal to be reviewed by both the client and the bank
- Data—specific client financial information, including assets

Figure 10.30 shows Bert's list for planning asset allocations.

Writing

Next, you have to write the draft of the document. Bert had to decide how to present his information to Sylvia. Bert wrote the following rough draft of the asset allocations in paragraph form (Figure 10.31).

Rewriting

You then revise and rewrite the rough draft. Bert realized that a paragraph would not communicate asset allocations easily to the customer, so he revised the paragraph into a pie chart for the presentation. Figure 10.32 shows Bert's pie chart of asset allocations.

FIGURE 10.30   Planning List of Asset Allocations

Reallocate funds as follows:

- $110,000 in an annuity
- $78,000 in an Individual Retirement Account (IRA)
- $45,000 in municipal bonds
- $37,000 in a stock fund
- $42,000 in certificates of deposit (CD)
- $3,500 in a checking account

FIGURE 10.31   Rough Draft of Asset Allocations

After reviewing your financial situation and considering your lifestyle, you should consider reallocating your assets in the following way. Put $110,000 in an annuity, $78,000 in an Individual Retirement Account (IRA), $45,000 in municipal bonds, $37,000 in a stock fund, $42,000 in certificates of deposit (CD), and $3,500 in a checking account. This will allow you to have a higher overall rate of return, yet remain almost risk free in regards to your capital. I know that preservation of capital is of paramount importance at this time in your life.

FIGURE 10.32   Revised Pie Chart of Asset Allocations

Figure 1 Portfolio Allocations

- Annuity
- IRA
- Bonds
- Stocks
- CDs
- Checking

The wedge is pulled out from the pie for emphasis and to show the conservative approach to the asset allocations.

CHAPTER HIGHLIGHTS

1. Use graphics to create a more concise document.
2. Graphics add cosmetic appeal to your text, breaking up wall-to-wall words.
3. Color and 3-D graphics can be effective. However, these two design elements also can cause problems. Your color choices might not be reproduced exactly as you planned, and a 3-D graphic could be misleading rather than informative.
4. Tables are effective for presenting numbers, dates, and columns of figures.
5. Figures, such as bar charts, pie charts, line charts, flowcharts, and organizational charts, highlight and supplement important parts of your text.
6. Online graphics are protected by U.S. and international copyright laws.
7. Icons can communicate effectively to multicultural audiences.
8. Boilerplate graphics, created within your company, can be altered and do not infringe upon copyright.
9. Graphics, such as line graphs, pie charts, and bar charts, show trends, percentages, and comparisons.
10. Cite the sources of your information to ensure ethical communication.

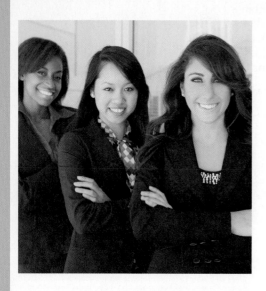

CASE STUDIES

1. Bert Lang is an investment banker at Country Commercial Bank, as noted in this chapter's beginning scenario. He is writing a proposal to a potential client, Sylvia Light, a retired public health nurse.

   She now has $315,500 in savings, allocated as follows: $78,000 in an individual retirement account (IRA); $234,000 in a low-earning certificate of deposit (CD); and $3,500 in her checking account.

   Sylvia contacted Bert, asking him to help her organize her portfolio for a comfortable retirement. Bert has studied Sylvia's various accounts and considered her lifestyle and expenditures. Now, he is ready to write the proposal. Bert plans to propose that Sylvia could reallocate her funds as follows:

   - $110,000 in an annuity
   - $78,000 in an IRA
   - $45,000 in municipal bonds
   - $37,000 in a stock fund
   - $42,000 in CDs
   - $3,500 in a checking account

   To make this proposal visually appealing and more readily understandable to Sylvia, Bert will use visual aids.

   ### Assignment

   - Create a table to show how he wants to invest her money.
   - Create a bar chart comparing her current allocations versus his proposed allocations.

2. Jim Goodwin owns an insurance agency, Goodwin and Associates Insurance (GAI). Letters are a major part of his technical communication with vendors, clients, and his insurance company's home office. However, many of Jim's employees fail to recognize the importance of communication. To prove how important their communication is to the company, Jim plans to summarize the amount of writing they do on the job. He has determined that they write the following each week:

   - **57 inquiry letters**—Insurance coverage changes constantly. To clarify these changes for customers, employees write inquiries, asking home office questions about new insurance laws, levels of coverage, coverage options, and rate changes.
   - **253 response letters**—A key to GAI's success is acquiring new customers. When potential customers call, e-mail, or write letters asking for insurance quotes, Jim and his employees write response letters.
   - **43 cover (transmittal) letters**—Once customers purchase insurance policies, Jim and his co-workers mail these policies, prefaced by cover letters. The transmittal letters clarify key points within the documents.
   - **12 good news letters**—Jim's clients receive discounts when they add new cars to an existing policy or include both car and home coverage. GAI writes good news letters to convey this information.
   - **25 adjustment letters**—When accidents occur or when losses take place, GAI writes adjustment letters stating that the claim is covered.
   - **2 bad news letters**—Like any business, GAI needs to communicate bad news to clients or vendors.
   - **7,000 e-mail messages**—Jim and his 20 employees send and receive a minimum of 50 e-mail messages each day.

Assignment

To show his employees the importance of technical communication, Jim wants to create a visual aid depicting this data. Create the appropriate graphic for GAI's employees.

INDIVIDUAL AND TEAM PROJECTS

1. Present the following information in a pie chart, a bar chart, and a table.

   In 2011, the Interstate Telephone Company bought and installed 100,000 relays. It used these for long-range testing programs that assessed failure rates. It purchased 40,000 Nestor 221s; 20,000 VanCourt 1200s; 20,000 Macro R40s; 10,000 Camrose Series 8s; and 10,000 Hardy SP6s.

2. Using the information presented in activity 1 and the following revised data, show the comparison between 2011 and 2012 purchases through two pie charts, a grouped bar chart, and a table.

   In 2012, after assessing the success and failure of the relays, the Interstate Telephone Company made new purchases of 200,000 relays. It bought 90,000 VanCourt 1200s; 50,000 Macro R40s; 30,000 Camrose Series 8s; and 30,000 Hardy SP6s. No Nestors were purchased.

3. Create a line chart. To do so, select any topic you like. The subject matter, however, must include varying values. For example, present a line chart of your grades in one class, your salary increases (or decreases) at work, the week's temperature ranges, your weight gain or loss throughout the year, the miles you've run during the week or month, amounts of money you've spent on junk food during the week, and so forth.

4. Analyze the following graphics and explain which ones succeed and which ones fail.

   a. Vertical Bar Chart

   b. Pie Chart

c. Horizontal Bar Chart

d. Table

TABLE 1 Comparison of Item Costs

	Price	Percentage
Item 1	$ 45.00	17%
Item 2	$ 67.00	25%
Item 3	$122.00	44%
Item 4	$ 38.00	14%
Total	$272.00	100%

5. Based on the criteria provided in this chapter, revise any of the poor graphics from the preceding assignment (number 4). Add any additional information necessary to correct the graphics.

PROBLEM-SOLVING THINK PIECES

1. Angel Guerrero, computer information systems technologist at HeartHome Insurance, was responsible for making an inventory of his company's hardware. He learned the following: The company had 75 laptops, 159 PCs, 27 printers, 10 scanners, 59 handheld computers (PDAs), 238 cell phones, and 46 digital cameras.

   To write his inventory report, Angel needs to chart the above data. Which type of visual aid should Angel use? Explain your answer, based on the information provided in this chapter. Create the appropriate visual aid.

2. Minh Tran works in the Marketing Department at Thrill-a-Minute Entertainment Theme Park (TET). Minh and her project team need to study entry prices, ride prices, food and beverage prices, and attendance of their park versus their primary competitor, Carnival Towne (CT).

   Minh and her team have found that TET charges $16.50 admission, while CT charges $24.95. Most of TET's rides are included in the entry price, but special rides (the Horror, the Bomber, the Avenger, and Peter Pan's Train) cost $2.50. At CT, the entry fee covers many rides, excluding Alice's Teacup, Top-of-the-World Ferris Wheel, and the Zinger, which cost $2.00 each. Food and beverages at TET cost $1.95 for a hot dog, $2.50 for a hamburger, $3.95 for nachos, and $1.50–2.50 for drinks. At CT, food and beverages cost $1.75 for hot dogs, $2.75 for hamburgers, $2.50 for nachos, and $1.50–2.50 for drinks. Attendance at TET last year was 250,000, while attendance at CT was 272,000.

What type of visual aid should Minh and her team use to convey this information? Explain your decision, based on the criteria for graphics provided in this chapter. Create the appropriate visual aid.

3. Toby Hebert is human resource manager at Crab Bayou Industries, the world's largest wholesaler of frozen Cajun food. Toby and her management team are concerned about the company's hiring trends. A prospective employee complained about discriminatory hiring practices at Crab Bayou.

To prove that the company has not practiced discriminatory hiring practices, Toby has studied the last ten years' hires by age. She found that in 2002, the average age per employee was 48; in 2003, the average age was 51; in 2004, the average age was 47; in 2005, the average age was 52; in 2006, the average age was 45; in 2007, the average age per employee was 47; in 2008, the average age was 42; in 2009, the average age was 39; in 2010, the average age rose to 42; in 2011 and 2012, the average age fell to 29 and 30 respectively (due to a large number of early retirements).

What type of visual aid should Toby use to convey this information? Explain your decision, based on the criteria for graphics provided in this chapter. Create the appropriate visual aid.

4. Yasser El-Akiba is a member of his college's International Students' Club. Yasser is a native of Israel. Other members of the club are from other countries: three from Australia, two from Ecuador, eight from Mexico, five from Africa, two from England, three from Canada, four from the Dominican Republic, and nine from China.

What kind of visual aid could Yasser create to show his club members' homelands? Defend your decision based on criteria for graphics in this chapter. Create the appropriate visual aid.

WEB WORKSHOP

Using an Internet search engine, type in phrases such as "automobile sales+line graph," "population distribution by age+pie chart," or "California+organizational chart." Create similar phrases for bar charts, pictographs, flowcharts, tables, and so forth. Open several links from your Web search and study the examples you have found. Which examples of graphics are successful and why? Which examples of graphics are unsuccessful and why? Explain your reasoning, based on this chapter's criteria.

QUIZ QUESTIONS

1. How can you effectively present large blocks of data or reveal comparisons?
2. How do graphics allow you to be concise?
3. What are four things graphics help readers to see?
4. What can happen when you use only unbroken text in a document?
5. When should you avoid using color in a graphic?
6. What are drawbacks to 3-D graphics?
7. When you use graphics, what are five criteria to follow to achieve successful graphics?
8. What are two things you achieve by using a table?
9. After a table, what explanatory information should you include in follow-up text?
10. In a table, should you use decimals or numerators and denominators for fractions?
11. Figures accomplish what two things in a document?
12. What are five types of figures?
13. What are two types of bar charts?
14. What is a pictograph?
15. What are three ways to achieve an effective pictograph?
16. When would you use a Gantt chart?
17. What do pie charts illustrate?
18. What type sequence can you reveal in a flowchart?
19. How are schematics useful?
20. What are two reasons for using icons?

CHAPTER ELEVEN

Communicating to Persuade

COMMUNICATION AT WORK

In this scenario, TechToolshop relies on persuasive writing to expand business.

TechToolshop provides sales, service, maintenance, installation, and data recovery for PC and Mac hardware and software tools. They install, repair, and maintain workstations, servers, printers, and peripherals. Through an online catalogue and storefront site, they sell printers, desktops, mainframes, minitowers, software, accessories, and encryption devices. Located in Big Springs, Iowa, this business of 1,200 employees began as a small computer store in 2005. Now, however, the company wants to expand its business.

As director of marketing communications at TechToolshop, Amanda Carroll's job is to communicate information to the public about product offerings, company product-line changes, environmentally responsible products, and warranty changes. Many of Amanda's written documents and oral presentations depend on persuasion. To be persuasive, she relies on the traditional methods of argument by appealing to the emotions, being logical, and maintaining ethical standards. She wants to communicate effectively to ensure that information contributes to "product branding" and, in turn, enhances the profitability of the company.

Amanda will use e-mail messages, text messages, instant messages, sales letters, brochures, and fliers to promote TechToolshop's expansion plans. To persuade new customers to

Objectives

When you complete this chapter, you will be able to

1. Understand the importance of argument and persuasion in technical communication

2. Recognize the traditional methods of argumentation

3. Use the *ARGU* technique to organize persuasive technical communication

4. Avoid logical fallacies in persuasive communication

5. Evaluate persuasive documents using the checklist

visit the store and online sites, Amanda is conducting a multifaceted viral marketing campaign. This will include

- **Full-color brochures** that focus on the company's primary functions: computer hardware and software sales and maintenance, data recovery, installation, and repair.

- **Fliers** to advertise what's on sale this month, recent industry news that affects the computer user, innovative software information, repair and replacement steps for software updates, best prices for used computer parts, and so on. In contrast to the brochures, the company's single-sided, postcard-sized fliers are more cost effective. These fliers, which can be produced quickly, will focus on individual topics where quick turnaround is important.

- **A corporate Facebook site.** On this social media site, TechToolshop will encourage customer involvement. The company's Facebook site will post photographs of customers who have used TechToolshop's products. In addition, TechToolshop will advertise time-sensitive events, such as holiday specials or limited-time sales, to encourage return visits to the site. Customers who invite their friends to visit the TechToolshop Facebook site will receive discounts on future purchases.

- **Twitter updates.** Through 140-character tweets, TechToolshop will update customers on the arrival of new products, advertise daily specials, and notify followers of upcoming events. Through retweets, messages sent by TechToolshop customers to their friends, the company benefits from an electronic mass marketing plan.

- **Web site and company blog.** The Web site and blog will allow customers to interact with the company. In the blog, customers can suggest additional parts or service the company can offer as it expands.

One of Amanda's major goals is maintaining excellent customer relations. Hard work and client referrals obviously help contribute to this goal. However, Amanda has found that continuous communication is another way to build client rapport and persuade people to do business with TechToolshop. Though TechToolshop's business is increasing, it wants to keep adding customers. Outstanding technical communication through persuasive sales letters, social media, brochures, and fliers equals more business for TechToolshop.

Check out our quarterly newsletters at www.pearsonhighered.com/gerson for dot.com updates, new case studies, insights from business professionals, grammar exercises, and facts about technical communication.

Web Resources

www.pearsonhighered
.com/gerson For more
information about
persuasive
communication, visit our
companion website.

The Importance of Argument and Persuasion in Technical Communication

When communicating technical information, you will write and speak for many reasons. You might write a memo, letter, e-mail message, tweet, text message, or instant message to *inform* about an upcoming meeting, a job opportunity, a new product release, or a facilities change. You might give an oral presentation to local businesses, government, or educational organizations in which you hope to *build rapport*. In writing a user manual, your goal will be to *instruct*. When you write a proposal, your goal is to *recommend* changes. If you write a technical specification, you will *analyze* components of a piece of equipment.

In addition to informing, building rapport, instructing, recommending, and analyzing, you also will need to communicate persuasively. Let's say you are a customer who has purchased a faulty product. You might want to write a letter of complaint or an e-mail message to the manufacturer of this product. To make your case strongly, you will need to convince your audience, clarifying how the product failed. If your argument is effective, then you will persuade the company to give you a refund or new product.

Professionally, you will need to use argument and persuasion daily. As a manager, you might need to argue the merits of a company policy to an unhappy customer. If your colleagues have decided that the department should pursue a course of action, you might need to persuade your boss to act accordingly. Maybe you are asking your boss for a raise or promotion, for office improvements, or for changes to the work schedule. Your task is to persuade the boss to accept your suggestions. In these instances, you will communicate persuasively using any of the following communication channels: routine correspondence (memos and letters); electronic communication (e-mail messages, instant and text messages, tweets, and blogs); reports; proposals; or oral presentations.

Figure 11.1 is a persuasive memo from a subordinate to a boss, documenting a problem and suggesting a course of action.

Frequently, you will write e-mail messages to supervisors and colleagues, persuading these readers to accept your point of view. Topics could range from requests for promotion, equipment needs, days off from work, assistance with projects, or financial assistance for job-related travel.

The "Before" and "After" e-mail messages seek to persuade a colleague to attend a work-related conference instead of the e-mail writer. The "Before" sample is not effectively persuasive. It reads more like a command and fails to consider the audience's reaction. The "After" sample is more persuasive.

You might write persuasively for some of the following reasons:

- **Automotives.** You are starting a new business for automobile service and sales. Now, you need to communicate this new venture to the public. To do so, you will write a sales letter, persuading the public to visit your service center and purchase your automobile products. You also will communicate with the public and potential customers through the company Web site and blog. In addition, you might send instant messages and text messages or tweets to advertise products and services.

- **Electrical/HVAC.** A new mall is being built in your community. Every store in the mall will require electrical work, including heating, ventilation, and air conditioning. To showcase your product, services, and credentials, you plan to send electronic fliers to potential clients.

- **Healthcare.** Your doctor's office or care center provides clients with brochures that market health services. Patients need easy-to-read literature to persuade them that they have chosen the correct physician, hospital, or laboratory.

Communication Goals

See Chapter 1 for more discussion of informing, instructing, and building rapport.
See Chapter 17 for more discussion of analyzing and recommending.

Routine Correspondence

See Chapter 6 for more discussion of e-mail messages.

FIGURE 11.1   Persuasive Memo

Date: March 22, 2012
From: Bob Ward
To: Lynn Richards
Subject: Maintenance of Photo Lab Machine

On March 21, our one-hour photo employee Brian Syoung reported that the photo lab machine was malfunctioning. It was cropping photos out of specification and producing blurred images. A service technician examined the machine and estimated that repairs would cost $1,000. Just as a reminder, Lynn, in one week, we were scheduled to have this machine replaced with an upgrade.

The problem is whether to repair the machine or wait a week for the replacement. Here are some facts that I've collected to help us decide.

- The lab makes $850 in profits each week. If we spend $1,000 to repair the machine, we will lose money.
- Closing the lab will save the company around $200 in supplies and electricity.
- If the photo lab is closed, then our higher-paid photo specialists will have to be assigned other jobs. We will lose approximately $2.00 an hour paying them to do work usually assigned to lower-paid employees.
- Closing the photo lab for a week to save repair expenses will inconvenience our repeat customers. In fact, they might shop elsewhere, which would impact us negatively long term.

Lynn, though repairing the machine costs $1,000, we will more than make up for this cost in customer satisfaction. We should repair the machine for the sake of our customers. Since this is a difficult decision, let's meet later this afternoon at 3:00 p.m. in your office. We need to resolve this issue today before we lose more money.

> The body objectively recognizes both sides of an issue and alternative approaches. The first two bullets focus on delaying the repair. The last two bullets explain why repair delays are bad for business.

> The body develops the argument persuasively by providing facts and figures. In doing so, the writer reveals his knowledge of the subject matter.

> The conclusion emphasizes persuasively the urgency of action.

BEFORE

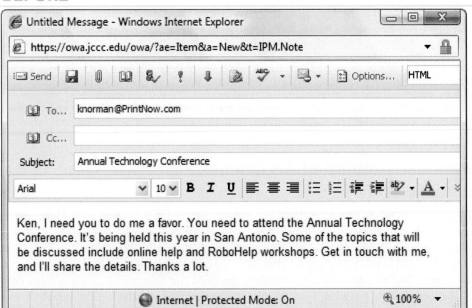

> This e-mail message is not persuasive for a number of reasons: The verb "need" is too commanding. The e-mail fails to arouse the reader's interest or provide sufficient detail to convince the reader of the conference's worth. It also fails to urge the reader to action.

The introductory paragraph arouses reader interest with an anecdote and a question.

The e-mail body begins by refuting any objections. Then, the following sentences show audience benefit.

The conclusion urges action by giving a due date and highlighting reader enjoyment.

AFTER

https://owa.jccc.edu/owa/?ae=Item&a=New&t=IPM.Note

Send | Options... | HTML

To... knorman@PrintNow.com

Cc...

Subject: Annual Technology Conference

Arial 10 **B** *I* <u>U</u>

Ken, remember last year when I took your place at the Dallas Technology Conference? I recall you saying, "Sue, I'll do the same for you sometime." Guess what? That time has arrived. I need your help. Please attend this year's conference in San Antonio.

I know that you are busily working on the Art.com contract, but I've already spoken to Harry. He says that your team is ahead of schedule. Harry says that he can finish the work on his own. In addition, I think that you would benefit more from this conference than I would. The keynote address is about online help, and that's one of your areas of interest. The agenda shows that the conference will provide hands-on RoboHelp workshops. This will help you learn more practical applications for our online help screens.

I'll need your answer by 5:00 p.m. tomorrow. After all, San Antonio is your favorite city.

Internet | Protected Mode: On | 100%

FAQs: Marketing and Technical Communication

Q: How does marketing fit in with technical communication?

A: Technical communication is more than hard, cold memos, letters, reports, user manuals, and technical descriptions. Technical communication also has a soft side—*marketing*. The bottom line is every company is in business to make money. Thus, every employee should perceive themselves as marketing personnel.

Examples: George Butler Associates (GBA), an architectural and engineering company in Kansas, Missouri, and Illinois, asks every one of its employees to take marketing classes on site. The goal is to help GBA employees work well with clients, vendors, and partners.

The Society for Technical Communication also recognizes the importance of marketing for technical writers. STC's Web site (http://stc.org) provides links to "Special Interest Groups." One such group is entitled "Marketing Communication." In addition, two recent links for general news included articles about the importance of marketing for technical communicators: "Tech writers as sales reps? Interface Software's award-winning docs boost brand, revenues, and customer satisfaction" and "Technical writers turn to marketing to survive." Persuasive marketing materials are essential to technical communication.

Traditional Methods of Argument and Persuasion

To argue a point persuasively, you can use any of the traditional methods of argumentation: *ethos* (ethical), *pathos* (emotional), and *logos* (logical) appeals. These three appeals to an audience are called the rhetorical triangle, as shown in Figure 11.2.

This equilateral triangle suggests that each part of a persuasive appeal is as important as any other part. In addition, it emphasizes the need for balance of all three appeals or types of proof. Excess emotion, for example, might detract from the logical appeal of your argument. Cold, hard facts might fail to persuade your audience.

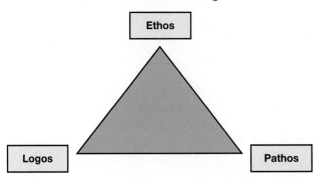

FIGURE 11.2 The Rhetorical Triangle

Ethical Argumentation

Arguments based on ethics (ethos) depend on your character. If you make arguments based on personal experiences, you must appear to be trustworthy and credible as a writer or speaker. To accomplish this goal, present information that is unbiased, reliable, and evenhanded.

LeBron James and Oprah Winfrey are classic examples of spokespeople known for their character. When LeBron speaks about basketball, you know he is an expert. In fact, he is so trustworthy in the eyes of the public that LeBron also can sell products outside of his primary areas of expertise, such as shoes and soft drinks for Nike and Coca Cola. Oprah Winfrey writes and speaks about health, education, entertainment, books, money, philanthropy, and more. When these individuals make arguments drawn from personal experiences, audiences respond favorably. Given their reputations for excellence and success, both are trustworthy and reliable experts.

Emotional Argumentation

Arguments based on emotion (pathos) seek to change an audience's attitudes and actions by focusing on feelings. If you want to move an audience emotionally, you would appeal to passion. You can do this either positively or negatively.

To sway an audience positively, you would focus on positive concepts like joy, hope, honor, pleasure, happiness, success, and achievement. You would use positive words to create an appealing message. In contrast, you also can appeal to emotions negatively. Fear, horror, anger, and unhappiness can be powerful tools in an argument.

Notice how the National Center for Environmental Health warns about the dangers of carbon monoxide: "Carbon monoxide is a silent killer. This colorless, odorless, poisonous gas kills nearly 500 U.S. residents each year, five times as many as West Nile virus" ("CDC and CPSC Warn of Winter Home Heating Hazards"). To highlight the dangers associated with carbon monoxide, the Centers for Disease Control uses emotional words, such as *killer* and *poisonous*, and compares this problem to the frightening "West Nile virus."

Logical Argumentation

Argumentation based on logic (logos) depends on rationality, reason, and proof. You can persuade people logically when you provide them the following:

- Facts—statistics, evidence, data, and research
- Testimony—citing customer or colleague comments, expert authorities, and results of interviews

FIGURE 11.3   E-mail Message Effectively Using Ethos, Pathos, and Logos

Ethics (ethos) is shown when the writer references his achievements at work.

Facts ("4.8 out of 5") and testimony from customer evaluations support the writer's case logically (logos).

The e-mail addresses the opposing point of view ("freeze wages") to present a "balanced" argument.

Note the use of positive words, such as "achievements," "highlight," "exceeded," and "outstanding." These are examples of pathos—emotional words to sway the reader.

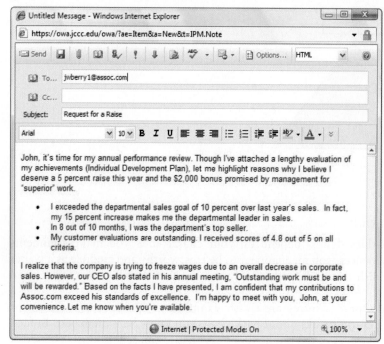

- Examples—anecdotes, instances, and personal experiences
- Strong, clear claims—including warrantees and guarantees
- Acknowledgment of the opposing points of view to ensure that information is balanced

Figure 11.3 is an e-mail message that argues a case for a raise by focusing on **ethics.** The writer refers to his work-related achievements to support his credibility. He uses **logic**—facts and testimony. Also note how the e-mail factors in an opposing point of view. In addition, the e-mail uses **emotion** to sway the reader through positive words.

ARGU to Organize Your Persuasion

Effective persuasive communication entails ethical, logical, and emotional appeals. Understanding the importance of this rhetorical triangle is only part of your challenge as a persuasive communicator. The next step is deciding how best to present your argument. Using our *ARGU* approach will help you organize your argument:

- *Arouse* audience involvement—grab the audience's attention in the introduction of your communication.
- *Refute* opposing points of view in the body of your communication.
- *Give* proof to develop your thoughts in the body of your communication.
- *Urge* action—motivate your audience in the conclusion.

Arouse Audience Involvement

You have only about five to eight seconds to grab your readers' attention in a sales letter, persuasive e-mail message, marketing brochure, speech, or any persuasive communication.

You must arouse the audience's interest imaginatively in the first few sentences of your document or oral presentation. Try any of the following attention grabbers in the **introduction** of your persuasive message:

- **Use an anecdote—a brief, dramatic story relating to the topic.** Stories engage your audience. You can involve your readers or listeners by recounting an interesting story to which they can relate. The story should be specific in time, place, person, and action. The drama should highlight an event. However, this must be a short scenario, since your time is limited. If you do not capture their interest quickly, the audience might lose interest.

E-mail Message Persuading a Manager to Take Action

Sam, last week, a customer fell down in our parking lot, cutting her knee, tearing her slacks, and requiring medical attention. We can't let this happen again. Please consider hiring a maintenance crew to salt and sand our lot during icy weather.

◀ **EXAMPLE**

This anecdote is specific in time, place, person, and action.

- **Start with a question to interest your audience.** By asking a question, you involve your audience. Questions imply the need for an answer. A question can make readers or listeners ask themselves, "How would I answer that?" This encourages their participation.

Sales Letter from a Financial Planner

"Where will I get money for my kid's college education?" "How can I afford to retire?" "Will my insurance cover all medical bills?" You have asked yourself these questions. Our estate planning video has the answers.

◀ **EXAMPLE**

By asking questions, you involve your audience.

- **Begin with a quotation to give your communication the credibility of authority.** By quoting specialists in a field or famous people, you enhance your credibility and support your assertions. A quote from Warren Buffett, world-renowned investor and businessman, about economics, for example, gives credence to a financial topic. Similarly quotes from Bill Gates, founder of Microsoft, about the computer industry are trustworthy.

An Oral Presentation about Stock Performance to Shareholders

As Warren Buffett says, "Our favorite holding period is forever." Though our stock prices are down now, don't panic and sell. The company will rebound.

◀ **EXAMPLE**

Use a quote from an authority to lend credence to your argument.

- **Let facts and figures enhance your credibility.** Factual information (percentages, sales figures, amounts, or dates) catches the audiences' attention and lays the groundwork for your persuasion.

Sales Letter about Computer Technology

Eighty-seven percent of all college students own a computer. Don't go off to college unprepared. Buy your laptop or PC at *CompuRam* today.

◀ **EXAMPLE**

Facts and figures help persuade your reader to your point of view.

- **Appeal to the senses.** You can involve your audience by letting them hear, taste, smell, feel, or visualize a product.

EXAMPLE ▶

An appeal to the senses can involve your reader.

Marketing Flier Advertising a New Restaurant

Hickory-smoked goodness and fire-flamed Grade A beef—let Roscoe's BBQ bring the taste of the South to your neighborhood.

- **Use comparison or contrast to highlight your message.** Comparison/contrast lets you make your point more persuasively. For example, you might want to show a client that your computer software is superior to the competition. You might want to show a boss that following one plan for corporate restructuring is better than an alternative plan. You could justify why an employee has not received a raise by comparing or contrasting his or her performance to the departmental norm. Comparison or contrast is useful in many different arguments for a variety of audiences.

EXAMPLE ▶

Comparing one year to another provides a benchmark for the writer's argument.

E-mail from Management to Subordinates Proposing Changes in Procedures

Last year our department fell short of corporate sales goals. This year we must surpass expectations. To do so, I propose a seven-step procedure.

- **Begin with poetic devices.** Advertising has long used alliteration (the repetition of sounds), similes (comparisons), and metaphors to create audience interest. Poetic devices are memorable, clever, fun, and catchy.

EXAMPLE ▶

Use alliteration, the repetition of sounds, to appeal to an audience.

Sales Letter from a Bank

Looking for a low-loan lease? Let US Bank take care of your car leases. Our 2 percent loans beat the best.

- **Create a feeling of comfort, ease, or well-being.** To welcome your audience, making them feel calm and peaceful, invoke nostalgia or good times.

EXAMPLE ▶

Interest your audience by making them feel comfortable and welcomed.

Slogan from a Mortgage Company Brochure

Come home again to Countryside Mortgage. We treat our customers like neighbors.

- **Create a feeling of discomfort, fear, or anxiety.** Another way to involve your audience is through stress. Beginning communication by highlighting a problem allows you to persuade the audience that you offer the solution.

EXAMPLE ▶

A negative example can incite an audience to action.

Sales Letter from an Apartment Complex

Why pay too much? If you paid over $300 a month for your apartment last semester, you got robbed. We'll charge 20 percent less—guaranteed.

Refute Opposing Points of View

You can strengthen your argument by considering opposing opinions. Doing so shows your audience that you have considered your topic thoroughly. Rather than looking at the subject from only one perspective, you have considered alternatives and discarded them as lacking in merit. In addition, by refuting opposing points of view, you anticipate negative comments an audience might make and defuse their argument.

To refute opposing points of view in the discussion (body) of your communication, follow these steps:

- Recognize and admit conflicting views.
- Let the audience know that you understand their concerns.
- Provide evidence.
- Allow for alternatives.

Figure 11.4 shows a letter of application that successfully uses refutation as part of its persuasion.

FIGURE 11.4   Letter of Application Using Refutation

1901 West King's Highway
San Antonio, TX 77910

February 27, 2012

Marissa Lee
Manager, Human Resources
PMBR Marketing
20944 Wildrose Dr.
San Antonio, TX 78213

Dear Ms. Lee:

In response to your advertisement in the *San Antonio Daily Register,* please consider me for the position of Marketing Accounts Associate. I have enclosed a resume elaborating on ways I can benefit your company.

Your advertisement requires someone with a BA in Marketing. Though I do not have this degree, my experience and skills prepare me for this position:

- Ten years experience in Marketing.
- Prepared press releases and created public service announcements.
- Created 30-second spots for local radio and television stations.
- Participated in press conferences providing corporate information about declining stock values and company layoffs.
- Created PowerPoint presentations for local governmental agencies.
- Currently enrolled in a Marketing class at the University.

My experience and abilities will make me an asset in your Marketing Department. I would be happy to meet with you to discuss ways in which I can benefit PMBR. Please contact me at arosa22@hotmail.com.

Sincerely,

Armondo Rosa

Attachment: Resume

The writer's refutation anticipates negative comments the audience might have and defuses the argument through proof and alternatives.

Refutes the job requirement for a BA.

The first five bullets provide evidence to show that the writer has skills appropriate for the job versus a degree.

The current enrollment provides an alternative to the degree requirements.

TABLE 11.1 Techniques for Supporting an Argument

Provide facts and figures to document your assertions.	Eighty-five percent of the homeowners contend that... or Seven out of ten buyers said they would...
Persuade through graphics. This pie chart not only conveys data, but also it makes a visual impact. It graphically highlights how many people own computers.	 15 85 ■ Homeowners with Computers □ Homeowners without Computers
Give testimony from satisfied customers, vendors, or co-workers.	The Job Corps of Blue Valley, Titan Co., and Amex Inc. have used our product for over ten years. or Julie Jones, sales associate, attests to my effective interaction with customers.
Document your credentials—years in business or certification of employees.	In business since 2000 . . . or All of our staff members are Certified Public Accountants.
Give examples.	For example, I attended five continuing education seminars this year. These included training workshops on Microsoft Word applications, Effective Business Writing, and Listening Skills. Because of my continuing education, I have the credentials needed to be an effective project manager. Please consider me for promotion.
Cite rules and regulations.	Section II.a of the State Waste Management Act specifies that wood debris, railroad ties, and used tires must be disposed of.
Attach an expert's name.	Linda Freeman, Certified Public Accountant, suggests that allocating $12,500 to a SEP (Self Employment Pension) account will lower your federal taxes.

Give Proof to Develop Your Thoughts

In the body of your communication, develop your argument with proof. Arousing an audience's interest and refuting opposing points of view will not necessarily persuade the audience. Most people require details and supporting evidence before making decisions. You can provide specific details to support your argument using any of the techniques in Table 11.1.

Urge Action—Motivate Your Audience

Throughout your communication, you have worked to persuade the audience to accept your point of view. In the conclusion, you need to motivate the audience to action. This could include any of the following: attend a meeting, purchase new equipment, invite you to interview for a position, vote on a proposition, promote you, give you a raise, allow you to work a flexible schedule, or change a company policy.

To urge the audience to action, consider the techniques in Table 11.2.

The example paragraph on page 336 concludes a persuasive letter from a governmental agency to a business owner.

TABLE 11.2 Techniques for Motivating an Audience to Action

Give due dates.	Please respond by January 15.
Explain why a date is important.	Your response by January 15 will give me time to prepare a quarterly review and meet with you if I have additional questions.
Provide contact information for follow-up.	Please submit your proposal to Hank Green, Project Director. You can e-mail him at hgreen@modernco.com.
Suggest the next course of action.	We need to plan our presentation before the next City Council meeting, so please attend Tuesday's meeting at 9:00 a.m.
Show negative consequences.	You must repair your sidewalk within 30 days to comply with city laws regarding pedestrian safety. Failure to do so will result in a $150 fine.
Reward people for following through.	Following these ten simple steps will help you load the software easily and effectively.

SPOTLIGHT

Writing Persuasively in Business and Industry

Logos, ethos, and *pathos*—these argumentation devices are ancient, but they're everyday communication tools in modern marketing. Here's what Ron Dubrov says about persuasive communication.

"Where words appeal primarily to rationality, the graphics and sound help to create emotional impact." To build trust, faith, and credibility in a product or service, Ron believes that words provide descriptive detail and supportive facts, while graphics, music, rhyme, and rhythm add interest.

Ron is CEO of RonDubrovCreative, an advertising/marketing company that utilizes creative copywriting, audio, video, and graphic design techniques to help clients develop their corporate identity and achieve their marketing communications objectives.

Words Provide Rational Proof

For example, in developing the Kansas Department of Transportation's "Kansas Clicks©" campaign to raise awareness and emphasize the importance of wearing seat belts, Ron created messages for radio, television, billboards, and print media based on the use of persuasive logic. To arouse the audience's interest, he used a variety of traditional techniques:

- **Questions** ("Hear it 'click'? Your seat belt is trying to tell you something. Listen. It's the first sound of a responsible driver.")
- **Facts** ("If you want your kids to be safe, let them hear your seat belt. Because when you click, they click!")
- **Startling images** (headline superimposed over a shattered windshield: "What's going to stop you from not wearing a seat belt?")

Statistical facts also served as the basis of the seat belt campaign message directed toward drivers of pickup trucks.

Research showed that over half the people in Kansas who drive pickup trucks fail to wear their seat belt—the one safety feature that could save their lives.

To urge action on the part of his audience, Ron uses "classic retail techniques"—reward for action, such as gifts or incentives. These include "limited time offers and coupon redemption." In the "Kansas Clicks" and "Click It or Ticket" campaigns, the rewards or benefits of taking positive action become even more emphatic—avoiding fines, being able to keep your license and continue to drive, and ensuring safety by surviving a car crash.

Graphics and Sound Create Emotional Context

To add interest and build emotional impact in his radio advertisements, Ron appeals to the "mind's eye" and memorable rhythm. For the "Kansas Clicks" campaign, these included a custom audio track with seat belt clicking sound; street/traffic sounds; driving on rough, gravel roads; automobile doors opening and slamming shut; and screeching tires and crashes.

Many of the radio commercials were written in Spanish and English, and the audio effects helped to maintain a cohesive campaign as well as maintain emotional impact. "It's a powerful tool," Ron says. "Sound transcends language."

Valuing the Audience

How does Ron judge the success of his persuasive documents? He offers value to his clients by being "the voice" of the customer, by reacting as the customer would, and by searching for and promoting the "real truth and value to the end user." These are attributes that not only abide by the Society for Technical Communication's ethical guidelines but also contribute to Ron's success as a professional.

EXAMPLE ▶

This conclusion provides a due date, explains why the date is important, shows negative consequences, and tells the reader what to do next.

Your business has not been in compliance for over 16 months. We have written three letters to give you guidance and information necessary to get the facility into compliance. You have not responded to our efforts, with the exception of last week's phone call. The deadline for the pending abatement order is 30 days. You must reply to this letter in a timely manner to avoid fines. We are open to having a meeting at our regional office to discuss the corrective measures you can take to address the pending order. Thank you for considering your options.

Avoiding Unethical Logical Fallacies

In a corporate environment, you must persuade your audience not only logically but also ethically. Your persuasive communication must be honest and reasonable. Honesty demands that you avoid the following logical fallacies.

Inaccurate Information

Facts and figures must be accurate. One way in which dishonest appeals are made in communication is through inaccurate graphics. Look at Figures 11.5 and 11.6. Figure 11.5 shows a $25,000 deficit in the third quarter. So does Figure 11.6. However, the size of the bar in Figure 11.6 is misleading and imprecise, visually suggesting that the loss is less. This is an inaccurate depiction of information.

FIGURE 11.5 Accurate Depiction of Information

FIGURE 11.6 Inaccurate Depiction of Information

Unreliable Sources

Being an expert in one field does not mean you are an expert in all fields. For example, quoting a certified public accountant to support an important healthcare issue is illogical. Not all specialists are reliable in all situations.

Sweeping Generalizations

Avoid exaggerating. Allow for exceptions. For instance, it is illogical to say that "*All* marketing experts believe that newsletters are effective." Either qualify this with a word like *some* or quantify with specific percentages.

Either . . . Or

Suggesting that a reader has only two options is deceitful if other options exist. Allow for other possibilities. It is wrong to write, "Either all employees must come to work on time, or they will be fired." This blanket statement excludes alternatives or exceptions. Is the "either . . . or" statement true if an employee has a car accident, if an employee's child is sick, or if the employee is caught in heavy traffic due to a snowstorm?

Circular Reasoning (Begging the Question)

"Some accountants are ambitious because they wish to succeed." This statement is illogical and uses circular reasoning because it states the same thing twice. *Ambitious* and *succeed* are essentially synonyms. The writer fails to prove the assertion.

Inaccurate Conclusions

When communicating persuasively, consider all possible causes and effects. Exact causes of events often are difficult to determine.

- A condition that *precedes* another is not necessarily the *cause* of it. This error is called *post hoc, ergo propter hoc*.

◀ EXAMPLE

> The contractor lost the bid, so he cannot expect to have increased revenues this fiscal year.

Yes, the contractor lost a bid, but this does not necessarily mean that the contractor won't increase revenues in some other way. To say so is to make a hasty conclusion based on too little information.

- A condition that *follows* is not necessarily the *effect* of another. This is called a *non sequitur*.

◀ EXAMPLE

> Because the manager is inexperienced, the report will be badly written.

Again, this is a hasty conclusion. Lack of experience in one area does not necessarily lead to a lack of ability in another area.

Red Herrings

If you focus on an irrelevant issue to draw attention from a central issue, this is called a red herring. For instance, you have failed to pay fines following citations for the mishandling of hazardous wastes. You contact the state environmental agency and complain about state taxes being too high. This is an irrelevant issue. By focusing on high taxes, you are merely avoiding the central issue.

To write effective persuasive documents, you must develop your assertions correctly, avoiding logical fallacies.

Types of Persuasive Documents

You have manufactured a new product (a mini digital camera, a paper-thin plasma computer monitor, a fiber optic cable, or a microfiber rain jacket). Perhaps you have just created a new service (home visitation healthcare, a mobile accounting business, a Web design consulting firm, or a gourmet food preparation and delivery service). Congratulations!

However, if your product sits in your basement gathering dust or your service exists only in your imagination, what have you accomplished? To benefit from your labors, you must market your product or service.

To convince potential customers to purchase your merchandise, you could write any of the following persuasive documents:

- Sales letters
- Fliers
- Brochures

Web Resources

www.pearsonhighered
.com/gerson See our
companion Website for
more sales letters, fliers,
and brochures.

Sales Letters

Routine Correspondence

See Chapter 6 for more
discussion of letters.

To write your sales letter, follow the format for letters discussed in Chapter 6. Include the letter essentials (letterhead address, date, reader's address, salutation, text, complimentary close, and signature). Your sales letter should accomplish the following objectives, relating to effective persuasion.

Arouse Reader Interest. The introductory paragraph of your sales letter tells your readers why you are writing (you want to increase their happiness or reduce their anxieties, for example). Your introduction should highlight a reader problem, need, or desire. If the readers do not need your services, then they will not be motivated to purchase your merchandise. The introductory sentences also should mention the product or service you are marketing, stating that this is the solution to their problems. Arouse your readers' interest with anecdotes, questions, quotations, or facts.

Refute Opposing Points of View. Your audience will not always agree with your assertions. To persuade the reader to accept your point of view, anticipate disagreements or alternative points of view. Consider opposing comments about your new product or service. Think about what your competitors offer as alternatives to your company. By mentioning competitors or alternative ideas and refuting them, you emphasize your product's value.

Give Proof to Develop Your Thoughts. In the discussion paragraph(s), specify exactly what you offer to benefit your audience or how you will solve your readers' problems. You can do this in a traditional paragraph. In contrast, you might want to itemize your suggestions in a numbered or bulleted list. Whichever option you choose, the discussion should provide data to document your assertions, give testimony from satisfied customers, or emphasize your credentials.

Urge Action. Make readers act. If your conclusion says, "We hope to hear from you soon," you have made a mistake. The concluding paragraph of a sales letter should motivate the reader to act.

Conclude your sales letter in any of the following ways:

- Provide your Web site URL; online help-desk e-mail address; and social media links to the company blog, Twitter account, or Facebook site.
- Give directions (with a map) to your business location.
- Provide a tear-out to send back for further information.
- Supply a self-addressed, stamped envelope for customer response.
- Offer a discount if the customer responds within a given period of time.
- Give your name or a customer-contact name and a phone number (toll free if possible).

Figure 11.7 provides a sample sales letters using the *ARGU* method of persuasion.

FIGURE 11.7 Sales Letter Using *ARGU* Method of Persuasion

4520 Shawnee Dr. Tulsa, OK 86221
721-555-2121 officestation.com

November 12, 2012

Bill Schneider
Office Manager
REM Technologies
2198 Silicon Way
Tulsa, OK 86112

Dear Mr. Schneider:

Are hardware and software upgrades making your profits plummet? Would you like to reduce your company's computer purchase and maintenance costs? Do computer breakdowns hurt your business productivity? Don't let technology breakdowns harm your bottom line. Many companies have taken advantage of **Office Station's** computer prices, service guarantees, and certified technicians.

Office Station, located in your neighborhood, offers you the following benefits:

 Purchase prices at least 10 percent lower than our competitors.

 IBM-trained technicians, available on a yearly contract or per-call basis.

 An average response time to service calls of under two hours.

 Repair loaners to keep your business up and running.

 Over 5000 satisfied customers, like IBM, Ford, Chevrolet, and Boeing.

 All-inclusive agreements that cover travel, expenses, parts, and shop work.

 State-of-the-art technologies, featuring the latest hardware and software.

Our service is prompt, our technicians are courteous, and our prices are unbeatable. For further information and a written proposal, please call us at **721-555-2121** or e-mail your sales contact, Steve Hudson (shudson@os.com). He's waiting to hear from you. Take advantage of our *Holiday Season Discounts!*

Sincerely,

Rachel Adams
Sales Manager
rachel.adams@os.com

Office Station
Authorized Sales and Service for
Gateway   3M   Microsoft   HP   Apple   Dell   Swingline

The introduction arouses reader interest by using alliteration ("profits plummet") and asking questions. The questions highlight reader problems: *profits, costs, productivity, and breakdowns.*

The last sentence shows how Office Station will solve the problems.

The letter uses positive words to persuade: *advantage, guarantees, certified, benefits, satisfied, prompt, and courteous.*

The body provides specific proof to sway the reader: *10 percent lower, IBM-trained technicians, two-hour response, and renowned satisfied customers.*

The conclusion urges action by giving contact names and numbers and seasonal discounts.

Fliers

Corporations and companies; educational institutions; religious organizations; museums, zoos, and amusement parks; cities and states—they all need to communicate with their constituencies (clients, citizens, members). An effective way for these organizations to communicate persuasively is with a one- or two-page flier, electronic or hard copy.

Fliers provide the following benefits:

- **Cost effective.** A flier costs less than an expensive advertising campaign and can be produced in-house.
- **Time efficient.** Creating a flier can take only a few hours of work or less by the company's employees.
- **Responsive to immediate needs.** Different fliers can be created for different audiences and purposes to meet unique, emerging needs.
- **Personalized.** Fliers can be created with a specific market or client in mind. Then, these fliers either can be mailed or e-mailed to that client or hand-delivered for more personalization.
- **Persuasive.** In a compact format, fliers concisely communicate audience benefit.

Flier Criteria. When writing your flier, follow these criteria:

Keep the Flier Short. Though one page might be preferable, you could create a two-page flier, using the front and back of 8½ × 11-inch paper. If you keep your flier to one page (front only), then you can save money by folding the flier in thirds, stapling it, and using the blank side for mailing purposes (addresses and stamp). A flier even could be smaller, the size of a postcard, for example. Many companies create electronic fliers, transmitted via their Web sites or through e-mail. This is even more cost effective than a hard-copy flier, since no postage is needed.

Focus on One Idea, Topic, or Theme per Flier. A flier should make one key point. This is how you make the flier's content relevant to your audience, fulfilling that audience's unique needs. For example, if your company's focus is automotive parts, avoid writing a flier covering every car accessory. Write the flier with one accessory in mind, such as windshield wipers, batteries, or custom rims.

Use a Title at the Top of Your Flier to Identify Its Theme. The title can be one or two words long, you could use a phrase, or you might want to write an entire sentence at the top of the flier. An effective persuasive approach is to begin your flier with a question to immediately arouse reader interest: "Is your software giving you a headache?" or "How usable is your Web site?"

Limit Your Text. Using few words, provide reader benefit, involve the audience, and motivate them to act. The action could be to purchase a product, attend an event, or contact you for additional information. By limiting your text, you avoid overwhelming either the flier's appearance or the audience's attention span. Getting to the point in a flier is a key concern. Limiting your text helps you achieve this goal.

Increase Font Size. In a flier you can use a 16-point font and up for text, and a 20-point font and up for titles. This will make the text more readable and dramatic. Your heading must be eye catching. To accomplish this goal, make sure the heading's font size and style are emphatic—at a glance, even from across a room.

Use Graphics. One graphic, at least, will emphasize your theme and visually make your point memorable. Another graphic could include your company logo (for corporate identity and namesake recognition). In addition, the logo should be accompanied by a street address, e-mail address, Web site URL, fax number, or phone number so clients can contact you or visit your site.

Use Color for Audience Appeal.   Pick one dominant color to emphasize key points. Use a color in your company logo to remind your reader of your company's identity. However, don't overuse color. Excess will distract your reader.

Use Highlighting Techniques.   Bullets, white space, tables, boldface, italics, headings, text boxes, or subheadings will help your reader access information. A little highlighting goes a long way, especially on a one-page flier. Too much makes a jumble of your text and distracts from the message.

Find the Phrase.   Select a catchy phrase, which you can use continually in all your fliers, to personify your company's primary focus. For example, McDonald's repeats the line "We love to make you smile"; Pepsi uses "The Joy of Cola"; Nike says "Just Do It"; and Ford's "Quality is Job #1." What phrase captures your company's personality?

Recognize Your Audience.   You want to show the readers how your product or service will benefit them. Understand your audience's needs and direct the flier to meet those concerns. In addition, you want to engage the reader. To do this, use pronouns which speak to the reader on a personal level and positive words which motivate the reader to action. Remember to speak at the reader's level of understanding, defining terms as necessary.

Figure 11.8 provides a sample persuasive flier.

FIGURE 11.8   Sample Persuasive Flier

Having trouble balancing your user support needs?

The question highlights the audiences need and arouses reader interest.

A graphic adds visual appeal to make the flier more persuasive. The graphic also develops how the company can meet the audience's need.

Software Manuals   Training Materials   Online Help   Help Desk Materials   Intranet Content

We can help.

This sentence answers the introductory question.

APPLIED COMMUNICATIONS GROUP, INC.

Call Mike Hanson at 913-362-9293
www.acgtech.com

We're here for you

The contact name and number are provided to motivate the reader.

The last sentence uses empathy to persuade the audience.

Brochures

A flier *must* be short—one or two pages. If you have more information to convey than can fit on a one-page flier, then a brochure might be a good option. Brochures offer you a detailed overview of products, services, options, and opportunities, complete with photographs, maps, or charts. Brochures are persuasive for the following reasons:

- Create awareness of your company, product, or service
- Increase understanding of a product, service, or your company's mission
- Advertise new aspects about your company, product, or service
- Change negative attitudes
- Show ways in which your company, product, or service surpasses your competition
- Increase frequency of use, visit, or purchase
- Increase market share

Criteria for Writing Brochures. Brochures come in many shapes and sizes. They can range from a simple front and back, four-panel, 8½ × 5½-inch brochure (one landscape 8½ × 11-inch page folded in half vertically) to six-, eight-, or even twelve-panel brochures printed on any size paper you choose. See Figure 11.9 for two ways to fold paper to create panels in a brochure. Brochures, like fliers, also can be transmitted electronically. Your topic and the amount of information you are delivering will determine your brochure's size and means of transmission.

To determine what you will write in each panel of your brochure and how the brochure should look, follow these criteria for writing an effective persuasive brochure.

Title Page (Front Panel). Usually, the title page includes at least three components:

- **Topic.** In the top third of the title panel, name the topic. This includes a product name, a service, a location (such as "Rocky Mountain Park" or "San Diego Zoo"), or the subject of your brochure.
- **Graphic.** In the middle third of the panel, include a graphic to appeal to your reader's need for a visual representation of your topic. The graphic will sell the value of your subject (its beauty, its usefulness, its location, or its significance) or visually represent the focus of your brochure.
- **Contact information.** In the bottom third of the panel, place contact information. Include your name, your company's name, street address, city, state, zip code, telephone number, Web site, or e-mail address. See Figure 11.10 for a sample brochure.

FIGURE 11.9 Four- and Six-Panel Brochures

FIGURE 11.10   Sample Brochure

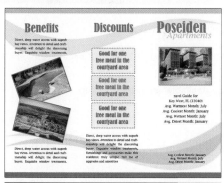

Body panel

Topic

Graphics

Contact information

Body panel

Graphic for
visual appeal

Back Panel.   The back panel could include the following:

- **Conclusion.** Summarize your brochure's content. Restate the highlights of your topic or suggest a next step for your readers to pursue.
- **Mailing.** The back panel could be used like the face of an envelope. On this panel, when left blank, you could provide your address, a place for a stamp or paid postage, and your reader's address.
- **Coupons.** As a tear-out, this panel could be an incentive for your readers to visit your site or use your service. Here you could urge action by providing discounts or complimentary tickets.
- **Location.** A final consideration would be to provide your readers with your address, hours of operation, phone numbers, e-mail, and a map to help them locate you.

Body Panels (Fold-in and Inside).   The following are some suggestions for creating the brochure's text:

- **Provide headings and subheadings.** These act as navigational tools to guide your readers, direct their attention, and help them find the information they need. The headings and subheadings should follow a consistent pattern of font type and size. First-level headings should be larger and more emphatic than your second-level subheadings. The headings must be parallel to each other grammatically.

   For example, if your first heading is entitled "Introduction," a noun, all subsequent headings must be nouns, like "Location," "Times," "Payment Options," and "Technical Specifications." If your first heading is a complete sentence, like "This is where it all began," then your subsequent headings must also be complete sentences: "It's still beautiful," "Here's how to find us," and "Prices are affordable."

Using Microsoft Word 2010 for Flier and Brochure Design Templates

You can access many design templates for fliers and brochures using your word processing software. For example, in Microsoft Word 2010, try this approach:

1. On the menu bar, click on the **File** tab and scroll to **New**. You will see **Available Templates** pop up.

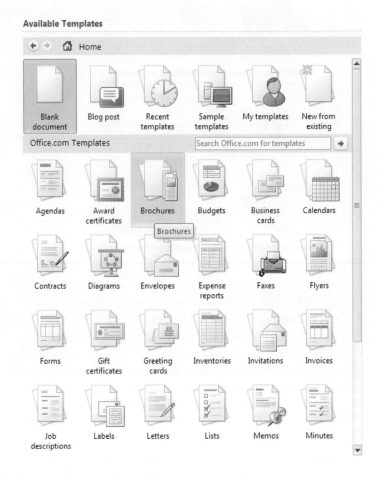

2. Click on the document of your choice (**Flyers** or **Brochures**). Word gives you 51 different brochure options for healthcare, business, event marketing, real estate, and more. These include templates for 8½ × 11 landscape with 3 folds, 8½ × 14 landscape with 3 and 4 folds, and 8½ × 11 portrait letter folds. When you click on fliers, you get templates for 67 different "Events" fliers, 64 "Marketing" options, 3 "Real Estate" choices, and numerous "Other Fliers."

Note: The templates will not tell you what to write in each panel. Instead, Word provides you a design layout, a shell for you to fill with your content.

Creating Your Own Brochure Layout

Instead of using a predetermined template, you can create your own as follows:

1. Change from **Portrait** to **Landscape** by clicking on the **Page Layout** tab and then **Orientation**.

2. Create three panels for your brochure by clicking on the **Columns** icon.

- **Use graphics.** Use photographs, maps, line drawings, tables, or figures to vary the page layout, for visual appeal, and to enhance your text.
- **Develop your ideas.** Consider including locations, options, prices, credentials, company history, personnel biographies, employment opportunities, testimonials from satisfied customers, specifications, features, uses of the product or service, payment schedules, or payment plans.
- **Persuade your audience.** Review the tips provided in this chapter for persuasive arguments. Use ethics, logic, and emotion to sway your reader.

Document Design.   Visual appeal helps to interest and persuade an audience. Compelling graphics, for example, can help to convince an audience. Use pie charts, bar charts, tables, or photographs to highlight key concerns. In addition to graphics, make your brochure visually appealing by doing the following:

- Limit sentence length to 10 to 12 words and paragraph length to four to six lines. When you divide paper into panels, text can become cramped very easily. Long sentences and long paragraphs then become difficult to read. By limiting the length of your text, you will help your readers access the information.
- Use white space instead of wall-to-wall words. Indent and itemize information so readers won't have to wade through too much detail.
- Use color for interest, variety, and emphasis. For example, you can use a consistent color for your headings and subheadings.
- Bulletize key points.
- Boldface or underline key ideas.
- Do not trap yourself within one panel. For variety and visual appeal, let text and graphics overlap two or more of the panels.
- Place graphics at angles (occasionally) or alternate their placement at either the center, right, or left margin of a panel. Panels can become very rigid if all text and graphics are square. Find creative ways to achieve variety.

See Figure 11.11 to get a better idea of what a typical brochure looks like.

DOT.COM UPDATES

For more information about creating brochures, check out the following links:

- http://www.hp.com/sbso/ productivity/howto/ marketing_main/ marketing_brochure/
- www.smallbusinessnotes. com/operating/marketing/ brochures.html

FIGURE 11.11   Brochure

A panel for mailing

Icon

Reporter's
questions
for
development

Title with
question
to arouse
reader
interest

Graphic
appropriate
to topic

Contact
information

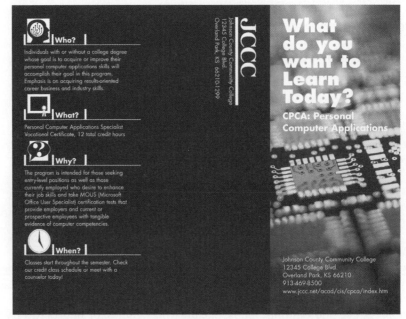

Fold-in   Back   Front

A heading
that overlaps
two panels

Bulleted
text

Graphics
for visual
appeal

Inside Panels

346

PERSUASIVE COMMUNICATION CHECKLIST

_____ 1. **Ethical argumentation:** Have you made an ethical argument based on character? You must be trustworthy and credible as a writer or speaker.

_____ 2. **Emotional argumentation:** Have you used emotion to change an audience's attitudes? You can appeal to an audience's emotions either positively or negatively.

_____ 3. **Logical argumentation:** Have you developed your persuasion by depending on rationality, reason, and proof? You can persuade people logically by providing facts, testimony, and examples.

_____ 4. **Arouse audience interest:** Have you used questions, quotes, anecdotes, comparison/contrast, poetic language, or an appeal to senses to interest your audience?

_____ 5. **Refute opposing points of view:** Have you presented a balanced argument?

To do so, recognize and admit conflicting views, let the audience know that you understand their concerns, and allow for alternatives.

_____ 6. **Give proof.** Have you provided evidence to prove your point?

_____ 7. **Urge action:** Have you motivated your audience to act? To do so, provide incentives, give discounts, mention warranties, provide contact information, or suggest a follow-up action.

_____ 8. **Highlighting/page layout:** Is your text accessible? To achieve reader-friendly ease of access, use headings, boldface, italics, bullets, numbers, underlining, text boxes, or graphics (tables and figures). These add interest and help your readers navigate your text.

_____ 9. **Conciseness:** Have you limited the length of your sentences, words, and paragraphs?

_____10. **Audience recognition:** Have you written appropriately to your audience?

This includes avoiding biased language, considering the multicultural/cross-cultural nature of your readers, and your audience's role (supervisors, subordinates, co-workers, customers, or vendors).

_____11. **Correctness:** Is your text grammatically correct? Errors will hurt your professionalism.

The Writing Process

Web Resources

www.pearsonhighered.com/gerson For more information about the writing process, visit our companion website.

Prewriting	Writing	Rewriting

• Determine your goals in writing persuasively.
• Try brainstorming, listing, and/or answering reporter's questions to clarify what content you will need to include.
• Decide whether your audience is high tech, low tech, or lay. This will help you decide which abbreviations and acronyms need to be defined.
• Choose the correct persuasive communication channel (routine correspondence, electronic communication, sales letter, brochure, or flier) for your audience and purpose.
	• Organize your content considering goals, audience, channel of communication, and data to include.
	• Use ethos, logos, and pathos in your writing to be persuasive.
	• Consider appropriate format to appeal to the audience and be persuasive.
		• Revise your draft by
		  • Adding details
		  • Deleting wordiness
		  • Simplifying words
		  • Enhancing the tone
		  • Reformatting your text
		  • Proofreading and correcting errors
		  • Adding visual appeal to attract the reader and enhance the persuasive content

The Writing Process at Work

Now that you know how to write persuasively, let's see how Georgia Nesselrode, director of government training for Mid-America Regional Council's (MARC) Government Training Institute, follows the writing process to create a flier. She plans to write a flier to advertise training workshops being offered to members of MARC's consortium of city and county employees. The best way to accomplish any technical communication challenge is by following a process: Prewrite to gather data and determine objectives, write a rough draft of your document, and rewrite to revise.

Prewriting

To create an effective persuasive document, first prewrite to determine goals, audience, channel of communication, and data. To plan her communication, Georgia considered the following:

- Goal—help communities meet their infrastructure needs
- Audience—small-town city councils
- Channels—informative and persuasive flier, mailed as hard-copy text and posted on MARC's Web site
- Data—specific workshop topics of discussion

The challenge at MARC is communicating to a diverse audience in different-sized cities. To prewrite, Georgia brainstormed and created a list (Figure 11.12).

Writing

After prewriting, write a draft of the document. Georgia's rough draft (Figure 11.13) included all of the information she needed to share with her intended audience. One goal was to provide answers to reporter's questions: who, what, when, where, why, and how.

FIGURE 11.12   Georgia's Brainstorming List

1. Brainstorm topics with internal staff and advisory committee (made up of planners and community development representatives).
2. Narrow the topics down to three or four workshops.
3. Draft course descriptions.
4. Recruit instructors.
5. Determine dates and times.
6. Secure locations.
7. Provide draft information to public affairs to create the flier.
8. Proofread and finalize marketing materials.
9. Send e-mail announcements with flier attachment to targeted audiences.
10. Hand out fliers at MARC-sponsored events.
11. Post the flier on our Web site.
12. Send fliers electronically to our customer list serves.

FIGURE 11.13   Georgia's Rough Draft for the Flier

Master Plans for Smaller Communities
Thursday, May 17, 2012
6:00 - 8:30 p.m.
Grandview Community Center – The View
13500 Byars Road
Grandview, MO

Moderator: Bob Linder, Planning Commissioner

Panelists: Chase Simpson, Real Estate Attorney; William Shane, Shane Appraisal, Inc.; Scott Mitchell, Vice President, Mitchell Engineering

Smaller communities throughout the Kansas City metropolitan area are facing growth challenges – residential, commercial, and industrial. Our smaller cities are eager to accommodate new residents and businesses, but are concerned about losing the small-town quality of life, impacting streams and other natural features, and paying for infrastructure to support new development. The preparation and adoption of a master plan or land use plan provides communities with an important citizen engagement opportunity and new tools to guide growth and development.

The two presenters will describe the conditions under which a master plan is useful or necessary; the elements that might be included in a master plan; the various ways that master plans might be developed, including ways to involve residents and other stakeholders; the adoption process; and how the planning commission uses the adopted plan in making decisions on development applications.

Rewriting

After prewriting and writing a draft of the persuasive document, you then rewrite by considering goals, audience, document design, and completeness of the content. You add highlighting techniques to emphasize key points in the flier. Because a flier is short, you want to offer the audience a visually appealing document filled with relevant content. See Georgia's rewritten flier in Figure 11.14.

FIGURE 11.14   Georgia's Revised Flier

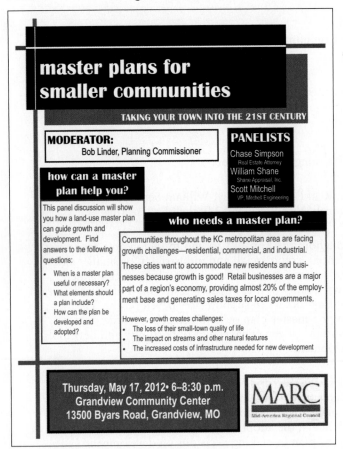

CHAPTER HIGHLIGHTS

1. Persuasive technical communication consists of a combination of ethos, logos, and pathos.
2. Use the *ARGU* technique to create persuasive routine correspondence, electronic communication, fliers, brochures, or sales letters.
3. Avoid logical fallacies when communicating persuasively.
4. Use many communication channels, including social media, routine correspondence, electronic communication, sales letters, brochures, and fliers to market services and products.
5. Fliers should be short, focused on one idea, titled, have limited text, and be visually appealing.
6. Recognize your audience and their needs when you write persuasive documents.
7. The front panel of a brochure contains a title and a graphic that depicts the topic.
8. The interior panels of a brochure contain information about the topic.
9. The back panel of a brochure can be used for mailing purposes, maps, or discounts.
10. Document design is important when you create persuasive routine correspondence, electronic communication, a flier, brochure, or sales letter.

CASE STUDY

A company has just expanded into your city. Having been in business in the Midwest since 2005, TechToolshop provides sales, service, maintenance, installation, and data recovery for PC and Mac hardware and software tools.

They install, repair, and maintain workstations, servers, printers, and peripherals. Through an online catalogue and storefront site, they sell printers, desktops, mainframes, mini-towers, software, accessories, and encryption devices. TechToolshop has proudly worked with many large companies, including McDonald's, Pepsi, Ford, Sprint, Transamerica, GE, JCPenney, and Chase Manhattan Bank. For their latest venture, this company will help you create your own corporate Web page and maintain the site, fees determined by number of screens, plug-ins, and graphics.

TechToolshop's home office is in Big Springs, Iowa, at 11324 Elm, where over 1,200 employees work. Their phone number at this site is 212-345-6666, and their e-mail address is ToolHelp@TechTools.com. This company's Web site can be found at www.TechTools.com. TechToolshop's new local address in your city is 5110 Nueces Avenue. The phone number is 345-782-8776.

TechToolshop is most proud of its product support and performance guarantees. The company offers free product support, 24 hours a day at 1-800-TechHelp, online help from the Web site, and 24/7 help through TweetForce. In addition, they warrant all products and services—money back—for 90 days, covering defects in material and workmanship.

The company is owned by James Wilcox Oleander (president) and his brother Harold Robert Oleander (CEO). They started this company after graduating from Midwestern State University, with degrees in IT, IS, and CPCA. Their first store had only three employees, but through hard work, their business grew 45 percent in the first two years of operation. Expert recruiting of the best State U. graduates increased their workforce, as did the Oleanders' philosophy of "employee ownership." TechToolshop's workers take pride in their work, since their success increases their stock options (TechToolshop's stock at NASDAQ opened in 2005 at $18 a share but has listed as high as $45). The Oleanders plan to open at least 12 new stores each year throughout the United States, as well as pursue franchise options. They have high hopes for their future success.

Write either a persuasive sales letter, flier, or brochure about TechToolshop, based on the information provided. Follow the criteria in this chapter.

INDIVIDUAL OR TEAM PROJECTS

ARGU

For the *ARGU* method of organization, you *A*rouse reader interest, *R*efute opposing arguments, *G*ive proof to support your argument, and *U*rge reader action. Read the following situations and complete the assignments.

- You plan to sell a flash drive. For a sales letter, write five different introductions to arouse reader interest using any of the options provided in the chapter.
- Write a body that refutes opposing arguments (too expensive, easily damaged, easily lost, etc.) and that gives proof to support your product claims.
- In the sales letter for the flash drive, conclude by urging reader action. Use at least two of the methods discussed in this chapter.

Analysis of Persuasive Writing

Find examples of persuasive writing (e-mail messages, sales letters, fliers, or brochures). Bring these to class, and in small groups or individually, accomplish the following:

1. Decide which methods of persuasion have been used. Where in the documents do the writers appeal to logic, emotion, and ethics? Give examples and explain your reasoning, either in writing or orally.
2. Have the writers aroused reader interest? Give examples and explain your reasoning, either in writing or orally. If the writers have not aroused reader interest, should they have? Explain why. Rewrite the introductions using any of the techniques discussed in this chapter to arouse reader interest.
3. Have the writers refuted opposing points of view? Give examples and explain your reasoning, either in writing or orally. If the writers have not negated opposing points of view, should they have? Explain why. Rewrite the text using any of the techniques discussed in this chapter to negate opposing points of view.
4. Have the writers developed their arguments persuasively? Give examples and explain your reasoning, either in writing or orally. If the writers have not provided persuasive proof, rewrite the text using any of the techniques discussed in this chapter to improve the arguments.
5. Do any of the examples you have found use logical fallacies to persuade the readers? Give examples and explain your reasoning, either in writing or orally. If the writers have used logical fallacies, rewrite the text using any of the techniques discussed in this chapter to improve the arguments.

PROBLEM-SOLVING THINK PIECES

Logical, Emotional, and Ethical Appeals

Read the following situations and determine whether the persuasive arguments appeal to logic, emotion, or ethics. These argumentation techniques can overlap. Explain your answers.

1. If you purchase this product, you can benefit from a healthier, happier, and longer life!
2. Seventy-two percent of SUV owners say that high gas prices will influence their next car purchases.
3. CEO Jim Snyder, an expert in the field of sports management, says, "Building a downtown sports arena enhances a city's image."
4. Style-tone Hair Gel improves your hair quality by preventing split ends, generating new hair growth, and inhibiting "frizzies."
5. Failure to recycle will cause 52 percent more dangerous hydrofluoric carbons to be released into the atmosphere, leading to harmful decreases in the ozone layer.

Logical Fallacies

Read the following logical fallacies and revise them, ensuring that the sentences provide logical, ethical, and correct argumentation.

1. All marketing experts believe that social media is an effective way to communicate persuasively.
2. Either all employees must come to work on time, or they will be fired.
3. Some accountants are ambitious because they wish to succeed.
4. The contractor lost the bid, so he cannot expect to have increased revenues this fiscal year.
5. Because the manager is inexperienced, the report will be badly written.

WEB WORKSHOP

You can find electronic brochures and fliers on the Internet. Using a search engine of your choice, type in phrases like "online brochure," "e-brochure," "online flier," "e-flier," "electronic brochure," or "electronic flier." Once you find examples of these online persuasive documents, do the following:

- Compare and contrast the electronic documents with hard-copy brochures and fliers.
- Compare and contrast the electronic documents with the criteria provided in this chapter.
- Decide how the online communication is similar to and different from the criteria and from hard-copy versions.
- If you decide that the online versions can be improved, print them out and revise them.

QUIZ QUESTIONS

1. What are the traditional methods of argumentation?
2. Which type of argumentation depends on character?
3. Which type of argumentation focuses on feelings?
4. Which type of argumentation depends on rationality, reason, and proof?
5. What are four ways to arouse reader interest?
6. Which steps do you follow to refute opposing points of view in the body of your argument?
7. How can you urge your audience to action?
8. What are three types of logical fallacies?
9. Why are brochures an effective marketing tool?
10. Why are fliers effective for marketing?

CHAPTER TWELVE

Technical Descriptions and Process Analyses

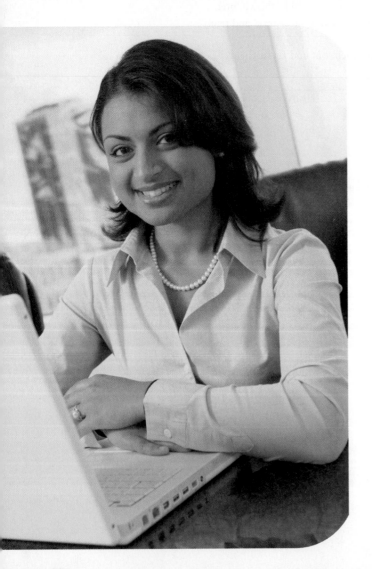

The Alpha Beta Consulting scenario shows the importance of technical descriptions and process analyses in the engineering profession.

Alpha Beta Consulting (A/B/C) specializes in creating software for public service engineering purposes, such as storm water, water distribution, transportation and traffic control, parks and recreation, facility and building management, and inventorying of fleets and equipment.

The city of Maple Valley, Texas, is interested in purchasing one of A/B/C's software suites related to storm water maintenance. The suite focuses on accurate and easy-to-use inventory and inspection tools for maintaining and improving storm water systems. A/B/C's software will help Maple Valley inspect and test all components of the city's storm water system including conduits, structures, detention basins, pump stations, and pumps.

Before A/B/C can sell the software to Maple Valley, the company's information technology director, Neeha Patel, must do the following:

Objectives

When you complete this chapter, you will be able to

1. Distinguish between technical descriptions and process analyses
2. Follow criteria for writing an effective technical description or process analysis
3. Use graphics with callouts effectively in your description or process analysis
4. Follow a process approach—prewriting, writing, and rewriting—to write your description or process
5. Evaluate your description or process by using the checklist

- Determine the city's unique needs—Neeha will interview city residents, city council members, and city government officials. To accomplish this goal, Neeha will create a questionnaire focusing on infrastructure needs. Then, through a combination of hard-copy mailers and online social media (Maple Valley's Facebook site and Twitter account), Neeha will collect constituent responses.

- Research software options—Based on the interviews, Neeha will determine if the city's current software is failing to meet the city's needs and what A/B/C software options exist.

- Create or adapt software to meet the city's technical specifications—If new software is needed, Neeha will work with her colleagues in IT to customize software for Maple Valley.

- Write technical descriptions of the hardware—Based on her analysis of hardware requirements, Neeha will write the technical descriptions of hardware needed to run the software. These descriptions will accompany the operations manual A/B/C will give Maple Valley employees.

- Provide process analyses—To clarify how the software will solve Maple Valley's infrastructure problems, Neeha will write a process analysis. This process analysis will show how the software will improve storm water, transportation, and traffic flow.

Once Neeha creates this documentation, she can provide it to Maple Valley's Planning Commission, which will decide whether or not to purchase the software and work with A/B/C.

Defining Technical Descriptions

A technical description is a part-by-part depiction of the components of a mechanism, tool, or piece of equipment. Technical descriptions are important features in several types of technical communication.

Types of Technical Descriptions

Operations Manuals

Manufacturers often include an operations manual in the packaging of a mechanism, tool, or piece of equipment. This manual helps the end user construct, install, operate, and service the equipment. Operations manuals also can contain technical descriptions.

Technical descriptions provide the end user with information about the mechanism's features or capabilities. For example, this information may tell the user which components are enclosed in the shipping package, clarify the quality of these components, specify what function these components serve in the mechanism, or allow the user to reorder any missing or flawed components. Following is a brief technical description found in an operations manual:

EXAMPLE ▶

> The Modern Electronics Tone Test Tracer, Model 77A, is housed in a yellow, high-impact plastic case that measures 1¼ inch × 2 inch × 2¼ inch, weighs 4 ounces, and is powered by a 1604 battery. Red and black test leads are provided. The 77A has a standard four-conductor cord, a three-position toggle switch, and an LED for line polarity testing. A tone selector switch located inside the test set provides either solid tone or dual alternating tone. The Tracer is compatible with the EXX, SetUp, and Crossbow models.

Product Demand Specifications

Sometimes a company needs a piece of equipment that does not currently exist. To acquire this equipment, the company writes a product demand specifying its exact needs. The product demand specification example shown below is written for a high-tech audience. It assumes knowledge on the part of the reader, such as definitions for "high-speed" and "deep-hole," which a lay reader would not understand. In addition, the high-tech abbreviation "AISI" is not defined.

EXAMPLE ▶

> Subject: Pricing for EDM Microdrills
> Please provide us with pricing information for the construction of 50 EDM Microdrills capable of meeting the following specifications:
>
> - Designed for high-speed, deep-hole drilling
> - Capable of drilling to depths of 100 times the diameter using 0.012-inch to 0.030-inch diameter electrodes
> - Able to produce a hole through a 1.000-inch-thick piece of AISI D2 or A6 tool steel in 1.5 minutes, using a 0.020-inch diameter electrode
>
> We need your response by January 13, 2012.

Study Reports Provided by Consulting Firms

Companies hire a consulting engineering firm to study a problem and provide a descriptive analysis. The resulting study report is used as the basis for a product demand specification requesting a solution to the problem. One firm, when asked to study crumbling cement walkways, provided the following technical description in its study report.

> The slab construction consists of a wearing slab over a ½ inch-thick waterproofing membrane. The wearing slab ranges in thickness from 3½ inches to 8½ inches and several sections have been patched and replaced repeatedly in the past. The structural slab varies in thickness from 5½ inches to 9 inches with as little as 2 inches over the top of the steel beams. The removable slab section, which has been replaced since original construction, is badly deteriorated and should be replaced. Refer to Appendix A, Photo 9, and Appendix C for shoring installed to support the framing prior to replacement.

◀ **EXAMPLE**

DOT.COM UPDATES

For examples of technical descriptions, check out the following links:

- Palm Pre Smartphone technical specifications http://www.palm.com/us/products/phones/pre/index.html
- iPod Nano technical specifications http://www.apple.com/ipodnano/specs.html
- Technical specifications for Garmin's GPS system http://www.garmin.com/manuals/425_TechnicalSpecification.pdf
- Technical specifications for the Boeing 737 http://www.boeing.com/commercial/737family/specs.html
- Original Xbox technical specifications http://www.xbox.com/en-US/support/familysettings/resources/consolespecs.htm
- Nintendo Wii technical specifications http://www.nwiizone.com/nintendo-wii/nwii/nintendo-wii-technical-specifications/

Construction Design

Prior to building any structure, architectural companies must clarify the construction design for the city and the clients. This requires a legal technical description of the property limits (location and dimensions). Once this description is written and drawn, the architectural company will use the text and graphics for a variety of purposes. These can include presentations of the design to a city council or homeowner's association, requests for a change of zoning, requests for special use permits, and re-platting of the site if a property line must be relocated. Finally, based on the legal technical description, the architectural firm can obtain city approval. See Figure 12.1 for a layout and landscaping plan and the legal technical description that accompanies it.

Sales Literature

Companies want to make money. One way to market equipment or services is to describe the product. Such descriptions are common in sales letters, proposals, and on Web sites. Figure 12.2 is a technical description in sales literature from Hewlett-Packard.

Defining Process Analysis

What is a process analysis? Let us define a process analysis by comparing it to an instruction. Instructions provide a step-by-step explanation of how to do something. For example, in an instruction, you explain how to change the oil in a car, how to connect a printer to a computer, how to make polenta, or how to put together a child's toy. In an instruction, the audience wants to know how to do a job. A process analysis, in contrast, focuses not on how to do something but on how something works. For instance, a process analysis might explain how viruses attack our bodies, how airbags save lives, how metal detectors work, or how e-mail messages are transmitted.

Examples of Process Analyses

A discussion of process is a common part of many technical descriptions:

- **Engineering.** Dubai, second largest emirate of the United Arab Emirates, plans to develop several islands off its coast. To increase tourism, these islands will have

Instructions

See Chapter 13 for more discussion of instructions.

FIGURE 12.1 Layout and Landscaping Plan with Legal Technical Description for High-Tech Audience

This landscape plan describes the construction design for section 10, township 14, range 25, located in the city of Hartford, Johnson County, Washington. The plot of land includes the following coordinates and distances:

- north 87°38'42" east, along the north line of the northwest quarter, a distance of 539.17 feet
- south 2°21'18" east, along a line perpendicular to the north line of the northwest quarter, a distance of 60.00 feet
- west 51°29'37", a distance of 314.56 feet, to a point of tangency
- south 09°56'20" west, along the westerly right-of-way line, a distance of 264.15 feet, to a point of curvature
- east 59°29'39", a distance of 129.80 feet, to the most easterly corner of tract "c" and east of the right-of-way line of Rolling Hills Drive

Source: Courtesy of LandPlan

both private residences and hotels. Your engineering company has been hired to build bridges connecting the islands to the coastline. Your options include beam bridges, arch bridges, and suspension bridges. To clarify which type of bridge will best meet your client's needs, you provide a PowerPoint presentation explaining how each bridge works.

- **Automotive sales.** A potential customer wants to buy a new car. This customer is concerned about the environment and is interested in buying a hybrid, knowing

FIGURE 12.2   Technical Description for a Lay Audience

HP Officejet Pro K550 series color printer
The title identifies the equipment to be described.

Impressionistic words such as "quick access," "easy troubleshooting," and "User-friendly control panel" are used for sales purposes to the lay audience.

HP Officejet Pro K550dtwn color printer shown

1 Ink cartridge cover provides quick access to four snap-in ink cartridges.

2 Top cover flips open for easy troubleshooting or maintenance.

3 User-friendly control panel includes graphical icons on the buttons and lights. It provides printer and wireless networking status at-a-glance and simplifies wireless configuration with SecureEasySetup™.

4 150-sheet output tray with extender.

5 250-sheet and 350-sheet input trays for high-volume printing.

6 Automatic two-sided printing accessory for professional, two-sided documents.

7 Built-in (Full Speed) USB host connector enables wireless configuration through Windows® Connect Now.

8 Built-in wired Ethernet networking.

9 Built-in Hi-Speed USB 2.0 port enables fast and easy direct connections.

Numbered callouts refer to the text and help the readers locate parts of the equipment.

Source: Courtesy of Hewlett-Packard

that these cars are energy efficient and ecologically friendly. However, the customer doesn't understand how hybrids work. Your job is to explain the process of how gasoline-electric hybrid cars work.

- **Biomedical technology.** You and a team of scientists have been working on new treatments for diabetes, a health problem in the United States that affects approximately 20.8 million Americans. One possible treatment is pancreatic islet transplantation. Islets are injected through a catheter into a diabetic's liver. In time, the islets begin releasing insulin. This treatment has great potential for diabetics, but research is ongoing. You are writing a proposal requesting additional funding for research. In this proposal, you provide a process analysis of how diabetes affects a patient and how the pancreatic islet transplantation will work to treat this disease.

How to Use Technical Descriptions in Landscape Architecture

Landplan, a multidisciplinary firm that specializes in surveying, civil engineering, landscape architecture, and planning, designs sites that conform to local city and county codes. As a landscape architect with Landplan, **Andrew Buchwitz** works with one of their larger corporate accounts to develop commercial sites. According to Andrew, "I lay out the best possible location for the building, parking, and landscaping while conforming to the local municipalities' City Development Code."

Communicating effectively is a major part of Andrew's job. He writes technical descriptions that accompany his architectural renderings, and he makes oral presentations at city planning commission meetings, city council meetings, and neighborhood meetings.

Andrew says that Landplan has "to write legal descriptions depicting the property size and location." Technical and legal descriptions are important for every project Andrew designs to

- Clarify the plan for the city and the client
- Meet legal description/property limits
- Meet city requirements for landscaping
- Obtain city approval so the client can build

Andrew's challenge is conforming to the local cities' development codes while still meeting his clients' needs. "Many times it becomes a compromise when the client wants the project to look a certain way, but the city does not," states Andrew.

For example, the city planners could review the renderings, technical specifications, technical descriptions, and documents Andrew designs but refuse to accept the plans as they have been drafted. The city might suggest site revisions including additional landscape screening around trash enclosures and additional upgrades to the exterior such as brick rather than stucco. In those instances, Andrew has to revise the plans to meet both the city and the client's needs.

Every site is unique in that it requires a different plan of attack. This could include requesting a change of zoning, requesting a special use permit, or having to re-plat the site because a property line needs to be relocated. "Some of these require public notice. In this case, Landplan would have to send a mail notification to residents living in a 200–500 foot buffer area surrounding the site."

According to Andrew, "Mail notices also usually require a neighborhood meeting for the general public. This is basically a Q & A session notifying the public of the changes that are being proposed for the particular area and hearing about any concerns they may have."

At these neighborhood meetings, Andrew often makes oral presentations. According to Andrew, "Some of the challenges may be getting the general public to back our plan. If the city commission sees that the general consensus of the plans for the buffered public area is against this development, it may be even harder to get approval. The city may not like the proposed development and want us to have another meeting with city planners to come up with another layout or design."

Once the City Planning Commission approves all of Andrew's site revisions, the Commission takes the plans to the City Council Community Development Committee for final approval. Andrew then returns the approved documents to the client who gives the "go-ahead" for the project. Throughout the multi-step process, Andrew faces many challenges communicating both orally and in writing with the client, the city, and other concerned parties. Carefully written technical and legal descriptions, as well as effective oral presentations, help to ensure approval of Landplan's proposed site development.

Web Resources

www.pearsonhighered .com/gerson See our companion website for more technical description samples.

Figure 12.3 is a process analysis of why, when, and how automobile air bags inflate.

Criteria for Writing Technical Descriptions and Process Analyses

As with any type of technical communication, there are certain criteria for writing technical descriptions and process analyses.

Title

Preface your text with a title precisely stating the topic. This could be the name of the mechanism, tool, piece of equipment, landform, product, or service you are describing or analyzing.

FIGURE 12.3   Why, When, and How Does an Air Bag Inflate?

Air bags save lives. Driver and passenger air bags are designed to inflate in frontal or side crashes. Steering wheel, right front instrument panel, or side-panel air bags will not inflate on all occasions. If a car drives over a bump or if a crash is "minor," such as hitting an object while driving at a slow speed, an air bag will not deflate. However, when cars hit walls (or trucks or cars or trees), air bags inflate to minimize injury and to save people's lives.

The introduction explains what the topic is, why it is important, and where the mechanism is located.

Before Inflation

How does the air bag detect whether the car has hit a bump versus being involved in a collision?

The airbag system's crash sensor can differentiate between head-on collisions and simple bumps in the road as follows:

1. A steel ball slides inside a smooth bored cylinder.
2. The ball is secured by a magnet or by a stiff spring. This inhibits the ball's motion when the car encounters minor motion changes, such as bumps or potholes.
3. If the car comes to a dramatic and rapid stop, a force equal to running into a brick wall at about 10 to 15 miles per hour, such as in a head-on crash, the ball quickly moves forward. This closes a contact and completes an electrical circuit, which initiates the process of inflating the airbag.

Parts of an Air Bag

Air bag systems consist of the following:

- **Air bag:** made of thin nylon. A nontoxic powder (cornstarch or talcum powder) inside the air bag keeps it flexible and ensures the parts of the air bag do not stick together.
- **Air bag holding compartment:** the steering wheel, dashboard, seat, or door.
- **Sensor:** a device that tells the bag to inflate.
- **Inflator canister:** consisting of sodium azide (NaN_3), potassium nitrate (KNO_3), and silicon dioxide (SiO_2), which produce nitrogen gas (N_2).

This technical description discusses the parts of the mechanism, the materials used, the location of these components, and the chemical compounds required for activation.

During Inflation

In a severe impact, the air bag sensing system will deploy in milliseconds. The following occurs:

1. The air bag's crash sensor triggers a switch that energizes a wire, sending electricity into a heating element in the propellant that releases gas from the inflator canister.
2. A pellet of sodium azide (NaN_3) is ignited, generating nitrogen gas (N_2).
3. A nylon air bag, folded into the dashboard, steering wheel, and/or side panels of the door, inflates at a speed of 150–250 mph, taking only about 40 milliseconds (about 1/20 of a second) for the inflation to be complete.

Here, providing the mechanism's process, the text explains how an air bag works. Note the specificity of detail: 150–250 mph and 1/20 of a second.

(Continued)

FIGURE 12.3   Why, When, and How Does an Air Bag Inflate? (*Continued*)

4. The sodium azide inside the air bag produces sodium metal, which is highly reactive, potentially explosive, and harmful when in contact with eyes, nose, or mouth. To render this harmless, the sodium azide reacts with potassium nitrate (KNO_3) and silicon dioxide (SiO_2)—also inside the air bag— to produce silicate glass, a harmless and stable compound.

This illustration graphically depicts the process in action. When the crash sensor is triggered, it activates the inflator. Nitrogen gas explodes, inflating the air bag.

After Inflation

Air bag vents, minute holes in the bag, allow the deployed air bags to deflate immediately after impact. This ensures that the car's inhabitants do not smother.

The conclusion sums up the process analysis by quantifying the significance of air bags as a means of saving lives.

   Though air bags were first used in 1973, they have only been mandatory in all cars since 1998. Have air bags made a difference? Absolutely! The National Highway Traffic Safety Administration says air bags saved an estimated 1,043 lives in 1998 alone.

Overall Organization

In the **introduction**, specify and define your topic, and explain the topic's functions, capabilities, or processes. You can also include a list of components, parts, or equipment.

EXAMPLE ▶

Technical Description

The Apex Latch (#12004), a mechanism used to secure core sample containers, is composed of three parts: the hinge, the swing arm, and the fastener.

EXAMPLE ▶

Technical Description and Process Analysis

The DX 56 DME (Distance Measuring Equipment) is a vital piece of aeronautical equipment. Designed for use at altitudes up to 30,000 feet, the DX 56 electronically converts elapsed time to distance by measuring the length of time between your transmission and the reply signal. The DX 56 DME contains the transmitter, receiver, power supply, and range and speed circuitry.

Why write process analyses?

Stacy Gerson is a technical writer for GBA Master Series, a software company that creates computerized maintenance management systems (CMMS) for public works and utilities organizations. Their clients include street, water, and sewer departments for cities, counties, and states, helping these constituencies track their inventory and assets. As a documentation specialist, Stacy writes online user manuals and training guides for end users such as office workers, dispatchers, maintenance supervisors, field workers, engineers, and employees of citizen complaint call centers.

These manuals help her clients use GBA Master Series' software related to public works preventative maintenance scheduling, work orders, asset inventory, asset inspections, employee timesheets, and administrative modules for city, county, and state planning and budgeting.

As part of these instructions, Stacy writes process analyses, complete with technical descriptions. She includes these process analyses for at least two reasons:

- **FYIs.** Many of Stacy's clients will e-mail her company asking "How did that work?" Her process analyses give these inquiring minds a view "behind the scenes."

- **Troubleshooting.** More importantly, process analyses help her clients complete a task. The software asks clients to input data regarding their asset management. If the clients get an error reading, knowing how the software works can give them a better understanding of how a problem might have occurred and how to fix it.

Here's a good example of one of Stacy's process analyses. GBA Master Series' "Accident Manager" software allows a city's public works department to determine if an intersection or stretch of road needs improvement. If many traffic accidents are occurring in one location, perhaps the city needs to add more signals, remove roadway obstructions, add speed bumps, change traffic patterns, and so on.

Stacy writes, "The screens in this ['Accident Manager'] are fully customizable, allowing you to tailor the module to fit your accident or case report forms. Additionally, the accident records are validated by the system to create an accurate collision diagram report."

To help her clients better understand the software and use it effectively, Stacy then writes, "For additional information on the validation performed by the system, consult the *Accident Validation Process* help guide." This process analysis says the following:

"The GBA Accident Manager™ Data Management module allows you to record details of your traffic accidents. Based on the information found in your accident records, the system conducts a validation process to verify the location of the accident, direction of the accident, and the type of accident that took place. These validated results are then compiled in an Intersection Collision Diagram. This diagram allows you to accurately project accident patterns and make recommendations for street improvements. Listed below are the steps taken by the system to validate the user-entered accident data.

1. The system checks to make sure there is at least one vehicle involvement record for the accident. This is found on the Involvement tab on the Data Management module.

   Note: You must have at least one vehicle record associated with the accident before it can be saved.

2. The system will then perform validations for the following: pedestrian involvement, type of accident (rear-end collision, head-on collision, sideswipe), pre-crash maneuver, and direction of vehicle travel."

In addition to her instructions and technical descriptions, a process analysis gives Stacy another way to meet her end user's needs for effective technical communication.

In the **discussion,** describe each of the mechanism's components and how the mechanism works—its process. Focus on details, word usage, and an organizational pattern that helps your audience easily understand the text.

- **Details.** To develop your discussion, include the following details:

Weight	Materials (composition)
Size (dimensions)	Identifying numbers
Color	Make/model
Shape	Texture
Density	Capacity
Parts	Procedural steps

TABLE 12.1 Photographic versus Impressionistic Word Usage

Photographic	Impressionistic
6'9"	tall
350 lb	heavy
gold	precious metal
6,000 shares of United Can	major holdings
700 lumens	bright
0.030 mm	thin
1966 XKE Jaguar	impressive car

Organization

See Chapter 3 for more discussion of types of organization.

- **Word usage.** Your word usage in the technical description and process analysis depends on your purpose. For factual, objective technical descriptions or process analyses, use photographic words. For subjective, sales-oriented descriptions or process analyses, use impressionistic words. Photographic words are denotative, quantifiable, and specific. Impressionistic words are vague and connotative. Table 12.1 shows the difference between photographic and impressionistic words.

- **Internal organization.** When describing your topic in the discussion portion of the technical description or process analysis, itemize the topic in some logical sequence. For a technical description, use spatial organization. When a topic is spatially organized, you literally lay out the components as they exist in space. You describe the components as they are seen either from left to right, from right to left, from top to bottom, from bottom to top, from inside to outside, or from outside to inside. In contrast, when writing your process analysis, you will use chronological organization to show how the product or service works in sequence.

Your **conclusion** depends on your purpose in describing the topic or analyzing how the process works. Some options are as follows:

Sales—"Implementation of this product will provide you and your company . . ."	**Guarantees**—"The XYZ carries a 15-year warranty against parts and labor."	**Comparison/contrast**—"Compared to our largest competitor, the XYZ has sold three times more . . ."
Uses—"After implementation, you will be able to use this XYZ to . . ."	**Testimony**—"Our satisfied customers include . . ."	**Reiteration of introductory comments**—"Thus, the XYZ is composed of the above interchangeable parts and works according to the process explained."

Highlighting Techniques

To aid understanding and allow for easy access, use highlighting techniques, including headings, itemization, and graphics. You can use line drawings, schematics, CAD/CAM drawings, photographs, architectural renderings, clip art, exploded views, or sectional cutaway views of your topic, each accompanied by callouts (labels identifying key components of the mechanism). An example of an exploded view with callouts can be seen in Figure 12.4.

Visual Aids

See Chapter 10 for more discussion of visual aids.

FIGURE 12.4 Exhalation Valve with Labeled Callouts

TECHNOLOGY TIP

How to Make Callouts Using Microsoft Word 2010

Callouts are a great way to label mechanism components in your technical description or process analyses. Follow these steps to create callouts.

1. Click on **Insert** in your Menu bar.

2. Click on **Shapes**. You will see a dropdown box.

3. Select the type of callout you want to use, such as the **Rounded Rectangular Callout** as shown.

4. When you choose the callout, a crosshair will appear. Use it to draw your Callout, as shown.

Note: When you draw the callout, Word 2010 opens up the **Format** tab, as shown.

From the **Format** tab, you can change the color of the text box (from **Text Box Styles**) change line type and density (from **Shape Outline**), create shadow effects, 3-D effects, and add content.

_____ 1. Does the technical description or process analysis have a title noting your topic's name (and any identifying numbers)?

_____ 2. Does the technical description or process analysis's introduction (a) state the topic, (b) mention its functions or the purpose, and (c) list the components?

_____ 3. Does the technical description or process analysis's discussion use headings to itemize the components for reader-friendly ease of access?

_____ 4. Do you need to define the mechanism and its main parts?

_____ 5. Does the discussion portion of the description or process analysis specify the following?

Colors	Capacities
Sizes	Textures
Materials	Identifying numbers
Shapes	Weight
Density	Make/model
Parts	Process

_____ 6. Are all of the calculations, measurements, or process steps correct?

_____ 7. Do you sum up your discussion using any of the optional conclusions discussed in this chapter?

_____ 8. Does your technical description or process analysis provide graphics that are correctly labeled, appropriately placed, and appropriately sized?

_____ 9. Do you write clearly and concisely, using a personalized tone?

_____10. Have you avoided biased language and grammatical and mechanical errors?

The Writing Process

Prewriting	Writing	Rewriting

- Decide whether you are writing to inform, instruct, or persuade.
- Determine whether your audience is high tech, low tech, or lay. This will help you decide which abbreviations and acronyms need to be defined.
- Gather your information through a prewriting technique, such as brainstorming/listing or flowcharting.

- Organize your description spatially and your process analysis chronologically.
- Include visual aids and callouts.
- Use headings and subheadings for easy navigation.

- Revise your draft by
  - Adding details
  - Deleting wordiness
  - Simplifying words
  - Enhancing the tone
  - Reformatting your text
  - Proofreading and correcting errors

Web Resources

www.pearsonhighered .com/gerson For more information about the writing process, visit our companion website.

The Writing Process at Work

The following student-written technical description shows how he used the writing process to create his document.

Prewriting

We asked students to describe a piece of equipment, tool, or mechanism of their choice. One student chose to describe a cash register pole display. To do so, he first had to gather data using brainstorming/listing. He provided the list shown in Figure 12.5.

FIGURE 12.5   Prewriting List

Brainstorming List of Parts in a Pole Display

1. Three parts
2. Pole printed circuit board (PCB)
3. Case assembly
4. Filter
5. Plastic materials

FIGURE 12.6   Student's Expanded Brainstorming/Listing

Pole PCB

- length—15 mm
- width—5.1 mm
- tube length—10.8 mm
- face plate width—2.3 mm
- thick—1.7 mm
- PCB thickness—0.2 mm
- 10 inch stranded wire with female connectors
- fiberglass and copper construction

Pole Case Assembly

- long—15.5 mm
- bottom width—2.5 mm
- top width—0.9 mm
- mounting pole—5 mm high × 3.2 mm diameter
- tounge for mounting—3.1 mm
- lower mounting tounge—1.5 mm
- side mounting tounge—0.8 mm high
- high—6.1 mm
- almond-colored plastic

Filter

- long—15.6 mm
- high—6.2 mm
- thick—0.7 mm
- plastic, blue

Next, we asked students to review their initial lists and edit them. Because this student's list was so sketchy, he did not need to omit any information. Instead, he had to add missing detail, as in Figure 12.6.

Writing

After the prewriting activity, we asked students to draft a technical description. Focusing on overall organization, highlighting, detail, and a hand-drawn graphic, this student wrote a rough draft (Figure 12.7).

FIGURE 12.7 Student's Rough Draft of a Pole Display

The QL169 Customer Pole Display provides the viewing of all transaction data for the customer. The display consist of a printed circuit board, a case assembly, and a filter display.

Display Circuit Board

- length—15 mm
- width—5.1 mm
- tube length—10.8 mm
- face plate width—2.3 mm
- thick—1.7 mm
- PCB thickness—0.2 mm
- 10 inch stranded wire with female connectors
- fiberglass and copper construction

Display Case Assembly

- long—15.5 mm
- bottom width—2.5 mm
- top width—0.9 mm
- mounting pole—
  5 mm high × 3.2 mm diameter
- tounge for mounting—3.1 mm
- lower mounting tounge—1.5 mm wide
- side mounting tounge—0.8 mm high
- high—6.1 mm
- almond-colored plastic

Display Filter

- long—15.6 mm
- high—6.2 mm
- thick—0.7 mm
- plastic, blue

Rewriting

Once students completed their rough drafts, we used peer review groups to help each student revise his or her paper. In Figure 12.8, students

- Added new detail for clarity
- Corrected any errors
- Perfected graphics

The student incorporated these suggestions and prepared the finished copy (Figure 12.9).

FIGURE 12.8   Rough Draft with Suggested Revisions

No Title   _define   — awkward —

The QL169 Customer Pole Display provides the viewing of ~~all transaction data~~ for the customer. The display consist of a printed circuit board, a case assembly, and a filter display. (SP)

Display Circuit Board
Make All Caps
- length—15 mm
- width—5.1 mm
- tube length—10.8 mm
- face plate width—2.3 mm
- thick—1.7 mm   thickness
- PCB thickness—0.2 mm
- 10 inch stranded wire with female connectors
- fiberglass and copper construction

Number your components

Too vague (specify)

Display Case Assembly
Make All Caps
- long—15.5 mm   length
- bottom width—2.5 mm
- top width—0.9 mm
- mounting pole—
  5 mm high × 3.2 mm diameter
- tounge for mounting—3.1 mm
- (SP) lower mounting tounge—1.5 mm wide
- side mounting tounge—0.8 mm high
- high—6.1 mm   height
- almond-colored plastic
  vague

Number your components

Display Filter
- long—15.6 mm   length
- high—6.2 mm   height
- thick—0.7 mm   thickness
- plastic, blue
  what

Add a conclusion

FIGURE 12.9   Student's Revised Technical Description

QL169 Pole Display

The QL169 pole is an electronic mechanism that provides an alphanumeric display for customer viewing of cash register sales. The display consists of a printed circuit board (PCB) assembly, a case assembly, and a display filter.

Item Description

1. PCB ASSEMBLY
2. CASE ASSEMBLY
3. DISPLAY FILTER

Figure 1. QL169

1.0 PCB Assembly

The printed circuit board, containing the display's electrical circuitry, is constructed of fiberglass with copper etchings.

The board consists of the following features:

1.1 Length—15 mm
1.2 Width—5.1 mm
1.3 Tube length—10.8 mm
1.4 Tube faceplate
width—2.3 mm
1.5 Tube total width—2.8 mm
1.6 Tube thickness—1.7 mm
1.7 PCB thickness—0.2 mm
1.8 20 conductor 10" 22-gauge
stranded wire with two AECC
female connectors
(AECC part #7214-001)
1.9 American Display Company blue phosphor display tube
(ADC part #1172177)

Figure 2. PCB

Include a definition in the introduction.

Exploded graphics help readers visualize the topic.

Specificity of detail adds clarity.

FIGURE 12.9 (Continued)

2.0 Case Assembly

Almond-colored ABS plastic, used to construct the case assembly, protects the PCB.

2.1 Length—15.5 mm
2.2 Bottom width—2.5 mm
2.3 Top width—0.9 mm
2.4 Mounting pole—5 mm high and 3.2 mm in diameter
2.5 Mounting tongue inside width—3.1 mm from side of assembly
2.6 Lower mounting tongue—1.5 mm wide
2.7 Side mounting tongue—0.8 mm high
2.8 Tongue thickness—0.2 mm
2.9 Height—6.1 mm

Figure 3. Case Assembly

3.0 Display Filter
3.1 Length—15.6 mm
3.2 Height—6.2 mm
3.3 Thickness—0.7 mm

Figure 4. Display Filter

Transparent blue plastic, used to construct the display filter, allows the customer to view the readings. The QL169 pole provides easy viewing of clerk transactions and ensures cashier accuracy.

The conclusion focuses on the end-user benefits.

CHAPTER HIGHLIGHTS

1. Technical descriptions are often used in user manuals, product specifications, construction designs, reports, and sales literature.
2. Process analyses tell how a tool, mechanism, piece of equipment, product, or service works.
3. The introduction of a technical description or process analysis specifies and defines the topic; explains the topic's functions, capabilities, or processes; and might include a list of components, parts, or equipment.
4. Graphics with callouts make a technical description and process analysis easier to understand for the reader.
5. Follow a spatial pattern of organization in a technical description.
6. Follow a chronological order of organization when describing how a process works.
7. Word usage for descriptions and processes will be photographic and/or impressionistic.
8. Use headings and subheadings to organize the content of your description or process analysis.
9. Use line drawings, schematics, CAD/CAM drawings, photographs, architectural renderings, clip art, exploded views, or sectional cutaway views of your topic for the technical description or process analysis.
10. Preface your text with a title precisely stating the topic. This could be the name of the mechanism, tool, piece of equipment, landform, product, or service you are describing or analyzing.

APPLY YOUR KNOWLEDGE

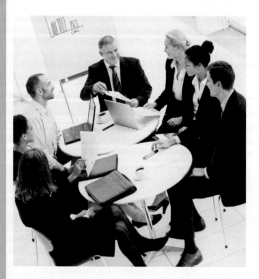

CASE STUDY

Nitrous Systems Biotechnology, Inc. plans to expand its company. Part of this expansion includes the need to hire 50 new biotechnicians. Each will need his or her own desktop computer, complete with monitor. But what kind of monitor should the company buy—CRTs or LCDs? To meet their computer technology needs, Nitrous has hired Alpha/Beta Consulting (A/B/C). Nitrous wants a brief, comparison/contrast, complete with a product description and a process analysis explaining the differences between these two monitor types.

In addition, which monitor type provides the best resolution and viewing quality, refresh rates, and screen-viewable size options? Are CRTs cheaper, or has that price break changed? Even energy savings are a consideration when purchasing either CRTs or LCDs. Finally, how do CRTs work versus LCDs?

As a computer technician who work for A/B/C, you need to research this topic and write the technical description, complete with process analysis, to meet the client's request.

INDIVIDUAL OR TEAM PROJECTS

1. Write a technical description, either individually or as a team. To do so, first select a topic. You can describe any tool, mechanism, or piece of equipment. However, don't choose a topic too large to describe accurately. To provide a thorough and precise description, you will need to be exact and minutely detailed. A large topic, such as a computer, an oscilloscope, a respirator, or a Boeing airliner, would be too demanding for a two- to four-page description. On the other hand, do not choose a topic that is too small, such as a paper clip, a nail, or a shoestring. Choose a topic that provides you with a challenge but that is manageable. You might write about any of the following topics:

USB flash drive	Computer disk	DVD remote control
Wrench	Computer mouse	Mechanical pencil
Screwdriver	Light bulb	Ballpoint pen
Pliers	Calculator	Computer monitor
Wall outlet	Automobile tire	Cell phone

2. Write a process analysis, either individually or as a team, on one of the following topics:

   - How does blood coagulate?
   - How does an x-ray machine work?
   - How are viruses spread?
   - How do Wi-Fi connections work?
   - How do rotary engines work?
   - How do fuel gauges work?
   - How is metal welded?
   - How does a metal detector work?
   - How do browsers work?
   - How do computer viruses work?

3. Find examples of professionally written technical descriptions or process analyses in technical books and textbooks, professional magazines and journals, or on the Internet. Bring these examples to class and discuss whether they are successful according to the criteria presented in this chapter. If the descriptions or process analyses are good, specify how and why. If they are flawed, state where and suggest ways to improve them. Rewrite the flawed descriptions or process analyses.

4. Select a simple topic for description, such as a pencil, coffee cup, toothbrush, or textbook (you can use brainstorming/listing to come up with additional topics). Describe this item without mentioning what it is or providing any graphics. Then, read your description to a group of students/peers and ask them to draw what you have described. If your verbal description is good, their drawings will resemble your topic. If their drawings are off base, you haven't succeeded in providing an effective description.

5. In degree-specific teams, choose a topic from your area of interest and expertise to describe (including a process analysis). For example, students majoring in HVAC could describe a humidifier and explain how it works. Biotech and nursing students could describe and provide a process analysis for a nebulizer, blood pressure monitor, or glucometer. EMT students could write about defibrillators. Automotive technicians could describe and include a process analysis for jumpstarters, battery chargers, or hydraulic jacks. Welding students could write about MIG, TIG, or stick welding.

PROBLEM-SOLVING THINK PIECE

Read the following process analysis and reorganize the numbered sentences to achieve a clear, chronological order. If necessary, research this process to learn more about how blood clots.

How does blood clot?

1. In our bodies, blood can clot due to platelets and the thrombin system.
2. When bleeding occurs, chemical reactions make the surface of the platelet "sticky." These sticky platelets adhere to the wall of blood vessels where bleeding has occurred.
3. Platelets, tiny cells created in our bone marrow, travel in the bloodstream and wait for a bleeding problem.
4. Soon, a "white clot" is formed, so called because the clotted platelets look white.
5. Blood clots consist of both platelets and fibrin. The fibrin strands bind the platelets and make the clot stable.
6. In the thrombin system, several blood proteins become active when bleeding occurs. Clotting reactions produce fibrin, long, sticky strings. These sticky strands catch red blood cells and form a "red clot."
7. Primarily, arteries clot due to platelets, while veins clot due to our thrombin system.

WEB WORKSHOP

1. You are ready to purchase a product. This could include printers, monitors, digital cameras, scanners, PCs, laptops, speakers, cables, adapters, automotive engine hoists, generators, battery chargers, jacks, power tools, truck boxes, screws, bolts, nuts, rivets, hand tools, and more. A great place to shop is online. By going to an online search engine, you can find not only prices for your products but also technical descriptions or technical specifications. These will help you determine if the product has the size, shape, materials, and capability you are looking for.

   Go online to search for a product of your choice and review the technical description or specifications provided. Using the criteria in this chapter and your knowledge of effective technical communication techniques, analyze your findings.

   - How do the online technical descriptions compare to those discussed in this textbook?
   - Are graphics used to help you visualize the product?
   - Are callouts used to help you identify parts of the product?
   - Are high-tech terms defined?
   - Is the use of the product explained?
     a. Report your findings, either in an oral presentation or in writing (e-mail message, memo, letter, or report).
     b. Rewrite any of the technical descriptions that need improvement according to the criteria provided in this textbook.

2. HowStuffWorks (http://www.howstuffworks.com/index.htm) is an extensive online library of process analyses. Revise a process analysis in this Web site by improving its layout, content, organization, and/or graphics.

QUIZ QUESTIONS

1. What is a technical description?
2. Why would you use a technical description?
3. What are the components of a technical description?
4. What is a process analysis?
5. Give three examples of topics for a process analysis.
6. What is a callout?
7. What is the best organizational pattern for a process analysis?
8. What is the best way to organize a technical description?
9. What information is included in a technical description?
10. What is the difference between photographic and impressionistic words?

CHAPTER THIRTEEN

Instructions, User Manuals, and Standard Operating Procedures

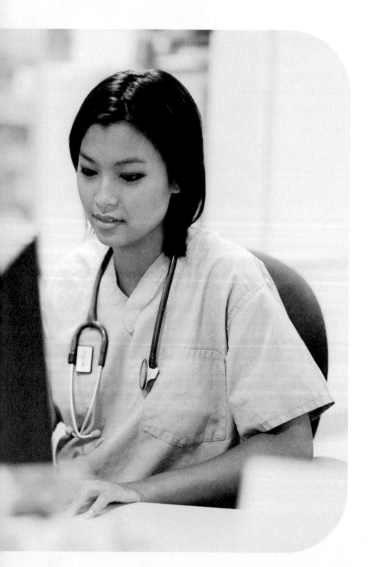

COMMUNICATION AT WORK

This scenario shows the importance of effective instructions in the biomedical industry and various communication channels used to convey step-by-step procedures.

As the baby-boom generation ages, medical needs are expanding—sometimes faster than medical care facilities and medical professionals can manage. One area in which this has been felt most acutely is in medical laboratories. Thousands of medical technicians need to be trained to accommodate increased demand.

PhlebotomyDR is a medical consulting firm seeking to solve this problem. Its primary area of concern is training newly hired technicians responsible for performing blood collection. PhlebotomyDR facilitates training workshops to teach venipuncture standards and venipuncture procedures.

PhlebotomyDR focuses on the following venipuncture instructions:

- Proper patient identification procedures
- Proper equipment selection, sterilization, use, and cleaning
- Proper labeling procedures

Objectives

When you complete this chapter, you will be able to

1. Understand when to write an instruction
2. Follow criteria in order to write an effective instruction
3. Recognize your audience when writing an instruction
4. Include key components for an effective instruction
5. Use videos for instructions
6. Write standard operating procedures (SOPs)
7. Follow the writing process to create instructions
8. Evaluate instructions using the effective instruction checklist and the usability checklist

- Order of phlebotomy draw
- Patient care before, during, and following venipuncture
- Safety and infection control procedures
- Procedures to follow when meeting quality assurance regulations

Each of the instructions mentioned requires numerous steps, complete with visual aids.

PhlebotomyDR offers its audience various communication channels. Hospitals, labs, and treatment centers can access PhlebotomyDR's instructions as follows:

- Hard-copy instructional manuals
- Online instructions, accessible at http://www.phlebotomydr.com
- Videos showing step-by-step performances of blood collection, complete with case studies enacted by technicians, patients, and supervisors
- Standard operating procedures used to achieve uniformity and efficiency

- Computer-aided instruction (CAI) for individual tutorials
- One-on-one tutorials with trained phlebotomists
- Instructional workshops designed for groups of seminar participants

PhlebotomyDR's outstanding staff realizes that trained technicians make an enormous difference. Training, achieved through instructional manuals, electronic aids, and individual facilitation, ensures the health and safety of patients. Of equal importance, excellent training also benefits many stakeholders. Untrained technicians make errors that cost us all. Medical errors create insurance problems, the need to redo procedures, increased medical bills, the potential involvement of regulators and legislators, and dangerous repercussions for patients.

In contrast, effective communication, achieved through successful instruction, saves lives, time, and money.

Check out our quarterly newsletters at www.pearsonhighered.com/gerson for dot.com updates, new case studies, insights from business professionals, grammar exercises, and facts about technical communication.

Why Write an Instruction?

Almost every manufactured product comes complete with instructions. You will receive instructions for baking brownies; making pancakes; assembling children's toys; or changing a tire, the oil in your car, or the coolant in your engine. Instructions help people set up bookshelves, use electronic equipment, maintain computers, and operate fighter planes.

We frequently see short instructions about computer-related problems. If you are at home or in the office, for example, and you can't figure out how to operate one of your computer's applications, you can access online help or dial a 1-800 number to speak to a computer technician. Figure 13.1 is an e-mail message giving a customer the steps to follow for computer technical support.

Include instructions or user manuals whenever your audience needs to know how to

Operate a mechanism	Collect lab specimens
Install equipment	Use an autoclave
Manufacture a product	Service equipment
Package a product	Troubleshoot a system
Perform lab experiments	Use software
Test components	Set up a product
Maintain equipment	Implement a procedure
Clean a product	Assemble a product
Draw blood	Repair a system

FIGURE 13.1   Instructions for Computer Technical Support

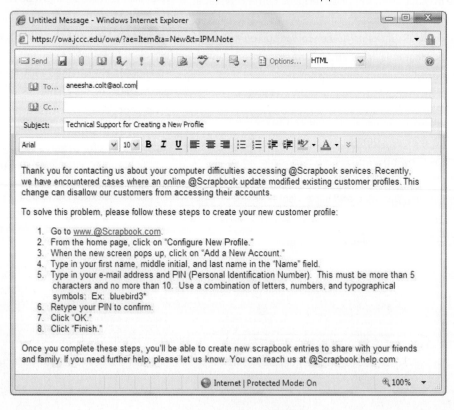

Criteria for Writing Instructions

Odds are good that you've read a badly written instruction. You've probably encountered an instruction that failed to define unfamiliar terminology, didn't tell you what equipment you needed to perform the task, didn't warn you about dangers, didn't clarify how to perform certain steps, and didn't provide you enough graphics or a video to complete the job successfully. Usually, such poorly written instructions occur because writers fail to consider their audience's needs. Successful instructions, in contrast, start with audience recognition.

Audience Recognition

The instruction tells you to "place the belt on the motor pulley," but you don't know what a "motor pulley" is, nor do you know how to "place the belt" correctly. Or, you are told to "discard the used liquid in a safe container," but you don't know what is safe for this specific type of liquid. You are told to "size the cutting according to regular use," but you have never regularly performed this activity before.

Here is a typical instruction written without considering the audience.

Audience

See Chapter 4 for more discussion of audience recognition.

> To overhaul the manual starter, proceed as follows: Remove the engine's top cover. Untie the starter rope at the anchor and allow the starter rope to slowly wind onto the pulley. Tie a knot on the end of the starter rope to prevent it from being pulled into the housing. Remove the pivot bolt and lift the manual starter assembly from the power head.

◀ EXAMPLE

Although many high-tech readers might be able to follow these instructions, many more readers will be confused. How do you remove the engine's top cover? Where is the anchor? Where is the pivot bolt, and how do you remove it? What is the power head?

The problem is caused by writers who assume that their readers have high-tech knowledge. This is a mistake for several reasons. First, even high-tech readers often need detailed information because technology changes daily. You cannot assume that every high-tech reader is up to date on these technical changes. Thus, you must clarify. Second, low-tech and lay readers—and that's most of us—carefully read each and every step for clear and thorough assistance.

As the writer, you should provide your readers with the clarity and thoroughness they require. To do this, recognize accurately who your readers are and give them what they want, whether that amounts to technical updates for high-tech readers or precise, even simple, information for low-tech or lay readers. The key to success as a writer of instructions is the following: Don't assume anything. Spell it all out—clearly and thoroughly.

Ethical Instructions

Your job as technical writer is to meet your audience's needs. This not only entails clarity in an instruction, but also it demands ethical behavior. In Chapter 3 we discuss legalities and practicalities. User manuals and instructions are perfect examples of the importance of ethical communication.

Legalities in User Manuals.   As a corporate employee, you never want your writing to lead to customer injuries or to cause a customer to damage equipment. Instead, to avoid issues of liability, your communication must identify the potential for harm. Clearly stated hazard notations, warranties, and disclaimers, discussed later in this chapter, allow you to warn your audience of potential dangers and to set limits and exceptions for product guarantees. By doing so, you adhere to your legal responsibilities to the company and to the client.

Practicalities in User Manuals. Ethical writing doesn't just keep customers safe. It also satisfies your customer's need for up-to-date information, and that's just good business. For example, let's say that a decade ago, your company wrote a technical specification for a product. The product has changed over the years, improving through numerous iterations. However, your company's customer support (training manuals, online help, user manuals, and troubleshooting guides) have not been revised to reflect these changes. Management says that the company doesn't have the time or budget to update the customer support. This can lead to many problems in the manuals, including inconsistencies, inaccuracies, irregularities, and customer complaints. If a client stops buying your products because the supporting documents are flawed, the company loses money. The practical solution, then, is to revise the manuals, bringing them up to date. Doing so is a company's ethical responsibility to its customers.

Components of Instructions

Not every instruction will contain the same components. Some very short instructions will consist of nothing more than a few, numbered steps. Other instructions, however, will consist of the components shown in Figure 13.2.

Title Page

Preface your instruction with a title page that consists of the *topic* about which you are writing, the *purpose* of the instruction, and a *graphic* depicting your product or service. For example, to merely title your instruction "iPod" would be uninformative. This title names the product, but it does not explain why the instruction is being written. Will the text discuss operating instructions, troubleshooting, service, or maintenance? A better title would be "Operating Instructions for the iPod." Adding a graphical representation lets your audience see what the finished product will look like and helps you market your product or service.

Safety Requirements

You can place safety requirements anywhere throughout your text. If a particular step presents a danger to the reader, call attention to this hazard just before asking the reader to perform the step. Similarly, to help your audience complete an action, place a note before the step, suggesting the importance of using the correct tool, not overtightening a bolt, or wearing protective equipment.

FIGURE 13.2 Key Components of Instructional Manuals

Title Page Topic Graphic Purpose	Hazards ⚠	Table of Contents	Introduction	List of Required Tools / Equipment
Glossary of Terms	Steps 1. 2.	Steps 3. 4. Etc.	Additional Components	Corporate Contact Information

In addition to placing safety requirements before a step, consider prefacing your entire instructions with hazard notations. By doing so, you make the audience aware of possible dangers, warnings, cautions, or notes in advance of performing the instructions. This is important to avoid potentially harming an individual, damaging equipment, and avoiding costly lawsuits. Correctly using hazard notations will avoid expensive liability issues and adhere to ethical communication standards. Include the following in your safety requirements:

- **Access.** Make the hazard notations obvious. To do so, vary your typeface and type size, use white space to separate the warning or caution from surrounding text, box the warning or caution, and call attention to the hazards through graphics.
- **Definitions.** What does *caution* mean? How does it differ from *warning, danger,* or *note?*

Four primary organizations that seek to provide a standardized definition of terms are American National Standards Institute (ANSI), the U.S. military (MILSPEC), the Occupational Safety and Health Administration (OSHA), and the MSDS Search National Repository (material safety data sheets).

To avoid confusion, we suggest the following hierarchy of definitions, which clarifies the degree of hazard:

1. **Note.** Important information, necessary to perform a task effectively or to avoid loss of data or inconvenience.
2. **Caution.** The potential for damage or destruction of equipment.
3. **Warning.** The potential for serious personal injury.
4. **Danger.** The potential for death.

Our goal is to avoid confusion, which can occur if one writes, "Cautionary Warning Notice!"

- **Colors.** Another way to emphasize your hazard message is through a colored window or text box around the word. Usually, *Note* is printed in blue or black, *Caution* in yellow, *Warning* in orange, and *Danger* in red.
- **Text.** To further clarify your terminology, provide the readers text to accompany your hazard alert. Your text should have the following three parts:
  1. **A one- or two-word identification alerting the reader.** Words such as "High Voltage," "Hot Equipment," "Sharp Objects," or "Magnetic Parts," for example, will warn your reader of potential dangers, warnings, or cautions.
  2. **The consequences of the hazards, in three to five words.** Phrases like "Electrocution can kill," "Can cause burns," "Cuts can occur," or "Can lead to data loss," for example, will tell your readers the results stemming from the dangers, warnings, or cautions.
  3. **Avoidance steps.** In three to five words, tell the readers how to avoid the consequences noted: "Wear rubber shoes," "Don't touch until cool," "Wear protective gloves," or "Keep disks away."
- **Icons.** Equipment is manufactured and sold globally; people speak different languages. Your hazard alert should contain an icon—a picture of the potential consequence—to help everyone understand the caution, warning, or danger.

Figure 13.3 shows an effective page layout and the necessary information to communicate hazard alerts.

DOT.COM UPDATES

For samples of hazard icons, check out the following link:

- http://www.speedysigns .com/signs/osha _readymade.asp

FIGURE 13.3 Hazard Alert

1	Hazard alert word
2	Hazard alert color
3	Alert symbol
4	Identification—what is the specific hazard?
5	Consequence—what are the potential results?
6	Avoidance steps—what should you do?
7	Hazard icon

Table of Contents

Table of Contents

See Chapter 17 for more discussion of a table of contents.

Your instruction might have several sections. In addition to the actual steps, the instructional manual could include technical specifications, warranties, guarantees, FAQs, troubleshooting tips, and customer service contact numbers. An effective table of contents will allow your readers to go to any of these sections individually on an as-needed basis.

Introduction

Companies need customers. A user manual might be the only contact a company has with its customer. Therefore, instructions often are reader-friendly and seek to achieve audience recognition and audience involvement. The manuals try to reach customers in a personalized way. Look at the following introduction from a user manual.

Thank you for your purchase. Installation and operating procedures are contained in this manual about your new High-Def TV. The minutes you spend reading these instructions will contribute to hours of viewing pleasure.

This introduction uses pronouns (*you* and *your*) to personalize the manual. The introduction also uses positive words, such as *Thank you* and *pleasure*, to achieve positive customer contact. An effective introduction promotes good customer-company relationships.

Glossary

If your instruction uses the abbreviations *BDC*, *CCW*, or *CPR*, will the readers know that you are referring to "bottom dead center," "counterclockwise," or "continuing property records"? If your audience is not familiar with your terminology, they might miss important information and perform an operation incorrectly.

To avoid this problem, define your abbreviations, acronyms, or technical terms. You can define your terms early in the instruction, throughout the manual, or in a glossary located at the end of your manual. The following example defines terms alphabetically in a glossary.

GLOSSARY

BDC	Bottom dead center
CCW	Counterclockwise
Danger	This hazard alert designates the possibility of death.
	Be extremely careful when performing an operation.
RMS	Root mean square

Required Tools or Equipment

What tools or equipment will the audience need to perform the procedures? You don't want your audience to be in the middle of performing a step and suddenly realize that a missing piece of equipment is needed. Provide this important information either through a list or graphics depicting the tools or equipment necessary to complete the tasks.

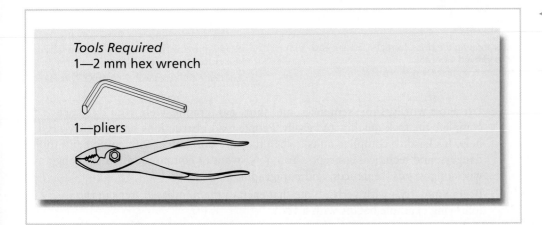

Tools Required
1—2 mm hex wrench

1—pliers

Instructional Steps

The instructional steps are the most important part of your manual—the actual actions required of the audience to complete a task. To successfully write your instruction, follow these steps:

1. **Organize the steps chronologically,** as a step-by-step sequence. You cannot tell your readers to do step 6, then go back to step 2, then accomplish step 12, then do step 4. Such a distorted sequence would fail to accomplish the task. To operate machinery, monitor a system, troubleshoot equipment, or perform a standard operating procedure, your readers must follow a chronological sequence. Be sure that your instruction is chronologically accurate.

2. **Number your steps.** Do not use bullets or the alphabet. Numbers, which you can never run out of, help your readers refer to the correct step. In contrast, if you used bullets, your readers would have to count to locate steps—seven bullets for step 7, and so on. If you used the alphabet, you'd be in trouble when you reached step 27.

3. **Use highlighting techniques.** Boldface, different font sizes and styles, text boxes, emphatic warning words, color, and italics call attention to special concerns. A danger, caution, warning, or specially required technique must be evident to your reader. If this special concern is buried in a block of unappealing text, it will not be read. This could be dangerous to your reader or costly to you and your company. To avoid lawsuits or to help your readers see what is important, call it out through formatting.

4. **Limit the information within each step.** Don't overload your reader by writing lengthy steps.

BEFORE	AFTER
Overloaded Steps	**Separated Steps**
Start the engine and run it to idling speed while opening the radiator cap and inserting the measuring gauge until the red ball within the glass tube floats either to the acceptable green range or to the dangerous red line.	1. Start the engine. 2. Run the engine to idling speed. 3. Open the radiator cap. 4. Insert the measuring gauge. 5. Determine whether the red ball within the glass tube floats to the acceptable green range or up to the dangerous red line.

5. **Develop your points thoroughly.** Avoid vague content by clarifying directions, needed equipment, and cautions.

This example clarifies what is meant by correct rotating and approved lubricant. The steps are also separated for enhanced readability.

BEFORE	AFTER
After rotating the discs correctly, grease each with an approved lubricant.	1. Rotate the disks clockwise so that the tabs on the outside edges align. 2. Lubricate the discs with 2 oz of XYZ grease.

6. **Use short words, short sentences, and short paragraphs.** Help your audience complete the task quickly and easily. People read instructions because they need help. It's hard to complete a task when they are unfamiliar with equipment, tools, dangers, and technical concepts. You don't want to compound their challenge with long words, sentences, and paragraphs.

7. **Begin your steps with verbs.** Note that each of the numbered steps in the following example begins with a verb.

Verbs Begin Steps

1. *Organize* the steps chronologically.
2. *Number* your steps.
3. *Use* highlighting techniques.
4. *Limit* the information within each step.
5. *Develop* your points thoroughly.
6. *Use* short words and phrases.
7. *Begin* your steps with verbs.

8. **Personalize your text.** Remember, people write to people. Involve your readers in the instruction by using pronouns.
9. **Do not omit articles.** Articles, such as *a, an,* and *the,* are part of our language. Although you might see instructions that omit these articles, please don't do so yourself. Articles do not take up much room in your text, but they make your sentences read more fluidly.

Audience Involvement

See Chapter 4 for more discussion of audience involvement through personalization.

BEFORE	AFTER
1. Press right arrow button to scroll through list of programs.	1. Press <u>the</u> right arrow button to scroll through <u>a</u> list of programs.
2. Select program to scan.	2. Select <u>the</u> program you want to scan.
3. Place item to scan face down on scanner glass in upper left corner.	3. Place <u>an</u> item to scan face down on <u>the</u> scanner glass in <u>the</u> upper left corner.

Additional Components

Your instruction might include the following additional components.

Technical Descriptions. In addition to the step-by-step instructions, many manuals contain technical descriptions of the product or system. A description could be a part-by-part explanation or labeling of a product or system's components. Such a description helps readers recognize parts when they are referred to in the instruction. For example, if the user manual tells the reader to lay shingles with the tabs pointing up, but the reader doesn't know what a "tab" is, the step cannot be performed. In contrast, if a description with appropriate callouts is provided, then the reader's job has been simplified.

Perhaps the user manual or standard operating procedure (SOP) will contain a list of the product's specifications, such as its size, shape, capacity, capability, and materials of construction. Specifications allow the user to decide whether the product meets his or her needs. Does it have the desired resolution? Is it the preferred weight and size? The specification for a PDA answers these questions.

Technical Descriptions

See Chapter 12 for more discussion of technical descriptions.

PDA Specifications

Height:	114 mm (4.48 inches)
Width:	66 mm (2.6 inches)
Depth:	15 mm (0.59 inches)
Weight:	136 grams (4.8 ounces)
Monitor Resolution:	480 × 320 pixel color display

Finally, a user manual might include a schematic, depicting the product or system's electrical layout. This would help the readers troubleshoot the mechanism.

Warranties.   Warranties protect the customer and the manufacturer. Many warranties tell the customer, "This warranty gives you specific legal rights, and you may also have other rights that vary from state to state." A warranty protects the customer if a product malfunctions sooner than the manufacturer suggests it might: "This warrants your product against defects due to faulty material or installation." In such a case, the customer usually has a right to free repairs or a replacement of the product.

However, the warranty also protects the manufacturer. No product lasts forever, under all conditions. If the product malfunctions after a period of time designated by the manufacturer, then the customer is responsible for the cost of repairs or a replacement product. The designated period of time differs from product to product. Furthermore, many warranties tell the customers, "This warranty does not include damage to the product resulting from normal wear, accident, or misuse."

Disclaimers.   Disclaimers are another common part of warranties. You can ensure that your company is legally protected by clearly stating what your company is *not* responsible for. This could include risks inherent to using the product. You also might want to clarify timeframes for legitimate repair requests, who can and should work on a product, equipment limitations, copyright laws applicable to the product or service, and explanations of the intended use of the product. Sometimes, before a customer can use a product, the company will require that the end user click on the "I Agree" dialogue box, therefore demanding that customer accept the disclaimer. This acts as a contract between the end user and the company.

Typical disclaimers read as follows:

- Note: Any changes or modifications to this system not expressly approved in this manual could void your warranty.
- Proof of purchase with a receipt clearly noting that this unit is under warranty must be presented.
- The warranty is only valid if the serial number appears on the product.
- The manufacturer of this product will not be liable for damages caused by failure to comply with the installation instructions enclosed within this manual.

EXAMPLE ▶

> ### Disclaimer
>
> All downloadable ring tones and wallpapers are used at the user's own risk. No warranty or guarantee is included with purchase of this product. We are unaware of any copyright infringement regarding ring tones or wallpapers. Downloads are assumed to be public domain. However, if you think there is a copyright infringement, please contact us at techsupport@pdadisplay.com so we can correct this problem immediately. Ring tones and wallpapers are not for commercial use.

Accessories.   A company always tries to increase its revenue. One way to do so is by selling the customer additional equipment. A user manual promotes such equipment in an accessories list. This equipment isn't mandatory, required for the product's operation. Instead, an accessories list offers customers equipment such as extra-long cables or cords, carrying cases, long-life rechargeable batteries, belt clips and cases, wireless headsets, automotive chargers, CD-ROM capabilities, and memory cards. Often the specifications are also provided for these accessories.

Frequently Asked Questions.   Why take up your customer support employees' valuable time by having them answer the same questions over and over? By including a frequently asked questions (FAQs) page in the user manual, common consumer concerns can be addressed immediately. This will save your company time and money while improving customer relations.

Corporate Contact Information. Conclude your manual by providing your company's street, city, and state address; 1-800 number; URL; e-mail address; fax number; social media links; or other ways your audience could contact the company with questions or requests for more information.

Graphics

Clarify your points graphically. Use drawings, photographs, and screen captures that are big, simple, clear, keyed to the text, and labeled accurately. Not only do these graphics make your instructions more visually appealing, but also they help your readers and you. What the reader has difficulty understanding, or you have difficulty writing clearly, your graphic can help explain pictorially.

A company's use of graphics often depends on the audience. With a high-tech reader, the graphic might not be needed. However, with a low-tech reader, the graphic is used to help clarify. Look at the examples of instructions for the same procedure. Figure 13.4 uses text with graphics to help the low-tech audience better understand the required steps. Table 13.1 uses text without graphics for the high-tech audience.

FIGURE 13.4 Instructions with Large, Bold Graphics for Low-Tech Audience

Presetting Ventilation Controls and Alarms

Begin by setting ventilator controls and alarms as follows, prior to connecting patient to ventilator.

(1) Set Mode Selector Switch at the "Power Off" position.

TABLE 13.1 Text without Graphics for High-Tech Readers

Location Item	Action	Remarks
	Note For additional operating instructions, refer to Companion 2800 operator's manual.	
1. MODE switch	Set to POWER OFF.	
2. Rear panel	Ensure that instructions on all labels are observed.	
3. Patient tubing circuit	Assemble and connect to the unit.	See Table 2, step 3.
4. Power cord	Connect to a 120/220 V AC grounded outlet.	If the power source is an external battery, connect the external battery cable to the unit per Table 2, step 4.
5. Cascade I humidifier	Connect to the unit.	See Table 2, step 2.

Source: Courtesy of Puritan Bennett Corp.

Using Screen Captures to Visually Depict Steps

Here's a good, simple way to visually depict steps in your instruction—screen captures. A screen capture lets you copy any image present on your computer monitor. To make screen captures, follow these steps:

Capturing the Image

1. Find the graphic you want to include in your instructions.
2. Press the **Print Screen** key on your computer keyboard.

Cropping the Image

Pressing the print screen key captures the entire image seen on the monitor. This might include images that you do not want to include in your instruction. To crop your image, follow these steps:

1. Paste the **Print Screen** image into a graphics program like MS Paint.
2. Click on the **Select** icon.

3. Click and drag the **Select** tool to choose the part of the captured graphic you want to crop.

4. Copy and paste this cropped image into your user manual.

"Selected" portion of the graphic chosen for cropping

Note: When capturing screenshots, be careful to avoid infringing upon a company's copyright to the image. Make sure that your use of the image meets the principle of "fair use" (which permits the use of images for criticism, comment, news reporting, teaching, scholarship, or research—Copyright Act of 1976, 17 U.S.C. § 107).

Instructional Videos

Another way to depict your instructional steps graphically is through videos. Screencasting, a mixture of animated screen visuals, voiceovers, and captioned text, will benefit your audience in the following ways:

- Animated videos show the end user how to perform a step more successfully than static text might.
- Videos are an excellent communication channel for visual learners.
- Through podcasting, instructional videos can be downloaded anywhere, anytime by mobile learners.
- A company can enhance its marketing strategy by advertising its instructional videos as an alternative communication channel to hard-copy text. YouTube is an excellent social networking tool for sharing instructional videos and for marketing a company's product.
- Instructional videos allow for end user interaction. The audience can fast forward, slow down, go backwards, or freeze screens to learn how to perform tasks at their own pace.
- A video can be more informative than static text and photography. A video screencast shows how an end user interacts with machinery, equipment, tools, and more.

> **Social Media**
>
> See Chapter 7 for more discussion of social media channels.

SPOTLIGHT

How GBA Master Series Uses Screencasting for Instructions

Stacy Gerson is a technical writer for GBA Master Series (gbaMS), a software company that creates computerized maintenance management systems (CMMS) for public works and utilities organizations. Clients include street, water, and sewer departments for cities, counties, and states. gbaMS's instructions help clients track inventory and assets. As a documentation specialist, Stacy writes online user manuals and training guides in hard-copy and online video formats for end users such as office workers, dispatchers, maintenance supervisors, field workers, engineers, and employees of citizen complaint call centers.

These manuals and videos help her clients use gbaMS's software related to public works preventative maintenance scheduling, work orders, asset inventory, asset inspections, employee timesheets, and administrative modules for city, county, and state planning and budgeting.

To write her user manuals and online videos, Stacy

- Interacts with her programmers, asking questions to clarify challenging content.
- Watches WebEx movies created by the software developers. These movies walk through programs and provide voiceover explanations of the content.

Stacy uses the movie for the basis of her text for both the hard-copy manuals and video scripts.

- Studies the complex coding, provided by the subject matter experts to find what will be most useful for her lay readers.
- "Learns how to use the software by using it."

Stacy focuses on the big picture when she creates her online manuals and videos. Her goals are to create easy-to-read text, easy-to-follow steps, and text that "is not just well written but content that will help the end user." As she adapts her scripts to online video instruction, she adds screen shots and provides a voiceover narration to explain which fields to fill in on the screens and what step to perform next.

Stacy has found that she needs to write well, have grammatical abilities, and empathize with the end-users' needs. She also has to consider whether or not her online videos use enough narration to explain the screen shots.

Stacy's number one task is to make things easily understandable for gbaMS's customers. The software can be complex at times, so if an instruction is unclear, "The customer can't do the job." To make the instructions understandable, "I try to write text for the manuals and the videos that is logical, clear, simple, and visually appealing. If I look at something and say, 'What's that?' then I have to assume that it won't be clear to the client either."

FIGURE 13.5   Video Instruction with Voiceover Narration

Voiceover Narration:
The first step in assessing street pavement field conditions is to determine the level of riding comfort. Click on **Riding Comfort** to select this option. The **Set Severity/Density** box drops down. Now, fill in the **Severity** and **Density** fields and click **OK**. This takes you to the next step.

At the bottom of the video instruction, GBA Master Series provides a WebEx Player control box. With these buttons and slides, the end user can control the sound, fast forward, rewind, or pause the video.
The video counter allows the end user to skip ahead or return to the other portions of the video.

Source: Courtesy of GBA Master Series

- Video instructions help end users see the consequences of hazard alerts. For example, the user can see a short animation showing the results of pushing the wrong button, using the incorrect equipment, or putting the wrong chemical in a test tube.
- You can make instructional videos using a variety of tools that are supported by Windows, Macs, or Linux. These include WebEx, Camtasia, Captivate, CamStudio, iShowU, SnapzPro, Wink, and Encorder.

(Tietjen; Sharp)
Figure 13.5 shows a screen capture of one step in a video instruction.

Standard Operating Procedure (SOP)

Why Write an SOP?

An SOP is a set of written instructions that documents routine or repetitive technical or administrative activities followed by business and industry. SOPs ensure accurate job performance and consistent quality and integrity of the end product. SOPs also ensure that governmental regulations are followed.

SOPs must be reviewed and enforced by management. Employees need to reference SOPs for accuracy of procedures, so SOPs must be available as hard-copy documents and in electronic formats. Annual reviews of SOPs by both management and employees are necessary to ensure that the SOPs are being followed.

SOPs are used in many fields, including science, healthcare, biomedical technology, government, military, the computer industry, and others. These industries use SOPs for the following reasons:

- Calibrate and standardize instruments
- Collect lab samples
- Handle and preserve food
- Analyze test data
- Troubleshoot equipment, machinery, and procedures
- List mathematical steps to follow for acquiring data and making calculations
- Assess hardware and software analytical data

Components of SOPs

SOPs contain many of the same components as user manuals already discussed in this chapter, such as procedural steps. However, several of the parts of an SOP are different. Figure 13.6 shows the key components of an SOP.

Title Page.   Preface your SOP with a title page that consists of the topic about which you are writing, the date of the SOP, the purpose of the SOP, and the routing list of people who have to sign off on this procedure.

◀ **EXAMPLE**

Standard Operating Procedure

January 14, 2012

Preparing and Processing Algae Samples to Ensure Proper pH

Author: _____

Date Approved: _____

Manager: _____

Date Approved: _____

Quality Assurance Manager: _____

Date Approved: _____

FIGURE 13.6   Key Components of SOPs

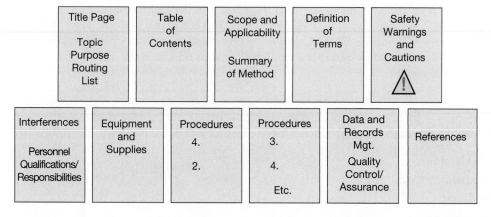

Scope and Applicability.   This section of the SOP can be limited to approximately three to five sentences. However, some SOPs, due to the complexity of the subject matter, can have Scope and Applicability sections that are several paragraphs long. This section of the SOP provides an overview for the audience. To do so, explain why the SOP has been written, show how it meets regulatory requirements, specify any limits for the use of the procedure, and state the applicability of the procedure.

EXAMPLE ▶

Scope

The purpose of this SOP is to establish uniform procedures for water compliance inspections (WCIs) performed by the Georgia Science and Assessment Division (GSAD). The WCIs evaluate the effectiveness and reliability of the state agency's inspection procedures to meet regulations of the Clean Water Act (CWA). The SOP will ensure thoroughness of all water compliance inspections and reports. The inspector may alter SOPs due to unexpected or unique problems in the field. Deviations from the SOP must be reported.

Applicability

The policies and procedures of the SOP apply to all personnel who take part in the WCIs.

Summary of Method.   In this brief summary of the procedure (three to five sentences or short paragraphs), you might focus on any of the following, determined by the topic of the SOP:

- Title of people involved
- Their roles and duties
- Sequence of their involvement
- Sequence of activities performed

EXAMPLE ▶

Summary of the Method

During the WCI, the state inspector is the lead person supervising all tests. The state inspector manages a staff of scientists who gathers water samples either manually or automatically with a portable compositor. Once the samples are collected, the state scientists divide the collection into containers, preserve the samples in ice, and ship the collection to a state laboratory for analysis. The laboratory technicians compare the test results to acceptable limits to determine compliance. The state inspector must sign off on all parts of the procedure.

Once these tests are completed, the WCI inspector and the United States EPA inspector discuss procedures, findings, variances, and corrective actions. The U.S. EPA inspector fills out the Water Compliance Overview Inspection Checklist. Both inspectors are required to sign this form, which is kept on file in the state and national archives.

Interferences.   Interferences include any components of the process that may interfere with the accuracy of the final product. This can include but is not limited to the following:

- Humidity
- Temperature
- Depth
- Altitude

- Weather
- Cleanliness
- Carelessness
- Sample size
- Contamination

◀ EXAMPLE

Interferences

If the water has high concentrations of chlorine, this can interfere with test results. Therefore, if the inspector determines that chlorine is present, add sodium sulfide to the sample bottle before autoclaving.

Personnel Qualifications/Responsibilities. This section lists the required experience and certifications of the individuals performing the SOP. These qualifications can include education, years of experience, courses taken, continuing education hours completed, and more.

◀ EXAMPLE

Personnel Qualifications/Responsibilities

This SOP is written specifically for water compliance inspections (WCIs) performed by the Georgia Science and Assessment Division (GSAD). All personnel who perform this activity must have the Basic Inspector Training for Water Compliance certification and have taken the state-required, 8-hour continuing education class. Additional training qualifications include GSAD regulation certification as well as on-the-job training.

Personnel performing this SOP have the following responsibilities:

1. **WCI inspector.** This person must have a minimum of a bachelor of science degree in chemistry (or related discipline), five years field work experience, three years laboratory work, and be certified in state water compliance regulations. This inspector is responsible for coordinating staff and conducting the WCI test to meet deadlines. The inspector must be able to enforce the policies and procedures outlined in this SOP.
2. **Scientists and laboratory technicians.** These employees must have a minimum of an associate of science degree in chemistry, biotechnology, or a related field. These employees also must have passed state-level training in water compliance regulations. Scientists and laboratory technicians must meet deadlines for the WCI, submit all samples in accordance with regulatory policies, and adhere to the SOP.

Equipment and Supplies. This section can include, among others, the following items necessary to perform the SOP:

- Tools
- Equipment
- Reagents needed for the procedure
- Standards for the chemicals
- Biological specimens

If the steps in the process break down into multiple sections, you can divide the equipment and supplies into the units in a chronological pattern of organization. Remember to include quantities of items if needed to perform different tests in the process.

Equipment and Supplies

1.1 Sample Collection
    1.1.1 Sterile sample jars
    1.1.2 Syringes with carbon filters
    1.1.3 Sterile screw-top lids
1.2 Lab Analysis
    1.2.1 Air incubator
    1.2.2 Water bath
    1.2.3 Disposable sterile pipettes
    1.2.4 Sterile applicator sticks
    1.2.5 Sterile culture tubes
    1.2.6 Sterile screw-cap tops
    1.2.7 Pyrex graduated cylinders
    1.2.8 Culture tube racks

Data and Records Management. Include the following information in this unit of the SOP:

- Calculations to be performed during the procedure
- Forms for the reports
- Required reports
- Reporting intervals
- Report recipients
- Process to follow for recording and storing data and information generated by the SOP

EXAMPLE ▶

Records Management

1. Appropriate SOPs will be kept in the laboratory library in black binders. All SOPs will be available to management and employees. When an SOP is removed from the library, sign your name and add the checkout date on the posted SOP inventory sheet.
2. No SOPs can leave the laboratory area.
3. The laboratory director will update the SOP binders as procedures or regulations change.
4. Staff will read all revised SOPs within 5 working days, sign, and date the updated SOP list.

Quality Control and Quality Assurance. Quality control activities help you ensure that you are checking for the highest quality and consistency of the procedure:

- Explain how you will verify your work through quality control. What self-checks will you implement? Will you retest, recount, or recalibrate findings?
- Include the specific process to follow for retesting the procedure, recounting to determine the performance level of the procedure, or recalibrating the machinery.

- Include the self-check intervals for retesting, recounting, or recalibrating. Detail the procedure for dealing with divergent results found through the self-checks.
- State how and to whom you will report the results of the quality control activities.

◀ EXAMPLE

> *Quality Control*
>
> To achieve quality assurance, the laboratory will follow this procedure:
>
> - **Sterility.** Each sample is incubated for 24 hours at 0.5°C and examined for contaminants. If a sample tests positive for contaminants, the entire lot is discarded.
> - **pH.** If the pH does not meet manufacturer's specifications prior to autoclaving, pH can be adjusted according to EPA specifications. If the pH does not meet specifications after autoclaving, the entire sample must be discarded.
> - **Testing.** The laboratory will test samples twice a year (January and June).

References. In this unit, cite the documents or procedures used in or referred to in your SOP. You should include documentation of additional SOPs referenced, literature used for research, and any additional instruction manuals used for reference. Attach copies of the documents if they are not readily available to your potential audience.

> **Research and Documentation**
>
> See Chapter 5 for more discussion of research.

Test for Usability

You've written the instruction or SOP, but is it usable? If the instruction doesn't help your audience complete the task, what have you accomplished? To determine the success of your manual, test the usability of the instruction or SOP as follows:

1. **Select a test audience.** The best audience for usability testing would include a representative sampling of individuals with differing levels of expertise. A review team composed only of high-tech readers would skew your findings. In contrast, a review team consisting of high-tech, low-tech, and lay readers would give you more reliable feedback.
2. **Ask the audience to test the instructions.** The audience members would attempt to complete the instructions, following the procedure step by step.
3. **Monitor the audience.** What challenges do the instructions seem to present? For example, has the audience completed all steps easily? Are the correct tools and equipment listed? Are terms defined as needed and presented where they are necessary? Has the test audience abided by the hazard notations? Are any steps overloaded and, thus, too complicated? Has each step been explained specifically?
4. **Time the team members.** How long does it take each member to complete the procedure? More important, why has it taken some team members longer to complete the task?
5. **Quantify the audience's responses.** Once your test audience has completed the procedure, debrief these individuals to determine what problems they encountered.

Use the Instruction, User Manual, or SOP Usability Checklist to help gather quantifiable information about the instruction's usability.

Audience Recognition

_____ 1. Are technical terms defined?

_____ 2. Are graphics used to explain difficult steps at the reader's level of understanding?

_____ 3. Do the graphics depict correct completion of difficult steps at the reader's level of understanding?

_____ 4. Are the tone and word usage appropriate for the intended audience?

_____ 5. Does the introduction involve the audience, clarify how the reader will benefit, or provide the scope and applicability of the procedure?

Development

_____ 1. Are steps precisely developed?

_____ 2. Is all required information provided, including hazards, technical descriptions, warranties, accessories, and required equipment or tools?

Conciseness

_____ 1. Are words, sentences, and paragraphs concise and to the point?

_____ 2. Does each step specify one action versus overloading the reader?

Consistency

_____ 1. Is a consistent and parallel hierarchy of headings used?

_____ 2. Are graphics presented consistently (same location, same use of figure titles and numbers, similar sizes, etc.)?

_____ 3. Does wording mean the same throughout (technical terms, cautions, warnings, notes, etc.)?

_____ 4. Is the same system of numbering used throughout?

Ease of Use

_____ 1. Can readers easily find what they want through the following:

- Table of contents
- Glossary
- Hierarchical headings
- Headers and footers
- Index
- Cross-referencing or hypertext links
- Frequently asked questions (FAQs)

Document Design

_____ 1. Do graphics depict how to perform steps?

_____ 2. Is white space used to make information accessible?

_____ 3. Does color emphasize hazards, key terms, or important parts of a step?

_____ 4. Do numbers or bullets divide steps into manageable chunks?

_____ 5. Are hazards (dangers, warnings, cautions, and notes) boldfaced, typed in appropriate colors, and/or placed in textboxes to emphasize importance?

Sample Instructions

See Figure 13.7 (Crate&Barrel Elements Bookcase) for sample instructions.

The Writing Process at Work

Web Resources

www.pearsonhighered.com/gerson For more information about the writing process, visit our companion website.

Now that you know what should be included in an instruction or standard operating procedure, it is time to write. The best way to accomplish your technical communication challenge is by following a process: Prewrite to gather data and determine objectives, write a rough draft of your instruction, and rewrite to revise.

Following is an example of an instruction written by Stacy Gerson, a technical writer at gbaMS. To write this instruction, Stacy gathered data, determined objectives and sequenced the instruction through prewriting, drafted an instruction, and finally revised the draft.

The Writing Process

Prewriting	Writing	Rewriting

- Use the reporter's questions to clarify what content you will need to include.
- Determine whether your audience is high tech, low tech, or lay. This will help you decide which abbreviations and acronyms need to be defined.
- Visualize the steps by flowcharting.

- Organize your instructions chronologically.
- Include visual aids and callouts.
- Use headings and subheadings for easy navigation.

- Revise your draft by
  - Adding details
  - Deleting wordiness
  - Simplifying words
  - Enhancing the tone
  - Reformatting your text
  - Proofreading and correcting errors
  - Testing for usability

Prewriting

Flowcharts chronologically trace the stages of an instruction, visually revealing the flow of action, decision, authority, responsibility, input/output, preparation, and termination of process. Flowcharting is not just a graphic way to help you gather data and sequence your instruction, however. It is two-dimensional writing. It provides your reader content as well as a panoramic view of an entire sequence. Rather than just reading an instruction step by step, having little idea what's next, with a flowchart your reader can figuratively stand above the instruction and see where it's going. With a flowchart, readers can anticipate cautions, dangers, and warnings, since they can see at a glance where the steps lead.

To create a flowchart, use the International Organization for Standardization (ISO) flowchart symbols shown in Figure 13.8.

Figure 13.9 illustrates Stacy's flowcharting to gather data and to arrange it chronologically.

Writing

Once you've graphically depicted your instruction sequence using a flowchart, the next step is to write a rough draft of the instruction. After completing the prewriting, Stacy composed a rough draft. The callouts were provided to us by Stacy after she reviewed her rough draft and considered revisions (Figure 13.10).

Rewriting

Rewriting, the last stage in the process, is the time to revise your instruction. Stacy revised the rough draft. The revision is shown in Figure 13.11.

FIGURE 13.7 Crate&Barrel Elements Bookcase Instructions

ELEMENTS BOOKCASE - LEFT FACING

Crate&Barrel

Before beginning assembly, decide if you want to have a right or left facing bookcase. **THESE INSTRUCTIONS ARE FOR ASSEMBLING A LEFT FACING BOOKCASE.**

Thank you for purchasing the **Elements Left Facing Bookcase.** This page lists all the contents included in the box. Please take time to identify the hardware as well as the individual components of this product. As you unpack and prepare for assembly, place the contents on a carpeted or padded area to protect them from damage.
Note: 2 people are required to safely assemble this product.

HARDWARE

24 - bolts   6 - cams   6 - cam screws   2 - screws

24 - plastic caps   4 - feet   2 - metal brackets   1 - wall strap

Missing hardware? Please call **800.606.6387** for replacements.

TOOLS REQUIRED

Allen Wrench (included)

Phillips Head Screwdriver

Hammer

COMPONENTS

1 - top shelf
(additional hole on long edge)

5 - shelves

2 - upright panels

1 - long kickplate

1 - short kickplate

① Lay 1 **upright panel** face down on a padded surface. Insert 2 **cam screws** in the holes as shown, securing each with a **phillips head screwdriver**. Do not over tighten the screws.

Made in China - r03

Page 1 of 3

Customer Service 800-967-6696 | http://www.crateandbarrel.com

FIGURE 13.7 Continued

ELEMENTS BOOKCASE - LEFT FACING

Crate&Barrel

A

B

C

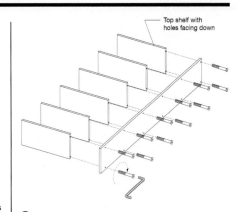

Top shelf with holes facing down

(2) A. Insert 2 **cams** into the large holes on the **long kickplate**. Make sure arrow on each cam points towards nearest edge with holes.
B. Insert 2 **cam screws** into the smaller holes of the long kickplate and secure with a screwdriver.
C. Insert 4 **cams** into the **short kickplate**, making sure each arrow points to the nearest edge with holes.

(3) With the assistance of another adult, align holes in the **upright panel** (panel without cam screws) with holes in the side of each **shelf** and **top shelf**. Attach each shelf with 2 **bolts**, fully tightening each bolt with the **allen wrench**.

Cam in unlocked position

Cam in locked position

(4) With the assistance of another adult, turn shelf assembly over. Attach remaining **upright panel** (panel with cam screws) with 12 **bolts**, fully tightening each bolt with the **allen wrench**. Make sure the cam screws face down.

(5) Turn assembly back over. Fit **short kickplate** onto cam screws as shown. Secure by turning both **cams** clockwise with a **screwdriver** until the cam securely engages the cam screw.

Made in China - r03

Page 2 of 3

Customer Service 800-967-6696 | http://www.crateandbarrel.com

(Continued)

FIGURE 13.7 Crate&Barrel Elements Bookcase Instructions (*Continued*)

ELEMENTS BOOKCASE - LEFT FACING

Crate&Barrel

(6) Fit cam screws of **long kickplate** into short kickplate and upright panel. Secure all **4 cams** by turning each clockwise with a **screwdriver**.

(7) A. Attach 4 **feet** to the bottom of the assembly as shown, gently tapping into place with a **hammer**.
B. Conceal all exposed bolt heads by inserting plastic caps. Assembly is now complete.

OPTIONAL WALL ATTACHMENT

No wall anchor hardware is included with this unit. For safe installation it is essential to use anchor hardware appropriate for your wall type. Mount to wood wall studs whenever possible. If you are unsure of what type of fasteners to use, consult your local hardware store or a qualified professional. This product is only deterrent, it is not a substitute for proper adult supervision.

Locate the pilot hole on the back of the shelf as shown.
A. Attach 1 **metal bracket** with 1 **screw** to the shelf.
B. With the assistance of another adult, place bookcase in desired location against the wall. Mark wall directly opposite of pilot hole in shelf.
C. Attach 1 **metal bracket** to the wall at the mark. Secure with 1 wall anchor (not included), preferably into a wall stud. If not attaching to a stud, be sure to use hardware appropriate to your wall type (drywall, masonry, etc).
D. Place bookcase back into position against the wall. Thread the end of the **wall strap** through each bracket and into the notched end of the strap. Pull until tight.

Pilot hole

Hole in Wall

CLEANING AND CARE

Clean surfaces with a dry or damp soft cloth. Do not use abrasive cleaners.

Made in China - r03

Page 3 of 3

Customer Service 800-967-6696 | http://www.crateandbarrel.com

Source: Courtesy of Crate&Barrel

FIGURE 13.8 Flowcharting Symbols

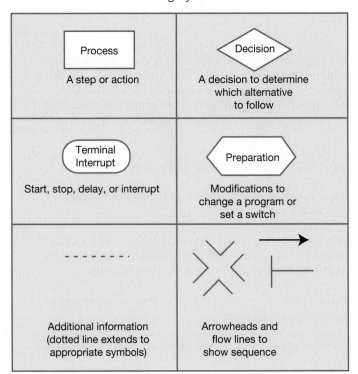

Process A step or action	**Decision** A decision to determine which alternative to follow
Terminal Interrupt Start, stop, delay, or interrupt	**Preparation** Modifications to change a program or set a switch
– – – – – – – Additional information (dotted line extends to appropriate symbols)	Arrowheads and flow lines to show sequence

FIGURE 13.9 Flowchart for gbaMS Parks Inventory

FIGURE 13.10   gbaMS Instruction Rough Draft with Stacy's Commentary

After I reviewed my rough draft, I knew I needed to add a title page, a table of contents for reader ease of access, a glossary to define terms the audience might not know, and a welcoming introduction.

My first step in the rough draft is actually three, separate steps. The use of the bullet and the numbering confused my test audience, so I'll revise that setup.

I added unnecessary transitional words, such as "then," "next," and "finally." I'll remove these words since the numbered steps act as transition.

Last, I'll add screen captures to help my audience understand the procedure.

Park Inventory

The *Parks* module is the most important module in the *Park Master*. All other park asset modules are dependant upon the *Parks* module. While you can access the other asset modules through the gbaMS main menu, it's best to add park assets directly from the *Parks* module. This allows the system to carry over linking information and details about the park. In the following step-by-step example, we'll show you how to create a park, record its location and attributes, and associate park assets.

1. Open **Trees/Parks>>Park Master>>Parks** from the gbaMS main menu.

   • Click *GO* to bypass the filter. When the module is opened, click the Add button in the toolbar to access a blank record.

2. Then, in the module header, create a unique Park Name and Alternate Park ID. Make sure to choose a name and ID that are recognizable; you'll need to be able to select the appropriate park name from a pick list in the park asset modules.

3. Next, record the park location, park supervisor, park type, district, region, owner, acquisition date, and opening date on the General tab.

4. Finally, select a Default WO Category. The category you select here will be carried over to any work order created from this record.

Inspecting Park Assets

All of our park asset inventory modules have associated inspection modules. In the following example, we'll show you how to add an inspection record for your park asset.

1. On the Inspections/Associations tab of the *Park Playground Equipment Inventory* module, right click in the Inspections grid and select *Add Record*.

2. The *Park Playground Equipment Inspection* module will open. There, you'll see the Equipment Number, Alternate Equipment ID, and Equipment Name carried over from the related inventory record.

3. Next, you'll need to enter the inspection date, inspection crew, and next scheduled inspection date.

4. Two Condition tabs are provided. You'll use these to record the condition codes (excellent, good, fair, and poor) for the various aspects of the playground equipment. These include the overall condition, stability, concrete exposure, protective surface, clearance, supports, and fastenings, among other things.

FIGURE 13.11 gbaMS Parks Inventory Completed Instruction

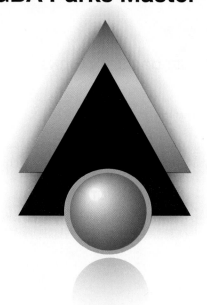

GBA MASTER SERIES® SOFTWARE APPLICATIONS

GBA Parks Master™

10561 Barkley #500
Overland Park, Kansas 66212
Phone 800.492.2468 • Fax 913.341.3128
www.gbams.com • info@gbams.com

Source: Courtesy of gbaMS

TABLE OF CONTENTS

LIST OF ILLUSTRATIONS

INTRODUCTION

In this workbook, we'll introduce you to the modules in the *GBA Parks Master*™. Using a series of step-by-step examples, we'll show you how to track and maintain your park assets. These include

- Parking lots
- Paths
- Fields and courts
- Landscaping
- Playground equipment
- Pools
- Furniture (and more)

Our software modules will help you manage your parks effectively and efficiently. While the *Parks Master's* multiple inventory and inspection modules track unique park assets, the steps provided in this workbook will apply equally to all park asset modules.

REQUIRED HARDWARE AND SOFTWARE

The following requirements are based on the simplest configurations with the best performance.

Hardware

CPU	450 MHz
RAM	256
Storage Space	20MB
Minimum Video	800 × 600

Software

Windows XP Professional

Internet Explorer Version 7.0

GLOSSARY

ID	Identification number (provided from the dropdown list)
Mb	Megabytes
MHz	Megahertz
WO	Work Order number (you choose your own for your records)

2

PARK INVENTORY

The Parks module is the most important module in the Parks Master. All other park asset modules are dependent upon the Parks module. While you can access the other asset modules through the gbaMS main menu, it's best to add park assets directly from the Parks module. This allows the system to carry over linking information and details about the park.

   The following step-by-step instructions show you how to create a park record and report park inspections.

1. Open Trees/Parks>>Park Master>>Parks from the gbaMS main menu.
2. Click GO to bypass the filter.
3. When the module is opened, click the Add button in the toolbar to access a blank record.
4. In the module header, create a unique Park Name and Alternate Park ID.

Note: Make sure to choose a name and ID that are recognizable. You'll need to be able to select the appropriate park name from a pick list in the park asset modules.

FIGURE 1   Name and ID

Note: In our example above, we've created a record for the Crossroads District Park and given it the Alternate Park ID, 4.

5. Record the park location, park supervisor, park type, district, region, owner, acquisition date, and opening date on the general tab.
6. Select a Default WO Category.

Note: The category you select here will be carried over to any work order created from this record.

3

INSPECTING PARK ASSETS

All of our park asset inventory modules have associated inspection modules. The following instructions show you how to add an inspection record for your park's swing set. This is an important consideration for liability reasons.

1. On the Inspections/Associations tab of the Park Playground Equipment Inventory module, right click in the Inspections grid and select Add Record.

FIGURE 2   Add Record

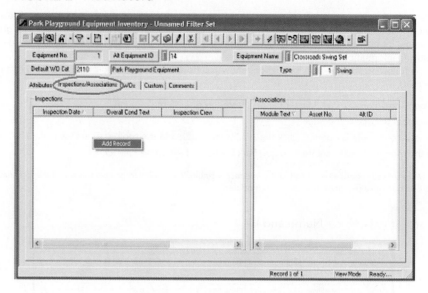

Note: The Park Playground Equipment Inspection module will open. Here, you see the Equipment Number, Alternate Equipment ID, and Equipment Name carried over from the related inventory record.

FIGURE 3   Equipment Name and Number

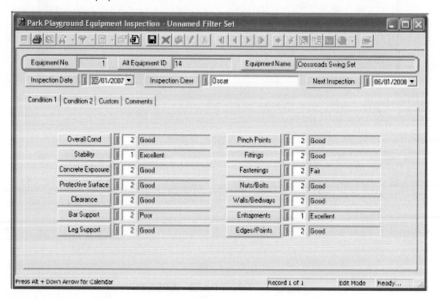

2. Enter the inspection date, inspection crew, and next scheduled inspection date.

3. Two Condition tabs are provided. Record the condition codes (excellent, good, fair, and poor) for the various aspects of the playground equipment. These include the overall condition, stability, concrete exposure, protective surface, clearance, supports, and fastenings, among other things.

4

FIGURE 4 Equipment Condition

Congratulations. Once you've saved the record, the Inspections grid in the Inventory module will be updated.

5

CHAPTER HIGHLIGHTS

1. To organize your instruction or SOP effectively, include an introduction, a table of contents, and a discussion of sequenced steps.

2. Number your step, and start each step with a verb.

3. Hazard alerts are important to protect your reader and your company. Place these alerts early in your instruction or SOP or before the appropriate step.

4. In the instruction or SOP, follow a chronological sequence.

5. Screencasting is a mixture of animated screen visuals, voiceovers, and captioned text. Screencasting benefits your audience by allowing the audience to see the procedure in action.

6. An SOP is a set of written instructions that documents routine or repetitive technical or administrative activities followed by business and industry. SOPs ensure accurate job performance and consistent quality and integrity of the end product. SOPs also ensure that governmental regulations are followed.

7. Highlighting techniques emphasize important points, thus minimizing damage to equipment or injury to users.

8. SOPs are used in many fields, including science, healthcare, biomedical technology, government, military, the computer industry, and others.

9. The writing process, complete with usability testing, will help you construct an effective instruction, user manual, or SOP.

10. Consider usability testing to determine if your procedure is valid.

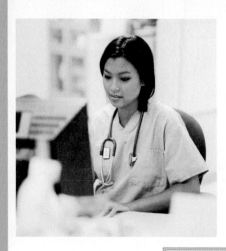

CASE STUDIES

1. PhlebotomyDR provides instructions for its staff (nurses, doctors, and technicians) regarding the correct procedures to ensure cleanliness and employee protection. These include hand washing procedures and the correct use of sterile equipment, such as gloves, masks, aprons, and shoe coverings.

   Following is a rough draft of one set of instructions for personnel safety. Revise the rough draft according to the criteria provided in this chapter. Correct the order of information, the grammar, and the instruction's content. In addition, improve the instruction by including appropriate graphics.

Hand washing

- Lather hands to cover all surfaces of hands and wrists.
- Wet hands with water.
- Rub hands together to cover all surfaces of hands and fingers. Pay special attention to areas around nails and fingers. Lather for at least 15 seconds.
- Dry thoroughly.
- Rinse well with running warm water.
- Avoid using hot water. Repeated exposure to hot water can lead to dermatitis.
- Use paper towels to turn off faucet.

Gloves

- Replace damaged gloves as soon as patient safety permits.
- Don gloves immediately prior to task.
- Remove and discard gloves after each use.

Masks

Wear masks and eye protection devices (goggles or eye shields) to avoid droplets, spray, or splashes and to prevent exposure to mucous substances. Masks are also worn to protect nurses, doctors, and technicians from infectious elements during close contact with patients.

Aprons and Other Protective Clothing

- Wear aprons or gowns to avoid contact with body substances during patient care procedures
- Remove and discard aprons and other protective clothing before leaving work area.
- Some work areas might require additional protective clothing such as surgical caps and shoe covers or boots.

2. BurgerNet Online Order System is a software product in development. As a technical communicator, write the instruction to accompany this online order system.

BurgerNet Online Order System

1. **Description.** The BurgerNet application will be used to order burgers for delivery to a customer's home or work.

2. **Dropdown Menu Bar.** The dropdown menu bar includes a Task, Edit, View, and Help menu.
   - Task menu. On the "Task" menu customers will find a "Send Order" command, used to electronically send the completed order to the restaurant. The "Print Order" command on the "Task" menu lets the customer print a copy of the order. The "Exit" command terminates the order session.
   - Edit menu. On the "Edit" menu, an "Add Favorites" command allows customers to add a frequently ordered combination of food items for make future online ordering quick and easy. The "Edit Favorites" command lets customers modify an existing favorite order.
   - View menu. On the "View" menu, the "Preferences" command is used to modify default customer. In addition, when favorites have been defined, they are listed under "View." Customers can select a favorite for display before sending the order.
   - Help menu. On the "Help" menu, customers can access FAQs.

3. **Key Components of BurgerNet.**
   - Customer Info. frame: Customers type in their name and telephone number in the "Name" and "Extension" fields.
   - Burgers frame. In the "Burgers/Of" fields customers specify the number of burgers desired. The "Cheese" dropdown list lets the customer select either American cheese, Pepperjack, Mozzarella, or Swiss cheese to top their burgers. "Bun" lets customers choose either plain or onion buns from the dropdown list.
   - Toppings. Here, customers check what to put on their burgers.
   - Sides: Select which side dish or dishes and specify the quantity for each.
   - Beverages: Select drink(s) and specify the quantity for each.
   - Total field. The total cost of the customer's order is automatically displayed.

INDIVIDUAL AND TEAM PROJECTS

1. Write an instruction or SOP. To do so, first select a topic. You can write an instruction telling how to monitor, repair, test, package, plant, clean, operate, manage, open, shut, set up, maintain, troubleshoot, install, use software, and so on. Choose a topic from your field of expertise or one that interests you. Follow the writing process techniques to complete your instruction. Prewrite (using a flowchart), write a draft (abiding by the criteria for instructions or SOPs presented in this chapter), and rewrite to perfect your text.

2. Find examples of instructions, user manuals, or SOPs written in the work environment. Bring these to class. Using the criteria for instructions and SOPs presented in this chapter, decide whether the instructions and SOPs are successful or unsuccessful. If the instructions and SOPs are good, show how and why. If they are flawed, explain the problem(s). Then rewrite the instructions or SOPs to improve them.

3. Good writing demands revision. Following is a flawed instruction. To improve it, rewrite the text, abiding by the criteria for instructions and the rewriting techniques included in this chapter.

Date: November 1, 2012
To: Maintenance Technicians
From: Second Shift Supervisor
Subject: Oven Cleaning

The convection ovens in kiln room 33 need extensive cleaning. This would consist of vacuuming and wiping all walls, doors, roofs, and floors. All vents and dampers need to be removed and a tack cloth used to remove loose dust and dirt. Also, all filters need replacing. I am requesting this because when wet parts are placed in the ovens to cure the paint, loose particles of dust and dirt are blown onto the parts, which causes extensive rework. I would like this done twice a week to ensure cleanliness of product.

---

4. One reason flawed instructions fail is lack of audience recognition. For example, an instruction uses high-tech terms but is intended for low-tech readers.

    Find an example of an instruction or SOP designed for a high-tech audience that correctly uses high-tech terms. Then find an example of an instruction geared to low-tech readers that incorrectly uses high-tech words. Next, find an instruction geared to low-tech or lay readers that has correctly defined its terms and explained its terminology.

5. Find examples of instructions that do not provide effective hazard alerts. Using the criteria for effective hazard alerts provided in this chapter, decide where the alerts are flawed. Then improve the instructions or user manuals by creating effective hazard alerts.

PROBLEM-SOLVING THINK PIECE

Read the following instructional steps. Are they in the correct chronological order? How would you reorder these steps to make the instruction more effective?

Changing Oil in Your Car

Run the car's engine for approximately 10 minutes and then drain the old oil.
Park the car on a level surface, set the parking brake, and turn off the car's engine.
Gather all of the necessary tools and materials you might need.
Open the hood.
Jack up and support the car securely.
Place the funnel in the opening and pour in the new oil.
Replace the cap when you have finished pouring in new oil.
Locate the oil cap on top of the engine and remove the cap.
Tighten the plug if you find leakage.
Run the engine for a minute, then check the dipstick. Add more oil if necessary.
Pour the used oil into a plastic container and dispose of it safely and legally.

WEB WORKSHOP

1. Review any of the following Web site's online instructions. Based on the criteria provided in this chapter, are the instructions successful or not?

   - If the answer is yes, explain why and how the instructions succeed.
   - If the answer is no, explain why the instructions fail.
   - Rewrite any of the flawed instructions to improve them.

Web Sites	Topics
http://www.hometips.com/diy.html	Electrical systems, plumbing, kitchen appliances, walls, windows, roofing, and siding
http://www.hammerzone.com	Kitchen projects, tubs, sinks, toilets, showers, and water heaters
http://dmoz.org/Home /Home_Improvement/	Links to step-by-step procedures for painting, welding, soldering, plumbing, walls, windows, and door repair and installation
http://directory.google.com/Top /Home/Home_Improvement/	Links to sites for instructions on decorating, electrical, flooring, furniture, lighting, painting, plumbing, welding, windows, and doors
http://www.quakerstate.com/#/car-care	Instructions for car care

2. The U.S. Department of Labor article, "Hazard Communication: A Review of the Science Underpinning the Art of Communication for Health and Safety," can be found at http://www.osha.gov/dsg/hazcom/hc2inf2.html. This article focuses on many important aspects of writing instructions, including the use of icons, readability, and audience variables. Access the article and report your findings either orally or in a memo, letter, or e-mail message.

3. YouTube provides thousands of instructional videos for topics including creating Origami shapes, playing World of Warcraft, playing pool and billiards, snowboarding and

paragliding, playing guitars, tossing pizzas, using Macs and PCs, installing software and hardware, and downloading music for iPods. Research any of these videos (or others). Based on the criteria provided in this chapter, assess your findings. Were the videos successful? How could they be improved? Share your findings with your teacher in an e-mail message or present your findings to your class in a brief oral presentation.

QUIZ QUESTIONS

1. What are five reasons to write instructions?
2. What should you include in the introduction of an instruction or SOP?
3. How can you create an effective hazard alert?
4. What is screencasting?
5. What is an SOP?
6. What colors highlight a warning, a danger, and a caution?
7. What are five components of an effective SOP?
8. What do graphics achieve in instructions?
9. In what way should you begin each step in an instruction?
10. What are some features of usability testing?

CHAPTER FOURTEEN

Online Help and Web Sites

COMMUNICATION AT WORK

In the following scenario, Future Promise is creating a cross-functional team to develop a Web site.

Future Promise is a not-for-profit organization created to help at-risk high school students. This agency realizes that to reach its target audience (teens aged 15–18), it needs an Internet presence.

Future Promise's CEO, Brent Searing, has decided to form a cross-functional team to create the agency's Web site. Brent will encourage the team to work collaboratively to determine the Web site's content, its level of interactivity, and its design features. Brent wants the Web site to include

- College scholarship opportunities
- After-school intramural sports programs
- Job-training skills (resume building and interviewing)
- Service learning programs to encourage civic responsibility
- An FAQ page

Objectives

When you complete this chapter, you will be able to

1. Recognize the importance of online technical communication

2. Know the characteristics of e-readers

3. Understand the characteristics of online communication

4. Understand online help

5. Follow criteria for effective online help

6. Follow criteria to create a successful Web site

7. Distinguish among Web-accessibility problems, such as cognitive, hearing, and visual impairments

8. Apply the usability checklist to create effective Web sites

- Future Promise's 800-hotline (for suicide prevention, STD information, depression, substance abuse, and peer counseling)

- Social media links to Future Promise's Facebook, Twitter, and blog sites

- Additional links (for donors, sponsors, educational options, job opportunities, etc.)

To accommodate these Web components, the Future Promise Web team will consist of the agency's accountant, sports and recreation director, public relations manager, counselor, training facilitator, graphic artist, and computer and information systems director. In addition to these Future Promise employees, Brent also has asked two local high school principals, two local high school students, and a representative from the mayor's office to serve on the committee. Future Promise's public relations manager Jeannie Kort will chair the committee.

Jeannie has a big job ahead of her. First, she must coordinate everyone's schedules. The two principals and high school students, for example, can attend meetings only after school hours. Next, she must meet Brent's deadline; he wants the Web site up and running within three months. Jeannie also must manage this diverse team (with varying ages, levels of responsibility, and levels of knowledge).

Another challenge involves hardware and software. Jeannie realizes that not all of her Web readers will have state-of-the-art computers. Therefore, she must ask her Web design team to create a Web site that is accessible on many platforms, even low-end hardware. Thus, though Jeannie wants the site to be colorful, interactive, and entertaining, the site also must load quickly, avoid frames, accommodate many different monitor resolutions, and be both PC and Mac friendly.

The task is daunting, but the end product will be invaluable for the city and the city's youth. Jeannie and Brent know that by conveying information about jobs, training, scholarships, and counseling to their end users (at-risk teens), Future Promise can improve the quality of many people's lives. Jeannie's Web design team has an exciting project on its hands.

Check out our quarterly newsletters at www.pearsonhighered.com/gerson for dot.com updates, new case studies, insights from business professionals, grammar exercises, and facts about technical communication.

Overview—the Importance of Electronic Communication

The written word has undergone significant changes, leaping from the printed page into cyberspace. Correspondence, once limited to hard-copy letters, reports, and memos, is now often online as Web sites and online help. Corporate brochures and newsletters, once paper bound, now are online. Product and service manuals, once paper bound, now are online. Technical communication in the twenty-first century is increasingly electronic.

This trend toward electronic communication is increasing with the growing presence of WiFi (wireless fidelity) and MiFi (mobile fidelity) hotspots. At WiFi hotspots, people can connect to the Internet by way of laptops, tablet PCs, and smart phones. MiFi allows for WiFi on the go, hotspots that fit in a user's pocket. These technologies allow people to work offsite and to communicate with co-workers, clients, vendors, and corporations anywhere and anytime.

The Characteristics of E-Readers

Electronic communication is an entirely different mode of communication than hard-copy text. Not only do Web sites and online help screens differ from paper text, but also the "e-reader" approaches online communication differently than he or she will tackle hard-copy text. What are the differences between paper text and online communication?

E-Readers Are Topic Specific

In libraries or bookstores, we wander up and down aisles, looking for any book that interests us. In contrast, e-readers tend to access the Web and online help with specific goals in mind. You go online to search for specific information found in specific Web sites: CD prices, automobile loan rates, hotel room availability, restaurant menus, technical specifications for laser printers, the start date for your college's spring semester, and so on. You access online help to find solutions to specific work-related problems.

E-Readers Want Information Quickly

E-readers often scan, skim, and skip over text, looking just for the information they want and ignoring the rest of the text. In fact, readers want to find information in "ten seconds. Your web site visitors love skim-readable pages. They're not lazy. They're just in a hurry. They don't want to waste time and money reading the wrong web page, when there are millions of other pages to choose from" (McAlpine).

E-Readers Access Electronic Communication Using Diverse Platforms

Another difference between an e-reader and one who reads a hard-copy novel is the way in which the reader accesses the document. Most hard-copy text is printed on either book-sized pages or on 8½ × 11 inch pieces of paper. In contrast, e-readers can access electronic communication "using cell phones, PDAs, and other wireless devices" (Moore 17). Thus, screen resolution and the size of type font are key elements for online readability. Furthermore, e-readers will be using Firefox, Internet Explorer, AOL, Macs and PCs, and a host of other electronic platforms to view your text. Each of these electronic platforms differs in subtle ways.

The Characteristics of Online Communication

Because online text is read differently than hard-copy text, you must alter the way you write online communication. When you write a Web site or online help, you must reconsider

- **The screen versus the page.** Forget 8½ × 11. Online help screens are smaller, and good Web screens rarely require scrolling.
- **Skimming versus linear reading.** We read books "linearly," line by line. In contrast, online help screens and Web screens are skimmed and scanned.
- **Hypertext links versus chronological reading.** We read books from beginning to end, sequentially. Web sites, however, allow us, even encourage us, to leap randomly from screen to screen.

Page Layout

Most hard-copy text is 8½ × 11 inches, with 1-inch margins (top, sides, and bottom). When you get to the bottom of the page, you turn to the next page. In contrast, the size of online help screens and Web sites varies.

Page Length.   On-screen, less is best. When writing online, "you must be aware that computer screens display smaller amounts of information than a printed page" (Hemmi 11). A successful Web page should limit pertinent information to one screen, without requiring the reader to scroll. Online help screens are even smaller. Thus, the "material that fits on one printed page might require three to six screens online" (Hemmi 11).

Margins.   Hard-copy text, when read from margin to margin, is approximately 100 characters long (a character is every letter, punctuation mark, and space). The size and type of your font, of course, affects the number of characters you will type per line. Online, horizontal margins vary from monitor to monitor (21 inches, 19 inches, 17 inches, 15 inches, and even smaller on hand-held devices).

On a Web page, your audience is forced to read more words per line if you use the entire screen. That's difficult. Readers have trouble maintaining their focus (visually and mentally) when expected to read long lines of horizontal text. Vertically, the problem increases. Regardless of the monitor size, cybertext can run forever. The reader can scroll, and scroll, and scroll. Whereas paper sizes are controlled by convention and printing companies, you must control the size of a Web page.

A successful Web page should limit lines of text to perhaps two-thirds of the screen, with a graphic, white space, or hypertext links placed in the remaining third. This breaks up the monotony of long lines into manageable chunks. Thus, the reader can maintain focus more easily (Eddings et al.).

Font.   Your computer's word processing software probably defaults to Times New Roman, 12 point, or Calibri, 11 point. These are considered the best types and size fonts for readable hard-copy text. You can enhance your hard-copy text with designer fonts (stylized and cursive), boldface, or underlining for visual effect. In contrast, Arial or Verdana seem to work best for online reading. "Stylized, cursive, italicized, and decorative typefaces are difficult to read online" (Hemmi 11).

Underlining is especially troublesome online, because online, hypertext links can be shown as underlined text. To avoid confusing your readers online, you must avoid using underlining as a highlighting technique.

Structure

Paper text is read sequentially. We read from page 1 through the end of the text. Think of a book: You would never consider reading page 102 before you have read page 18 or Chapter 7 before you have read Chapter 4. In fact, you are expected to read sequentially—page after page after page. After all, the book (or newsletter or manual) is bound; each page is physically stitched, glued, or stapled to the next. In contrast, online help screens and Web sites are composed of randomly "stacked" screens. For online help screens, you access a search mechanism, find the topic you want, and click on a link to access it. For a Web site, you scan the home page, and then you decide what you want to read and when you want to read it. Once you make this decision, you click on hypertext links, access any screen, and read content in any order. Thus, you receive "just in time" or "as needed" information. That is, if you want to access the linked text, you can. If you don't want to read this screen, you don't have to. Again, on-screen, less is best for the random reader who plans to skim and scan (Goldenbaum and Calvert). Hard-copy documentation follows a page-by-page format, whereas online text is nonlinear. The structure of online text is modular and flexible (Hemmi 11).

Noise

Though hard-copy text is rigidly linear, it can be easier to read than online text. Paper is dull and absorbs light. Most hard-copy text is colorless and motionless; a book is composed of black ink on white paper. For online text, however, the screen is composed of glass, which reflects light, creating visual glare. Many of our screens are smudged where we have used our fingertips to point out interesting images.

To help e-readers, you need to consider a key challenge of reading online: computer "noise"—sound and visual distractions. Extended viewing of a computer screen is more demanding than continued reading of paper text. Web sites often contain lots of color, blinking text, animated graphics, frames of layered text, and sound and video. Noise—multiple distractions—inundates the screen. In such a busy communication channel as a Web site, the less we give the reader on one screen, the better.

Online Help

Imagine this possibility. Your boss calls you into a meeting and says, "As of today, we are no longer producing paper documents" (McGowan 23). It's not as farfetched as it sounds. Online Web help is a major part of the technical communicator's everyday job for several reasons, including

- The increased use of computers in business, industry, education, and the home
- The reduced dependence on hard-copy manuals by consumers
- The need for readily available online assistance
- Proof that people learn more effectively from online tutorials than from printed manuals (Pratt 33)

In addition, online help screens provide the practical value of portability. Rather than take up computer archival space with thousands of pages of PDF files or print out hundreds of pages of instructional help, online help screens allow companies to save space and provide readers ease of use.

What is an online help system? Online help systems, which employ computer software to help users complete a task, include procedures, reference information, wizards, and indexes (see Figure 14.1). Typical online help navigation provides readers hypertext links, tables of contents, and full-text search mechanisms.

Help menus on your computer are excellent examples of online help systems. As the computer user, you pull down a help menu, search a help list, and click on the topic of your choice, revealed as hypertext links When you click on a link, you could get a *pop-up* (a small window superimposed on your text), or your computer might *link* to another full-sized screen layered over your text. In either instance, the pop-up or hyperlink gives the following information about your topic:

- **Overviews**—explanations of why a procedure is required and what outcomes are expected
- **Processes**—discussions of how something works

FIGURE 14.1 Microsoft Word 2010 Help Screen with a Search for "Margins" and the Resulting Pop-up Help Screen

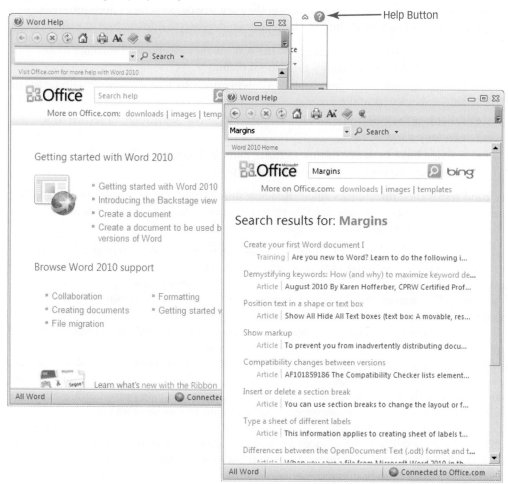

- **Definitions**—an online glossary of terms
- **Procedures**—step-by-step instructions for completing a task
- **Examples**—feedback verifying the completion of a task or graphic depictions of a completed task, including a screen capture or a description with callouts
- **Cross-references**—hypertext links to additional information
- **Tutorials**—opportunities to practice online

Online help systems allow the technical communicator to create interactive training tools and informational booths within a document. These online systems can be created using a wide variety of help authoring tools (HATs). Some popular HATs include Adobe RoboHelp, Author-it, Doc-To-Help, DocArchitector, HelpScribble, HyperText Studio, MadCap Flare, and Sandcastle. The value of online help is immense. Because online help offers "just in time" learning or "as needed" information, readers can progress at their own pace while learning a program or performing a task.

Techniques for Writing Effective Online Help

To create effective online help screens, consider the following suggestions.

Organize Your Information for Easy Navigation. Poorly organized screens lead to readers who are "lost in cyberspace." Either they cannot find the information they need, or they have accessed so many hypertext links that they are five or six screens deep into text.

You can avoid such problems and help your readers access information in various ways:

- Allow users to record a history of the screens they've accessed through a bookmark or a Help Topics pull-down menu.
- Provide an online Contents menu, allowing users to access other, cross-referenced help screens within the system.
- Provide a Back button or a Home button to allow the readers to return to a previous screen.
- Provide links such as "In this section" and "Related topics" on each screen for immediate access to topic-specific content.

A good test is the "three clicks rule." Readers should not have to access more than three screens to find the answer they need. Similarly, readers should not have to backtrack more than three screens to return to their place in the original text (Timpone).

Recognize Your Audience. Online help must be user oriented. After all, the only goal of online help is to help the user complete a task. Thus, a successful online help system must be "designed at the same level of detail as the user's knowledge and experience" (Wagner). This means that technical communicators have to determine a user's level of knowledge.

If the system is transmitted by way of an intranet or extranet, you might be tempted to write at a high- or low-tech level. Your readers, you assume, will work within a defined industry and possess a certain level of knowledge. That, of course, is probably a false assumption. Even within a specific industry or within a specific company, you will have co-workers with widely diverse backgrounds: accountants, engineers, data processors, salespeople, human resource employees, management, technicians, and so on.

Don't assume. Find out what information your readers need. You can accomplish this goal through usability testing, focus groups, brainstorming sessions, surveys, and your company's hotline help desk logs. Then don't scrimp on the information you provide. Don't just provide the basic or the obvious. Provide more detailed information and numerous pop-ups or links. Pop-up definitions are an especially effective tool for helping a diverse audience. Remember, with online help screens, readers who don't need the information can skip it. Those readers who do need the additional information will appreciate your efforts. They will more successfully complete the tasks, and your help desk will receive fewer calls.

Audience

See Chapter 4 for more discussion of definitions and tone.

Achieve a Positive, Personalized Tone. Users want to be encouraged, especially if they are trying to accomplish a difficult task. Thus, your help screens should be constructive, not critical. Your text should be "written in the affirmative," a concept supported by human-computer interaction (HCI) concerns (Wagner). HCI-driven online help systems "coach" the users rather than "command" them ("Human-Computer Interaction" 2). The messages also should be personalized, including pronouns to involve the reader.

Document Design

See Chapter 9 for more discussion of document design.

Design Your Document. How will your help screens look? Document design is important because it helps your readers access the information they need. (See Figure 14.2.)

To achieve an effective document design, consider the following points:

- **Use color sparingly.** Color causes several problems in online documents. Bright colors and too many colors strain your reader's eyes. Furthermore, a color that looks good on a high-resolution monitor might be difficult to read on a monitor with poor resolution. Your primary goal is contrast. To help your audience read

FIGURE 14.2 Help Screen

Hypertext link

Pop-up definition

Source: Courtesy of Earl Eddings, technical communications specialist.

your text, maximize the contrast between the text color and the background color, as seen in Figures 14.1 and 14.2.

- **Be consistent.** Pick a color scheme and stick with it. Your headings should be consistent, along with your word usage, tone, placement of help screen links and pop-ups, graphics, wizards, and icons. Readers expect to find things in the same place each time they look. If your help screens are inconsistent, readers will be confused.

- **Use an easy-to-read font.** A 12-point type size is standard for most printed documents, but 10-point type will save you valuable space online. Serif fonts are the standard for most technical writing. A serif font, with small, horizontal "feet" at the bottom of each letter, helps guide the eye while reading printed text. However, on lower resolution monitors, serif text is more difficult to read. Many online help screens, such as those provided by Microsoft Word, use a sans serif font, such as Arial or Calibri. Avoid "designer fonts," such as Algerian, Bauhaus 93, or Brush Script MT, which are hard to read.

- **Use white space.** Don't clutter your help screens. Avoid excessive emphasis techniques. For example, typical PowerPoint "fly-in" effects not only are distracting, but also they negatively impact the online screen's performance, slowing load time down dramatically. Minimize your reader's overload by adding ample horizontal and vertical white space. Online, less is best.

Be Concise. "It takes 20 to 30 percent longer to read on-screen than in print, so you must minimize text" (Timpone). In addition to limiting word and sentence length, a help screen, just like any Web page, should limit horizontal and vertical scrolling. Each screen should include one self-contained message. If readers need additional information, use "In this section" and "Related topics" links.

> **Conciseness and Clarity**
>
> See Chapter 3 for more discussion of conciseness and clarity.

Be Clear.   Your audience reads the help screen only to learn how to perform a task. Thus, your only job is to meet the reader's needs—clearly. To accomplish this goal, be specific. In addition, clarity online could include the following:

- Tutorials to guide the reader through a task. Microsoft provides clear, step-by-step instructions with embedded screen captures to help its audience complete activities.
- Pop-up definitions with expanding blocks for longer definitions.

Provide Access on Multiple Platforms.   You will want every reader to be able to access your online help, not just those using a PC or a Mac or UNIX. After all, if readers can't access online help, how have you helped them? To ensure that all readers can access your system,

- Do not use platform-specific technologies, such as ActiveX, which only runs on 32-bit Windows.
- Create content that can be viewed on multiple browsers, including Internet Explorer and Firefox.
- Create online help for the oldest version that you will support—the lowest common denominator. Remember, all users won't have the latest version browser, software, or hardware.
- Test your online help on at least three different browsers and platforms. Though you might be creating your online help screens on a 30" widescreen, your audience might be accessing the text on a 15" panel or smaller smart phone. Make sure they can see what you see, even if they are viewing text on smaller screens.

Correct Your Grammar.   As in all technical communication, incorrect grammar online leads to two negative results: a lack of clarity and a lack of professionalism. Don't embarrass your company or confuse your reader with grammatical errors. Proofread.

Web Sites

Web sites can convey any amount of information about any topic. Web sites are created by companies, organizations, schools, and government agencies, not to mention individuals. That's why a Web site's URL (Uniform Resource Locator—the Web site's address) reads *.com* (commercial), *.org* (organization), *.edu* (education), *.gov* (government), and so on.

ONLINE HELP CHECKLIST

_____ 1. Have you used your online help screens to explain why a procedure is required, show how something works, define terms, give examples, allow opportunities to practice tasks, and/or provide access to additional information?

_____ 2. Have you helped readers access information with pull-down menus, tables of content, back and/or home buttons, and links to related topics?

_____ 3. Have you tried to ensure that readers do not need more than three screens to find answers to their questions?

_____ 4. Have you met your audience's needs by designing content at the level of their knowledge and expertise?

_____ 5. Have you written your content using a positive tone, striving to "coach" rather than "command"?

_____ 6. Did you design your help screens effectively by limiting color, using a consistent color scheme, choosing an easy-to-read font, and incorporating white space for easy access?

_____ 7. Is your online help concise, limiting word length, sentence length, paragraph length, and horizontal and vertical scrolling?

_____ 8. Have you developed your text thoroughly?

_____ 9. Can your audience access the online help using different platforms and browsers?

_____ 10. Is your online help grammatically correct?

The Process of Creating Effective Web Sites

Shannon Conner is a freelance technical writer and Web designer. When asked what his most important job is as a freelance technical writer and Web designer, Shannon says, "Finding the perfect balance between content and design to give a Web site user reason to come back to the site."

One of the biggest challenges Shannon faces in his work is balancing clients' expectations with audience need while achieving concise, clear design.

Define the client's need. "I always ask clients, 'What are the top three ideas you want your visitors to take with them when they leave this page or Web site? Undoubtedly, the common response is, 'It's all important,' but when the clients put some serious thought to it and pick the top three, they get a Web site with clear intent and satisfied users," Shannon says.

Define your target audience. Because of the mass availability of the Internet, it can be a challenge really knowing who the audience is for a Web site. Defining the target audience ensures you are communicating the right way to the right people. "Your vernacular would be vastly different for a Web site geared toward retirement-aged users compared to a Web site geared toward high school-aged users," explains Shannon.

Employ clarity, conciseness, and consistency. "Consider the three Cs the keystone of your communication goals," Shannon says. "Always communicate with clarity, conciseness, and consistency. Without these three traits—especially when your medium is a Web site—the thoughts and ideas you worked hard to produce are worthless. What is the value of your content if no one reads it? Remember, users of Web sites typically scan the text; they don't read it. Provide them clean content so they can find exactly what they're looking for with as little effort as possible."

A final challenge for Shannon involves the global audience. Users can easily translate Web site content without the Web site owner's knowledge or participation, and many clients request their Web sites be available in multiple languages. "This fact not only impacts the words I use and my sentence structure, but it also affects how graphics are used," Shannon says.

Shannon, like all good technical communicators, knows the value of graphics. A screen capture of a Web application or a graphical representation of a workflow will be more effective than hundreds of words in helping an end user perform a task. In the United States, a screen capture of a Window's dialog box will read "Save," "Exit," or "Next." This will not communicate to all members of a multilingual audience. The dialog box has to be translated into many languages, and then new visuals must be made for each application. If an online help screen has many visuals, suddenly, Shannon's task becomes more challenging and more costly for the client.

Usability for the Web audience is paramount for Shannon. The audience must be able to say, "I can use this Web site, and I can benefit from it." To accomplish these goals, Shannon must consider global customers, their varied knowledge of the subject matter, and their diverse computer skills.

The Democratic National Party has a Web site. So do McDonald's, Ford Motor Company, the city of Las Vegas, the University of Texas, and Greenpeace. There are hundreds of Web sites dedicated to Elvis Presley. Whatever topic you are interested in researching, you can find it online in someone's Web site.

Web Accessibility

An international goal is Internet accessibility for persons with disabilities. The Web Accessibility Initiative (WAI), in coordination with organizations around the world, pursues accessibility of the Web. The following are Web accessibility problems ("Internet Accessibility").

Cognitive

Cognitive problems include learning disabilities, reading disorders, and attention-deficit/hyperactivity disorder (ADHD). Web access can be enhanced with illustrations, graphics, and headings, which provide visual cues for easier Web understanding.

Hearing

Web users who are hearing impaired may need assistive technology to read Web audio or captioned text for multimedia content.

Color-blindness causes problems on the Internet. To combat this challenge, Web writers need to choose colors correctly, perhaps avoiding green and red font. The United States Government recognizes the importance of the Internet and the need for inclusiveness for both employees and external Web readers. In 1998, Congress amended the Rehabilitation Act, requiring federal agencies to make electronic and information technology accessible to people with disabilities. Section 508 for federal employees and the public eliminates barriers in information technology ("508 Law").

Following are Web sites that provide additional information about Web accessibility:

- Web Accessibility Initiative (WAI)—http://www.w3.org/WAI/
- U.S. Government Section 508 Website Accessibility Guidelines—http://www.section508.gov/index.cfm
- The Society for Technical Communication Accessibility Special Interest Group—http://www.stcsig.org/sn/internet.shtml
- National Center for Accessible Media—http://ncam.wgbh.org/
- International Web Accessibility Concerns—http://www.stcsig.org/sn/international.shtml

Criteria for a Successful Web Site

Follow these criteria when creating your site.

Home Page

The home page on a Web site is your welcome mat (Figure 14.3). It is the first thing your reader sees; thus, the home page sets the tone for your site. A successful home page should consist of the following components:

Identification Information.   Who are you? What is the name of your company, product, or service? How can viewers get in touch with you if they want to purchase your product? A good home page should clearly name the company, service, or product. In addition, a successful home page should provide the reader access to contact information, such as a corporate phone number, an e-mail address, a fax number, and an address.

Graphics.   Don't just tell the reader who or what you are. Show the reader. An informative, attractive, and appealing graphic depicting your product or service will convey more about your company than words. Check out various Web sites to prove this point. At the writing of this textbook, McDonald's Web site (www.mcdonalds.com) creates a family-friendly environment, complete with a sparkling golden arch and delicious food images. iPod's home page (http://www.apple.com/itunes/) shows its iPod touch, with the tag line "Next level fun." Starbuck's Web site (www.starbucks.com) reads "People in 156 countries joined together to sing all you need is love." This is accompanied by photographs of people from around the world and links to Starbuck's "Community" (MyStarbucksIdea.com, Starbucks V2V, Starbucks Shared Planet, Twitter, Facebook, and YouTube Channel). Your home page's graphic, like a corporate logo, should represent your company's product, service, and ideals.

Lead-In Introduction.   In addition to a graphic, provide a phrase or sentence that tells your reader who and what you do. What's your company's focus? To clarify its goals to prospective clients, Creative Courseware (http://www.creativecourseware.com/) adds the following lead-in introduction on its Web site homepage.

FIGURE 14.3 Web Site Home Page with Lead-in Information

Source: Courtesy of George Butler Associates, Inc.

> Creative Courseware helps employers **hire, train, develop,** and **manage** their employees within the values and goals of the organization.

◀ EXAMPLE

Navigation Bar.   By providing the reader with hypertext links, the home page's navigation bar acts as an interactive table of contents or index. The reader selects the topic, clicks on that link, and jumps to a new screen. There is no one way to present these hypertext links on the home page. You could provide a vertical or horizontal navigation bar listing your hypertext links.

Linked Pages

Once your readers click on the hypertext links from the home page, they will jump to the designated linked pages. These linked pages should contain the following information.

Headings and Subheadings.   Readers click on the home page link and arrive at a new screen. Will readers know where they are now? Or will they be lost in cyberspace? To ensure that readers know where they are in the context of the Web site, you need to use headings. These give the readers visual reminders of their location. Successful headings on linked pages are in the same location, font size, and font type. Consider creating a story-board so you can envision the linked pages for the Web site. See Figure 14.4 for a sample storyboard.

Development.   As in all correspondence, you should develop ideas thoroughly. Answer the reporter's questions (who, what, when, where, why, and how). In addition, avoid vague

FIGURE 14.4   Web Site Storyboard

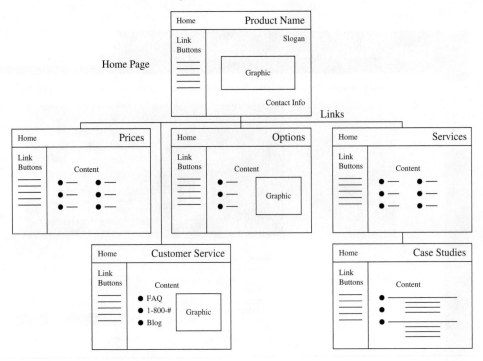

See Chapter 3 for more discussion of reporter's questions.

Gathering Information

words and phrases, like "some cities," "several products," or "quality customer support." Specify instead by writing "we're located in Miami, Atlanta, Los Angeles, and Houston"; "we offer over three dozen product accessories"; and "our customer support is available 24/7 at www.helpforce.net." Each linked page will develop a new idea. Prove your points precisely.

Navigation

If readers are viewing one screen but want to return to a screen they looked at before, how do they get there? There are no pages to turn. You need to help the readers navigate through cyberspace. You can do this in two ways:

- **Home buttons.** Readers must be able to return to the home page easily from any page of a Web site. Remember, the home page acts as a table of contents or index for all the pages within the site. By returning to the home page, readers can access any of the other pages. To ensure this easy navigation, provide a hypertext-linked Home button on each page.
- **Links between Web pages.** Why make readers return to the home page each time they want to access other pages within a site? If each page has a navigation bar with a hypertext link to all pages within the site, then readers can access any page, in any order of discovery.

See Figure 14.5 for an example of a successful linked page.

Document Design

You can do amazing things on a Web site. You can add distinctive backgrounds, colored fonts, different font faces and sizes, animated graphics, frames, and highlighting techniques (such as lines, icons, and bullets). However, just because you can doesn't mean you should. Document design should enhance your text and promote your product or service, not distract from your message.

FIGURE 14.5   MoDOT "Business" Page Link

MoDOT logo
used as a
"Home" button

Banner color
coded to match
heading

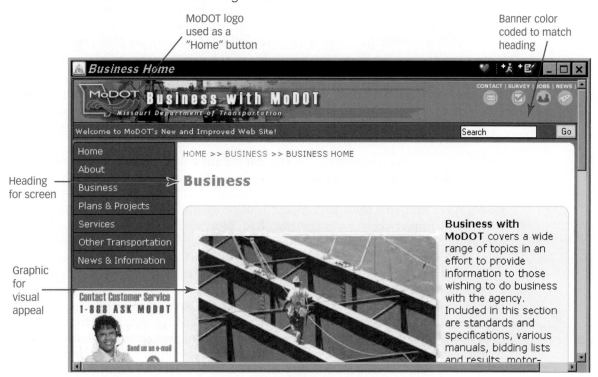

Heading
for screen

Graphic
for
visual
appeal

Source: Reprinted with the permission of the Missouri Department of Transportation.

Background.   On Web sites, you can add backgrounds, running the gamut from plain white, to various colors, to an array of patterns: fabrics, marbled textures, simulated paper, wood grains, grass and stone, psychedelic patterns, water and cloud images, and waves. They all look exciting. However, they are not all effective. When choosing a background, you want to consider your corporate image. If you are an engineering company, do you want a pink background? Perhaps a subtle stone or brick background would be more effective in portraying your corporate identity. If you are a child care center, do you want a dark background? Maybe this will be too depressing. In contrast, a white background with toys as a watermark might more successfully convey your center's mission statement.

In addition, remember that someone will attempt to read your Web site's text. Very few font colors or styles are legible against a psychedelic background—or against most patterned backgrounds, for that matter. Instead, to achieve readability, you want the best contrast between text and background. Despite the vast selection of backgrounds at your disposal, the best contrast is still black text on a white background.

Font.   It can be difficult to read text online. Text tends to flicker, glare impedes readability, and we can get eye strain. Thus, font selection is very important. Consider these suggestions (Williams 389):

- Use sans serif fonts, like Arial, or one specifically designed for the Web—Verdana. Conventional wisdom has always suggested that serif fonts, like Times New Roman, work best for correspondence. However, online, a serif (the "foot" at the bottom of text) displays irregularly on low-resolution screens.
- Use 12- to 14-point type, the size that most readers with normal vision find easiest to read. A smaller font will cause problems for readers. Then use a larger font for headings.

- Avoid overusing bold and italics. First, these should only be used for emphasis and then used sparingly. Boldface is often poorly formed on-screen because it must add pixels to a letter. Italics, due to the oblique orientation, do not always work well with the horizontal and vertical orientation of the pixel grid.
- Avoid using underlining, which readers could confuse with a hypertext link.
- Avoid using all caps. As with e-mail, text typed in all caps makes you look as if you are shouting. In addition, studies show that we read lowercase text about 13 percent faster than we read text in all caps. Uppercasing, thus, will add additional reading problems for your audience, already stressed by the challenges of reading online.

Color.   When you create a Web site, you can use any color in the spectrum, but do you want to? What corporate image do you have in mind? The font color should be suitable for your Web site. A yellow font on a blue background will cause headaches for your reader. A red font on a black background is nightmarish. The issue, however, is not just esthetics. A primary concern is contrast. Red, blue, and black font colors on a white background are very legible because their contrast is optimum. Other combinations of color don't offer this contrast; thus, the reader will have trouble deciphering your words.

In addition, when you use color on your Web site, the color you see on your monitor will *not* necessarily be the color your reader sees. As a webmaster, you cannot solve this problem because you cannot control other people's monitors. All you can do is try to reduce the problem. First, use restraint with color. Stick with black, red, and blue font color for text. Variations such as salmon, lime green, cornflower blue, medium aquamarine, and goldenrod won't be true. Next, test your use of color on several different monitors. You may be surprised at what you see, but now you can change your font color if needed for clarity (Eddings et al.).

Graphics

See Chapter 10 for more discussion of figures and tables.

Graphics.   Graphics affect download time. For example, a "small" graphic (approximately 90 × 80 pixels, 1½ × 1½ inches) consumes about 2.5 kilobytes. A "medium" graphic (approximately 300 × 200 pixels, 5 × 4 inches) consumes about 25 kilobytes. A "large" graphic (approximately 640 × 480 pixels, 10 × 8 inches) consumes about 900 kilobytes. When you use graphics with varying amounts of color depth, your file size increases. The larger your file, the longer it will take for the images to load. The longer it takes for your file to load, the less interested your reader will be in your Web site. Limit the size and number of your graphics.

Furthermore, use suitable graphics. If you work for an accounting firm, do you want graphics depicting a smiling calculator operated by an oversized flamingo? Create and insert your graphics carefully.

Highlighting Techniques.   Font sizes and styles, lines, icons, bullets, frames, Java applets, animation, video, audio—you can do it all on your Web page, but should you? Nothing is more distracting than a Web site full of highlighting techniques. Avoid overwhelming your Web site with an excessive number of design elements.

Style

Conciseness is important in all technical correspondence. Conciseness is even more important in a Web site. A successful Web page should be limited to one viewable screen, and a line of text should rarely exceed two-thirds of the screen. On a Web site, use pronouns, contractions, and positive words. Create a personalized tone that engages your readers.

Grammar

Your Web site represents your company; you don't want prospective clients to perceive your company as lacking in quality control. In fact, that's what poor grammar online denotes—poor attention to detail. It's easy to see why many sites are grammatically flawed. First, anyone can go online almost instantaneously. All you need to do is save a

Effective Online Highlighting	Ineffective Online Highlighting
1. *Lines* (horizontal rules) can separate headings and subheadings from the text.	1. *Frames* are considered to be one of the worst highlighting techniques. Jesse Berst, editorial director of ZDNet, considers frames to be one of the "Seven Deadly Web Site Sins." He says, "Too many frames . . . produce a miserable patchwork effect." If you want to achieve the frame look, but without the hassle, use tables instead. They are easier to create and revise, load faster, and are less distracting to your readers.
2. *Bullets and icons* enliven your text and break up the monotony of wall-to-wall words.	2. *Italics* and *underlining*. Both are hard to read online, and underlining looks like a hypertext link.
3. *First- and second-level headings*, achieved by changing your font size and style, separate key ideas.	3. *Java applets* take a long time to load.
4. *Boldface* also emphasizes important points and headings.	4. *Video* requires an add-on users must download before they can enjoy your creation.
5. *White space,* created by indenting, makes text more readable.	5. *Animation* can be sophomoric and distracting.

DOT.COM UPDATES

For Web design tips, check out these sites:

- Website Tips (http://www.websitetips.com/), complete with articles, tutorials, and resources.
- Grantastic Design (http://www.grantasticdesigns.com/tips.html) complete with design tips, rules for effective web layout, and readability and usability ideas.
- Web Design Features (http://www.ratz.com/features.html), complete with discussion of good Web design ideas and examples of problems associated with bad Web designs.
- HTML Goodies (http://www.htmlgoodies.com/), one of the most complete sites for all kinds of Web design hints.

Word document as an HTML file and transfer the file to your Web server. Speed is great for business, but it's bad for proofreading. Good proofreading takes time. When building your Web site, you must be vigilant. The integrity of your Web site demands correct grammar.

TECHNOLOGY TIP

Using Microsoft Word 2010 to Create a Web Site

You can use Microsoft Word 2010 to create your Web site. Follow these steps to do so:

1. From the **Insert** tab, insert a 3 × 2 table.

(Continued)

2. Merge the top right two columns.

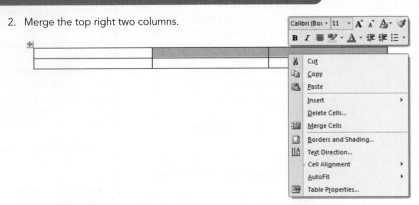

3. Place your cursor in the right rows and press the **Enter** key to add space. At this point, you have the beginning design for a Web page.
4. Add graphics, text, color, and font types just as you would with any Word document.

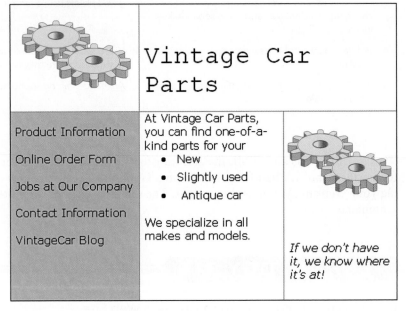

5. To create hypertext links, highlight selected text, right click, and scroll to **Hyperlink**.

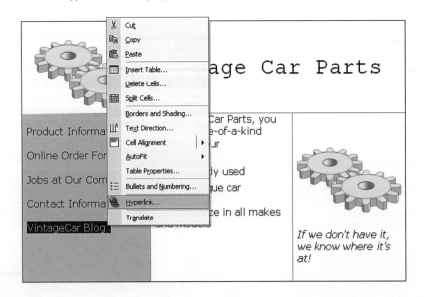

6. To create your Web site, click on the Word 2010 **File** tab and scroll to **Save As**.

7. In the pop-up **Save As** window, click on the **Save as type** down arrow and click on **Web Page**. This saves your work as an HTML (hypertext markup language) file which you can open in a Web browser.

WEB SITE USABILITY CHECKLIST

Audience Recognition and Involvement

_____ 1. Does the Web site meet your reader's needs?

_____ 2. Does the Web site give your audience a reason to return (tutorials, tips, comics, links to other interesting sites, regular updates, social media links, and so on)?

_____ 3. Does the Web site involve the audience by asking for user feedback, displaying customer comments in a blog, providing an FAQ page, or providing links to the company's Twitter or Facebook site?

_____ 4. Does the Web site make it easy for your audience to purchase online?

_____ 5. Does the Web site use pronouns to engage the reader?

Home Page

_____ 1. Does the home page provide identification information (name of service or product, company name, e-mail, fax, city, state, street address, and so on)?

_____ 2. Does the home page provide an informative and appealing graphic that represents the product, service, or company?

_____ 3. Does the home page provide a welcoming and informative introductory phrase, sentence, or paragraph?

_____ 4. Does the home page provide hypertext links connecting the reader to subsequent screens?

Linked Pages

_____ 1. Do the linked pages provide headings clearly indicating to the readers which screen they are viewing?

_____ 2. Do the linked pages develop ideas thoroughly (appropriate amount of detail, specificity, valuable information)?

_____ 3. Are the linked pages limited to one or two primary topics?

(Continued)

(Continued)

Navigation

_____ 1. Does the Web site allow for easy return from linked pages to the home page?

_____ 2. Does the Web site allow for easy movement between linked pages?

_____ 3. Does the Web site provide access to other Web sites for additional information?

Document Design

_____ 1. Does the Web site provide an effective background, suitable to the content and creating effective contrast for readability?

_____ 2. Does the Web site use color effectively, in a way suitable to the content and creating effective contrast for readability?

_____ 3. Does the Web site use a consistent document design (colors, background, graphics, font), carried throughout the site?

_____ 4. Does the Web site use headings and subheadings for easy navigation?

_____ 5. Does the Web site vary font size and type to create a hierarchy of headings?

_____ 6. Does the Web site use graphics effectively—suitable to the content, not distracting, and load quickly?

_____ 7. Does the Web site use highlighting techniques effectively (white space, lines, bullets, icons, audio, video, font size and type) in a way that is suitable to the content and not distracting?

Style

_____ 1. Is the Web site concise?

- Short words (1–2 syllables)
- Short sentences (10–12 words per sentence)
- Short paragraphs (4 typed lines maximum)
- Line length limited to approximately two-thirds of the screen (40–60 characters per line)
- Text per Web page limited to one viewable screen, minimizing the need to scroll

Accuracy

_____ 1. Does the Web site avoid grammatical errors?

_____ 2. Does the Web site ensure that information is current and correct?

_____ 3. Do the internal and external links work?

The Writing Process at Work

Web Resources

www.pearsonhighered
.com/gerson
For more information about the writing process, visit our companion website.

Shannon Conner, freelance technical writer and Web designer, used the writing process to create an effective and dynamic Web site for his wife's photography business.

The Writing Process

Prewriting	Writing	Rewriting
• Use the reporter's questions to clarify what content you will need to include.	• Create a navigation bar and home buttons.	• Revise your draft by

• Determine whether your audience is high tech, low tech, or lay. This will help you decide which abbreviations and acronyms need to be defined.	• Use headings and subheadings for easy navigation in cyberspace.	• Adding details
		  • Deleting wordiness
		  • Simplifying words

• Visualize the Web site's layout by storyboarding.	• Consider enhancing the Web site with elements of document design.	• Enhancing the tone
		  • Reformatting your text
		  • Proofreading and correcting errors

• To create focus for the Web site, use a statement of purpose.		• Testing for usability

Prewriting

See how Shannon used prewriting to begin his Web site.

To plan his Web site, Shannon considered the following:

- Goal—provide policy and procedures for the photography studio's Web site
- Audience—existing and potential photography studio clients
- Channels—statement of purpose for the Web site
- Data—policy and procedures including fees, ordering, pricing structure, and so on

Shannon used a Statement of Purpose to define the audience, communication channel, and shelf life of the content for the Web site. Figure 14.6 shows how Shannon used a Statement of Purpose to plan his Web site.

FIGURE 14.6 Statement of Purpose Used for Prewriting

Thank you for choosing J Conner Photography!

Session Fee
- Session fees are $125. This covers Jeanne's time, talent, and equipment and does not include any prints. Session fees are due at time of booking. Due to her busy schedule, session dates are only held once the session fee has been received.

Your Session
- Your session is individualized to fit your needs. I do not rush you through the session. We will take our time to get the best image possible. For small children and newborns, this may mean taking short breaks for feeding and snacks.

Ordering
- A minimum of 15 proofs will be provided in a private, password-protected online gallery within one week of your session date. J Conner will email you a link to your gallery. You're welcome to forward your gallery to family and friends.

Print Prices

Wall Portraits (mounted and coated)	Gift Prints (coated)	
30 × 40 $400	8 × 10	$40
20 × 24 $275	5 × 7	$25
16 × 20 $170	8 Wallets	$20
Add $150 for gallery-wrapped canvas		

(Continued)

FIGURE 14.6   Statement of Purpose Used for Prewriting (*Continued*)

Boutique

- Boutique items are available after a minimum purchase of $300, unless otherwise noted.
- Session Album - $100
- A small 5x7, spiral bound album that includes watermarked proofs from your session. It's perfect for carrying around to show to friends and family.

Final prints

- Prints will be delivered in 2–3 weeks. Larger images, canvas prints, and specialty items may take 4–6 weeks. J Conner Photography will contact you for the best time to deliver the prints. All sales are final.

Reprints

- Reprints may be ordered up to one year from the session date. They are subject to the $45 re-hosting fee and a minimum order of $250 applies.

Copyright Information

- All images copyrighted by photographer, Jeanne Conner.

Referrals

- Thank you for your business. As a small business owner, I rely heavily on referrals and word-of-mouth advertising. As a thank you for spreading the word, I'd like to offer you a $20 print credit for each referred client who completes a session.

Writing

You next have to write a draft of every Web page. Figure 14.7 shows Shannon's rough draft of a Web page.

Rewriting

Rewriting is the final part of the writing process. Using input from Jeanne, Shannon revised and reformatted the Web page considering how the design would be effective for the audience. Figure 14.8 is Shannon's revised Web pages.

FIGURE 14.7 Webpage Mockup with Input from the Client

Thank you for choosing J Conner Photography! Please take a moment to look through our policies and procedures. If you have any questions, please contact Jeanne at jconner@jconnerphotography.com.

Session Fee

Session fees are $125. This covers Jeanne's time, talent, and equipment and does not include any prints. Session fees are due at time of booking. Due to her busy schedule, session dates are only held once the session fee has been received.

Your Session

Your session is individualized to fit your needs. I do not rush you through the session. We will take our time to get the best image possible. For small children and newborns, this may mean taking short breaks for feeding and snacks.

Ordering

A minimum of 15 proofs will be provided in a private, password-protected online gallery within one week of your session date. J Conner will email you a link to your gallery. You're welcome to forward your gallery to family and friends. The ordering process is secure. Cash and credit/debit payments are made through the PayPal system.

Print Prices

Wall Portraits (mounted and coated)	Gift Prints (coated)
30 × 40 $400	8 × 10 $40
20 × 24 $275	5 × 7 $25
16 × 20 $170	8 Wallets $20
Add $150 for gallery-wrapped canvas	

Boutique

Boutique items are available after a minimum purchase of $300, unless otherwise noted. J Conner Photography also offers custom sterling silver photo jewelry and photo purses and handbags. Due to the handmade nature of these items, ask about current prices, styles, and availability.

Jeanne said to Shannon, "I see why you wanted me to pick the most important ideas for the potential customer to take away from the Web site. Let's make the Web site simpler, with far fewer words, and use lots more linked pages so that the customer can focus. I want the Web site to be as elegant as possible and illustrate some of my most treasured photographs so the customers can see what they will get when they hire me."

FIGURE 14.8 Revised Home Page and Linked Page

Home Page

(Continued)

FIGURE 14.8   Revised Home Page and Linked Page (*Continued*)

Linked Page

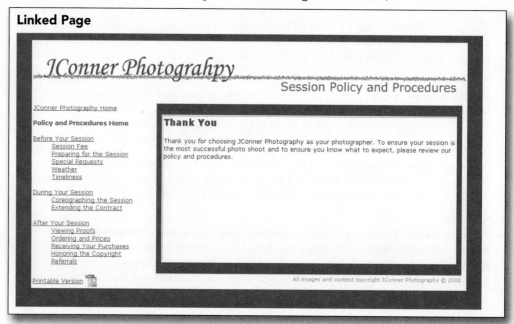

CHAPTER HIGHLIGHTS

1. To be an effective communicator in business and industry today, you have to communicate electronically.

2. A successful Web page should limit lines of text to perhaps two-thirds of the screen, with a graphic, white space, or hypertext links placed in the remaining third.

3. Hard-copy documentation follows page-by-page layout, whereas online text is nonlinear.

4. E-readers are topic specific and want information quickly.

5. A good home page should clearly name the company, service, or product. In addition, a successful home page should provide the reader access to contact information: a corporate phone number, an e-mail address, a fax number, and an address.

6. Online help systems allow the technical communicator to create interactive training tools and informational booths within a document. These online systems can be created using a wide variety of authoring tools. Some popular ones include Adobe RoboHelp, Author-it, Doc-To-Help, DocArchitector, HelpScribble, HyperText Studio, MadCap Flare, and Sandcastle.

7. Interactive online help systems provide "just in time" and "as needed" information to help your audience complete a task.

8. Online help screens provide the practical value of portability. Rather than take up computer archival space with thousands of pages of PDF files or print out hundreds of pages of instructional help, online help screens allow companies to save space and provide readers ease of use.

9. The usability checklist can ensure that you create an effective Web site.

10. The process approach to writing applies to all technical communication including Web site construction.

CASE STUDIES

1. Future Promise is a not-for-profit organization to help at-risk high school students. This agency realizes that to reach its target audience (teens aged 15–18), it needs an Internet presence.

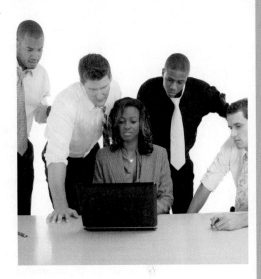

   To do so, it has formed a 12-person team, consisting of the agency's accountant, sports and recreation director, public relations manager, counselor, technical writer, graphic artist, computer and information systems director, two local high school principals, two local high school students, and a representative from the mayor's office. Jeannie Kort, the PR manager, is acting as team leader.

   The team needs to determine the Web site's content, design, and levels of interactivity. Jeannie's boss, Brent Searing, has given the team a deadline and a few components that must be included in the site:

   - College scholarship opportunities
   - After-school intramural sports programs
   - Job-training skills (resume building and interviewing)
   - Service learning programs to encourage civic responsibility
   - Future Promise's 800-hotline (for suicide prevention, STD information, depression, substance abuse, and peer counseling)
   - Additional links (for donors, sponsors, educational options, job opportunities, etc.)

   Jeannie has a big job ahead of her.

Assignment

Form a team and design Future Promise's Web site. To do so, follow the criteria for Web design provided in this chapter. Then, prewrite, write, and rewrite as follows:

Prewrite:

- Research the topics listed above (either in a library, online, or through interviews) to gather details for your Web site.
- Consider the audience and their respective needs and interests.
- Focus on your Web site's purpose—are you writing to inform, persuade, instruct, or build rapport?
- Draw a storyboard of how you would like the Web site to look or use an organizational chart to lay out the Web design.
- Divide your labors among the team members (who will research, write, create graphics, etc.?).

Write:

- Draft your text in a word processing program (you can save as an HTML file).

Rewrite:

- Review the Web site criteria in this chapter.
- Add, delete, simplify, move, enhance, and correct your Web site.
- Test the site to make sure all links work and that it meets your audience's needs.

2. The following report recommends that upper management approve the construction of a corporate Web site. The company's management has agreed with the recommendation report. You and your team have received a memo directing you to build the Web site. Using the report, build your company's Web site.

Date: November 11, 2012
To: Reuben Ortega, CEO
From: Sandy Warner, Information Technology Manager
Subject: Recommendation Report for New Corporate Web Site

Studies show that most companies with an online presence increase their corporate profits. Our company, Java Lava, is lagging behind. We need to create a corporate Web site to maximize our sales opportunities. Please consider the following facts.

1. **International bean sales:** Currently, coffee bean sales account for only 27 percent of our company's overall profit. These coffee bean sales depend solely on walk-in trade. The remaining 72 percent stems from over-the-counter beverage sales. If we go global with a Java Lava Web site, research shows that we can expand our coffee bean sales by almost 15 percent. Potential clients from every continent will be able to order our coffee beans online.

2. **International promotional product sales.** Mugs, T-shirts, boxer shorts, jean jackets, leather jackets, key chains, paperweights, and calendars could be marketed online. Currently, we give away items imprinted with our logo. By selling these items online, we could make money while marketing our company.

3. **International franchises.** We now have three coffeehouses, located at 1200 San Jacinto, 3897 Pecan Street, and 1801 Paloma Avenue. Let's franchise. On the Internet, we could offer franchise options internationally.

4. **Online employment opportunities.** Once we begin to franchise, we'll want to control hiring practices. This will ensure that Java Lava's standards are met. Through a Web site, we could post job openings internationally and list job requirements. Then potential employees could submit their resumes online.

In addition to this information, used to increase our income and expand our brand, we could provide the following:

- A map showing our three current sites.
- Our company's history—founded in 1976 by Hiram and Miriam Coenenburg, with money earned from their import/export bean business.
- Sources of our coffee beans—Guatemala, Costa Rica, Colombia, Brazil, Sumatra, France, and the Ivory Coast.
- Freshness guarantees—posted ship dates and ground-on dates; 100 percent money-back guarantees.
- Corporate contacts—street and city addresses, phone numbers, e-mail address, and fax number.
- Social media options—today's client wants personal contact with companies. If we provided a Facebook and Twitter link on our Web site, we can achieve greater involvement with our customers.

Coffee is a "hot" commodity now. The international market is ours for the taking. We can maximize our profits and open new venues for expansion. I'm awaiting your approval for this exciting new venture.

INDIVIDUAL AND TEAM PROJECTS

Online Help

1. Study the help menus in your word processing package. Distinguish between the pop-up windows and the hyperlinks. What are the differences? What are the similarities?
2. In a small group, have each individual ask a software-related question regarding a word processing application. These could include questions such as "How do I print?" or "How do I set margins?" Then, using a help menu, find the answers to these questions. Are the answers provided by way of pop-up windows or hyperlinks? Rewrite any of these examples to improve their conciseness or visual layout.
3. Take an existing document (one you have already written in your technical writing class or writing from your work environment) and rewrite it as online help. To do so, find key words, phrases, or concepts that need to be defined. Then, expand on them as either pop-up windows or hyperlinks.
4. Using our suggestions in this chapter, rewrite the following text as an effective online help screen. To do so, reformat the information for access on a smaller screen. Next, determine which of the words, phrases, or concepts can be expanded with a short pop-up window and which require a longer hyperlink.

Because your company is located in the East Side Commercial Park, the outside power is subject to surges and brownouts. These are caused when industrial motors and environmental equipment frequently start and stop. The large copiers in your office also cause power surges that can damage electronic equipment.

As a result of these power surges and outages, your three network servers have become damaged. Power supplies, hard drives, memory, and monitors have all been damaged. The repair costs for damages totaled $55,000.

Power outages at night also have caused lost data and failed backups and file transfers. Not being able to shut down the servers during power failures is leading to system crashes. Data loss requires time-consuming tape restorations. If the failures happen before a scheduled backup, the data are lost, which requires new data entry time. Ten percent of your data loss can not be restored. This will lead to the loss of customer confidence in the reliability of your services.

Finally, staff time losses of $40,000 are directly related to network downtime. Data-entry staff are not able to work while the servers are being repaired. These staff also must spend extra time reentering lost data. Fifteen percent of the data management staff time is now spent recovering from power-related problems.

To solve your problems, you need to install a BACK-UPS 9000 system. It has the configuration and cost-effectiveness appropriate for your current needs.

Web Sites

1. Create a corporate Web site. To do so, make up your own company and its product or service. Your company's service could focus on dog training, computer repair, basement refinishing, vent cleaning, Web site construction, child care, auto repair, personalized aerobic training, or online haute cuisine. Your company's product could be paint removers, diet pills, interactive computer games, graphics software packages, custom-built engines, flooring tiles, or duck decoys. The choice is yours. To create this Web site, follow our writing process.

2. Create a Web site for your technical communication, business writing, or professional writing class. To create this Web site, follow our writing process.

3. Create a Web site for your high school, your college, your fraternity or sorority, your church or synagogue, or your professional organization. To create this Web site, follow our writing process.

4. Create a personal Web site for yourself or for your family.

5. Research several Web sites, either corporate or personal. Use our Web Site Usability Checklist to determine which sites excel and which sites need improvement. Then write a report justifying your assessment. In this report, clarify exactly what makes the sites successful. To do so, you could use a table, listing effective traits and giving examples from the Web sites to prove your point. Next, explain why the unsuccessful sites fail. Finally, suggest ways in which the unsuccessful sites could be improved.

PROBLEM-SOLVING THINK PIECES

1. FlyHigh Travel specializes in "adventure trips . . . in the air, sea, and land." If it's shark watching off the Barrier Reef, paragliding from a cliff in Acapulco, or mountain climbing in Nepal, FlyHigh is the traveler's answer. FlyHigh books adventures like feeding stingrays in the Caribbean, animal photo shoots in Kenya, and kayaking down the Colorado River. FlyHigh also makes travel arrangements for scuba diving in Hawaii, spelunking in French caves, and skydiving anywhere in the world.

   To reach as wide an audience as possible, FlyHigh is building its Web site (www .WeFlyHigh.com). FlyHigh's CEO wants this Web site to include trip information, testimonials from satisfied travelers, pricing, and photos that highlight each trip's excitement.

Assignment

Create the Web site for FlyHigh including the following content:

- Specific trip information travelers need to plan their trips (costs, accommodations, options, and so on)
- Information international customers will need (passports, visas, or inoculations)
- Information for FlyHigh's clients who travel with families (alternative, age-specific activities)
- Information for FlyHigh's "fit and fearless" clients whose travels will involve extreme sports

2. Web site design is challenging. Some Web sites are outstanding; others are not as well designed. Check out http://www.webpagesthatsuck.com/ and http://www.webpractices.com/samplesites.htm, two sites that assess flawed Web pages. Do you agree with their assessments? Write a memo to your instructor or give an oral presentation explaining your decisions.

WEB WORKSHOP

1. Access any company's Web site and study the site's content, layout (color, graphics, headings, use of varying font sizes and types, etc.), links (internal and external), ease of navigation, tone, and any other considerations you think are important. Then, determine how the Web site could be improved if you were the site's webmaster. Once you have made this determination, write a memo or e-mail message recommending the changes that you believe will improve the site. In this memo or e-mail,

   - Analyze the Web site's current content and design, focusing on what is successful and what could be improved.
   - Recommend changes to improve the site.

2. Audience recognition involves a reader's level of knowledge as well as gender, religion, age, culture, and language. No communication channel allows for a more diverse audience than the Internet, where anyone can click on a link and enter a site.

   Still, effective Web sites recognize and appeal to specific audiences. To better understand how successful workplace communication recognizes and involves audience, click on the following sites:

   - http://www.mcdonalds.com/
   - http://www.gap.com/
   - http://www.harley-davidson.com/

   Who are the intended audiences for each Web site, and what techniques do the Web sites use to involve and recognize their unique audience types?

QUIZ QUESTIONS

1. What are the characteristics of e-readers?
2. How does online communication differ from hard-copy documents?
3. Why is the use of online help growing?
4. How can you create effective online help screens?
5. How can you achieve a pleasant tone in your help screen?
6. In what three ways can you design an effective screen?
7. What are characteristics of successful Web sites?
8. Why should you avoid "noise" on your Web site?
9. What information is usually included on a home page in a Web site?
10. What information should be included in a linked page, and how is navigation important?

CHAPTER FIFTEEN

Summarizing Technical Communication

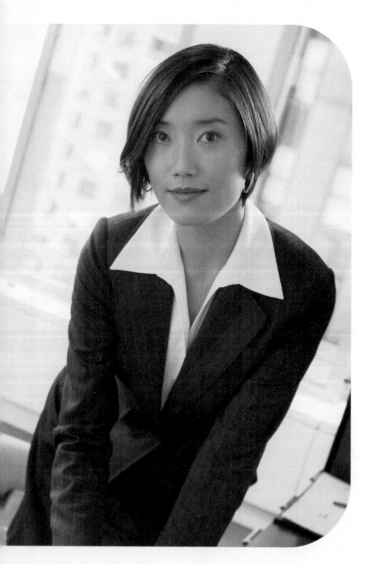

COMMUNICATION AT WORK

The "Deer Creek Construction" scenario shows how poor communication, both oral and written, can negatively impact a company's success and how a consulting firm used summaries to share researched findings.

Deer Creek Construction had a wonderful reputation in the city, county, and region. The company was known for providing construction services—primarily roads, bridges, on- and off-ramps, and cloverleafs—that were built on time and within budget. Both civic leaders and community end users praised Deer Creek's quality. However, lately, Deer Creek has not been getting selected for new projects. Their proposals continually were coming in second and third to competitors.

To determine why this was occurring, Deer Creek's management hired a consulting firm. This firm studied Deer Creek's written proposals and oral presentations. The consultants also performed primary research by interviewing clients who had worked with the construction company. Finally, the consulting firm conducted secondary research on best practices for proposal writing.

After doing so, the consulting firm found that Deer Creek's written proposals were as follows:

Objectives

When you complete this chapter, you will be able to

1. Understand why you write summaries
2. Apply the criteria for effective summaries
3. Use an appropriate style for writing summaries
4. Write an objective summary
5. Use the checklist to evaluate your summary
6. Follow the writing process to write a summary

- Stiff and dry, instead of creating a more personalized tone that achieved audience involvement.
- Gender insensitive, using phrases like "when an engineer visits your site, *he* will. . . ."
- Visually unappealing, leaning too heavily on long, dense paragraphs instead of graphics and other highlighting techniques.
- Poorly organized with little attention to placement of content for ease of access and readability.

Furthermore, the consulting firm found that Deer Creek's presenters had the following problems:

- Used hand-written flip charts instead of more professional PowerPoint slides.
- Fidgeted and slumped against the lecterns while they spoke.
- Read text to the audience.

- Failed to engage the audience through conversational discussion or Q and A.
- Did not provide an abstract or summary of the written material for the audience.

The consultants also discovered from their research that Deer Creek was not adhering to several best practices for proposal writing.

Despite Deer Creek's quality work, the firm was losing business due to poor communication skills: oral, written, and nonverbal. Something had to change—*fast*. To share their findings with Deer Creek, the consultants wrote a summary of the problems the company was experiencing, based on their research. By presenting a summary of their findings to management, the consulting firm was able to show how to correct the problems with proposals and presentations. The consulting firm helped to ensure that Deer Creek's proposals would be accepted by potential customers.

Web Resources

www.pearsonhighered
.com/gerson For more
information about
summaries, visit our
companion website.

Why Write Summaries?

You might be required to write a summary if, for example, your boss is planning to give a presentation at a civic meeting (Better Business Bureau, City Council, or Businesswomen's Association) and needs some up-to-date information for his or her presentation. Bosses often are too busy to perform research or attend a conference at which new information might be presented. Therefore, your boss asks you to research the topic and provide a summary of your findings.

Similarly, your boss might need to give a briefing to upper-level management. Again, time is a problem. How can your boss get the appropriate information and digest it rapidly? The answer is for *you* to read the articles, attend the conference or meeting, and then summarize your research. At other times, you will be asked to summarize your own writing. For example, if you are writing a long report, you might include an executive summary, an overview of the report's key points.

Perhaps your boss is trying to decide on a course of action regarding new facilities, new software, new hiring practices, new client possibilities, or changes to existing procedures. Before he or she makes the decision, however, the boss has asked you to research the topic and then summarize your findings.

Finally, as a class assignment, your teacher might want you to write a summary of a formal research report. The summary, though shorter than a research report, would still require that you practice research skills, such as reading, note taking, and writing paraphrases. To write a summary, you'll need to study the research material. Then, in condensed form (a summary is no more than 5 to 15 percent of the length of the original source), you'll report on the author's main points. Preparing a summary puts to good use your research, analyzing, and writing skills.

Executive Summaries Versus Summaries

Executive Summaries

See Chapter 17 for more
discussion of executive
summaries.

An executive summary is found at the beginning of either a formal report or a proposal and summarizes the major topics covered in the document. By reading the executive summary, your audience gets an overview of the much longer report or proposal. Executive summaries can even take the place of the much longer formal report or proposal. In this way, you save your reader's time and communicate only the most relevant parts of the report or proposal.

Criteria for Writing Summaries

A well-constructed summary, though much shorter than the original material being summarized, highlights the author's important points. A summary is a condensed approach to writing. Although the summary will not cover every fact in the original, after reading the summary you should have a clear overview of the main ideas from the original speech or document.

Locating Your Periodical Article, Book Chapter, or Report

If you are summarizing a meeting or seminar, take notes. If your boss or instructor gives you the source of information, one major obstacle has been overcome. If, however, you need to visit a library or go online to research your topic, refer to Chapter 5 for helpful hints on research skills.

Works Cited or References Information

Once you've located your research material, immediately write down the author's name, the article's title, the name of the periodical or book in which this source was found, the date of publication, and the page numbers. Doing so will save you frustration later. For example, if you lose your copy of the article and have not documented the source of your information, you will have to repeat your research. Instead, spend a few minutes at the beginning of your summary documenting your sources. This will ensure that you do not waste time searching for the missing article. For correct documentation format, follow the CSE, APA, or MLA style sheets found in Appendix B.

Overall Organization

As in any well-written document, a summary contains an introduction, a discussion, and a conclusion.

Introduction.   Begin with a topic sentence clearly stating the author's main idea. The topic sentence will present the primary focus of the original source and list the two or three major points to be discussed. You must also tell your reader what source you are summarizing. You can accomplish this in one of three ways:

- List the author's name and article title in the topic sentence.
- Preface your summary with a works cited or references notation providing the author's name, title, publication date, and page numbers.
- Follow your topic sentence with a footnote and give the works cited or reference information at the bottom of the page.

Providing accurate documentation of the source is an ethical consideration and an important reflection of your professional skills.

◀ EXAMPLE

Name and Title in Topic Sentence

In "What the Literature Says About Using Game Worlds and Social Worlds in Cyberspace for Communicating Technical and Educational Content" (*Technical Communication*, August 2008), Marci Araki and Saul Carliner study the history of virtual worlds and analyze the basic differences between social and game worlds.

◀ EXAMPLE

Works Cited Prefacing Topic Sentence

Araki, Marci and Saul Carliner. "What the Literature Says About Using Game Worlds and Social Worlds in Cyberspace for Communicating Technical and Educational Content." *Technical Communication*. August 2008. 251–258. Print.
   Marci Araki and Saul Carliner study the history of virtual worlds and analyze the basic differences between social and game worlds.

◀EXAMPLE

Topic Sentence Followed by Explanatory Footnote

Marci Araki and Saul Carliner study the history of virtual worlds and analyze the basic differences between social and game worlds.*

_____

*This paragraph summarizes Marci Araki and Saul Carliner's article "What the Literature Says About Using Game Worlds and Social Worlds in Cyberspace for Communicating Technical and Educational Content." *Technical Communication*. August 2008. 251–258.

Paraphrasing

See Chapter 5 for more discussion of how to paraphrase.

Discussion. In this section, briefly summarize the main points covered in the original material. To convey the author's ideas, you can paraphrase, using your own words to restate the author's point of view. Organize the summary to parallel the author's method of organization (comparison/contrast, argument/persuasion, problem/solution, cause/effect, analysis, chronology, spatial, importance, and so forth). Also use transitional words and phrases to create a coherent and unified summary.

Conclusion. To conclude your summary, you can do one of the following:

- Reiterate the focus statement, reminding the reader of the author's key ideas.
- Highlight the author's conclusions regarding the topic.
- State the author's recommendations for future activity.

Internal Organization

How will you organize the body of your discussion? Because a summary is meant to be objective, you should present not only what the author says but also how he or she organizes the information. For example, if the author has developed ideas according to a problem/solution format, your summary's discussion should also be organized as a problem/solution. This would give your audience the author's content and method of presentation. Similarly, if the author's article is organized according to cause/effect, comparison/contrast, or analysis (classification/division), this would determine how you would organize your summary.

Organization

See Chapter 3 for more discussion of methods of organization.

Development

To develop your summary, focus on the following:

- **Most important points.** Because a summary is a shortened version of the original, you can't include all that the author says. Thus, you should include only the two or three key ideas within the article. Omit irrelevant details, examples, explanations, or descriptions.
- **Major conclusions reached.** Once you've summarized the author's key ideas, state how these points are significant. Show their value or impact.
- **Recommendations.** Finally, after summarizing the author's major points and conclusions, tell your audience if the author recommends a future course of action to solve a problem or to avoid potential problems.

Style

The summary, like all technical communication, must be clear, concise, accurate, and accessible. Watch out for long words and sentences. Avoid technical jargon unless the terminology is essential to the author's message. Most important, be sure that your summary reflects the author's content. Your summary must be an unbiased presentation of what the author states and include none of your opinions.

Length

As mentioned earlier, the summary will be approximately 5 to 15 percent the length of the original material. To achieve this desired length, omit references to the author (after the initial reference in the works cited or topic sentence). You also can omit material such as the following:

- Past histories
- Definitions (unless needed by the audience for understanding of concepts)
- Complex technical concepts
- Statistics
- Tables and figures
- Tangential information (such as anecdotes and minor refutations)

- Lengthy examples
- Biographical information

Deleting is especially important in a summary. Your major challenge is brevity. Therefore, review your draft to see if you have included any of the author's points that are not essential. Such nonessential information may include side issues that are interesting but not mandatory for your reader's understanding.

Another type of information you'll want to delete will be complex technical theories. Although such data might be valuable, a summary is not the vehicle for conveying this kind of information. If you have included such theories, delete them. These deletions serve two purposes. First, your summary will be stronger since it will focus only on key ideas and not on tangential arguments. Second, the summary will be more concise due to your deletions.

Delete biased comments. If you inadvertently have included any of your attitudes toward the topic, remove these biases. A summary should not present your ideas or your subjective comments and opinions; it should reflect only what the author says.

Finally, reread your draft and delete any unnecessary words and phrases. Strive for an average of 15-word sentences and one- or two-syllable words. If an author has written about telecommunications or microbiology, for example, you shouldn't simplify these words. You should, however, avoid long words that aren't needed and long sentences caused by wordy phrases.

Audience Recognition

Consider your audience when deciding whether to omit or include information. Although you usually would omit definitions from a summary, this depends on your audience. Technical communication is useless if your reader does not understand your content. Therefore, determine whether your audience is high tech, low tech, or lay. Then define your terms accordingly.

Audience

See Chapter 4 for more discussion of types of audiences.

Grammar and Mechanics

As always, flawed grammar and mechanics will destroy your credibility. Not only will your reader think less of your writing and research skills, but also errors in grammar and mechanics might threaten the integrity of your summary. Your summary will be inaccurate and, therefore, invalid.

Use the following Summary Checklist to help you write an effective summary.

SUMMARY CHECKLIST

_____ 1. Does your summary provide the works cited or reference information (documentation) for the article that you're summarizing?

_____ 2. Does your summary begin with an introduction clearly stating the author's primary focus?

_____ 3. Does your summary's discussion section explain the author's primary contentions and omit secondary side issues?

_____ 4. In the discussion section, do you explain the author's contentions through pertinent facts and figures while avoiding lengthy technicalities?

_____ 5. Is your content accurate? That is, are the facts that you've provided in the summary exactly the same as those the author provided to substantiate his or her point of view?

_____ 6. Have you organized your discussion section according to the author's method of organization?

_____ 7. Did you use transitional words and phrases?

_____ 8. Have you omitted direct quotations in the summary, depending instead on paraphrases?

_____ 9. Does your conclusion either reiterate the author's primary contentions, reveal the author's value judgment, or state the author's recommendations for future action?

_____ 10. Is your summary completely objective, avoiding any of your own attitudes?

_____ 11. Have you used an effective style, avoiding long sentences and long words?

_____ 12. Are your grammar and mechanics correct?

_____ 13. Have you avoided biased comments and omitted personal opinions?

The Writing Process at Work

To write a summary, follow the writing process of prewriting, writing, and rewriting. One of our students wrote a summary using the writing process.

The Writing Process

Prewriting	Writing	Rewriting

• Determine whether your audience is high tech, low tech, or lay. This will help you decide which abbreviations and acronyms need to be defined.
• Conduct your research and determine which information to summarize.
• Consider how your summary will be used by the audience.
	• Organize your content using modes such as problem/solution, cause/effect, comparison/contrast, argument/persuasion, analysis, chronology, etc. Your summary's organization will reflect the original document's organization.
	• Abide by rules for correct paraphrasing.
	• Document your sources correctly.
		• Revise your draft by
		  • Adding details
		  • Deleting wordiness
		  • Deleting biased comments and technical terms
		  • Simplifying words
		  • Reformatting your text
		  • Proofreading and correcting errors
		  • Determining if all material taken from the source is documented accurately

Prewriting

To prewrite the summary, the student made marginal notes as he read the article about virtual worlds.

Writing

After studying the article and his marginal notes, the student wrote a rough draft. The following is the student's rough draft with editing suggestions (Figure 15.1).

Rewriting

The student revised the draft of the summary as seen in Figure 15.2.

FIGURE 15.1   Rough Draft of Summary with Revision Suggestions

Araki and Carliner study the history of virtual worlds and analyze the basic differences between social and game worlds. They apply their analysis to the uses of both social and game worlds for communicating educational and technical content. Literature began being written in the early 1990s about game and social virtual worlds. The distinction between game worlds and social worlds is that game worlds require a participant (as an avatar) to role play to reach certain goals. In contrast, an avatar in a social virtual world such as Second Life shops, socializes, and develops and shares content. Both game and social virtual worlds have similar features including shared space where avatars share one environment and another feature of virtual worlds is GUI in 2D or 3D. A third distinguishing characteristic is that users interact in real time, and this interactivity allows users to roleplay with other participants. A fourth trait of the virtual world is that it exists 24/7 whether the individual is logged on or not and in the social and game virtual worlds, users socialize and form a sense of community with other participants.

Social worlds have some distinct characteristics that differ from game worlds including an environment that is open ended with no specific rules or a storyline that participants have to follow. There exists no distinct separation from the real world of rules; in contrast, specific rules and guidelines exist in the game world and in the social world avatars live in residences similar to homes and have human-type identities. These avatars also have jobs that are similar to the real world. Often avatars are involved in commercial activities just as is a real-world counterpart. Technical communicators might be able to use the virtual world to disseminate both technical and educational materials if they consider carefully the developers role in creating the virtual environment. Technical communicators must also consider how to share their content with the end user. Virtual communities of users of technical and educational content must also be created. Last, the technical communicator should create an avatar and then maintain a professional, legitimate reputation and identity with users to be effective.

Insert the source citation.

Correct the spelling error versus "role play."

Revise these paragraphs, making both shorter.

Revise this sentence to make it more concise.

Revise this sentence to make it more concise.

Correct the apostrophe error, writing "developer's."

FIGURE 15.2   Completed Summary

In "What the Literature Says About Using Game Worlds and Social Worlds in Cyberspace for Communicating Technical and Educational Content" (*Technical Communication,* August 2008), Marci Araki and Saul Carliner study the history of virtual worlds and analyze the basic differences between social and game worlds. Literature began being written in the early 1990s about game and social virtual worlds. The distinction between game worlds and social worlds is that game worlds require a participant (as an avatar) to role play to reach certain goals. In contrast, an avatar in a social virtual world such as Second Life shops, socializes, and develops and shares content. Colleges have begun offering courses in Second Life, and some organizations and businesses are involved in the virtual world to advertise and market products and services.

(Continued)

FIGURE 15.2 Completed Summary (Continued)

Both game and social virtual worlds have similar features including shared space where avatars share one environment. Another feature of virtual worlds is graphical user interface (GUI) in 2D or 3D. A third distinguishing characteristic is that users interact in real time, and this interactivity allows users to role play with other participants. A fourth trait of the virtual world is that it exists 24/7 whether the individual is logged on or not. Finally, in the social and game virtual worlds, users socialize and form a sense of community with other participants.

Social worlds have some distinct characteristics that differ from game worlds including an environment that is open ended with no specific rules or a storyline that participants have to follow. There exists no distinct separation from the real world of rules; in contrast, specific rules and guidelines exist in the game world. Finally, in the social world avatars live in residences similar to homes and have human-type identities. These avatars also have jobs that are similar to the real world. Often avatars are involved in commercial activities just as is a real-world counterpart.

Technical communicators might be able to use the virtual world to disseminate both technical and educational materials if they consider carefully the developer's role in creating the virtual environment. Technical communicators must also consider how to share their content with the end user. Virtual communities of users of technical and educational content must also be created. Last, the technical communicator should create an avatar and then maintain a professional, legitimate reputation and identity with users to be effective.

CHAPTER HIGHLIGHTS

1. A summary is a compressed version of a much longer document or speech.
2. You can summarize chapters, books, speeches, reports, material from the Internet, and more.
3. A well-written summary is about 5 to 15 percent the length of the original.
4. Include an introduction, a discussion, and a conclusion.
5. Include source citations when appropriate.
6. The writing process helps you create an effective summary.
7. Delete biased comments from a summary because a summary should not reflect your ideas.
8. Avoid using long words and sentences in a summary.
9. Organize your summary in the same way the author organized the original source.
10. In a summary, focus only on key ideas and not on tangential arguments.

CASE STUDY

Due to Deer Creek's problems getting new clients (refer to the scenario at the beginning of this chapter), Deer Creek management has decided to hire a public relations specialist/manager of corporate communication to do the following:

- Act as a liaison between Deer Creek's engineers, city/county/state government agencies, and end-user clients.
- Prepare proposals (gather information through meetings and write the reports).
- Prepare summaries of proposals for management as updates on potential new business.
- Make oral presentations to potential clients.

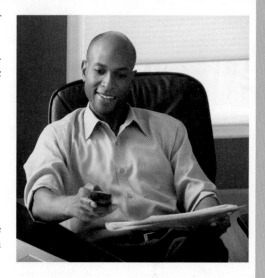

You have been asked to write the job description for this position, to be advertised in the local newspaper, in Deer Creek's employment office, and online in the company's Web site: DeerCreekConstruction.com.

Assignment

Research the position of public relations specialist in the *Occupational Outlook Handbook*. Find out the salary range for this job and educational requirements. What additional job responsibilities will this position entail? What skills should this public relations specialist have? Based on your research, write a summary of your findings. Your audience is your boss.

INDIVIDUAL AND TEAM PROJECTS

1. Locate an article that interests you (one within your field of expertise, your degree program, or an area that you would like to pursue). Study this article and summarize it according to the criteria provided in this chapter.
2. Locate an article that interests you. After reading it, take marginal notes (one- to three-word notations per paragraph) highlighting the article's key points. Then, write either a topic or a sentence outline (discussed in Chapter 5).
3. Read three to five articles. Then determine what method of organization the authors have used. Have the authors used analysis, problem/solution, comparison/contrast, argument/persuasion, cause/effect, and so on? Share your findings with your class or professor either in an oral presentation or e-mail message.
4. After attending a lecture, meeting, or conference, summarize its content. Provide the speaker's name, the location of the presentation, and the date of presentation for the source citation.

WEB WORKSHOP

1. Using the Internet, research 10 companies in your major field. Write a summary of the company's mission statement, types of products or services they provide, job opportunities, case studies, and so on.
2. Many companies provide annual stockholder's reports online. Find a company's online stockholder's report. Using the techniques discussed in this chapter, summarize the report.
3. Many companies issue press releases or news releases online to update their employees or stakeholders of important corporate information. Find a company's online press release. Using the techniques discussed in this chapter, summarize the press release.
4. Research articles on current issues, such as *outsourcing, telecommuting, globalization, multiculturalism, flextime, job sharing, virtual teams, best practices,* and *green technology.* Write a summary of the article using the techniques presented in this chapter.

QUIZ QUESTIONS

1. What constitutes a well-constructed summary?
2. Explain the function of a topic sentence in a summary.
3. In what ways can you organize a summary?
4. What types of material are omitted from a summary?
5. What key aspects of writing style are evident in a summary?
6. Approximately how long is a summary?
7. What is an executive summary?
8. Why should you avoid biased comments in a summary?
9. What two primary goals do you achieve when you delete unnecessary information or biased comments from a summary?
10. Why should you avoid complex technical terms in a summary?

CHAPTER SIXTEEN

Short, Informal Reports

COMMUNICATION AT WORK

In this scenario, a director of information technology must write a variety of reports, including meeting minutes, trip reports, feasibility/recommendation reports, and progress reports.

Cindy Katz is director of information technology at WIFINation, an electronics communication company with branches in 35 cities and a home office in Austin, Texas.

Cindy has been traveling to Austin from her office site in Miami, Florida, one week each month for a year. She is part of a team being trained on WIFINation's new corporate software. Her team consists of staff members from human resources, information technology, accounting, payroll, and administrative services. The software they are learning will be used to manage WIFINation's electronic communications systemwide. It will allow for

- Electronic payroll reports
- Employee benefits
- Paystubs
- Record of cumulative sick leave and vacation days available

Objectives

When you complete this chapter, you will be able to

1. Understand the purposes of a report

2. Distinguish the differences between short, informal reports and long, formal, researched reports

3. Follow guidelines for writing short reports

4. Use headings and talking headings to organize your short reports

5. Write different types of short reports including incident reports, investigative reports, trip reports, progress reports, lab reports, feasibility/recommendation reports, and meeting minutes

6. Evaluate your reports using the report checklist

- Corporate blogging
- WIFINation's intranet
- WIFINation's Web site
- WIFINation's e-mail system

In addition, with the help of the new software, WIFINation is creating an intracorporate Web site called "WIFILand." This social media site, based on a Facebook model, will provide WIFINation employees an online location to share stories, update their personal profiles, and inform their colleagues about community activities. The site will also allow for the posting of instructional videos and videos of corporate activities. WIFILand will be an online meet and greet site designed to build corporate rapport. Cindy's team has first-level responsibility for this new social media venture. In her travels to Austin, she is gathering information from her co-workers about what they want to build into this social media site.

Cindy must document her activities monthly and biannually. First, as the project team's recording secretary, Cindy keeps her team's meeting minutes. Next, she must record her travel expenses and the team's achievements, necessitating monthly trip reports submitted to her manager. As a member of her team's technology impact task force subcommittee, Cindy also has been asked to study technology options and emerging technology concerns. This means that she will write a feasibility/recommendation report following her study to justify the implementation of the new technology systems the team decides on.

Finally, when her project is completed, Cindy will collaborate with her team members to write a progress report for WIFINation's board of directors. Though Cindy's area of expertise is computer information systems, her job requires much more than programming or overseeing the networking of her corporation's computer systems. Cindy's primary job has become communication with colleagues and administrators. Writing reports is a major component of this job requirement.

Check out our quarterly newsletters at www.pearsonhighered.com/gerson for dot.com updates, new case studies, insights from business professionals, grammar exercises, and facts about technical communication.

FAQs: Reports

Q: What's a report? I've seen "reports" that look like memos and "reports" that look like letters. What's the difference?

A: Reports come in all shapes and sizes, and different companies define the word *report* differently. For one company, a report could consist of an informal, half-page e-mail message, a page-long memo, or a three-page letter. Other companies might define a report as a formal, long document, such as a 30-page report on acquiring new clients or a 100-page annual report to stockholders.

Few companies use the same terms for their reports or even write the same kinds of reports. For example, note how the following industries use different terminology for their reports:

- Banking supervisors might write a **Job in Jeopardy** report to document employee problems.
- If a customer's loan request is denied, the bank will report this by writing an **Adverse Action Letter** to the client, documenting this unfavorable finding.
- Accountants write **Financial Statements,** reports that document if a client's finances are normal, if some aspect of the client's finances does not meet industry standards, or if rules have not been followed and violations have occurred.
- Accounting **Valuation Reports** are sent to the IRS for estate planning and gift tax reporting.
- Software developers write **Flash Notification** reports informing clients of vital flaws discovered in software that could lead to financial loss.
- Software **Release Notes,** sent electronically via a company's Web site or on a CD accompanying software, are reports that explain corrections or enhancements and include how-to instructions for testing software functionality.

Each industry sector writes different kinds of reports using different communication channels. However, all reports have certain characteristics in common, as this chapter will illustrate.

What Is a Report?

Report is a simple word that is hard to define. Reports come in different lengths and levels of formality, serve different and often overlapping purposes, and can be conveyed to an audience using different communication channels. You could write a short, informal progress report that conveys information about job-related projects. For a short report, an e-mail format might be sufficient, directed to either a high-tech, low-tech, or lay audience. However, you might write a longer, more formal progress report, in a letter format for an external lay audience, which provides facts, analyzes these findings, and recommends follow-up action.

Your reports will satisfy one or all of the following needs:

- Supply a record of work accomplished
- Record and clarify complex information for future reference
- Present information to a large number of people
- Record problems encountered
- Document schedules, timetables, and milestones
- Recommend future action
- Document current status
- Record procedures

Long? Short? Formal? Informal? To clarify unique aspects of different kinds of reports, look at Table 16.1.

TABLE 16.1 Unique Aspects of Reports

Report Features	Distinctions	Definition of Unique Characteristics
Length and Scope	Short	A typical short report is 1–5 pages. Short reports focus on topics with limited scope. This could include a limited timeframe covered in the report, financial impact, personnel, and impact on the company.
	Long	Long reports are more than 5 pages long. If a topic's scope is large, including a long timeframe, significant amounts of money, research, many employees, and a momentous impact on the company, a long report might be needed.
Formality (tone)	Informal	Most short reports are informal, routine messages, written as letters, memos, or e-mail.
	Formal	Formal reports are usually long and contain standardized components, such as a title page, table of contents, list of illustrations, abstract, appendices, and works cited/references.
Audience	Internal (high tech or low tech)	Colleagues, supervisors, or subordinates within your company are an internal audience. Usually you would write an e-mail or memo report.
	External (multiple audience levels)	An external audience is composed of vendors, clients, customers, or companies with whom you are working. Usually you would write a report in letter format.
	Internal and external	If a report is being sent to both an internal and external audience, you would write either an e-mail, memo, or letter report.
Purpose	Informational	Informational reports focus on factual data. They are often limited in scope to findings: "Here's what happened."
	Analytical	Analytical reports provide information but analyze the causes behind occurrences. Then, these analytical reports draw conclusions, based on an interpretation of the data: "Here's what happened and why this occurred."
	Persuasive	Persuasive reports convey information and draw conclusions. Then, these reports use persuasion to justify recommended follow-up action: "Here's what happened, why this occurred, and what we should do next."
Communication Channels	E-mail	E-mail reports, written to internal and external audiences, are short and informal.
	Memo	Memo reports are written to internal audiences and are usually short and informal.
	Letter	Letter reports are sent to external audiences. These reports can be either long or short, formal or informal, depending on the topic, scope, purpose, and audience.
	Electronic (online)	Many reports can be accessed via a company's Web site. These reports can be downloaded and printed out and often are "boilerplate"—text that can be used repetitively. Electronic reports also provide interactivity, allowing end users to fill out the report online and submit it to the intended audience.

Types of Reports

Many reports fall into the following categories:

- Incident reports
- Investigative reports
- Trip reports
- Progress reports
- Lab reports
- Meeting minutes
- Feasibility/recommendation reports
- Research reports
- Proposals

> **Proposals**
>
> See Chapter 18 for more discussion of internal and external proposals.

How much time do you spend writing reports and why are the reports important in the accounting field?

Linda M. Freeman, CPA, says she and her colleagues throughout the office of Marks, Nelson, Vohland & Campbell spend approximately "20 to 25 percent of our day on report writing, including background research, drafting, and proofreading."

On a daily basis, Linda and her colleagues write Compilation Reports, Review Reports, and Audit Reports. Because these reports are "subject to guidelines of the American Institute of Certified Public Accountants," the reports, in large part, are boilerplate with prescribed wording.

Linda also writes Valuation Reports on a daily basis. Some Valuation Reports are sent to the IRS for estate planning and gift tax reporting. Others "go to the courts for settling disputes between business owners or divorcing couples." Because Valuation Reports are judgment calls on the part of the CPAs, these reports are not boilerplate and entail less prescribed language.

Private Letter Ruling Requests are less-frequently written reports. They are "high-level tax-planning" documents that require intensive research. In these formal requests sent to the IRS, a client proposes a new business or activity structured in a unique way. The accountant projects that the venture will be taxed according to one standard and asks the IRS if they agree with this assumption.

Times are changing. Communication is "going more electronic," Linda says. Reports to federal agencies, for example, are "template-driven." Accountants are encouraged to use online forms and software from federal agencies to help the government cut back on paperwork.

Writing takes up much of Linda's time. Her reports must address many precise accounting standards. Linda's ultimate goal in report writing is quality assurance: adherence to legalities.

This chapter will focus on short, informal reports (incident reports, investigative reports, trip reports, progress reports, lab reports, feasibility/recommendation reports, and meeting minutes).

Criteria for Writing Reports

Although there are many different types of reports and individual companies have unique demands and requirements, certain traits, including format, development, audience, and style, are basic to all report writing.

Organization

Every short report should contain five basic units: identification lines, headings and talking headings, introduction, discussion, and conclusion/recommendations.

Identification Lines. Identify the date on which your report is written, the names of the people to whom the report is written, the names of the people from whom the report is sent, and the subject of the report. As discussed in Chapter 6, the subject line should contain a *topic* and a *focus*. In a short internal report written using a memo format, the identification lines will look like the following example.

Subject Lines

See Chapter 6 for more discussion of subject lines.

EXAMPLE ▶

Date: March 15, 2012
To: Rob Harken
From: Stacy Helgoe
Subject: Report on Usenet Conference

Headings and Talking Headings. To improve page layout and make content accessible, use headings and talking headings. Headings—words or phrases such as "Introduction," "Discussion," "Conclusion," "Problems with Employees," or "Background Information"— highlight the content in a particular section of a document. Talking headings, in contrast, are more informative than headings. Talking headings, such as "Human Resources Committee Reviews 2012 Benefits Packages," informatively clarify the content that follows.

Headings

See Chapter 9 for more discussion of headings, talking headings, and hierarchy of headings.

Introduction. The introduction supplies an overview of the report. It can include three or more optional subdivisions, such as the following:

- **Purpose**—a topic sentence(s) explaining why you are submitting the report (rationale, justification, objectives) and the subject matter of the report.
- **Personnel**—names of others involved in the reporting activity.
- **Dates**—what period of time the report covers.

In this introductory section, use headings or talking headings to summarize the content. These can include headings for organization, such as "Overview" or "Purpose" or more informative talking headings, such as "HVAC Conference Dates Set for 2012."

◄ **EXAMPLE**

> ## Introduction
>
> *Report Objectives*: I attended the Southwest Regional Conference on Workplace Communication in Fort Worth, Texas, to learn more about how our company can communicate effectively. This report addresses the workshops I attended, consultants I met with, and pricing for training seminars.
>
> *Conference Dates*: August 5–8, 2012
>
> *Committee Members*: Susan Lisk and Larry Rochelle

Some businesspeople omit the introductory comments in reports and begin with the discussion. They believe that introductions are unnecessary because the readers know why the reports are written and who is involved.

These assumptions are false for several reasons. First, it is false to assume that readers will know why you're writing the report, when the activities occurred, and who was involved. Perhaps if you are writing only to your immediate supervisor, there's no reason for introductory overviews. Even in this situation, however, you might have an unanticipated reader for the following reasons:

- Immediate supervisors change—they are promoted, fired, retire, or go to work for another company.
- Immediate supervisors aren't always available—they're sick for the day, on vacation, or off-site for job-related travel.

Second, avoiding introductory overviews assumes that your readers will remember the report's subject matter. This is false because reports are written not only for the present, when the topic is current, but also for the future, when the topic is history. Reports go on file—and return at a later date. At that later date, the following may occur:

- You won't remember the particulars of the reported subject matter.
- Your colleagues, many of whom weren't present when the report was originally written, won't be familiar with the subject.
- You might have outside, lay readers who need additional detail to understand the report.

An introduction, which seemingly states the obvious, satisfies multiple and future readers.

Discussion. The discussion section of the report can summarize many topics, including your activities, the problems you encountered, costs of equipment, warranty information, and more. This is the largest section of the report requiring detailed development, illustrated in the different report types discussed throughout this chapter.

Conclusion/Recommendations. The conclusion section of the report allows you to sum up, to relate what you have learned, or to state what decisions you have made regarding the activities reported. The recommendation section allows you to suggest future action, such as what the company should do next. Not all reports require recommendations.

EXAMPLE ▶

Headings, such as "Benefits of the Conference" and "Proposed Next Course of Action" provide focus.

The conclusion shows how the writer benefited.

The recommendation explains what the company should do next and why.

Conclusion/Recommendation

Benefits of the Conference:

The conference was beneficial. Not only did it teach me how to improve my technical communication but also it provided me contacts for technical communication training consultants.

Proposed Next Course of Action:

To ensure that all employees benefit from the knowledge I acquired, I recommend hiring a consultant to provide technical communication training.

Development

Now that you know what subdivisions are traditional in short reports, the next questions might be, "What do I say in each section? How do I develop my ideas?"

First, answer the reporter's questions.

- **Who** did you meet or contact, who was your liaison, who was involved in the accident, who was on your technical team, and so on?
- **When** did the documented activities occur (dates of travel, milestones, incidents, etc.)?
- **Why** are you writing the report and why were you involved in the activity (rationale, justification, objectives)? Or, for a lab report, for example, why did the electrode, compound, equipment, or material act as it did?

- **Where** did the activity take place?
- **What** were the steps in the procedure, what conclusions have you reached, or what are your recommendations?

Clarity

See Chapter 3 for more discussion of how to develop ideas.

Second, when providing information, *quantify.* Do not be vague or imprecise. Specify to the best of your abilities with photographic detail.

The following justification is an example of vague, imprecise writing.

BEFORE

Installation of the machinery is needed to replace a piece of equipment deemed unsatisfactory by an Equipment Engineering review.

Which machine are we purchasing? Which piece of equipment will it replace? Why is the equipment unsatisfactory (too old, too expensive, too slow)? When does it need to be replaced? Where does it need to be installed? Why is the installation important? A department supervisor will not be happy with the preceding report. Instead, supervisors need information *quantified,* as follows:

AFTER

The *exposure table* needs to be installed by *September 10, 2012* so that we can *manufacture printed wiring products with fine line paths and spacing (down to 0.0005 inch).* The table will replace the *outdated printer* in *Department 76.* Failure to install the table *will slow the production schedule by 27 percent.*

Note that the italicized words and phrases provide detail by quantifying.

Audience

Since reports can be sent both internally and externally, your audience can be high tech, low tech, lay, or multiple. Before you write your report, determine who will read your text. This will help you decide if terminology needs to be defined and what tone you should use. In a memo report to an in-house audience, you might be writing simultaneously to your immediate supervisor (high tech), to his or her boss (low tech), to your colleagues (high tech), and to a CEO (low tech). In a letter report to an external audience, your readers could be high tech, low tech, or lay. To accommodate multiple audiences, use parenthetical definitions, such as cash in advance (CIA) or continuing property records (CPR).

In reports, audience determines tone. For example, you cannot write directive reports to supervisors mandating action on their part. It might seem obvious that you can write directives to subordinates or a lay audience, but you should not use a dictatorial tone. You will determine the tone of your report by deciding if you are writing vertically (up to management or down to subordinates), laterally (to co-workers), or to multiple readers.

Style and Highlighting Techniques

See Chapter 3 and Chapter 9 for more discussion of conciseness and document design.

Style

Style includes conciseness and highlighting techniques. Achieve conciseness by eliminating wordy phrases. Use *consider* rather than *take into consideration;* use *now* rather than *at this present time.* Headings, subheadings, and graphics can be used to help communicate content. Note how the following text in the "before" example fails to communicate effectively.

BEFORE

Johnson County is expected to add 157,605 persons to its 1980 population of 270,269 by the year 2012. That population jump would be accompanied by a near doubling of the 96,925 households the county had in 1980. The addition of 131,026 jobs also is forecast for Johnson County by 2012, more than doubling its employment opportunity.

The information is difficult to access readily. We are overloaded with too much data. A table in the "after" example allows for easier access to the data.

Johnson County Predicted Growth by 2012

	Population	Households	Employment
1980	270,269	96,925	127,836
2012	427,874	192,123	258,862
% change	+58.3%	+98.2%	+102%

Types of Short, Informal Reports

All reports include identification information, an introduction, a discussion, and conclusion/recommendations. However, different types of short reports customize these generic components to meet specific needs. Let's look at the criteria for seven common types of reports: incident reports, investigative reports, trip reports, progress reports, lab reports, feasibility/recommendation reports, and meeting minutes.

Incident Reports

Purpose and Examples. An *incident report* documents an unexpected problem that has occurred. This could be an automobile accident, equipment malfunction, fire, robbery, injury, or even problems with employee behavior. In this report, you will document what happened. If a problem occurs within your work environment that requires analysis (fact finding, review, study, etc.) and suggested solutions, you might be asked to prepare an incident report (also called a trouble report or accident report), as follows:

- **Sales.** One of your sales representatives has been involved in a car accident while on job-related travel. The sales representative or his or her supervisor must document this problem.
- **Biomedical technology.** A CAT scan in the radiology department is not functioning correctly. This has led to the department's inability to read x-rays. To avoid similar problems, you need to report this incident.
- **Hospitality management.** An oven in your restaurant caught fire. This not only injured one of your cooks but also damaged the oven, requiring that it be replaced with more fire-resistant equipment.
- **Retail.** A customer was hurt while in your showroom. Incidents also could include employees who are not abiding by company policy. Maybe one of your retail locations has experienced a burglary. The police have been contacted, but as site manager, you believe the problem could have been avoided with better in-store security. Your incident report will document the event and show how to avoid future problems.

Criteria. To write an incident report, include the following components:

1. Introduction
   Purpose. In this section, document when and where the incident occurred. What motivated your visit to the scene of the problem?
   Personnel. *Who* was involved, and *what* role do you play in the report? That could entail listing all of the people involved in the accident or event. These might be people injured, as well as police or medical personnel answering an emergency call.

In addition, *why* are you involved in the activity? Are you a supervisor in charge of the department or employee? Are you a police officer or medical personnel writing the report? Are you a maintenance employee responsible for repairing the malfunctioning equipment?

2. Discussion (body, findings, agenda, work accomplished)

   Using subheadings or itemization, quantify what you saw (the problems motivating the report). Organize your content using problem/solution, chronology, cause/effect, or other modes of organization. Develop your content by including the following information:
   - Make or model of the equipment involved
   - Police departments or hospitals contacted
   - Names of witnesses
   - Witness testimonies (if applicable)
   - Extent of damage—financial and physical
   - Graphics (sketches, schematics, diagrams, layouts, etc.) depicting the incident visually
   - Follow-up action taken to solve the problem

3. Conclusion/recommendations

   Conclusion. Explain what caused the problem.

   Recommendations. Relate what could be done in the future to avoid similar problems.

Often, on the job you will write reports to colleagues and supervisors with whom you work on a day-to-day basis. Using e-mail allows the audience to quickly and conveniently access this short, informal report and respond to any issues. Figure 16.1 presents an example of an incident report, written as an e-mail.

Investigative Reports

Purpose and Examples. As the word *investigate* implies, an *investigative report* asks you to examine the causes behind an incident. Something has happened. The report does not just document the incident. It focuses more on why the event occurred. You might be asked to investigate causes leading up to a problem in the following instances:

- **Security.** You work in a bank's security department. You are responsible for investigating theft, burglary, fraud, vandalism, check forging, and other banking illegalities. One of your customers, a college student at the local university, reports losing her purse at a campus party. Within hours of the theft, checks bearing her name are showing up across the city. Your job now is to investigate the incident and report your findings.

- **Engineering.** A historic, 100-year-old bridge crossing your city's river is buckling. The left lane is now two inches higher than the right lane, and expansion joints are separating beyond acceptable specifications. You must visit the bridge site, inspect the damage, and report on the causes for this construction flaw.

- **Medicine.** As radiographic technologist, you have administered a bone scan to a patient. This scan detected a shadowy area in the right medial humerus. This medical problem could be arthritis or inflammation. However, the shadow also could indicate a metastatic condition. With the help of a pathologist and radiologist, you must submit an investigative report explaining the causes for this aberration and suggesting a follow-up procedure.

DOT.COM UPDATES

For more information about online incident reports, check out the following links:

- US Consumer Product Safety Commission's online "Consumer Product Incident Report" form https://www.cpsc.gov/cgibin/incident.aspx
- National Fraud Information Center's "Online Incident Report Form" http://www.fraud.org/info/repoform.htm
- Health & Safety Executive's "Incident Reporting" forms https://www.hse.gov.uk/forms/incident/index.htm

FIGURE 16.1   E-mail Incident Report

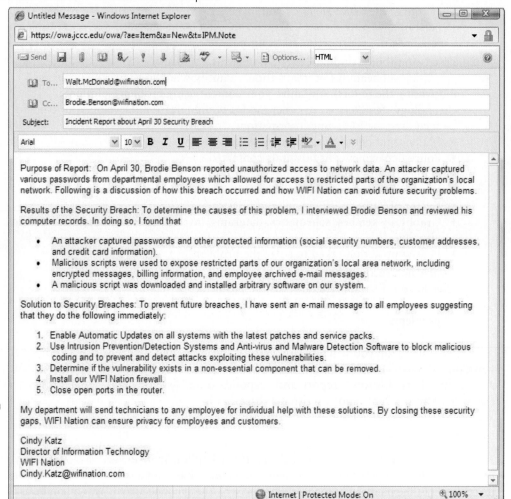

The writer relies on a "problem/solution" analytical pattern of organization for this short, informal, internal e-mail report.

The writer includes a "signature file" that acts like an online business card, clearly identifying the writer and providing contact information.

- **Computer technology.** You work in a college's technology department. Primarily, your job is to help faculty members with their technology applications. The college requires that all student grades be kept on a newly installed, campuswide database and then submitted electronically when the semester ends. For some reason, faculty cannot access their students' records for grade inputting. You must investigate the causes behind this technology glitch and solve the problems—*now!* The semester grades are due within 24 hours.

Criteria.   Following is an overview of what you might include in an effective investigative report:

1. Introduction (overview, background)

   Purpose. In the purpose section, document the date(s) of the incident. Then comment on your objectives or rationale. What incident are you reporting on and what do you hope to achieve in this investigation?

   You might also want to include these following optional subheadings:

   Location. Where did the incident occur?

   Personnel. Who was involved in the incident? This could include those with whom you worked on the project or those involved in the situation.

   Authorization. Who recommended or suggested that you investigate the problem?

2. Discussion (body, findings, agenda)

This is the major part of the investigation. Using subheadings, document your findings. This can include the following:

- A review of your observations. This includes physical evidence, descriptions, lab reports, testimony, and interview responses. Answer the reporter's questions: who, what, when, where, why, and how.
- Contacts—people interviewed
- Difficulties encountered
- Techniques, equipment, or tools used in the course of the investigation
- Test procedures followed, organized chronologically

3. Conclusion/recommendations

Conclusion. What did you accomplish? What did you learn? What discoveries have you made regarding the causes behind the incident? Who or what is at fault?

Recommendations. What do you suggest next? Should changes be made in personnel or in the approach to a particular situation? What training is required for use with the current technology, or should technology be changed? What is the preferred follow-up for the patient or client? How can the problem be fixed?

Use letter formats to write short, informal reports to an external audience. Figure 16.2 illustrates an investigative report, written in letter format. This example is written at a high-tech level because the city council needs to understand the scientific information

Letters

See Chapter 6 for more discussion of letter formats.

FIGURE 16.2   Investigative Report in Letter Format (High-Tech Level)

Follow letter format when you write a short, informal report for an external audience.

Frog Creek Wastewater Treatment Plant
9276 Waveland Blvd.
Bowstring City, UT 86554

September 15, 2012

Bowstring City Council
Arrowhead School District 234
Bowstring City, UT 86721

Subject: Investigative Report on Frog Creek Wastewater Pollution

On September 7, 2012, teachers at Arrowhead Elementary School reported that over a five-day period (September 2–6), approximately 20 students complained of nausea, lightheadedness, and skin rashes. On the fifth day, the Arrowhead administration called 911 and the Arrowhead School District (ASD 234) in response to this incident.

Bowstring City paramedics treated the children's illnesses, suggesting that the problems might be due to airborne pollutants. The Bowstring City Council contacted the Frog Creek Wastewater Treatment Plant (FCWTP) to investigate the causes of this problem. This report is submitted by Mike Moore (Director of Public Relations), Sue Cottrell (Wastewater Engineer), and Thomas Redburn (Wastewater Engineer), in response to Bowstring City's request.

The subject line provides a topic (Frog Creek Wastewater Pollution) and a focus (Investigative Report).

In the letter's introduction, reporter's questions clarify when and where the event occurred, who was involved, and why the report is being written.

(Continued)

Include a new-page notation on every page, after page 1. In this notation, provide the writer's name, page number, and date. You can use a header to create this notation.

The findings use chronological organization to document the incidents.

Mike Moore
Page 2
September 15, 2012

Committee Findings

Impact on School Children: Arrowhead Elementary School administrators reported the following:

- Monday, September 2—two children reported experiencing nausea.
- Tuesday, September 3—two children reported experiencing nausea, and one child experienced lightheadedness.
- Wednesday, September 4—three children experienced skin rashes.
- Thursday, September 5—two children complained of nausea, one child was lightheaded, and two children showed evidence of skin rashes.
- Friday, September 6—two children reported nausea, three reported skin rashes, and two lightheadedness.

After Friday's occurrences, Arrowhead Elementary School administrators called 911. Bowstring paramedics reported that (with parental approval) the children were treated with antacids for nausea, antihistamines for skin rashes, and oxygen for their lightheadedness. No other incidents were reported in the neighborhood surrounding the school.

Report on Frog Creek Pollutants: Wastewater Engineers Sue Cottrell and Thomas Redburn took samples of Frog Creek on September 7–10 and found the following:

Typical Frog Creek Readings	Readings from September 7–10
Low alkalinity: (generally <30 mg/l) (milligrams per liter)	High alkalinity readings: <45 mg/l
Low inorganic fertilizer nutrients (phosphorous and nitrogen): <20 mg/l	High phosphorous and nitrogen readings: <25 mg/l
Limited algae growth: 1 picometers	High algae readings: 4 picometers

Explanation of Findings:
Despite normally low readings, in late summer, with heat and rain, these readings can escalate. Higher algae-related odors above the 3–6 picometer thresholds, along with increased alkalinity (<50 mg/l) can create health problems for youth, elderly, or anyone with respiratory illnesses.

These studies showed that algae, alkalinity, and fertilizer nutrients were higher than usual.

The above elevated readings were caused by three factors (heat, rain, and northeasterly winds).

- Heat—On the days of the Arrowhead Elementary School incident, the temperature ranges were 92 degrees F to 95 degrees, unusually high for early September. Algae and chemical growth increases in temperatures above 84 degree F.

FIGURE 16.2 (Continued)

Mike Moore
Page 3
September 15, 2012

- Rain—In addition, on September 4–6, Bowstring City received 2″ of rain, swelling Frog Creek to 3′ above its normal levels. Studies show that rain-swollen creeks and rivers lead to increased pollutants, as creek bottom silt rises.
- Wind—On September 4–5, a prevailing northeasterly wind blew from Frog Creek toward Arrowhead Elementary School's playground.

Follow-up Studies: On September 11–14, our engineers rechecked Frog Creek, finding that the chemical levels had returned to a normal, acceptable range.

Conclusion about Incident

Frog Creek normally has acceptable levels of algae, alkalinity, and fertilizer nutrient levels. The heat and higher water levels temporarily led to elevated pollutant readings. These levels subsequently returned to normal. Wind directions during the school incidents also had an impact on the children's illnesses. On follow-up questionnaires, FCWTP employees found that the school children's ailments had subsided.

The Arrowhead Elementary School situation appears to have been an isolated incident due to atmospheric changes.

Recommendations for Future Course of Action

Though we constantly monitor Frog Creek for safety, FCWTP HAZMAT employees would be happy to work with parents and teachers to provide additional health information. In a one-hour workshop, presented during the school day or at a Parent-Teacher Organization meeting, FCWTP could offer the following information:

- Scientific data about stream and creek pollutants
- The effects of rain, wind, and heat on creek chemicals
- Useful preventive medical emergency techniques

This information would explain real-world applications for science classes, as well as provide valuable health tips for parents and teachers. We have enclosed for your review information about our proposed training sessions.

Please let us know if you would like to benefit from this free-to-the-public workshop. We would be happy to schedule one at your convenience.

Mike Moore
Frog Creek Wastewater Treatment Plant Director of Public Relations

Enclosure

The findings not only investigate the causes of the incident, but also document with specific details. To achieve a readable format, the text is made accessible through highlighting techniques—underlined and italicized subheadings, bulleted details, and a comparison/contrast table.

The letter's conclusion provides options for the readers and a positive tone appropriate for the intended audience.

FIGURE 16.3   Investigative Report in Letter Format (Lay Audience)

Frog Creek Wastewater Treatment Plant
"Safety is Our Number 1 Concern."
9276 Waveland Blvd.
Bowstring City, UT 86554

September 15, 2012

Attention: Parents and Arrowhead Elementary School Teachers and Administrators

Subject: Report on September 2–6 Frog Creek Incident

Last week, your children at Arrowhead Elementary School experienced nausea, lightheadedness, and skin rashes. This was due to an unusual environmental situation at Frog Creek, a rare case of airborne pollutants caused by high winds and rain.

How Did This Happen?

Algae is a good thing. The crayfish, snails, and minnows that your children love seeing in Frog Creek thrive on algae and lichen (the small, green plants that are the food base for most marine life). However, when temperatures rise above 84 degrees, algae can grow to an unhealthy level.

The same thing applies to the acid level in water. When acid is regulated by alkalinity, algae growth is controlled. When acid levels rise, however, algae can bloom or marine creatures can die. Both of these problems lead to unusual odors.

That's what happened last week. Our studies showed that algae and alkalinity were higher than usual. The causes were increased heat, rain, and wind:

- <u>Heat</u>—On the days of the Arrowhead Elementary School incident, the temperature ranges were 92 degrees F to 95 degrees, unusually high for early September.
- <u>Rain</u>—In addition, on September 4–6, Bowstring City received 2" of rain, swelling Frog Creek to 3' above its normal levels. The rain forced the creek bottom silt to rise, which led to increased pollutants.
- <u>Wind</u>—On September 4–5, a northeasterly wind blew from Frog Creek toward Arrowhead Elementary School's playground.

Because this letter report is sent to multiple audiences, no reader address is given. Instead, an "attention line" is used.

To communicate effectively with the public, the writer used pronouns and a pleasant tone. Words like "unusual" and "rare" are used to lessen the parents' concerns.

For headings, the writer used questions that an audience of concerned parents and teachers might have.

Because the audience is lay, the writer uses simple sentence structure ("Algae is a good thing"; "The same applies to acid levels in water.") and avoids complex scientific discussion, focusing on temperature, rain levels, and wind direction.

about the incident. Moreover, this report will be kept on file not only for documentation but also for potential litigious situations.

In contrast to Figure 16.2, the high-tech investigative report, Figure 16.3 is written at a lay level for an external audience consisting of the elementary school administration, the teachers, and the parents of the school children affected by the incident in Frog Creek. The highly technical and scientific information is omitted, the material is condensed, and the writer includes sufficient details to alleviate the concerns of the audience. In addition, the writer emphasizes actions taken and assurances of how and why this incident will not recur.

Trip Reports

Purpose and Examples.   A *trip report* allows you to report on job-related travel. When you leave your work site and travel for job-related purposes, your supervisors not only

FIGURE 16.3 (Continued)

page 2
September 15, 2012

What Can We Do to Help?

Could heat, rain, and wind lead to similar situations in the future? Yes. But . . . the incidents from last week were very rare. Please do not expect a repeat occurrence any time soon. Tell your children to enjoy Frog Creek for its beauty and natural resources.

We would be happy to meet with you and your children to explain the science of this environmental event. Plus, we'd like to provide techniques for managing simple ailments like nausea, skin rashes, and lightheadedness.

In a one-hour workshop, presented during the school day or at a Parent-Teacher Organization meeting, Frog Creek Wastewater Treatment Plant (FCWTP) could offer the following information:

- Scientific data about stream and creek pollutants—with hands-on tutorials for your students.
- The effects of rain, wind, and heat on creek chemicals—complete with graphics and an age-appropriate PowerPoint presentation.
- Useful preventive medical emergency techniques—which every parent and child should know.

This information would explain real-world applications for science classes, as well as provide valuable health tips for parents and teachers. We have enclosed for your review information about our proposed training sessions.

Please let us know if you would like to benefit from this free-to-the-public workshop. We would be happy to schedule one at your convenience. FCWTP wants to assure you that your child's "safety is our number 1 concern."

Mike Moore
Frog Creek Wastewater Treatment Plant Director of Public Relations

Enclosure

The writer emphasizes the positive by using words such as "please," "enjoy Frog Creek for its beauty," and "We would be happy to meet with you."

The writer focuses on the audience's future concerns and suggests age-appropriate material to explain the incident to the elementary school children. This attempt at community outreach connects the wastewater treatment plant to concerned constituents.

require that you document your expenses and time while off site, but they also want to be kept up to date on your work activities. For example, you might be engaged in work-related travel as follows:

- **Information technology.** You go to a conference to learn about the latest hardware and software technologies for the workplace. There, you meet with vendors, participate in hands-on technology workshops, and learn what other companies are doing to manage their technology needs. When you return, you write a trip report documenting your activities.
- **Heating, ventilating, and air conditioning.** One of your clients is building an office site. Your company has been hired to install their heating, ventilating, and air conditioning (HVAC) system. You travel to your client's home office to meet with other contractors (engineering and architectural) so all team members can agree on construction plans. At the conclusion of your job-related travel, you will write a trip report about your meeting.

Creating Headers and Footers in Microsoft Word 2010

To create headers and footers (useful for new-page notations in reports), follow these steps:

1. Click on the **Insert** tab on your toolbar. You will see the following ribbon.

2. Click on either **Header** or **Footer**. When you click on your choice of either **Header** or **Footer**, you will see a drop-down menu, such as shown below.

3. Choose the type of **Header** or **Footer** you want to use in your document and type your content (date, name, page number, etc.), as shown in the following example.

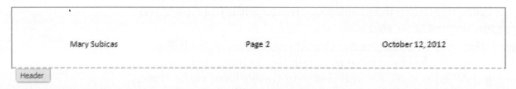

- **Engineering.** To be a responsible member of your community, your company has partnered with the local school district. One of your jobs is to visit the city's high schools and discuss engineering job preparedness. To do so, you give an oral presentation about the engineering job market, desired job skills, and the importance of technical communication in the workplace. Upon returning to your office, you write a trip report not only to document how the company is working with the community but also to explain to your supervisor how the company can meet community needs.
- **Biomedical equipment sales.** Four days a week, you are on the road making sales calls. Each month, you must document your job-related travel to show that you are making your quota and to receive reimbursement for travel expenditures.

Criteria. Following is an overview of what you will include in an effective trip report:

1. Introduction (overview, background)
   Purpose. In the purpose section, document the date(s) and destination of your travel. Then comment on your objectives or rationale. What motivated the trip, what did you plan to achieve, what were your goals, why were you involved in job-related travel?
   You might also want to include these following optional subheadings:
   Personnel. With whom did you travel?
   Authorization. Who recommended or suggested that you leave your work site for job-related travel?
2. Discussion (body, findings, agenda)
   Using subheadings, document your activities. This can include a review of your observations, contacts, seminars attended, or difficulties encountered.
3. Conclusion/recommendations
   Conclusion. What did you accomplish—what did you learn, whom did you meet, what sales did you make, what of benefit to yourself, colleagues, or your company occurred?
   Recommendations. What do you suggest next? Should the company continue on the present course (status quo) or should changes be made in personnel or in the approach to a particular situation? Would you suggest that other colleagues attend this conference in the future, or was the job-related travel not effective? In your opinion, what action should the company take?

Figure 16.4 presents an example of an informal trip report written in memo format. The memo format, providing identification lines (date, to, from, subject), is written to an internal audience and creates a hard-copy document.

Progress Reports

Purpose and Examples. A *progress report* lets you document the status of an activity, explaining what work has been accomplished and what work remains. Supervisors and customers want to know what progress you are making on a project, whether you are on schedule, what difficulties you might have encountered, and what your plans are for the next reporting period. Because of this, your audience might ask you to write progress (or activity or status) reports—daily, weekly, monthly, quarterly, or annually:

- **Biomedical technology.** You and your team are developing a new heart monitor. This entails researching, patenting, building, testing, and marketing. You have been working on this project for months. What is your status? A progress report will tell your investors and supervisors where you stand, if you are on schedule, and when the project will conclude.

DOT.COM UPDATES

For more information about trip reports, check out the following links:

- This Disney Trip Report Archive lets you read reports for all trips taken during an indicated period, plus you can add your own trip report. http://www.mouseplanet.com/dtp/trip.rpt/
- Vanguard, an independent consulting company specializing in customer contact and convergence, provides their employees a sample trip report template. http://www.vanguard.net/documents/CCT%20Trip%20Report%206-02.doc

FIGURE 16.4   Trip Report in Memo Format

Date: February 26, 2012
To: Debbie Rulo
From: Oscar Holloway
Subject: Trip Report—Unicon West Conference on Electronic Training

Introduction

The introduction section answers the "reporter's questions"—who, what, when, where, and why.

Purpose of the Meeting: On Tuesday, February 23, 2012, I attended the Unicon West Conference on Electronic Training, held in Ruidoso, New Mexico. My goal was to acquire hands-on instruction and learn new techniques for electronic training, including the following:

- Online discussion groups
- E-based tutorials
- Intranet instruction
- Videoconference lecture formats

Conference Participants: My co-workers Bill Cole and Gena Sebree also attended the conference.

Discussion

Different "heading levels" and highlighting techniques are used to make the information more accessible, to help the readers navigate your text.

Presentations at the Conference:
Gena, Bill, and I attended the following sessions:

- *Online Discussion Groups*
  This two-hour workshop was presented by Dr. Peter Tsui, a noted instructional expert from Texas State University, San Marcos, Texas. During Dr. Tsui's presentation, we reviewed how to develop online questions for discussion, post responses, interact with colleagues from distant locales, and add to streaming chats. Dr. Tsui worked individually with each seminar participant.

- *E-based Tutorials*
  This hour-long presentation was facilitated by Debbie Gorse, an employee of Xenadon E-Learning Inc. (Colorado Springs, Colorado). Ms. Gorse used videos and screen captures displayed on overhead projectors to give examples of successful E-based, computer-assisted instructional options.

- *Intranet Instruction*
  Dr. Randy Towner and Dr. Karen Pecis led this hour-long presentation. Both are professors at the University of Nevada, Las Vegas. Their workshop focused on course development, online instructional methodologies, customizable company-based examples, and firewall-protected assessment. The professors provided workbooks and hands-on learning opportunities.

- *Videoconferencing*
  Denise Pakula, Canyon E-Learning, Tempe, Arizona, spoke about her company's media tools for teleconferenced instruction. These included lapel microphones, multi-directional pan/tilt video cameras, plasma display touch screens, wideband conference phones, recessed lighting ports with dimming preferences, and multimedia terminals.

FIGURE 16.4 (Continued)

Oscar Holloway
Page 2
February 26, 2012

A "new-page notation" helps your readers avoid losing or misplacing pages.

Conclusion

Presentation Benefits:

Every presentation we attended was beneficial. However, the following information will clarify which workshop(s) would benefit our company the most:

The conclusion focuses on the primary findings to give the audience direction.

1. Dr. Tsui's program was the most useful and informative. His interactive presentation skills were outstanding and included hands-on activities, small-group discussions, and individual instruction. In addition, his online discussion techniques offer the greatest employee involvement at the most cost-effective pricing. We met with Dr. Tsui after the session, asking about his fees for onsite instruction. He would charge only $90 per person (other people we researched charged at least $150 per person). Dr. Tsui's fees should fit our training budget.

2. E-based tutorials will not be a valid option for us for two reasons. First, Xenadon's products are prepackaged and allow for no company-specific examples. Second, Ms. Gorse's training techniques are outdated. Videos and overhead projections will not create the interactivity our employees have requested in their annual training evaluations.

3. The Towner/Pecis workshop was excellent. Intranet instruction would be ideal for our needs. We will be able to customize the instruction, provide participants individualized feedback, and ensure confidentiality through firewall-protection. Furthermore, Drs. Towner and Pecis used informative workbooks and hands-on learning opportunities in their presentation.

4. Ms. Pakula's presentation on videoconferencing focused more on state-of-the-art equipment rather than on instruction. We believe that the price of the equipment both exceeds our budget and our needs. Our current videoconference equipment is satisfactory. If the Purchasing Department is looking for new vendors, they might want to contact Canyon.

Recommendations:

Gena, Bill, and I suggest that you invite Dr. Tsui to our site for further consultation. We also think you might want to contact Drs. Towner and Pecis for more information on their training.

The recommendation suggests the next course of action.

- **Hospitality management.** The city's convention center is considering new catering options. Your job has been to compare and contrast catering companies to see which one or ones would best be suited for the convention center's needs. The deadline is arriving for a decision. What is the status? Whom have you considered, what are their prices and food choices, what additional services do they offer, and so forth? You need to submit a progress report so management can determine what the next steps should be.

- **Project management.** Your company is renovating its home office. Many changes have occurred. These include new carpeting, walls moved to create larger cubicles, the construction of larger conference rooms, a new cafeteria and fitness center, and improved lighting. Other changes are still in progress, such as increased parking spaces, exterior landscaping, a child care center, and handicapped accessibility. The supervisor wants to know when these renovations will be concluded. You need to write a progress report to quantify what has occurred, what work is remaining, and when work will be finished.

- **Automotive technology.** Your company recently suffered negative publicity due to product failures. As manufacturing supervisor, you have initiated new procedures for automotive manufacturing to improve your product quality. How are these procedural changes going? Your company CEO needs an update. To provide this information, you must write a progress report.

Criteria.   Following is an overview of what you will include in an effective progress report:

1. Introduction (overview, background)
   Objectives. These can include the following:
   - Why are you working on this project (what's the rationale)?
   - What problems motivated the project?
   - What do you hope to achieve?
   - Who initiated the activity?

   Personnel. With whom are you working on this project (i.e., work team, liaison, contacts)?

   Previous activity. If this is the second, third, or fourth report in a series, remind your readers what work has already been accomplished. Bring them up to date with background data or a reference to previous reports.

2. Discussion (findings, body, agenda)
   Work accomplished. Using subheadings, itemize your work accomplished either through a chronological list or a discussion organized by importance.

   Problems encountered. Inform your reader(s) of any difficulties encountered (late shipments, delays, poor weather, labor shortages) not only to justify your possibly being behind schedule but also to show the readers where you'll need help to complete the project.

   Work remaining. Tell your reader what work you plan to accomplish next. List these activities, if possible, for easy access. A visual aid, such as a Gantt chart or a pie chart, fits well after these two sections. The chart will graphically depict both work accomplished and work remaining.

3. Conclusion/Recommendations
   Conclusion. Sum up what you've achieved during this reporting period and provide your target completion date.

   Recommendations. If problems were presented in the discussion, you can recommend changes in scheduling, personnel, budget, or materials that will help you meet your deadlines.

Figure 16.5 presents an example of a progress report.

Graphics

See Chapter 10 for more discussion of how to create successful graphics, like Gantt and pie charts.

Web Resources

www.pearsonhighered .com/gerson See our companion website for more progress report samples.

Lab Reports

Purpose and Examples.   A *lab report* lets you document the status of and findings from a laboratory experiment, procedure, or study. Professionals in electronics, engineering, medical fields, the computer industry, and other technologies often rank the ability to communicate as highly as they do their technical skills. Conclusions drawn from a technical procedure are worthless if they reside in a vacuum. The knowledge acquired from a laboratory activity *must* be communicated to colleagues and supervisors so they can benefit from your discoveries. You write a lab report after you have performed the lab to share your findings:

- **Biomedical technology.** You have performed a pathology study on tissue, reviewed a radiological scan, or drawn blood. What have you found? To help nurses and doctors provide the best patient care, you must write a lab report documenting your findings.

FIGURE 16.5   Progress Report in Memo Format

To: Buddy Ramos
From: Pat Smith
Date: April 2, 2012
Subject: First Quarterly Report—Project 80 Construction

Purpose of Report

In response to your December 20, 2011, request, following is our first quarterly report on Project 80 Construction (Downtown Airport). Department 93 is in the start-up phase of our company's 2012 build plans for the downtown airport and surrounding site enhancements. These construction plans include the following:

1. *Airport construction*—terminals, runways, feeder roads, observation tower, parking lots, maintenance facilities.

2. *Site enhancements*—northwest and southeast collecting ponds, landscaping, berms, and signage.

The introduction explains why *the report has been written and* what *topic will be discussed.*

Work Accomplished

In this first quarter, we have completed the following:

1. *Subcontractors*: Toby Summers (Project Management) and Karen Kuykendahl (Finance) worked with our primary subcontractors (Apex Engineering and Knoblauch and Sons Architects). Toby and Karen arranged site visitations and confirmed construction schedules. This work was completed January 12, 2012.

2. *Permits*: Once site visitations were held and work schedules agreed upon, Toby and Wilkes Berry (Public Relations) acquired building permits from the city. They accomplished this task on January 20, 2012.

3. *Core Samples*: Core sample screening has been completed by Department 86 with a pass/fail ratio of 76.4 percent pass to 23.6 percent fail. This meets our goal of 75 percent. Sample screening was completed January 30, 2012.

4. *Shipments*: Timely concrete, asphalt, and steel beam shipments this quarter have provided us a 30-day lead on scheduled parts provisions. Materials arrived February 8, 2012.

5. *EPA Approval*: Environmental Protection Agency (EPA) agents have approved our construction plans. We are within guidelines for emission controls, pollution, and habitat endangerment concerns. Sand cranes and pelicans nest near our building site. We have agreed to leave the north plat (40 acres) untouched as a wildlife sanctuary. This will cut into our parking plans. However, since the community will profit, we are pleased to make this concession. Our legal department also informs us that we will receive a tax break for creating this sanctuary. EPA approval occurred on February 15, 2012.

The discussion provides quantified data and dates for clarity, such as "76.4 percent pass" and "January 30, 2012." The discussion also clarifies who worked on the project and lists other primary contacts.

(Continued)

FIGURE 16.5 Progress Report in Memo Format (Continued)

Pat Smith
Page 2
April 2, 2012

Problems Encountered

Core samples are acceptable throughout most of our construction site. However, the area set aside for the northwest pond had a heavy rock concentration. We believed this would cause no problem. Unfortunately, when Anderson Brothers began dredging, they hit rock, which had to be removed with explosives.

Since this northwest pond is near the sand crane and pelican nesting sites, EPA told us to wait until the birds were resettled. The extensive rock removal and wait for wildlife resettlement have slowed our progress. We are behind schedule on this phase.

This schedule delay and increased rock removal will affect our budget.

Work Remaining

To complete our project, we need to accomplish the following:

1. *Advertising*: Our advertising department is working on brochures, radio and television spots, and highway signs. Advertising's goal is to make the construction of a downtown airport a community point of pride and civic celebration.

2. *Signage*: With new roads being constructed for entrance and exit, our transportation department is working on street signage to help the public navigate our new roads. In addition, transportation is working with advertising on signage designs for the downtown airport's two entrances. These signs will juxtapose the city's symbol (a flying pelican) with an airplane taking off. The goal is to create a logo that simultaneously promotes the preservation of wildlife and suggests progress and community growth.

3. *Landscaping*: We are working with Anderson Brothers Turf and Surf to landscape the airport, roads, and two ponds. Our architectural design team, led by Fredelle Schneider, is selecting and ordering plants, as well as directing a planting schedule. Anderson Brothers also is in charge of the berms and pond dredging. Fredelle will be our contact person for this project.

4. *Construction*: The entire airport must be built. Thus, construction comprises the largest remaining task.

Project Completion/Recommendations

Though we have just begun this project, we have completed approximately 15 percent of the work. We anticipate a successful completion, especially since deliveries have been timely.

Only the delays at the northwest pond site present a problem. We are two weeks behind schedule and $3,575.00 over cost. With approximately 10 additional personnel to speed the rock removal and with an additional $2,500, we can meet our target dates. Darlene Laughlin, our city council liaison, is the person to see about corporate investors, city funds, and big-ticket endowments. With your help and Darlene's cooperation, we should meet our schedules.

A "Problems Encountered" section helps justify delays and explain why more time, personnel, or funding might be needed to complete a project.

Numbered points with italicized subheadings help readers access the information more readily. This is especially valuable when an audience needs to refer to documents at a later date.

The conclusion sums up the overall status of the project: "15 percent" complete.

The recommendation explains how the problems discussed in the "Problems Encountered" section can be solved and what is needed to complete the job: "additional personnel" and "increased funds." It also states who to contact for help, "Darlene Laughlin."

FIGURE 16.5 (Continued)

Pat Smith
Page 3
April 2, 2012

The Gantt chart in Figure 1 clarifies our status at this time.

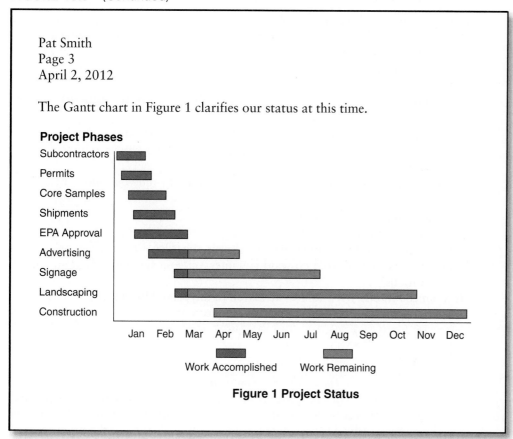

Figure 1 Project Status

A graphic not only adds interest to the report but also helps quantify information and clarify details visually.

- **Electronics.** Your company manufactures a global positioning system (GPS) to correctly inform a user of an exact location. The GPS receiver must compare the time a signal is transmitted by a satellite with the time it is received. Your company's receptors are malfunctioning, as are the units' electronic maps. Why? Your job is to study the electronic systems on randomly selected GPS units and write a lab report documenting your findings.
- **Information technology.** Customers are calling your company's 1-800 hotline almost daily, complaining about hard drive error readings. This is bad for business and profitability. To solve these hard drive malfunctions, you must study units to find the problem. Then, you will write a lab report to document your discoveries.

You write a lab report after you've performed a laboratory test to share with your readers

- Why the test was performed
- How the test was performed
- What the test results were
- What follow-up action (if any) is required

The following are components of a successful lab report:

1. Introduction (overview, background)
   Purpose. Why is this report being written? To answer this question, provide any or all of the following:
   - Rationale (What problem motivated this report?)
   - Objectives (What does this report hope to prove?)
   - Authorization (Under whose authority is this report being written?)
2. Discussion (body, methodology)
   How was the test performed? To answer this question, provide the following:
   - Apparatus (What equipment, approach, or theory have you used to perform the test?)
   - Procedure (What steps—chronologically organized—did you follow in performing the test?)
3. Conclusion/recommendations
   Conclusion. The conclusion of a lab report presents your findings. Now that you've performed the laboratory experiment, what have you learned or discovered or uncovered? How do you interpret your findings? What are the implications?
   Recommendations. What follow-up action (if any) should be taken?
      You might want to use graphics to supplement your lab report, as shown in Figure 16.6.

Feasibility/Recommendation Reports

Purpose and Examples. A *feasibility/recommendation report* accomplishes two goals. First, it studies the practicality of a proposed plan. Then, it recommends action. Occasionally, your company plans a project but is uncertain whether the project is feasible. Will the plan work, does the company have the correct technology, will the idea solve the problem, or is there enough money? One way a company determines the viability of a project is to perform a feasibility study, document the findings, and then recommend the next course of action:

- **Manufacturing.** Your company is considering the purchase of new equipment but is concerned that the machinery will be too expensive, the wrong size for your facilities, or incapable of performing the desired tasks. You need to research and analyze the options, determining which equipment best suits your company's needs. Then, you will recommend purchase.
- **Accounting.** Your company wants to expand and is considering new locations. The decision makers, however, are uncertain whether the market is right for expansion. Are interest rates good? Are local property taxes and sales taxes too high? Will the city provide tax rebate incentives for your company's growth? You need to study the feasibility of expansion and report your recommendations.
- **Web design.** Your company wants to create a Web site to market your products and services globally. The CEO wants to be sure that online checkout is easy, pricing is cost effective, products are depicted in a visually appealing way, and the site loads quickly. How will you make your Web site stand out from the competition? You must write a feasibility report to present the options as well as to offer your recommendations.
- **Health management.** It is time to update your health information system. With increasingly complex insurance and regulatory challenges, your current system is outdated. What are your options? You could install software to help code and classify patient records. You could hire consultants to help comply with in-patient and outpatient regulations. You could outsource your patient load to home health

FIGURE 16.6 Lab Report in Memo Format

Date: April 12, 2012
To: Dr. Lee Wang
From: Cassidy Poston
Subject: Lab Report on Anticancer Characteristics of Anthocynanins in SW480
           Cancer Cells

Introduction

Biosystems Inc. is testing new drugs in different applications to prevent colon
cancer. This lab report will evaluate colon cancer cell growth rate and cell cycle
apoptosis in response to varying concentrations of anthocynanin. The study was
authorized by Dr. Amanda Wharton, laboratory supervisor at Biosystems, and
conducted by me, Andrew Boston, and Cynthia Ruiz, laboratory technicians at
Raston Pharmaceuticals.

Discussion

Apparatus—To conduct the lab work, we used the following chemicals and equipment:

Chemicals	Equipments
<ul><li>Anthocynanin</li><li>Peonidin 3-glucose</li><li>Cyanidin 3-glucose</li><li>SW480 colon cancer cells</li><li>Fetal bovine serum</li><li>Penicillin</li></ul>	<ul><li>Spectroscope</li><li>Petri dishes</li><li>96-well microtiter plate</li></ul>

Procedure—To conduct the lab work, we followed this procedure:

1. *HPLC analysis*: Ms. Ruiz purified anthocynanin samples extracted from a
   sweet potato using HPLC (high-performance liquid chromatographic)
   analysis.
2. *Cells cultured*: SW480 colon cancer cells were cultured in a standard growth
   medium supplemented with 10 percent fetal bovine serum and 1 percent
   penicillin. The cells were then placed in 96-well microtiter plates for drug
   administration; approximately $2.0*10^4$ cell density per well.
3. *Anthocynanin application*: The standard anthocynanins along with the
   sweet potato extracts were applied to the cell cultures in varying
   concentrations: 0.5, 1.0, 5.0 and 10.0 microgram/ml.
4. *Results*: Cancer cell growth inhibition was measured using spectroscopy,
   absorbance at 490 nm with a 96-well plate reader. The analysis of
   anthocynanin standards were recorded graphically. We are awaiting results
   on the cells treated with sweet potato extract. (See graphs below.)

(Continued)

FIGURE 16.6   Lab Report in Memo Format (Continued)

Cassidy Poston
Page 2
April 12, 2012

Colon Cancer Cell Growth vs. Anthocynanin Concentration

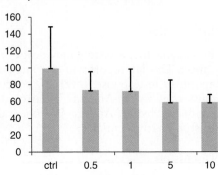

Standard: Peonidin 3-glucose   Standard: Cynanin 3-glucose

Conclusion

The application of the anthocynanin standards gave undesirable results. The cell growth inhibition was similar for all concentrations of drug treatment: 0.5 to 10 microgram/ml. Our team of lab techs suspects that more concentrated applications would have a greater inhibitory effect on colon cancer cell growth or at least varying growth recordings for each concentration. We believe this discrepancy may be due to differences in cell densities per well. We did not invert our cell suspension often or follow a randomized application. This delayed analysis of the cell cultures treated with sweet potato extract.

Recommendation

With the permission of Dr. Wharton, we suggest a follow-up procedure:

1. *Reapplication of cells*: This would include re-culturing and inoculating SW480 cells on the 96 well plates. To do so, we suggest applying similar amounts of cell density to each well by inverting the cell suspension and applying the cells in a zig-zag pattern down the plate.
2. *Reapplication of anthocynanin*: Both the standard and sweet potato extracts should be reapplied to the new plates in the same concentrations discussed above.
3. *Purchase of new equipment*: In addition, we suggest the purchase of a digital pipette. This would allow for precise and quick application of cell suspension, medium and anthocynanin to the 96-well plates.
4. *Analysis*: The cell growth could then be measured using spectrometry with the same conditions of prior trials.
5. *Cytoflowmetry*: Once all values have been recorded for cell growth, the samples should be sent to Biosystems's laboratory to be assessed using cytoflowmetry. This will detect the specific stage of cell apoptosis.

Although our results have not been optimal, treated colon cancer cells did have inhibited growth compared to control values (which had no anthocynanin applied). This is an encouraging lab result.

care agencies. You could upgrade your intranet system to provide decision makers with more accurate information. A feasibility report will study the options before you recommend changes.

Criteria.   One way a company determines the viability of a project is to perform a feasibility study and then write a feasibility report documenting the findings. The following are components of an effective feasibility report:

1. Introduction (overview, background)
   Objectives. Under this subheading, answer any of the following questions:
   - What is the purpose of this feasibility report? One of your responsibilities is to provide background data. To answer the question regarding the report's purpose, provide a clear and concise statement of intent.
   - What problems motivated this study? To clarify for your readers the purposes behind the study, explain what problems cause doubt about the feasibility of the project (is there a market, is there a piece of equipment available that would meet the company's needs, is land available for expansion?). You can also explain what problems led to the proposed project (current equipment is too costly or time consuming, current facilities are too limited for expansion, current net income is limited by an insufficient market).
   - Who initiated the feasibility study? List the name(s) of the manager(s) or supervisor(s) who requested this report.
   Personnel. Document the names of your project team members, your liaison between your company and other companies involved, and your contacts at these other companies.

2. Discussion (body, findings)
   Under this subheading, provide accessible and objective documentation:
   Criteria. State the criteria upon which your recommendation will be based. Criteria are established so you have a logical foundation for comparison of personnel, products, vendors, costs, options, schedules, and so on.

   Analysis. In this section, compare your findings against the criteria. In objectively written paragraphs, develop the points being considered. You might want to use a visual aid such as a table to organize the criteria and provide easy access.

3. Conclusion/recommendations
   Conclusion. In this section, state the significance of your findings. Draw a conclusion from what you have found in your study. For example, state that "Tim is the best candidate for director of personnel" or "Site 3 is the superior choice for our new location."
   Recommendations. Once you have drawn your conclusions, the next step is to recommend a course of action. What do you suggest that your company do next? Which piece of equipment should be purchased, where should the company locate its expansion, or is there a sufficient market for the product?

Figure 16.7 presents an example of a feasibility/recommendation report.

Meeting Minutes

Purpose and Examples.   *Meeting minutes* document the results of a meeting—what was discussed, proposed, voted on, and planned for future meetings. As the recording secretary for your project team, a community organization, your city council, or a departmental meeting, your job is to record the meeting's minutes. In meeting minutes, you record who attended the meeting, when it began, when it ended, and where it took place. You report on the topics discussed, decisions, and plans for the future.

FIGURE 16.7 Feasibility Report in Memo Format

From: Cindy Katz, Director of Information Technology
To: Shamir Rammalah, Accounts Payable
Date: August 13, 2012
Subject: Feasibility Study for Technology Purchases

Purpose of the Report

The purpose of this report is to study which technology will best meet your communication needs and budget. After analyzing the feasibility of various technologies, we will recommend the most cost-effective technology options.

The purpose reminds the reader why this report has been written and what the report's goal is.

Technology Problems

According to your memo dated August 1, 2012, your department needs new communication technologies for the following reasons:

- Your department has hired three new employees, increasing your headcount to ten.
- Currently, your department has only five laptops.
- We also must update training because WIFINation has purchased new software.
- Accounts Payable's current printer is also insufficient due to your increased headcount.

The problem details what issues have led to this report.

Vendor Contacts

Our vendor contacts for the laptops, printers and software are as follows:

Electek	**Tech On the Go**	**Mobile Communications**
Steve Ross	Jay Rochlin	Karen Allen
stever1@electek.com	jrochlin@tog.com	karen.allen@mobcom.net

The vendor section provides contact information (names and e-mail addresses).

Criteria for Vendors

The following criteria were considered to determine which communication technology would best meet your department's needs:

1. *Trainers*—Because of the unique aspects of the software, we need trainers who are familiar with password creation, privacy laws, malware, spyware, and corporate e-mail policy.
2. *Maintenance*—We need to purchase equipment and software complete with either quarterly or biannual service agreements (at no extra charge).
3. *Service Personnel*—The service technicians should be certified to repair and maintain whatever hardware we purchase. In addition, the vendors must also be able to train our personnel in hardware usage.
4. *Warranties*—The warranties should be for at least one year with options for renewal.
5. *Cost*—The total budgeted for your department is $15,000.

The criteria states the topics used to research the report and includes precise details. In this way, the audience can understand the rationale for later decisions.

FIGURE 16.7 (Continued)

Cindy Katz
Page 2
August 13, 2012

Needs Assessment

Purchasing agrees that the Accounts Payable hardware and software needs exceed their current technology. Not only are the department's laptops and printer insufficient in number, but also they do not allow the personnel to access corporate e-mail, the Internet, or word-processing packages. Updated equipment is necessary.

Vendor Evaluation

- **Electek**—Having been in business for 10 years, this company is staffed by highly trained technicians and sales staff. All Electek employees are certified for software training. The company promises a biannual maintenance package and subcontracted personnel if employees cannot repair hardware problems. They offer manufacturers' guarantees with extended service warranties costing only $100 a year for up to 5 years. Electek offers 20 percent customer incentives for purchases of over $2,000.

- **Tech On the Go (TOG)**—This company has been in business for two years. TOG provides only subcontracted service technicians for hardware repair. TOG's employees are certified in software training. The owners do not offer extended warranty options beyond manufacturers' guarantees. No special customer pricing incentives are offered though TOG sells retail at a wholesale price.

- **Mobile Communications**—Having been in business for 5 years, Mobile has certified technicians and sales representatives. All repairs are provided in house. The company offers quarterly maintenance at a fee of $50 ($200 per year). Mobile offers a customer incentive of 10 percent discounts on purchases over $5000.

The organizational mode comparison/contrast is used to analyze the strengths and weaknesses of each vendor.

Cost Analysis

- *Laptops*—Accounts Payable requests one laptop per departmental employee. Our analysis has determined that the most affordable laptops we can purchase (with the requested software and wireless Internet connections) would cost $1,500 per unit. Thus, 10 laptops would cost $15,000. Even with discounts, this exceeds your department's budget.

- *Printers*—Accounts Payable requests three additional printers, with the capability to print double-side pages, staple, collate, print in color, and print three different sizes of envelopes and three different sizes of paper. Our analysis has found that the most affordable printers meeting your specifications would cost $2,500 each. Again, when combining this cost with that of laptops, you exceed your budget.

- *Training*—The manufacturer can provide training on our new software. Training can be offered Monday through Friday, starting in September. The manufacturer says that effective training must entail at least 20 hours of hands-on practice. The cost for this would be $5,000.

The discussion provides specific details to prove the feasibility of the plan or project.

(Continued)

FIGURE 16.7 Feasibility Report in Memo Format (Continued)

Cindy Katz
Page 3
August 13, 2012

The following table compares the three vendors we researched on a scale of 1–3, 3 representing the highest score.

Graphics depict the findings more clearly and more concisely than a paragraph of text.

Table 1: Criteria Comparison			
Criteria	Electek	TOG	Mobile
Maintenance	3	2	3
Personnel	3	3	3
Warranties	3	2	2
Cost	3	2	2
Total	12	9	10

Summary of Findings

We cannot purchase the number of laptops and printers you have requested. Doing so exceeds the budget. Training is essential. Your department must adjust its budget accordingly to accommodate this need.

All three vendors have the technology you require. However, TOG and Mobile do not meet the criteria. In particular, these companies do not provide either the maintenance packages, warranties, or pricing required.

The conclusion sums up the findings, explaining the feasibility of a course of action and why a plan should or should not be pursued.

Recommended Action

Given the combination of cost, maintenance packages, warranties, and service personnel, Electek is our best choice.

We suggest the following options for printers and laptops: purchasing five laptops instead of ten; purchasing one additional printer instead of three; and/or sharing printers with nearby departments.

The recommendation explains what should happen next and provides the rationale for this decision.

Criteria.   Meeting minutes should include the following key components:

1. Introduction
   At the beginning of the minutes, list the date on which the meeting is held, what time the meeting began, and where the meeting was located. List the names of those who attended the meeting.
2. Discussion (findings, agenda)
   This is the most important part of your minutes. In this long section, you report on the agenda items in chronological order.
3. Conclusion
   Conclude the meeting minutes by listing any unresolved topics needing further discussion. Highlight new topics for future discussion. Report when the committee will meet next, providing the date, time, and location. Finally, report when the meeting ended.

Figure 16.8 shows an example of meeting minutes.

FIGURE 16.8   Meeting Minutes

**Employee Benefits Project Team
Meeting Minutes**

Date: April 12, 2012

Time: 7:00 P.M.–9:30 P.M.

Objectives: In this second project team meeting, the goal was to discuss employee benefits options that could be added to the company's cafeteria plan.

Attendees: Stacy Helgoe, Christie Pieburn, Darren Rus, Andrew McWard, Bill Lamb, Jessica Studin

Team Leader: Phyllis Goldberg, Director of Employee Benefits

Agenda:

1. Flextime—Stacy and Christie reported on flextime options. They stated that in a survey dated March 15, 76% of employees favored flextime benefits. These included beginning our workdays at 7:00 A.M. and extending work hours until 6:30 P.M. Doing so would allow employees to either arrive for work earlier than standard and/or work later than usual. This work-schedule flexibility would allow employees to manage child care needs, appointments, and other concerns that might fall within the traditional workday.

   The downside of flextime, according to management, includes increased security coverage, increased utilities (heating, lighting, etc.), and inconsistent coverage of office hours. The benefits of flextime include increased employee morale, less congestion in the parking lots at traditionally prime times, and employee empowerment. Our project team voted 6-1 to promote flextime to management at next month's meeting.

2. Stock Options—At last month's meeting, Darren was asked to study the possibility of adding stock options to our cafeteria plan. He met with Accounting, our external Auditor, and three proprietary stock brokerage companies (Bull and Bear, Market Trend, and BuyItNow). Please see the attached brokerage firm proposals. The proposals offered exciting information.

   However, both Accounting and our Auditor convinced Darren that this year would not be the best time to offer our employees stock options. Our profit margin ratio is down, we cannot meet our quarterly forecasts, and we cannot distribute dividends to stockholders.

   Future European expansion plans should increase our revenues. Therefore, our project team voted 5-2 to table this issue until the end of next quarter.

3. Personal Leave Days—Currently, our employees are allotted five sick leave days a year. Our proposal asks for an additional two personal leave days. These days would have to be used within the calendar year. The personal leave days would not roll over, nor could they be banked. Andrew McWard surveyed the employees regarding this topic. Ninety-eight percent of those surveyed favored the personal leave proposal.

(Continued)

FIGURE 16.8   Meeting Minutes (Continued)

Page 2
Meeting Minutes
April 12, 2012

Our project team voted 6-1 to promote personal leave days as a cafeteria plan option.

4. Sick Bank—Last year, a survey suggested that employees might be willing to share their unused sick leave days with other employees facing catastrophic illnesses or family emergencies. This is a common practice at other corporations of our size. This not only gives employees a feeling of ownership, but also it is a method of team building.

Bill revisited this issue with employees. He attended departmental meetings across the company, asking for input. Bill reported that employees do not seem to favor this proposal at the moment. His informal survey shows only a 30 percent interest. Unless these numbers change, our committee sees no reason to pursue a sick bank. We voted 4-3 against proposing a sick bank option.

5. Short- and Long-Term Disability—Jessica visited three insurance carriers (In-need Insurance, Evergreen Insurance Inc., and Employee Associates General). Her goal was to get quotes on short- and long-term disability insurance. Our employee survey ranked this as a number one priority.

Our committee agreed with the survey and highly recommended this as an add-on to our employee cafeteria plan (7-0 vote). In fact, without this benefit, employees with illnesses longer than five days or any banked hours have no salary or job insurance. This fact will negatively impact both employee retention and recruitment. Most major firms offer this benefit. To stay competitive, we must offer this benefit also.

Old Business—Andrew reminded us that the company holiday party and family picnic needed to be planned. We will contact the hospitality committee to determine their status.

New Business—In next month's meeting, we will focus on ways to implement our recommendations.

Submitted by Jessica Studin, Recording Secretary

Attachment:
Brokerage Firm Proposals
Insurance Proposals

The Writing Process at Work

Look at how one student used the writing process (prewriting, writing, and rewriting) to construct her progress report.

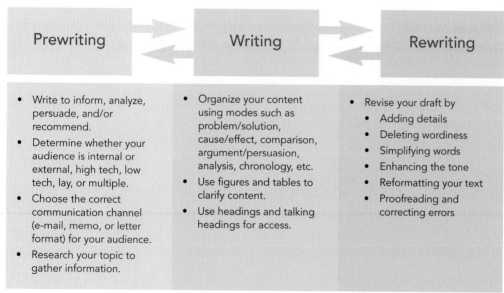

The Writing Process

Prewriting	Writing	Rewriting

- Write to inform, analyze, persuade, and/or recommend.
- Determine whether your audience is internal or external, high tech, low tech, lay, or multiple.
- Choose the correct communication channel (e-mail, memo, or letter format) for your audience.
- Research your topic to gather information.

- Organize your content using modes such as problem/solution, cause/effect, comparison, argument/persuasion, analysis, chronology, etc.
- Use figures and tables to clarify content.
- Use headings and talking headings for access.

- Revise your draft by
  - Adding details
  - Deleting wordiness
  - Simplifying words
  - Enhancing the tone
  - Reformatting your text
  - Proofreading and correcting errors

Prewriting

The student used a simple topic outline to gather data and determine her objectives (Figure 16.9).

FIGURE 16.9   Topic Outline for Prewriting

Sales Proposal to Round Rock Hospital

I. Introduction
        A. When—Oct. 10
        B. What—Sales Proposal to hospital
        C. Why—in response to your memo

II. Discussion
        A. What points to cover?
            1. Containers
            2. Sterilization
            3. Documentation
            4. Cost
            5. Disposal
        B. What's next?
            1. Future work requirements

III. Conclusion/Recommendations
        A. Conclusion—When? On schedule, 50 percent completed
        B. Recommendations—Comply with regulations

Writing

The student drafted her progress report, focusing on the information she discovered in prewriting. After drafting this report, the student submitted her copy to a peer review group, which helped her edit and revise her text (Figure 16.10).

Rewriting

After discussing the suggested changes with her peer review group, the student revised her draft and submitted her finished copy (Figure 16.11).

FIGURE 16.10   Rough Draft with Peer Group Revision Suggestions

November 6, 2012
To: Carolyn Jensen
From: Shuan Wang.
Subject: Progress Report.

Omit the period and clarify the topic of this report.

Purpose: This is a progress report on the status of a sales proposal you requested in your memo of October 10, 2012. The objective of this proposal is the sale of a total program for infectious waste control and disposal to a Round Rock Hospital. The proposal will cover the following five <u>things</u>.

1. Containers
2. Steam sterilization
3. Cost savings
4. Landfill operating
5. Computerized documentation

The second sentence above is long. Remember, our teacher told us to limit sentences to 10–15 words.

"Things" = weak word!

Work Completed: The following is a list of items that are finished on the project.

Boldface the headings for emphasis.

Containers
I have finished a description of the specially designed containers with disposable biohazard bag.

Steam sterilization
I have <u>set a</u> instructions for using the Biological Detector, model BD 12130 as a reliable indicator of sterilization of infectious wastes.

Correct the typo = "a set of instructions"

Cost savings
Central Hospital sent me some information on installing a pathological incinerator as well as the constant personal and maintenance costs to operate them. I ran this information through our computer program "Save-Save". The results show that a substantial savings will be realized by the use of our services. I have data to support this, since more and more of our customer base is online with the system proving that expenses to other companies has been reduced.

Future Work: I have an appointment to visit Round Rock Hospital on November 7, 2012. At that time I will make on-site evaluations of present waste handling practices at Round Rock Hospital. After carefully <u>studing</u> this evaluation, I will report to you on needed changes to comply with governmental guidelines.

Correct the spelling error

Conclusion: The project is proceeding on schedule. Approximately 50% of work is done (see graph). I don't see any problem at this time and should be able to meet the target date. I will update you on our continuing progress in the near future.

Why is there an empty cell in the graphic?

FIGURE 16.11   Finished Progress Report in Memo Format

Date: November 6, 2012
To: Carolyn Jensen
From: Shuan Wang
Subject: Progress Report on Round Rock Hospital

Purpose of Report

In response to your October 10, 2012, request, following is a progress report on our Round Rock Hospital sales proposal. This proposal will present Round Rock Hospital our total program for infectious waste control and disposal. The proposal will cover the following five topics:

- Containers
- Steam sterilization
- Cost savings
- On-site evaluations
- Landfill disposal

Work Completed

1. *Containers*—On October 5, I finished a description of the specially designed containers with biohazard bags. These were approved by Margaret Chase, Round Rock's Quality Control Administrator.

2. *Steam sterilization*—On November 4, I wrote instructions for using our Biological Detector, model BD 12130. This mechanism measures infectious waste sterilization levels.

3. *Cost savings*—Round Rock Hospital's accounting manager, Lenny Goodman, sent me their cost charts for pathological incinerator expenses and maintenance costs. I used our computer program "Save-Save" to evaluate these figures. The results show that we can save the hospital $15,000 per year on annual incinerator costs, after a two year break-even period. I have data to support this from other customers.

Work Remaining

1. *On-site evaluation*—When I visit Round Rock Hospital on November 7, I will evaluate their present waste handling practices. After studying these evaluations, I will report necessary changes for governmental compliance.

2. *Landfill disposal*—I will contact the Environmental Protection Agency on November 13, 2012, to receive authorization for disposing sterilized waste at our sanitary landfill.

FIGURE 16.11 (Continued)

Wang
Page 2
November 6, 2012

Completion of Project

The project is proceeding on schedule. Approximately 50 percent of work is done (see Figure 1).

Figure 1: Completion Status of Planned Activities

Legend:
- Computerization
- Landfill
- Cost Savings
- Sterilization
- Containers

Y-axis: % Complete
X-axis: 0, 50, 100

I see no problems at this time. We should meet our December 31, 2012, target date. I will update you on our continuing progress in two weeks.

CHAPTER HIGHLIGHTS

1. Reports are used to document many different occurrences on the job and written to both internal and external audiences.
2. Use headings and talking headings when designing your report.
3. E-mail, letters, and memos are effective communication channels for short, informal reports.
4. Progress reports recount work accomplished and work remaining on a project.
5. Lab reports document the findings from a lab analysis.
6. Feasibility/recommendation reports are used to determine the viability of a proposed project.
7. Graphics help your audience understand and access content.
8. An incident report documents an unexpected problem that has occurred.
9. An investigative report asks you to examine the causes behind an incident.
10. Meeting minutes document the results of a meeting.

CASE STUDIES

Read the following case studies and write the appropriate report.

1. You manage an engineering department at Acme Aerospace. Your current department supervisor is retiring. Thus, you must recommend the promotion of a new supervisor to the company's executive officer, Kelly Adams. You know that Acme seeks to promote individuals who have the following traits:

   • Familiarity with modern management techniques and concerns, such as teamwork, global economics, crisis management, and the management of hazardous materials
   • An ability to work well with colleagues (subordinates, lateral peers, and management)
   • Thorough knowledge of one's areas of expertise

   You have the following candidates for promotion. Using the information provided about each and the criteria for feasibility/recommendation reports discussed in this chapter, write your report recommending your choice for a new supervisor.

   a. **Pat Jefferson.** Pat has worked for Acme for 12 years. In fact, Pat has worked up to a position as a lead engineer by having started as an assembler, then working in test equipment, quality control, and environmental safety and health (ESH). As an engineer in ESH, Pat was primarily in charge of hazardous waste disposal. Pat's experience is lengthy, although Pat has only taken two years of college coursework and one class in management techniques. Pat is well liked by all colleagues and is considered to be a team player.

   b. **Kim Kennedy.** Kim is a relatively new employee at Acme, having worked for the company for two years. Kim was hired directly out of college after earning an MBA degree from the Mountaintop College School of Management. As such, Kim is extremely familiar with today's management climate and modern management techniques. Kim's undergraduate degree was a BS in business with a minor in engineering. Currently, Kim works in the engineering department as a departmental liaison, communicating the engineering department's concerns to Acme's other departments. Kim has developed a reputation as an excellent co-worker who is well liked by all levels of employees.

   c. **Chris Clinton.** Chris has a BS degree in engineering from Poloma College and an MBA degree from Weatherford University. Prior to working for Acme, Chris served on the IEEE (Institute of Electrical and Electronic Engineering) Commission for Management Innovation, specializing in global concerns and crisis management. In 1991, Chris was hired by Acme and since then has worked in various capacities. Chris is now lead engineer in the engineering department. Chris has earned high scores on every yearly evaluation, especially regarding knowledge of engineering. Whenever you have needed assistance with new management techniques, Chris has been a valued resource. Chris's only negative points on evaluations have resulted from difficulties with colleagues, some of whom regard Chris as haughty.

   Whom will you recommend for supervisor? Write your feasibility/recommendation report stating your decision.

2. Your company, Telecommunications R Us (TRU), has experienced a 45 percent increase in business, a 37 percent increase in warehoused stock, and a 23 percent increase in employees. You need more room. Your executive officer, Polina Gertsberg, has asked you to research existing options. To do so, you know you must consider these criteria:

   • **Ample space for further expansion.** Gertsberg suggests that TRU could experience further growth upward of 150 percent. You need to consider room for parking, warehouse space, additional offices, and a cafeteria—approximately 20,000 square feet total.

- **Cost.** Twenty million dollars should be the top figure, with a preferred payback of five years at 10 percent.
- **Location.** Most of your employees and customers live within 15 miles of your current location. This has worked well for deliveries and employee satisfaction. A new location within this 15-mile radius is preferred.
- **Aesthetics.** Ergonomics suggest that a beautiful site improves employee morale and increases productivity.

After research, you've found three possible sites. Based on the following information and on the criteria for feasibility/recommendation reports discussed in this chapter, write your report recommending a new office site.

a. **Site 1 (11717 Grandview).** This four-story site, located 12 miles from your current site, offers three floors of finished space equaling 18,000 square feet. The fourth floor is an unfinished shell equaling an additional 3,000 square feet. As is, the building will sell for $19 million. If the current owner finishes the fourth floor, the addition would cost $4 million more. For the building as is, the owner asks for payment in five years at 12 percent interest. If the fourth floor is finished by the owner, payment is requested in seven years at 10 percent. The building has ample parking space but no cafeteria, although a building next door has available food services. Site 1 is nestled in a beautifully wooded area with hiking trails and picnic facilities.

b. **Site 2 (808 W. Blue Valley).** This one-story building offers 21,000 square feet that includes 100 existing offices, a warehouse capable of holding 80 storage bins that measure 20 feet tall × 60 yards long × 8 feet wide, and a full-service cafeteria. Because the complex is one story, it takes up 90 percent of the lot, leaving only 10 percent for parking. Additional parking is located across an eight-lane highway that can be crossed via a footbridge. The building, located 18 miles from your current site, has an asking price of $22 million at 8.75 percent interest for five years. Site 2 has a cornfield to its east, the highway to its west, a small lake to its north where flocks of geese nest, and a strip mall to its south.

c. **Site 3 (1202 Red Bridge Avenue).** This site is 27 miles from your current location. It has three stories offering 23,000 square feet, a large warehouse with four-bay loading dock, and a cafeteria with ample seating and vending machines for food and drink. Because this site is located near a heavily industrialized area, the asking price is $15 million at 7.5 percent interest for five years.

Which site do you recommend? Write your feasibility/recommendation report stating your choice.

INDIVIDUAL AND TEAM PROJECTS

1. Write a progress report. The subject of this report can involve a project or activity at work. If you haven't been involved in job-related projects, write about the progress you're making in this class or another course you're taking. Write about the progress you're making on a home improvement project (refinishing a basement, constructing a deck, or painting and papering a room). Write about the progress you're making on a hobby (rebuilding an antique car, constructing a computer, or making model trains). Follow the criteria presented in this chapter regarding progress reports.

2. Write a lab report. The subject of this report can involve a test you're running at work or in one of your classes. Use the criteria regarding lab reports presented in this chapter to help you write the report.

3. Write a feasibility/recommendation report. You can draw your topic either from your work environment or home. For example, if you and your colleagues were considering the purchase of new equipment, the implementation of a new procedure, expansion to a new location, or the marketing of a new product, you could study this idea and then write a report on your findings. If nothing at work lends itself to this topic, then consider plans at home. For example, are you and your family planning a vacation, the purchase of a new home or car, the renovation of your basement, or a new business venture? If so, study this situation. Research car and home options, study the market for a new business, and get bids

for the renovation. Then write a feasibility/recommendation report to your family documenting your findings. Follow the criteria for feasibility/recommendation reports provided in this chapter to help you write the report.

4. Write an incident report. You can select a topic either from work or home. If you have encountered a problem at work, write an incident report documenting the problem and providing your solutions to the incident. If nothing has happened at work lending itself to this topic, then look at home. Has your car broken down, did the water heater break, did

---

Site Visit—Alamo Manufacturing
November 1

Sam, I visited our Alamo site and checked on the following:

1. Our plant facilities suffered some severe problems due to the recent wind and hail storms. The west roof lost dozens of shingles, leading to water damage in the manufacturing room below. The HVAC unit was submerged by several feet of water, shorting out systems elsewhere in the plant. In addition, our north entryway awning was torn off its foundation due to heavy winds. This not only caused broken glass in our front entrance door and a few windows bordering the entrance, but also the entryway driveway now is blocked for customer access.

2. Maintenance and security failed to handle the problems effectively. Security did not contact local police to secure the facilities. Maintenance responded to the problems far too many hours late. This led to additional water and wind damage.

3. Luckily, the storm hit early in the day, before many of our employees had arrived at work. Still, a few cars were in the parking lot, and they suffered hail damage. Are we responsible?

4. The storm, though hurting manufacturing, will not affect sales. Our 800-lines and e-mail system were unaffected. But, we might have problems with delivery if we can't fix the entryway impediment. I think I have some solutions. We could reroute delivery to one of our new plants, or maybe we should consider direct delivery to our sales staff (short term at least). Any thoughts?

5. Meanwhile, I have gotten on Maintenance's case, and it is working on repairs. Brownfield HVAC service has given us a quote on a new sump pump system that has failsafe programs (not too bad an extra cost, given what we can save in the long run). Plainview Windows & Doors is already on the front entrance problem. It promises replacement soon.

6. As for future plans, I think we need to reevaluate both Maintenance and Security. Training is an option. We could also add new personnel, reconsider our current management in those departments, or maybe just a few, good, hard, strongly stated comments from you would do the trick.

Anyway, we're up and running again. Upfront repair costs are covered by insurance, and long-term costs are minimal because sales weren't hurt. Our only remaining challenges are preventive, and that depends on what we do with our Maintenance and Security staff. Let me know what you think.

---

you or any members of your family have an accident of any sort, or did your dog or cat knock over the vase your mother-in-law gave you for Christmas? Consider such possibilities, and then write an incident report documenting the incident. Follow the criteria for incident reports provided in this chapter.

5. Revision is the key to good writing. An example of a flawed progress report that needs revising is shown in the example on page 494. To improve this report, form small groups, decide what's missing according to this chapter's criteria for good progress reports, and then revise the report.

PROBLEM-SOLVING THINK PIECES

1. Angel Guerrero, computer information technologist at HeartHome Insurance, has traveled from his home office to a branch location out of town. While on his job-related travel, he encountered a problem with his company's remanufactured laptop computer. He realized that the problem had been ongoing not only for this laptop but also for six other remanufactured laptops that the company had recently purchased.

   Angel thinks he knows why the laptops are malfunctioning and plans to research the issue. When he returns to his home office from traveling, he needs to write a report. What type of report should he write? Explain your answer, based on the information provided in this chapter.

2. Minh Tran is a special events planner in the marketing department at Thrill-a-Minute Entertainment Theme Park. Minh and her project team are in the middle of a long-term project. For the last eight months, they have been planning the grand opening of the theme park's newest sensation ride—The Horror—a wooden roller coaster that boasts a 10 g (gravity) drop.

   During one of its weekly project meetings, the team has hit a roadblock. The rap group Bite R/B Bit originally slated to play at the midnight unveiling has cancelled at the last minute. The team needs to get a replacement band. One of Minh's teammates has researched the problem and presented six alternatives bands (at varying prices and levels of talent) for consideration. Minh needs to write a report to her supervisor.

   What type of report should she write? Explain your decision, based on the criteria for reports provided in this chapter.

3. Toby Hebert is human resource manager at Crab Bayou Industries (Crab Bayou, Louisiana), the world's largest wholesaler of frozen Cajun food. She and her staff have traveled to New Orleans to attend meetings held by five insurance companies that planned to explain and promote their employee benefits packages.

   During the trip, Toby's company van is sideswiped by an uninsured driver. When Toby returns to Crab Bayou, she has to write a report. What type of report should she write? Base your decision on the criteria for reports provided in this chapter.

4. Bill Baker, claims adjuster for CasualtyU Insurance Company (CUIC), traveled for a site visit. Six houses insured by his company were in a neighborhood struck by a hailstorm and 70 mile per hour winds. The houses suffered roof and siding damages. Trees were uprooted, with one falling on a neighbor's car. Though no injuries were sustained, over $50,000 worth of property damage occurred.

   When he returns to CUIC, what type of report should he submit to his boss? Defend your decision based on criteria for reports provided in this chapter.

WEB WORKSHOP

1. More and more, companies and organizations are putting report forms online. The reason for doing this is simple—ease of use. Go online and access online report forms. Use any Internet search engine, and type in "online _____ report form." (In the space provided, type in "trip," "progress," "incident," "investigative," or "feasibility/recommendation.") Once you

find examples, evaluate how they are similar to and different from the written reports shown in this chapter. Share your findings with others in your class, either through oral presentations or written reports.

2. Every company writes reports. You can find examples of them online. Use the Internet to access Report Gallery, found at http://www.annualreports.com/. Report Gallery bills itself as "the largest Internet publisher of annual reports." From this Web site, you can find the annual reports from 2,200 companies, including many Fortune 500 companies.

Study the annual reports from 5 to 10 different companies. What do these reports have in common? How do they differ from each other? Discuss the reports' page layout, readability, audience involvement, content, tone, and development. How are the reports similar to and different from those discussed in this chapter?

Share your findings with others in your class, either through oral presentations or written reports.

QUIZ QUESTIONS

1. What are purposes of a report?
2. What are common types of short, informal reports?
3. List the four basic units of a report.
4. What types of material are included in the discussion section?
5. Explain how the reporter's questions can help you prepare your report.
6. Why would you use e-mail, memo, or letter formats for short, informal reports?
7. What organizational modes are useful for writing short, informal reports?
8. When do you write a trip report?
9. When do you write a progress report?
10. When do you write a lab report?
11. When do you write a feasibility/recommendation report?
12. When do you write an incident report?
13. When do you write an investigative report?
14. What content do you include in meeting minutes?
15. How do visual aids assist the reader in a report?
16. What do you include in the identification lines?
17. What do you include in the conclusion?
18. What do you include in the recommendation?
19. What is the writing style appropriate for a report?
20. How do you satisfy multiple readers in the introduction?

CHAPTER SEVENTEEN

Long, Formal Reports

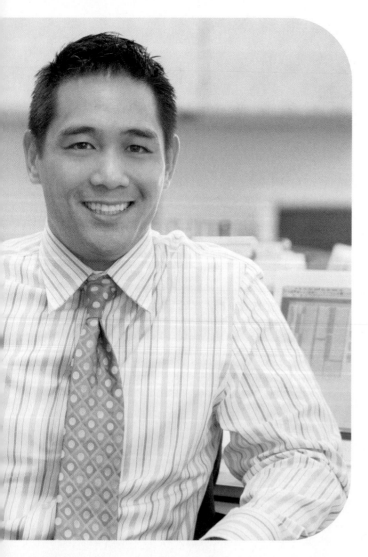

COMMUNICATION AT WORK

In the following scenario, an information technology consulting company needs to write a long, formal report recommending a backup data storage facility for a client.

Alpha/Beta Consulting (A/B/C) is going to work with Nitrous Solutions to help them choose an offsite backup data storage facility that will be safe, secure, dependable, prompt, and able to grow as the company grows. Nitrous Solutions, a multidiscipline architectural design company, is a relatively new but successful business with no offsite backup data storage location.

Nitrous knows that offsite data backup is important for many reasons. First, after the 9/11 attacks, many companies realized that their data, if only housed on site, could be endangered. What companies need is a defense-in-depth strategy to ensure data security. Offsite data storage provides companies remote backup for worst-case scenarios, such as terrorism. With offsite data storage, even if a company's primary facility is damaged, data stored in another location is preserved.

Objectives

In addition, weather problems, highlighted by the New Orleans Hurricane Katrina catastrophe, proved to companies that multiple levels of data redundancy were needed to secure their files. High waters and rain damage caused many Gulf Coast businesses to lose valuable data. Offsite storage, in contrast, would provide a higher level of security.

Data loss isn't just due to terrorism or natural disasters. Nitrous also realizes that their files are endangered from simple causes like software and hardware failure. When disasters occur, a company can lose days, weeks, or months of past work. Recovering these files takes more time, negatively impacts the company's credibility with clients, and can lead to financial losses.

Nitrous needs to factor in one more consideration when choosing an offsite storage option. The company does not want to trust couriers to transport files from their home site to a separate location. Nitrous doesn't want an employee to put data tapes in a car and drive hundreds or thousands of miles. This kind of transport could lead to additional problems—lost time, car wrecks, theft, etc. Instead, Nitrous wants a secure means of electronic file transfer that will be quick, safe, and efficient.

Management has hired A/B/C to help them determine their data storage needs. A/B/C has met with management and key employees at Nitrous to study their client's unique needs. Now, A/B/C will write a long, formal report to inform, analyze, and recommend. The report will provide factual information about backup solutions for archiving architectural materials and critical communications for Nitrous.

The long, formal report will then analyze options, focusing on pros and cons of various solutions. In this analysis, A/B/C will draw conclusions about pricing, support, scalability, reliability, ease of use, and archiving choices.

Based on their findings and analysis of the options, A/B/C will draw a conclusion. Then, the consulting company will recommend the best options to help Nitrous fulfill their company's mission statement.

Why Write a Long, Formal Report?

Short Reports

See Chapter 16 for more discussion of short, informal reports.

In some instances, your subject matter might be so complex that a short report will not thoroughly cover the topic. For example, your company asks you to write a report about the possibility of an impending merger. This merger will require significant commitments regarding employees, schedules, equipment, training, facilities, and finances. Only a long report, complete with research, will convey your content sufficiently and successfully. Long reports are not simply extensions of short reports. Long, researched reports require time, resources of people and money, and have far-reaching effects.

The following are titles of long, formal, research reports:

- The Effect of Light Rail on Wyandotte County: An Economic Impact Study on Infrastructure
- The Increasing Importance of Mobile Communications: Security, Procurement, Deployment, and Support
- Managing Change in a Technologically Advancing Marketplace at EBA Corp.
- Information Stewardship at The Colony, Inc.: Legal Requirements for Protecting Information
- Remote Access Implementation and Management for New Hampshire State University

Types of Long, Formal Reports: Informative, Analytical, and Recommendation

Reports can provide information, be analytical, and/or recommend a course of action. Occasionally, you might write a report that only informs. You might write a report that just analyzes a situation. You could write a report to recommend action persuasively. Usually, however, these three goals will overlap in your long, formal reports.

For example, let's say that your company is expanding globally and will need a WAN (a wide area network that spans a large geographical area). In a long, formal report to management, you might first write to inform the audience that this WAN will meet the strategic goals of storing and transmitting information to your global co-workers. The report also will analyze the ways in which this network will be safe, reliable, fast, and efficient. Finally, the report will recommend the best designs for providing secure communications to clients, partners, vendors, and co-workers. Ultimately, you must persuade the audience, through research, that your envisioned network's design, hardware specifications, software, and estimated budget will satisfy both internal and external needs. Such a report would have to be informative, analytical, and persuasive to convince the audience to act on the recommended suggestion.

Information

When you provide information to your audience, focus on the facts. These facts will help your readers better understand the situation, the context, or the status of the topic. For example, Nitrous Solutions asked Alpha/Beta Consulting to report on backup data storage options for software archives (see the opening scenario on page 498). To so do, A/B/C needed to present factual information about the importance of backup storage. In a long, formal report, A/B/C informed its client about the following: the threats to data, the cost of lost data, how often data should be backed up, and what data should be archived (see Figure 17.1).

FIGURE 17.1   Information about Backup Data Storage

Description of Backup Data Storage:

The following information provides facts about permanent storage to help Nitrous Solutions maintain a high degree of security in an offsite backup data storage location.

Facts:

Due to events such as 9/11 and Hurricane Katrina, the importance of maintaining mission-critical electronic data and backups has become evident for companies that wish to avoid a catastrophic outcome. Moving data to a separate offsite location dilutes risk and protects against data loss when, for example, the company's IT infrastructure or its critical electronic information is damaged by the following:

Threats to Your Data	Dangers and Costs of Data Loss
➤ Fire	➤ Loss of mission-critical files
➤ Flood	➤ Corrupted database
➤ Hurricane or tornado	➤ Corrupted operating system
➤ Lightning strike	➤ Loss of all files on hard drive
➤ Earthquake	➤ Laptop damage, theft or loss
➤ Heat, sunlight	➤ Damaged computer hardware
➤ Humidity, moisture, spilled liquids	➤ Backup media loss or theft
➤ Smoke, dust and dirt	➤ Total loss of all data at your site
➤ Electrical surge or power failure	➤ Compromised data security
➤ Media failure	➤ Competitor access to your data
➤ Hard drive failure	➤ Lost business records

When and What Data Should be Backed Up?

Some basic questions that are commonly asked when determining when and what data should be backed up are as follows:

How often should desktop PCs or Macs be backed up?

- Data on desktop PCs and Macs should be backed up at a minimum of at least once a week. This will prevent anyone losing more than one week's worth of information.

Should everything on my PC or Mac be backed up once a week?

- This is not necessary. Only data needs to be backed up once a week. Software, operating systems, and even static data do not need to be backed up that often. Backing up everything on your PC or Mac adds time to the process and requires substantially more backup tapes, backup disks, or other types of media.

How long should I keep computer backups?

- Some companies keep original backups up to seven years and have a set schedule that includes five different increments of backups: daily, weekly, monthly, quarterly, and yearly.

This is a topic for discussion with your key management, legal staff, information systems technicians, records management, and key personnel. Balancing the needs of a company's computer system protection and good records management takes informed decisions on the parts of all these departments.

Information is presented in a variety of ways. The consulting company details data in bulleted lists, asks questions and provides answers, and explains facts and findings in paragraphs.

Analysis

Visual Aids

See Chapter 10 for more discussion of tables and figures.

When you analyze for your audience, you begin with factual information. However, you expand on this information by interpreting it and then drawing conclusions. Once Alpha/Beta Consulting presented the informational findings about backup data storage facilities, they followed this information with a more in-depth analysis of backup options and drew a conclusion for their client (see Figure 17.2).

FIGURE 17.2 Analysis of Backup Data Storage Options

Analysis of Backup Data Storage Options

Industry analysts claim that two out of five businesses that experience a disaster will go out of business within five years of the event due to information and service loss as a consequence of disasters.

This is an alarming statistic considering the high cost of not expecting the unexpected. In the event of a disaster, a corporation could lose hundreds, thousands, or even millions of dollars through lost productivity. At worst the corporation could go out of business without the possibility of a second chance.

The two main types of offsite data storage include the following:

Offsite Data Storage Facility	Internet Data Storage Facility
Tapes, CDs, DVDs, or hard drives are sent to a predetermined offsite location for security and disaster recovery purposes. This allows a business to be safe in the knowledge that it will be able to recover quickly and seamlessly from any disaster. With simply a walk across the street or a drive downtown to the storage facility, the business can quickly begin to rebuild from the data it has stored.	Offsite Internet data storage, although relatively new, has quickly become one of the fastest growing arenas in the records and information management industry. A business can simply select the files and the time that it wishes to have its files backed up with the offsite backup software. Its files are compressed and encrypted and sent to a backup server over an existing Internet connection. The business can upload or download your files as often as needed, with usually no additional charges.

FIGURE 17.2   Continued

Offsite Location for Data Storage Options	
Offsite Residential Location	**Offsite Facility Location**
Costs	
• Completely Free	• $5–$650 per month depending on space and features
Reliability	
• Data are accessible when needed. • Save and archive as much as you need with no restrictions or extra fees.	• Insurance on data is automatic with contract. • Data are fully accessible during normal business hours, after hours by emergency only.

Security	
• More secure than leaving your data in your office building	• Vault and/or building monitored by 24-hour surveillance and alarm systems • Security patrolled
Scalability	
• Can store as little or as much data as needed, with no restrictions, guidelines, rules or regulations free of charge	• Just like the Internet storage, as your company grows, so can data storage. This increases monthly fees.
Ease of Use	
• Your house, your rules • Maintenance of data management can be completed in your spare time.	• Easy to set up and manage • Step-by-step instructions available • Some companies offer free technical help.

Conclusion

Regardless of how much or how little a business uses a computer, your company will create important and unique data. The unique data can include financial and project budget records, digital images, client profiles, and marketing sales. The data are priceless and constantly at risk.

Data loss is likely to occur for many reasons:

- Hardware failure = 42%
- Human error = 35%
- Software corruption = 13%
- PC viruses = 7%
- Hardware destruction = 3%

When analyzing information, A/B/C presents facts and figures, pros and cons, and specific details.

Recommendation

After providing information and analysis, you can recommend action as a follow-up to your findings. The recommendation allows you to tell the audience why they should purchase a product, use a service, choose a vendor, select a software package, or follow a course of action. Alpha/Beta Consulting presented findings and analyzed information for its client Nitrous Solutions. Based on the analysis, the consulting company then made a recommendation (see Figure 17.3).

Major Components of Long, Formal Reports

Because short reports run only a few pages, you can assume that your readers will be able to follow your train of thought easily. Thus, short reports merely require that you use headings such as "Introduction," "Discussion," and "Conclusion/Recommendation" to guide your readers through the document or talking headings to summarize the content more thoroughly. Long reports, however, place a greater demand on readers. Your readers could be overwhelmed with many pages of information and research. A few headings won't be enough to help your readers wade through the data.

In addition to the basic components of a short report, a long, formal report includes the following:

- **Front matter** (title page, cover letter, table of contents, list of illustrations, and an abstract or executive summary)
- **Text** (introduction including purpose, issues, background, and problems; discussion; and conclusion/recommendation)
- **Back matter** (glossary, works cited or references page, and an optional appendix)

Figure 17.4 shows the components of a long, formal report.

Title Page

The title page tells your reader the

- Title of the long report
- Name of the company, writer, or writers submitting the long report
- Date on which the long report was completed

If the long report is being mailed outside your company to a client, you also might include on the title page the audience to whom the report is addressed. If the long report is being submitted within your company to peers, subordinates, supervisors, or owners, you might want to include a routing list of individuals who must sign off or approve the report.

FIGURE 17.3   Analysis and Recommendation

Offsite Location for Permanent Storage:
Analysis and Recommendation

Weighted Analysis:

Type		Offsite Residential		Offsite Facility	
Criteria	Weight	*Rating	Score	*Rating	Score
Cost	15%	5	.75	3	.45
Performance	25%	-	-	-	-
Reliability	30%	4	1.20	4	1.20
Security	15%	3	.45	4	.60
Scalability	10%	5	.50	5	.50
Ease of use	5%	4	.20	4	.20
Total	100%		3.1		2.95

*Rating: The weighted analysis is based on a 1 to 5 scale: 1 is poor, 5 is excellent.

Recommendation:

Because Nitrous Solutions does not currently have a network infrastructure in place, the data storage decision should be considered a high priority since all new data that the company creates will be significant to the company's overall success. Alpha/Beta Consulting recommends that Nitrous Solutions use the residential offsite data storage option due to the slight advantage it has over a professional storage facility. The one real advantage between the two is cost. Whereas residential storage is free with no contracts or monthly charges, a storage facility will charge anywhere from $5 to $650 per month, depending on the size of data and equipment that needs to be accompanied with such data.

If Nitrous Solutions determines the need to have multiple backup solutions, rather than a single reinforcement to the backup data, another recommended means of storage would be an Internet storage facility. Due to the popular demand and increasing technology, many different companies are now offering free trials of their Internet storage facilities and software. Because of the unknown intentions of Nitrous Solutions, only the features for Internet storage will be researched. Almost all options for Internet companies are either compatible or identical.

Based on information and an analysis of the findings, provided in a weighted table, A/B/C recommends a backup solution for the client.

FIGURE 17.4   Components of a Long, Formal Report

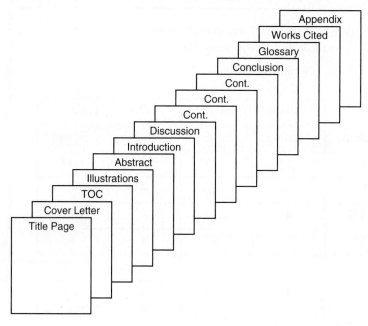

Following are two sample title pages. Figure 17.5 is for a long report with routing information. Figure 17.6 is for a long report without routing information.

Web Resources

www.pearsonhighered
.com/gerson For more
information about letters,
visit our companion website.

Cover Letter

Your cover letter prefaces the long report and provides the reader an overview of what is to follow. It tells the reader

- Why you are writing
- What you are writing about (the subject of this long report)
- What exactly of importance is within the report
- What you plan to do next as a follow-up
- When the action should occur
- Why that date is important

Letters

See Chapter 6 for more
discussion of cover letters.

Table of Contents

Long reports are read by many different readers, each of whom will have a special area of interest. For example, the managers who read your reports will be interested in cost concerns, timeframes, and personnel requirements. Technicians, in contrast, will be interested in technical descriptions and instructions. Not every reader will read every section of your long report.

Your responsibility is to help these different readers find the sections of the report that interest them. One way to accomplish this is through a table of contents. The table of contents should be a complete and accurate listing of the main and minor topics covered in the report. In other words, you don't want just a sketchy outline of

FIGURE 17.5 Title Page for Long Report (with Routing Information)

Report on Multicultural Workforce at StartCo Insurance

Prepared by: _____ Date: _____
           Pete Niosi
           Assistant Director, Human Resources

Reviewed by: _____ Date: _____
           Leah Workman
           Manager, Accounting

Recommended by: _____ Date: _____
           Greg Foss
           Department Supervisor, Customer Service

Recommended by: _____ Date: _____
           Shirley Chandley
           Director, Human Resources

Approved by: _____ Date: _____
           Ralph Houston
           Vice President

FIGURE 17.6 Title Page for a Long Report (without Routing Information)

Computer Networking Report:
The Need to Upgrade Intranet Capabilities

Title the report.

for
Acme Products, Inc.
2121 New Line Avenue
San Antonio, TX 78666

Provide the reader's name(s), title, and company name.

Submitted by
Thomas Brasher
CEO, Technology Upgrades
3254 West King's Highway
San Antonio, TX 78221

Give the writer's name(s), title, and company.

August 13, 2012

major headings. This could lead to page gaps; your readers would be unable to find key ideas of interest. In the "before" table of contents, the discussion section contains approximately 16 pages of data. What is covered in those 16 pages? Is anything of value discussed? We don't know.

BEFORE

This poorly written table of contents omits second-level headings and has too many page gaps.

In contrast, an effective table of contents fleshes out this detail so your readers know exactly what is covered in each section. By providing a thorough table of contents, you will save your readers time and help them find the information they want and need. The "after" example is a successful table of contents.

In the "after" example, note that the actual pagination (page 1) begins with the introductory section. Page 1 begins with your main text, not the front matter. Instead, information prior to the introduction is numbered with lowercase Roman numerals (i, ii, iii, etc.). Thus, the title page is page i, and the cover letter is page ii. However, you don't need to print the numbers on these two pages. Therefore, the first page with a printed number is the table of contents. This is page iii, with the lowercase Roman numeral printed at the foot of the page and centered.

List of Illustrations

If your long report contains several tables or figures, provide a list of illustrations. This list can be included below your table of contents, if there is room on the page, or on a separate page. As with the table of contents, your list of illustrations must be clear and informative. Don't waste your time and your reader's time by providing a poor list of illustrations like in the "before" example. Instead, provide a list like the one shown in the "after" example.

Web Resources

www.pearsonhighered .com/gerson For more information about graphics, visit our companion website.

Abstract

As mentioned earlier, a number of different readers will be interested in your long report. Because your readers are busy with many different concerns and might have little technical knowledge, they need your help in two ways: They need information quickly, and they need it presented in low-tech terminology. You can achieve both these objectives through an abstract or executive summary.

Table of Contents

iii

Second-level and third-level headings help the audience access all parts of the report in any order.

Creating a Hierarchy of Headings Using Microsoft Word 2010

When writing your long, formal reports, include headings and subheadings. To help your readers navigate the text, create a clear hierarchy of headings (also called cascading headings) that distinguish among first-level headings, second-level headings, third-level headings, and fourth-level headings.

Microsoft Word 2010 will help you create this hierarchy of headings. You can apply a style to your entire document in Word's "formatting text by using styles" tool.

For example, let's say that you want your headings and subheadings to look as follows:

Hierarchy of Headings	Description
HEADING ONE	First-level headings: Arial, all cap, boldface, 16 pt. font.
Heading two	Second-level headings: Times New Roman, italics, boldface, 14 pt. font.
Heading three	Third-level headings: Times New Roman, boldface, 12 pt. font.
Heading four	Fourth-level headings: Times New Roman, italics, 12 pt. font.

To create a hierarchy of headings, follow these steps:

1. Click on the **Home** tab and then click on the down arrow to the right of Heading 3.

The following styles and formatting box will pop up.

2. Right click on the **Heading 1** button and scroll to **Modify.** Here, you can change font type, size, and color.

BEFORE	AFTER

The abstract is a brief overview of the proposal or long report's key points geared toward a low-tech reader. To accomplish the required brevity, limit your abstract to approximately one to two pages.

Each long report you write will focus on unique ideas. Therefore, the content of your abstracts will differ. Nonetheless, abstracts might focus on the following: the *problems* necessitating your report, your suggested *solutions*, and the *benefits* derived when your suggestions are implemented.

For example, let's say you are asked to write a formal report suggesting a course of action (limiting excessive personnel, increasing your company's workforce, improving your corporation's physical facilities, etc.). First, your abstract should specify the problem requiring your planned action. Next, mention the action you are planning to implement. This leads to a brief overview of how your plan would solve the problem, thus benefiting your company.

Another approach to an abstract for a long report is to summarize the content being analyzed in the report, setting the stage for the major issues to be discussed. In this abstract, highlight all the main points in the long report, omitting any supporting facts and documentation (to be developed in the Discussion section of the report). You not only want to be brief, focusing on the most important issues, but also you should avoid high-tech terminology and concepts. The purpose of the abstract is to provide your readers with an easy-to-understand summary of the entire report's focus. Your readers want the bottom line, and they want it quickly. Therefore, either avoid all high-tech terminology completely, define your terms parenthetically, or refer readers to a glossary.

Executive Summary

An executive summary, similar to an abstract but generally longer, is found at the beginning of either a formal report or a proposal and summarizes the major topics covered in the document. By reading the executive summary, your audience (the decision makers) gets an overview of the much longer report or proposal. Executive summaries can even take the place of the much longer formal report or proposal. In this way, you save your reader's time and communicate only the most relevant parts of the report or proposal. To create an effective executive summary, follow these key points:

1. Begin with a purpose statement for the report or proposal.
2. Give an overview of the key ideas discussed in the report or proposal.
3. Inform the reader if any problems will affect the outcome of the report or proposal.

4. Suggest solutions to the problems, or encourage the reader to suggest solutions.

5. Conclude the executive summary and recommend a course of action.

Key Parts of an Executive Summary. To write an effective and concise executive summary, use only the most important details and supporting statistics or information. Omit details or technical content which your reader either does not need to know or would unnecessarily confuse the reader. Details are used in the discussion section of the report or the proposal. To limit the length of your executive summary to one or two pages, choose the most important elements, and omit those that are secondary. An effective summary can include the following:

- Purpose and scope of the report or proposal, stating the problem or need and your ability to improve the situation
- Research or methods used to develop your content
- Conclusions about your analyses of the topic
- Your qualifications showing that you can resolve the issue
- A project management plan and timetable
- The total project budget
- Recommendations based on your findings

Introduction

Your introduction should include the following: purpose, background, and problem.

Purpose. In one to three sentences or a short paragraph, tell your readers the purpose of your long report. This purpose statement informs your readers *why* you are writing or *what* you hope to achieve. This statement repeats your abstract to a certain extent. However, it's not redundant; it's a reiteration. Although numerous people read your report, not all of them read each line or section of it. They skip and skim.

The purpose statement, in addition to the abstract, is another way to ensure that your readers understand your intent. It either reminds them of what they have just read in the abstract or informs them for the first time if they skipped over the abstract. Your purpose statement is synonymous with a paragraph's topic sentence, an essay's thesis, the first sentence in a letter, or the introductory paragraph in a shorter report.

Background and Problem. Whereas the purpose statement should be limited to one to three sentences or a short paragraph for clarity and conciseness, your discussion of the problem must be much more detailed. For example, if you are writing a report about the need for a new facility, your company's current work and storage space must be too limited. Your company has a problem that must be solved.

Your introduction's focus on the problem and background, which could average one to two pages, is important for two reasons. First, it highlights the importance of your report and establishes a context for the reader. The introduction emphasizes for your readers the report's priority. In this section, you persuade your readers that a problem truly exists and needs immediate attention.

Second, by clearly stating the problem and background information, you also reveal your knowledge of the situation. This section reveals your expertise. Thus, after reading this section of the introduction, your audience should recognize the importance of the issue and trust you to solve it or understand the complexity of the topic being discussed.

Discussion

The discussion section of your long report constitutes its body. In this section, you develop the detailed content of the long report. As such, the discussion section represents the major portion of the long report, perhaps 85 percent of the text.

What will you focus on in this section? Because every report will differ, we can't tell you exactly what to include. However, your discussion can contain any or all of the following:

- Analyses
  - Existing situation
  - Solutions
  - Benefits
- Product specifications of mechanisms, facilities, or products
- Comparison/contrast of options
- Assessment of needs
- Features of the systems or products
- Optional approaches or methodologies for solving the problems
- Managerial chains of command (organizational charts)
- Biographical sketches of personnel
- Corporate and employee credentials
  - Years in business
  - Satisfied clients
  - Certifications
  - Previous accomplishments
- Schedules
  - Implementation schedules
  - Reporting intervals
  - Maintenance schedules
  - Delivery schedules
  - Completion dates
  - Payment schedules
  - Projected milestones (forecasts)
- Cost analyses
- Profit and loss potential
- Documentation and researched material
- Survey results
- Lab report results
- Warranties
- Maintenance agreements
- Online help
- Training options
- Impact on the organization (time, personnel, finances, customers)

You will have to decide which of these sections will be geared toward high-tech readers, low-tech readers, or a lay audience. Once this decision is made, you will write accordingly, defining terms as needed. In addition to audience recognition, enhance your discussion with figures and tables for clarity, conciseness, and cosmetic appeal.

Web Resources

www.pearsonhighered .com/gerson For more information about audience, visit our companion website.

Conclusion/Recommendation

Sum up your long report in a page or so, providing your readers with a sense of closure. The conclusion can restate the problem, the important implications of your analysis, your solutions, and the benefits to be derived. In doing so, remember to quantify. Be specific—state percentages and amounts.

Your recommendation will suggest the next course of action. Specify when this action will or should occur and why that date is important. The conclusion/recommendation section can be made accessible through highlighting techniques, including headings, subheadings, underlining, boldface, itemization, and white space.

Glossary

Because you will have numerous readers with multiple levels of expertise, you must be concerned about your use of high-tech language (abbreviations, acronyms, and terms). Although some of your readers will understand your terminology, others won't. However, if you define your terms each time you use them, two problems will occur: You might insult high-tech readers, or you will delay your audience as they read your text. To avoid these pitfalls, use a glossary. A glossary is an alphabetized list of high-tech terminology placed after your conclusion/recommendation.

Glossaries

See Chapter 4 for more discussion of glossaries.

A glossary is invaluable. Readers who are unfamiliar with your terminology can turn to the glossary and read your definitions. Those readers who understand your word usage can continue to read without stopping for unneeded information.

Ethical Considerations when Documenting Sources in a Long, Formal Report

Research

See Chapter 5 for more discussion of documenting sources.

If you use research to write your long report, include a works cited or references page. This page(s) documents the sources (books, periodicals, interviews, computer software, Internet sites, etc.) you have researched and from which you have quoted or paraphrased. Correct documentation and source citations are essential in your long report to enhance your credibility and demonstrate your ethical behavior. Remember that boilerplate content and templates, already created material in your company's document library, do not necessarily require documentation. However, if you research the material on the Internet or from any other published source, you must document this material or be guilty of unethical behavior.

Appendix

A final, optional component is an appendix. Appendices allow you to include any additional information (survey results, tables, figures, previous report findings, relevant letters or memos, etc.) that you have not built into your long report's main text. The contents of your appendix should not be of primary importance. Any truly important information should be incorporated within the report's main text. Valuable data (proof, substantiation, or information that clarifies a point) should appear in the text where it is easily accessible. Information provided within an appendix is buried, simply because of its placement at the end of the report. You don't want to bury key ideas. An appendix is a perfect place to file nonessential data that provides either documentation for future reference or further explanation or support of content in the report.

Using Research in Long, Formal Reports

You can use researched material, quotes, and paraphrases to support and develop content in your long, formal report. Technical communicators often ask how much of a long, formal report should be *their* writing, as opposed to researched information. A general rule is to lead into and out of every quotation or paraphrase with your own writing. In other words,

- Make a statement (your sentence).
- Support this generalization with a quotation or paraphrase (referenced material from another source).
- Provide a follow-up explanation of the referenced material's significance (your sentences).

Research Includes Secondary and Primary Sources

To conduct *secondary research*, you rely on printed and published information taken from sources including books, periodicals, newspapers, encyclopedias, reports, proposals, or other business documents. You might also rely on information taken from a Web site or a blog. All of this secondary research requires parenthetical source citations. In some instances, however, you will use already existing (boilerplate) material from your company's document files. This in-house material does not necessarily require documentation.

Primary research is research performed or generated by you. You do not rely on books or periodicals. Instead you create original research by preparing a survey or a questionnaire targeting a group of respondents, networking to discover information from other individuals, visiting job sites, or performing lab experiments. You may perform this research to determine for your company the direction a new marketing campaign should take, the importance of diversity in the workplace, the economic impact of relocating the company to a new office site, the usefulness of a new product, or the status of a project. You may also need to interview people for their input about a particular topic. For example, your company might be considering a new approach to take for increased security on employee computers. You could ask employees for a record of their logs which would highlight the problems they have encountered with their computer security. With primary research, you will be generating the information based on data or information from a variety of sources that might include observations, tests of equipment, interviews, networking, surveys, and questionnaires.

If you choose to use a survey for your primary research, consider doing the following to get other people's assistance:

- Ask politely for their help.
- Explain why you need the interview and information.
- Explain how you will use the information.
- Tell the survey participants approximately how long the survey should take.
- Make a convenient appointment for the interview or to fill in the survey.

- Come prepared. Research the subject matter so you will be prepared to ask appropriate questions. Write your interview questions or the survey before you meet with the person. Take the necessary paper, forms, pencils, pens, laptop, electronic notepads, handheld PDAs, or recording devices you will need for the meeting.
- Create survey questions that allow for quantifiable answers. In other words, try to avoid simple yes/no questions. Give survey participants a choice from a value-based scale, such as 4 = absolutely critical; 3 = important; 2 = somewhat important; 1 = not important.
- When the interview ends or the individuals complete the survey, thank them for their assistance.
- Ask the survey participant(s) for permission to use their responses in your research.

Figure 17.7 is a sample survey. An information technology manager at Design International, Inc., used this primary research to substantiate his long, formal report recommending that the company switch from a Microsoft Windows operating system to a Linux operating system.

Who writes long, formal reports?

City, state, and the federal government constitute the largest employer in the United States. The government hires engineers, financial analysts, accountants, scientists, administrative assistants, fire fighters, police officers, employees in public works and utilities, and parks and recreation workers.

Do these employees have to write long, formal reports? The answer is "yes." Vicki Charlesworth, assistant city manager/city clerk of Shawnee, Kansas, says that her city's engineers, police officers, fire chiefs, financial advisors, and parks and recreation employees often write formal reports to city council members and the city's mayor. These long reports help the city council members "make informed decisions" that impact the city's residents and business owners.

Topics for the reports bubble up from citizens. A citizen contacts one of the city council members, for example, to share a concern about streets, sidewalks, animal control, traffic, or taxes. Maybe a citizen wants to expand a business, improve safety guard training for school crosswalks, or build a skateboard park for resident recreation. City council members ask the city's staff to research the issues, write a report that provides information on relevant statutes and/or other cities' best practices for handling issues, analyze options, and then recommend follow-up actions.

In one recent instance, the city needed to expand an existing road. The two-lane, winding road, initially built in the early 1940s, no longer met the city's growing population needs. To accommodate city growth, the road had to be widened to four lanes and straightened to meet increased traffic and speed limits. This change would necessitate significant research and resources. City engineers needed to study the extent to which a road change would impact

- An adjacent animal refuge
- Current homeowners' land
- Sewage lines
- Easements
- Electrical and water lines
- New signage
- Traffic signals

The roadway project would be time consuming and costly. Only through a long, formal report could city engineers thoroughly provide the city council information about this project, analyze the options, and recommend appropriate construction steps.

Vicki says that poor communication damages the city employees' credibility and diminishes their professionalism. Flawed reports—ones that aren't well written or clearly developed—suggest that the writers "lack interest in the city's residents and business owners," are unresponsive to their constituents, or "just don't want to help" the city's taxpayers. That's bad PR, potentially leading to citizen complaints and inefficient government. It's not surprising that Vicki says, "One of the most important assets you can have in government is writing skills."

FIGURE 17.7   Survey Used to Conduct Primary Research

User Survey on Computer Operating Systems

Our company is considering migrating from a Windows-based computing system to a Linux-based system. To help us make this decision, please answer the following questions, using a 4 pt. scale.

1. Would you like to customize your desktop, choosing your own GUI links to computer operations (graphical images that you choose as iconic buttons to click on)?

___ 4 = absolutely critical
___ 3 = important
___ 2 = somewhat important
___ 1 = not important

2. Your DOS commands in Windows are a "text mode interface." Are you comfortable with our current system that relies on limited scripting/commands, or would you prefer a text mode interface that allows for multiple commands and is customizable?

___ 4 = absolutely critical
___ 3 = important
___ 2 = somewhat important
___ 1 = not important

3. Windows impacts your departmental budget because Microsoft allows only a single copy of Windows to be used on only one computer. In contrast, if we purchased a Linux system, you can run it on any number of computers for no additional charge. Such a purchase would require training to help computer users learn the new system. Realizing that a change in systems would necessitate training, how important is this budgetary issue to your department?

___ 4 = absolutely critical
___ 3 = important
___ 2 = somewhat important
___ 1 = not important

4. Windows comes with an operating system. Linux-based machines require that consumers install the operating system. Our company would provide you the software and hardware; an individual or individuals within your department would be required to perform this installation. Would this requirement make you vote *against* a purchase of a Linux system?

___ 4 = absolutely critical in a vote of "no"
___ 3 = important in a vote of "no"
___ 2 = somewhat important in a vote of "no"
___ 1 = not important—our department would not vote "no" due to this
requirement

(Continued)

FIGURE 17.7   Survey Used to Conduct Primary Research (Continued)

5. Has spyware or malware become a problem in your work environment?

___ 4 = absolutely critical
___ 3 = important
___ 2 = somewhat important
___ 1 = not important

6. Has the misuse of user IDs and passwords become a problem in your department, leading to security issues?

___ 4 = absolutely critical
___ 3 = important
___ 2 = somewhat important
___ 1 = not important

7. If your department has different types of and ages of computers, this could cause problems with operating systems with different hardware platforms. Windows, for example, does not run on different platforms. Linux systems can run on many different platforms. How important is it in your department to allow for different hardware platforms?

___ 4 = absolutely critical
___ 3 = important
___ 2 = somewhat important
___ 1 = not important

8. Does your department need computers to be clustered for multiple use and networking?

___ 4 = absolutely critical
___ 3 = important
___ 2 = somewhat important
___ 1 = not important

9. How important is storing data, backing it up, and moving stored data from one computer to another?

___ 4 = absolutely critical
___ 3 = important
___ 2 = somewhat important
___ 1 = not important

10. Windows supports many different kinds of printers, while Linux does not. If your department has different makes and models of printers, the operating system is an important choice. If your department does not have different makes and models of printers, the choice of an operating system is not important. How important is the choice of an operating system in your department, regarding printers?

___ 4 = absolutely critical
___ 3 = important
___ 2 = somewhat important
___ 1 = not important

Long, Formal Report Checklist

_____ 1. Have you included the major components for your long, formal report (front matter, text, and back matter)?

_____ 2. In your long, formal report, have you written to inform, analyze, and/or recommend?

_____ 3. Have you used research effectively, incorporating quotes and paraphrases successfully?

_____ 4. Have you used primary and/or secondary research?

_____ 5. Did you correctly give credit for the source of your research?

_____ 6. Have you correctly used a hierarchy of headings to help your audience navigate the text?

_____ 7. Did you use tables and/or figures to develop and enhance content?

_____ 8. Have you met your audience's need for definitions of acronyms, abbreviations, and high-tech terms by providing a glossary?

_____ 9. Have you achieved clarity and conciseness?

_____10. Is your text grammatically correct?

The Writing Process

Prewriting	Writing	Rewriting

• Write to inform, analyze, and/or recommend.
• Determine whether your audience is internal or external.
• Research your topic to gather information.
	• Organize your content using modes such as problem/solution, cause/effect, comparison, argument/persuasion, chronology, etc.
	• Develop your ideas by informing, analyzing, and recommending.
	• Use figures and tables to clarify content.
	• Use headings and talking headings for access.
		• Revise your draft by
		  • Adding details
		  • Deleting wordiness
		  • Simplifying words
		  • Enhancing the tone
		  • Reformatting your text
		  • Proofreading and correcting errors

The Writing Process at Work

Now that you know the criteria for long, formal reports and the importance of research, the next step is to write your report. To do so, follow the recursive, step-by-step process of _prewriting_, _writing_, and _rewriting_. Remember that the process is dynamic and the steps frequently overlap.

Web Resources

www.pearsonhighered
.com/gerson For more
information about process,
visit our companion website.

Prewriting

Throughout this textbook, we have presented several techniques for prewriting—reporter's questions, clustering/mind mapping, flowcharting, brainstorming/listing, and outlining. An excellent way to gather data and to determine your objectives in a long report is to

DOT.COM UPDATES

For samples of and more information about long, formal reports, check out the following links:

- World Wide Web Consortium (W3C) Reports: http://www.w3.org/TR/
- The National Institute on Aging's 2005 Report on Alzheimer's Disease: http://www.nia.nih.gov/Alzheimers/Publications/ADProgress2004_2005/
- The June 2004 Progress Report on the Federal Building and Fire Safety Investigation of the World Trade Center: http://wtc.nist.gov/progress_report_june04/progress_report_june04.htm
- The World Health Organization's Report on HIV/AIDS: http://www.who.int/hiv/fullreport_en_highres.pdf

brainstorm and list. One businessperson we met, John Staples, considered the following about his long, formal report:

- Goal—inform the audience of the need for new financial software, analyze findings from research, and recommend a preferred vendor
- Audience—lay readers (city manager and city council members)
- Channels—long, formal report
- Data—vendor options; software applications; cost of software, training, and implementation; payment schedules

To gather content for the long, formal report, John says he must do the following primary research:

- Meet with the accounting manager to determine the required software applications for the finance department.
- Seek proposals from software vendors.
- Determine how to allocate finances to pay for the software purchases.

To accomplish the above, John uses brainstorming to list the project requirements (Figure 17.8).

Writing

After prewriting, you then draft the document. Figure 17.9 shows the rough draft of John's executive summary.

FIGURE 17.8 John's Brainstorming List of End-User Requirements

Prioritized List of Needs

1. End-user ease of use, complete with extensive online help for inexperienced end users
2. Ability to accomplish specific finance tasks
   a. Automated interest allocations
   b. Pooled cash
   c. Seamless purchasing card interaction
   d. Superior budget module
   e. Project accounting
3. Lotus Notes compatibility
4. Seamless interaction with Excel and Word

Re: Procurement of new financial software

The financial software (ACS) that the Finance Department is currently utilizing is 16 years old. The current software is not Windows based and is extremely cumbersome to accomplish the simplest tasks. Many staff hours are wasted retyping information into Excel spreadsheets for reporting purposes; dual and even triple data entry is common due to the shortcomings of the current software. Additionally, the current software currently resides on an AS/400. Many technology experts believe the AS400 is becoming obsolete. In their management letter for year ended December 31, 2011, the City's auditors, Cottrell and Kreisler, LLC made several references to the inadequacy of the current software. Attached is an excerpt from the management letter referencing these weaknesses. They cite multiple internal control issues, inability to track vital information, and redundant data entry. During the 2012 budget discussions a Decision Package was approved for $90,000 for hardware or software that might be required for a new financial software package. It was determined at that time to wait to make the decision on how to proceed.

Requests for proposals for new financial software were sent to seven software vendors. The following table illustrates the five responses received on March 14, 2012.

New Land	Utica, NY	$254,295
MEGS	St. Louis, MO	$159,538
MoonRay	Jupiter Beach, FL	$129,884
ACS	Springfield, IA	$123,800
Roedell	Taos, NM	$109,318

After a review of the responses, three vendors were selected to provide on-site demonstrations. Staff from finance and from other departments spent many hours comparing these vendor's products. The highest priorities were identified as end-user ease of use, ability to accomplish specific finance tasks, and Lotus Notes compatibility. MEGS software met all of the department's needs including the following: automated interest allocation, pooled cash, seamless purchasing card interaction, superior budget module, project accounting, etc. Unfortunately MEGS is $70,000 over our allotted budget of $90,000. MEGS has offered to divide the payment over two years so the remaining $70,000 would be included in the General Fund in the 2012 budget. With the implementation of the software the Finance Department will have some cost reductions related to the constant need for temporary help to accomplish current requirement for redundant data entry.

Add headings to break the text into more manageable units. Use bullets to make the content accessible and to shorten long sentences. Avoid long paragraphs.

Rewriting

Next, revise and rewrite your draft by considering the revision suggestions. See Figure 17.10 for John's rewritten executive summary.

See Figure 17.11 for a sample long, formal report.

FIGURE 17.10 Rewritten Executive Summary for City Manager and City Council Members

Executive Summary

Background of Finance Department's Software Requirements
The Finance Department's current ACS software is 16 years old. Because the software is outdated, it has the following problems:

- The software is not Windows based.
- The software is not time efficient. Many staff hours are wasted retyping information into Excel spreadsheets for reporting purposes. Dual and even triple data entry is common due to the shortcomings of the current software.
- The ACS financial software resides on an AS/400 hardware system. Many technology experts believe the AS400 is becoming obsolete.
- In their management letter for year ended December 31, 2011, the City's auditors, Cottrell and Kreisler, LLC, made several references to the inadequacy of the current software, including the following: multiple internal control issues, inability to track vital information, and redundant data entry.

During the 2012 budget discussions, a Decision Package was approved for $90,000 for hardware or software that might be required for a new financial software package.

Comparison of Vendor Proposals
Requests for proposals for new financial software were sent to seven software vendors. Table 1 illustrates the five responses received on March 14, 2012.

Table 1 Vendor Comparison		
Vendors	**Locations**	**Proposed Fees**
New Land	Utica, NY	$254,295
MEGS	St. Louis, MO	$159,538
MoonRay	Jupiter Beach, FL	$129,884
ACS	Springfield, IA	$123,800
Roedell	Taos, NM	$109,318

FIGURE 17.11   Long, Formal Report Recommending a Linux Operating System

Linux Desktop in Enterprise Settings at Design International, Inc.

Submitted by
John Staples Manager of Information Technology

March 2, 2012

Date: March 2, 2012
To: Tiffany Steward, CEO
From: John Staples, Manager Information Technology
Subject: Recommendation for Migration to a Linux Operating System

Ms. Steward, thank you for allowing me to report on a possible solution to Design International's operating system challenges. This report is in response to a survey my team conducted and the Information Technology Department's focus on improving internal and external communication.

Currently, we use Microsoft Windows as our computing system. However, Windows has a few inherent challenges. These include limited security, limited customization, and increased cost. Based on substantial research, my team and I would like to present an option—Linux. This report will provide you the following information:

- A definition of Linux   page 2
- A comparison/contrast of Linux vs. Windows   page 2–3
- An explanation of how Linux works   page 3–4
- The benefits of migrating to Linux   page 13

Information Technology believes migrating to a Linux Operating System can benefit Design International. I appreciate your thoughts on this report and look forward to answering any questions you might have. If you could assess this report by March 20, we would be able to initiate a Linux migration before the next quarter begins.

(Continued)

Table of Contents

List of Illustrations

iii

FIGURE 17.11 Continued

Abstract

The Microsoft Windows desktop has established a dominant presence in the enterprise desktop marketplace. Currently, the Windows desktop has nearly total control over the enterprise desktop market primarily due to

- an aggressive marketing strategy
- its intuitive ease of use
- its apparent ease of installation and administration
- readily available support

Windows, however, is not the only viable desktop solution in the enterprise marketplace. Because of potential risks and benefits associated with any technology product, as IT manager I must be alert for alternatives that are secure, reliable, and cost effective.

This report provides technical information regarding the issue of migration to a Linux enterprise desktop solution, analyzes the pros and cons of migrating to a Linux solution, and recommends appropriate action for our company, Design International, Inc.

Introduction

Purpose

We need to consider changing from a Windows-based desktop to a Linux-based desktop for Design International.

Background of Enterprise Desktop Computing

Throughout the history of desktop computing solutions, Microsoft has prevailed in the marketplace. The near monopoly enjoyed by Microsoft over users in all aspects of desktop computing has been established and maintained for many reasons including the user-friendly graphic environment which has been at the center of virtually all Microsoft products.

Despite the dominance of the Microsoft desktop in the desktop marketplace, other alternatives are emerging. One promising alternative is the family of Linux desktop operating systems.

Challenges Created by Relying Only on Windows

Though Microsoft Windows is the standard in the marketplace, it presents certain challenges for end users. At Design International, we have encountered the following challenges:

- Security—In Windows, the operating system and key applications are combined, which can lead to security issues for e-mail clients. In addition, users of Windows are dependent upon the vendor for recognition and correction of security issues, as well as the implementation of security patches.
- Customization—Prepackaged IT systems, such as Windows, disallow Design International to customize our computerization.
- Cost—Windows requires that Design International pay costly software licensure fees. This cost is compounded by the fact that Windows is platform-dependent, limiting server selection.

1

Summarizing the report in the Abstract allows the reader to get a "snapshot" of the content. Since most people think that Microsoft Windows is the only option for desktop computing, this report discusses an alternative—Linux.

The IT manager lets the reader see that a recommendation will be made based on the facts in the report.

Providing a background for the report establishes and explains context for the audience.

Highlighting problems in the introduction shows the importance of your topic and the need to consider other options.

(Continued)

Discussion

What the Linux Desktop Environment Is

The current industry standard for the computing desktop is the Microsoft XP operating system. This system, like all other Microsoft products, is distributed to consumers, including enterprise consumers, in machine code only. The source code is kept as a tightly guarded secret. The Microsoft operating system comes to the consumer as an "end-to-end solution" that includes operating system, desktop, common applications, and utilities all bundled into one package.

In contrast, Linux is not necessarily distributed as machine code, nor is it necessarily an end-to-end solution. "Strictly speaking, Linux refers to the kernel maintained by Linus Torvalds and distributed under the same name through the main repository and various mirror sites. This code base includes only the kernel and no utilities whatsoever. The kernel provides the core system facilities" (Yaghmour).

The Linux kernel is open source. As such, a custom operating system may be built to suit a particular need based on the kernel. However, most distributions of Linux include common user applications and are marketed by one of several vendors in the open source community such as RedHat, SuSe, Mandrake, or Caldera. The products that are currently available come with differing levels of utilities, installation tools, support, and documentation. Linux is available either by way of machine code or in its native source code format. In this manner, and by virtue of the general public license, the user can modify the operating system, so long as the source code remains open.

Linux desktop environments including KDE and GNOME are utilities that are available for the user to interface with the other utilities that are driven by the Linux kernel. These tools also provide a graphic user interface (GUI) for the user to interact with the Linux operating system. Because a number of GUI desktop solutions are available in the marketplace, the user is free to choose from the ones that come with their Linux distribution or to search from the other available alternatives.

Linux vs. Windows Design Issues

As previously noted, the Windows operating system is distributed as an end-to-end solution. As such, it has to be designed to meet most of the needs of most of the users, most of the time.

The Microsoft Windows operating system follows the monolithic design. Most of the features of the system, including the GUI by way of the Internet Explorer, are built into one single unit, the kernel (Petreley 20).

While the Linux kernel is also characterized in the literature as being "monolithic" in relationship to traditional Unix microkernel design, fewer of the utilities and drivers are part of the kernel. What further differentiates the design of the Windows operating system and the design of Linux is that Linux is fully modular in nature (Petreley 21). Microsoft developed the Internet Explorer browser as an integrated part of the operating system, ostensibly to provide for more convenience for the user. However, because of the interdependencies between the Internet Explorer and the other parts of the operating system, the entire system is exposed to any flaws that exist in the interdependent system (Petreley 23).

2

Embedding secondary research helps to develop the analysis and gives authority to your argument.

Comparing and contrasting facts about the options helps to explain the topic to the reader. By focusing on pros and cons, you analyze the topic, clarifying the issue and supporting the conclusions you will draw later.

FIGURE 17.11 Continued

Unfortunately, the highly integrated graphic nature of Windows tends to cause the most problems. Because of this underlying architecture, a flaw in the graphics renderings can be detected and exploited to the extent that the entire computer can be taken over or destroyed by malicious code. According to Petreley, a common cause of the "blue screen of death" is something as innocuous as a bug in a graphics card driver (26).

In contrast, the Linux operating system does not for the most part allow graphics drivers to run inside the kernel. Again, according to Petreley, Linux runs almost all graphics drivers outside of the kernel. "A bug in a graphics driver may cause the graphical desktop to fail, but not cause the entire system to fail. If this happens, one simply restarts the graphical desktop" (Petreley 25).

Another architectural issue that has led to security issues for Windows, particularly in the networked environment, has been that Windows is heavily reliant on the remote procedural call (RPC), a situation where one computer initiates a call on another computer, over a network. While this is sometimes not possible to avoid, it is a procedure that should be kept to a bare minimum. This is the case with Linux (Petreley 27).

A final architectural issue that has rendered security issues with Windows is the high level of reliance on the GUI for performing administrative tasks. When the administrator is performing tasks as the administrative user, the GUI being used is essentially an instance of Internet Explorer. Because this browser is built so closely to the operating system, the underlying kernel, and in the case of a server, the entire network, is exposed to security risks that are inherent with the browser (Petreley 22).

How Linux Works

Proprietary software primarily exists for the purpose of generating revenue for the owner of the copyright to the software. Proprietary software is not usually sold to the user. Rather, its use is permitted in accordance with a licensing agreement. According to Kenston, software licenses serve three basic functions:

1. They create profit for the developer.
2. They provide a means for controlling the distribution of the product.
3. They protect the underlying source code from modification. (6)

Vendors of proprietary software have placed a premium on protecting the source code from modification so that product development remains under the tight control of the vendor's development group. This model of software development was described by Raymond as the "cathedral" model of software development (29). Raymond characterized software developed under the cathedral model as "carefully crafted by individual wizards or small bands of magi working in splendid isolation, with no beta released before its time" (29).

These concepts of business and development have two primary advantages:

1. They protect the vendor's profit motive.
2. They provide for centralized control of changes and development, along with a mechanism for funding and sustaining ongoing research and development.

Highlight facts and rely on research to emphasize important parts of the topic and strengthen the analysis.

3

(Continued)

One of the most compelling arguments in support of the commercial software development model is forwarded by Microsoft Senior Vice President Craig Mundie:

> Companies and investors need to focus on business models that can be sustainable over the long term in the real world economy. A common trait of many of the companies that failed is that they gave away for free or at a loss the very thing they produced that was of greatest value—in the hope that somehow they'd make money selling something else. (29)

In contrast, open source software exists for the benefit of the user community. It is distributed freely under the General Public License (GPL), which states that anyone may use or distribute the source code. The GPL says, "You may modify your copy or copies of the Program or any portion of it, thus forming a work based on the Program, and copy and distribute such modifications or work under the terms of Section 1 above" ("GNU").

In general, open source software products, including Linux, are developed by groups of users collaborating together to improve systems and applications. Eric Raymond provides a description of the culture in which open source software is developed. Some of the cornerstones of this culture that were described by Raymond are as follows:

Use bullets to draw the reader's attention to important research content that substantiates your analysis.

- Software development starts with some personal need of the developer.
- The developer remains in charge of the project, until it is passed along to some other competent developer.
- Reuse and improvement of code is preferable to coding from scratch.
- Treating users as co-developers helps to harness their creative energy.
- Releasing beta code often and listening to feedback leads to rapid debugging.
- Developing a large beta testing and co-developer base leads to rapid bug identification and resolution.
- Recognizing and using the ideas of co-developers leads to software that meets user's needs. (Raymond 27–28)

Raymond describes the open source software development environment as follows:

> The Linux community seemed to resemble a great babbling bazaar of differing agendas and approaches (aptly symbolized by the Linux archive sites, which would take submissions from anyone) out of which a coherent and stable system could seemingly emerge only by a succession of miracles. (30)

Clearly, the development models that drive commercial and open source software have differing strengths and weaknesses. While the commercial development model tends to assure future research and development, as well as the ongoing maintenance of its products, the open source development model uses a more broad range of expertise. As David Freedman says, Linux is a viable alternative to Windows because as an open-source software, Linux is not owned by any company, it is free to end users, and "it boasts a worldwide network of programmers constantly trying to perfect it" (69).

4

FIGURE 17.11 Continued

Hardware Support for a Linux-Based Platform

Support for hardware varies from device to device and configuration to configuration. This support is either provided by the Linux kernel or through drivers provided by the device manufacturer.

The Linux 2.4 kernel provides full support for the Intel x86 family of processors, the most common platform from which Linux is run. In addition, there is considerable, but not complete support for advanced RISC machine processors, IBM Motorola Power PC processors, MIPS, Hitachi Super H, and Motorola 6800 (Dalheimer et al. 42).

In terms of busses and interfaces, support is either contained in the Linux 2.4 kernel or is readily available for ISA, PCI, PCMCIA, PC104, VME, parallel port, USB, IEEE1394. Support is under development for other interfaces, including SCSI, and I2PC (Dalheimer et al. 42).

In terms of I/O (Input/Output) devices, the Linux kernel supports a broad range of serial port and parallel port devices, along with keyboards, pointing devices, printers, scanners, displays, and other common business tools. However, Linux does not support all such devices. In implementing a Linux desktop solution, as IT manager I had to research the transition of current devices and any new device purchases to assure that drivers are available and that performance is established.

File Systems for Linux

The Second Extended File system (ext2fs) was developed specifically for Linux. According to Dalheimer et al., "The ext2fs is one of the most efficient and flexible file systems" (35). This file system was designed to provide an implementation of UNIX file commands and offers advanced file system features. This system is currently the most common file system found in Linux distributions. Some of the features that are supported by ext2fs are

- Standard Unix file types, directories, device files, and links
- Large file systems, up to 32TB
- Long file names, usually 255 characters, but extendable to 1012
- Root user reserved files
- Root user block sizes that can improve I/O performance
- File system state tracking

   (Dalheimer et al. 6)

Perhaps the most important aspect of ext2fs is that it is a Journaling file system that keeps track of changes that are made so that a corrupted file system can be more easily restored after a crash (Dalheimer et al. 6).

5

Using subheadings organizes information for the reader and makes the content accessible. You will use subheadings to develop your table of contents.

Because long report topics are complex, help your reader navigate the text easily. To create a readable style in your long report, use short sentences, short paragraphs, bulleted lists, and headings and subheadings.

(Continued)

In addition to ext2fs, Linux supports a wide range of non-Linux file systems including

- Minix-1 and XENIX
- Windows NTFS (Windows 2000 and NT)
- VFAT (Windows 95/98)
- FAT (MS-DOS)
- Macintosh
- HPFS (OS/2)
- Amiga
- ISO 9660 (CD-ROM)
- Novell
- SMB
- Apple McIntosh
- BFS (SCO Unix)
- UFS
    (Dalheimer et al. 6)

This list is not exhaustive, but rather represents the broad range of systems that can work with Linux. Many other file systems are available, either fully supported or in the form of experimental beta drivers.

Network Support for Linux

The Linux kernel provides for a full implementation of the TCP/IP networking protocols. According to Dalheimer et al., some of the features included with the kernel are

- Device drivers for a wide range of Ethernet cards
- Parallel Line Internet Protocol (PLIP)
- IPv6 protocol suite
- DHCP
- Appletalk
- IRDA
- DECnet
- AX.25 for packet radio networks
- FTP
- Telnet
- NNTP
- Simple Mail Transfer Protocol (SMTP)
    (Dalheimer et al. 42)

In addition, Linux comes with support for a network firewall. This means that virtually any Linux computer can be configured as a router and a firewall.

6

FIGURE 17.11 Continued

How Linux Addresses the Enterprise Desktop Need at Design International

The User and the Enterprise Desktop

As previously noted, the available Linux desktops have developed independently. Nonetheless, there are two examples of desktops that are emerging as the front runners for enterprise use. These are the KDE and the GNOME desktops. Both provide a colorful graphic user environment that is highly customizable, even to the extent that it can be made to look a lot like the familiar Windows environment. Support for other features including drop and drag, multiple languages, and session management is now available.

Enterprise Desktop Management

While the Linux desktop has evolved in the past few years, until recently, there was no true enterprise desktop management solution for the Linux environment. While installing a Linux desktop in a network environment has been easy, it is much more difficult to manage an entire network of hundreds or thousands of desktops. Even the most basic desktop management issues such as common configuration management, software installation, software updates, printer configurations, network file configuration, and authentication management become a resource drain without an enterprise management solution.

In the past, some centralized administration tools were available. Notable examples include webmin, a browser-based Linux remote administration tool, and its less graphics-intensive and less feature-rich predecessor, linuxconf. Both of these are open source products which provided network management functionality for some of the more basic network services.

More recently, end-to-end enterprise desktop management software has been entering the market. While no open source version of a true enterprise management system has yet emerged, there are a number of promising proprietary solutions that have entered the Enterprise Linux Desktop arena. These include

1. ZENworks 7 Linux Management
2. CA Unicenter Management Software for Linux
3. Xandros xMD

What Are Design International's Options for Accessing Linux?

We can load Linux systems on our computers at Design International in the following ways (Freedman 69):

- **Live CD:** The most risk free approach to accessing Linux is achieved by inserting CDs in all our computers, rebooting, and circumventing Windows once the system restarts. Computers started this way will function as Linux machines. This solution, however, is a quick fix and not advisable for long-term functionality.

- **Parallel installation:** We can download and install Linux software and run it parallel to Windows. This way, our end users can choose between the two operating systems at start-up. Our employees will not be able to switch back and forth between Linux and Windows while the computers are running. Thus, this solution does not allow users to access both systems synchronously.

7

Tracing historical developments, such as desktop management, shows the reader the viability of changing desktop solutions.

Persuasion is a key component of recommendations. To communicate persuasively, you might need to refute opposing points of view. Notice how the writer analyzes information in this section but then rebuts the argument by pointing out its drawbacks.

(Continued)

- **Windows replacement:** If we choose to use Linux exclusively, we must remove all Windows files and install Linux. Then, if we would want access to Windows, we would need to perform a complete Windows reinstall.

Use of a Linux Environment at Design International

Is Design International ready for a migration to a Linux Environment? Will our employees benefit from such a computing change? To find answers to these questions, I conducted an online survey of our 3,000 nationwide employees. The employees were asked to assess 10 topics related to their computer usage. Specifically, the survey focused on issues such as security, printer usage, budget, installation of hardware, networking, archiving of files, and more. They ranked their responses on the following four-point scale:

___ 4 = absolutely critical
___ 3 = important
___ 2 = somewhat important
___ 1 = not important

I received a 75 percent response rate, equaling 2,250 responses to the survey (see Appendix). The results of this survey are shown in Figure 1.

Figure 1: Survey on Computer Operating Systems

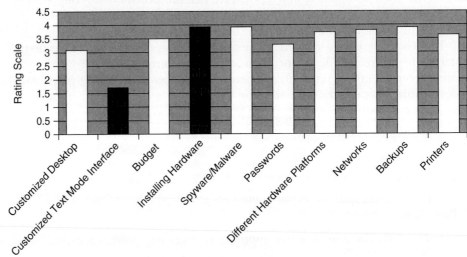

In Figure 1, notice that two bars in the bar chart are dark colored ("Customized Text Mode Interface" and "Installing Hardware"). The bar noting "Customized Text Mode Interface" scores low, suggesting that our employees are not interested in creating customized coding. The "Installing Hardware" scores high, but this is a strong "No" vote, suggesting that our employees *do not* want to be responsible for installing their own hardware. However, all of the other topics scored highly, suggesting that a Linux environment would meet and exceed our employees' computing needs, regarding cost, security, printer options, backup storage, and more.

8

Conducting primary research, such as through a survey, is an excellent way to add content that is customized for your particular audience and topic.

FIGURE 17.11   Continued

Explanation of Survey Results

Based on the 10 questions asked, the results strongly suggest that our employees would profit from a migration to a Linux environment. To clarify why this is true, the following table compares and contrasts key facts regarding Windows and Linux, as represented in the survey:

Table 1 Comparison/Contrast of Key Facts Regarding Windows vs. Linux	
Windows	**Linux**
Graphical User Interface	
Windows offers some customization of the GUI.	Linux allows advanced users to customize GUIs to their liking.
Text Mode Interface	
Each version of Windows has a single command interpreter.	Linux supports multiple command interpreters that can present a highly customized user interface employing the user interface and input/output facilities.
Cost	
Windows is expensive. Microsoft allows a single copy of Windows to be used on only one computer.	Linux is very cheap and/or free. Once you have purchased Linux, you can run it on any number of computers for no additional charge.
Getting the Operating System	
Windows comes with an operating system.	PC vendors sell Linux based machines only as servers, not to consumers, so you must buy a new computer with the operating system preinstalled or as a consumer you must install the Linux system yourself.
Viruses and Spyware	
Spyware is the worst problem affecting Windows-based computers.	One of the major reasons for using Linux is that viruses and spyware are far less troublesome.

Table 1 uses comparison/contrast to analyze Windows vs. Linux. This analysis provides information and then draws conclusions.

9

(Continued)

Table 1 Continued	
Windows	**Linux**
Users and Passwords	
Windows XP allows the omission of the password. Windows software is not designed to be used by a restricted user. Therefore, viruses and malware can access Windows machines easily.	Linux supports root and restricted users, which limits viruses and malware.
Hardware the OS Runs On	
Windows does not run on different hardware platforms.	Linux runs on many different hardware platforms, even very old personal computers such as 486 based machines.
Clustering, Multiple Users, and Networking	
Windows is designed to be used by one person at a time.	Linux has been used to make enormous clusters of computers. Linux is a multiuser system, designed to handle multiple concurrent users.
User Data	
Windows allows programs to store user information (files and settings) anywhere. This makes it hard to back up user data files and settings and to switch to a new computer.	Linux stores all user data in the home directory making it much easier to migrate from an old computer to a new one.
Printer Drivers	
Windows supports many printers.	Linux does not support as many printers as Windows.

Migration Planning to the Linux Environment

Migration planning is critical to deploying the Linux desktop. This is the process of identifying how the organization will move from its current desktop environment to the Linux environment. In general, the literature appears to advocate a "divide and conquer" approach. According to Almond et al., the migration plan will include the assessment of user patterns, establishing functional continuity, dealing with user issues, and retraining considerations.

During the initial process of planning the migration, a critical task will be identifying groups of users by way of patterns of IT usage. This accomplishes groupings by way of "role based client segmentation" (Almond et al. 28).

10

FIGURE 17.11 Continued

To accomplish this, you must assess the use of IT services and then group the users into functional usage patterns. Table 2 depicts a typical example of client segmentation.

Table 2 Client Segmentation				
Fixed Function	Technical Workstation	Transactional Workstation	Basic Office	Advanced Office
Limited Use of Business Apps	Limited Use of Business Apps	Applications Drive Business Process	Applications Drive Business Process	Applications Drive Business Process
Limited Office Productivity	Simple Office Productivity	Simple Office Productivity	Simple Office Productivity	Advanced Office Productivity
No E-mail	Simple E-mail	Simple E-mail	Simple E-mail	Advanced E-mail
No Instant Messaging	No Instant Messaging	No Instant Messaging	No Instant Messaging	Instant Messaging
Simple Browser Access	Simple Browser Access	Simple Browser Access	Simple Browser Access	Advanced Browser Applications
Filepoint Systems Management; Network Access; Host Emulation	Filepoint Systems Management; Network Access; Host Emulation	Filepoint Systems Management; Network Access; Host Emulation	Filepoint Systems Management; Network Access; Host Emulation	Filepoint Systems Management; Network Access; Host Emulation

A major concern with altering desktop solutions is how such a change will affect the end user. This table highlights key points of the analysis and clarifies pros and cons.

(Almond et al. 28)

Once general user roles have been identified along with the applications that they use, you must establish functional continuity by identifying potential technological solutions for the new network environment.

Solutions to Migration Issues

According to Almond et al., six potential technological solutions for solving migration issues are as follows:

- Bridging applications, or applications that work in both the old and new environment.
- Similar applications, or applications that work in the new environment that have the look and feel of the application used in the old environment.

11

(Continued)

- Server based applications, or applications that run solely on the server.
- Independent Software Vendor applications, or custom software purchased for the new environment.
- Web-based functional equivalents, or applications that run in the browser, regardless of the operating system.
- Custom porting of custom applications, or modifying the code of custom applications so that they operate in the new environment as they did in the old. (33–36)

An example of establishing a migration solution by way of bridging applications is shown in Table 3:

The table helps your reader quickly see the comparison/contrast between moving from a Windows-based application to a Linux-based environment.

Table 3 Bridging Applications in a Migrations Solution	
Application in Windows	**Linux/Open Source Equivalent**
Internet Explorer	Mozilla.org Firefox
Outlook	Mozilla.org Thunderbird
MS Office Word	OpenOffice.org Writer
MS Office Excel	OpenOffice.Org Spreadsheet
MS Office Power Point	OpenOffice.org Impress
Paint Shop Pro	Gimp
Messenger Client	Gaim

(Almond et al. 31)

Many custom application data systems that were designed for Microsoft systems will not run on a Linux client. An example of this type of situation is with Microsoft Visual Basic. In case of these sorts of applications, difficult decisions will have to be made whether to customize or to abandon the application. However, this issue could soon be moot. Microsoft has recognized the growing importance and acceptance of Linux. In late 2006, Microsoft announced a deal to promote the use of the "Linux operating system alongside Windows in mixed server environments. . . . Microsoft has taken steps toward Linux and other open source software before, but the tie-up with Novell is its strongest admission thus far of the increasing popularity of those technologies," according to Andi Mann, an analyst at Enterprise Management Associates (Lai 6). In addition, Centeris, a Washington-based company, has created software to bridge the Windows-Linux gap (Brown 22). Centeris' Likewise Management Suite 2.0 lets small and medium-size business network administrators

12

FIGURE 17.11 Continued

- Manage Linux systems using Windows interfaces
- Leverage Microsoft's Active Directory structure
- Create and remove shared files
- Set up printers
- Configure firewalls
    (Sturdevant 47)

Therefore, our company is no longer faced with an either/or dilemma. Centeris software and the impending partnership between Microsoft and Novell will help us manage the inherent "interoperability kinks" between the two operating systems (Khanna 10).

Managing User Satisfaction in Migration Planning

Perhaps the most critical issue in migration planning will be managing user satisfaction. As with any change, the reaction by users may be anything from enthusiastic acceptance to outright rebellion. Some strategies suggested by Almond et al. that can be used to help move users forward during the migration process are as follows:

- **Communicate.** Make sure that users understand what changes are upcoming, what the schedule is, and how the changes will have a positive impact on the organization.
- **Use bridging applications.** Some Linux based software such as OpenOffice runs well on Microsoft systems. Where possible, give users the opportunity to gradually become acclimated to the new environment prior to the migration.
- **Provide a similar look and feel.** The GNOME and KDE desktops can be customized to provide the user with a familiar Windows lookalike work environment.
- **Use similar actions.** Linux customization features allow desktop actions (for example, mouse double clicks) to be changed so that they can more closely resemble the familiar Windows actions.
- **Move file systems in a contiguous manner.** A common issue with users is the new Linux file structure. When moving file systems, the user's new file structure can be customized to closely resemble the familiar Windows file structure.
- **Provide users with the opportunity to try it before the migration.** Many Linux distributions are available in the form of a bootable CD. Users can test the operating system on their own computers without actually installing it by merely booting from the CD (Almond et al. 34–36).

13

To ensure enthusiastic acceptance of migration to Linux, use bullets to highlight the information and draw the reader's attention to the benefits.

(Continued)

Benefits of Using Linux at Design International

The Linux desktop operating system is a reliable and viable alternative to Microsoft Windows. The current release of the Linux kernel (2.6) contains features that include the following:

- Enhanced support for small embedded systems (PDAs, handhelds, etc.).
- Enhanced support for large, server systems.
- Full 32 bit user and group addressing capability, enabling enterprise size scalability.
- Improved hardware support, including improved USB support, and newly added support for touch screens, feature pointing devices, Braille teletype, as well as wireless devices.

Of course, Linux still retains its traditional attributes that make it a viable operating system. According to Dalheimer et al., some of the key attributes that make Linux a viable operating system for Design International include the following:

Focusing on the key benefits to using Linux helps your readers understand the implications of your topic and substantiates the recommendation.

- **Linux is free.** Organizations using the Linux desktop can eliminate the need for paying software licensure fees. In fact, an independent study by Enterprise Management Associates concluded that "Linux may, in many cases, be substantially less expensive to own than Windows" (Galli 16). This management firm based their conclusion on Linux's increased productivity. "Linux administrators often manage more servers than Windows administrators, and . . . Linux systems tend to handle greater workloads than Windows systems" (Galli 16).
- **Linux is reliable.** Linux systems rarely crash.
- **Linux is gaining popularity.** As previously noted, Linux is supported by user groups by way of the Internet through the open source community. In addition, commercial software vendors have seen the value of making software that is compatible with Linux.
- **Linux is powerful.** The Linux operating system is efficient and makes good use of hardware. Many companies are finding that older hardware can be kept in service for a longer period of time through the use of Linux.
- **Linux is robust.** Linux can be anything from a simple office application environment to a full featured development workstation with advanced networking capability.
- **Linux is compatible.** Linux can exist on the same computer as Windows and even access Windows files that are in a different partition on the same disk as the Linux installation. Additionally, Linux can access Windows files across a network.
- **Linux is supported.** Support and documentation for Linux users is widely available through the World Wide Web. In addition to the Linux Documentation Project, there are many independently posted "frequently asked questions" lists available. Additionally, a number of reputable books have been published.

14

FIGURE 17.11 Continued

- **Linux has inherently superior desktop security.** Its modular design separates the operating system from key applications that are at the root of many security problems, including the browser and the e-mail client.
- **Superior manageability of security issues.** The nature of the open source environment lends itself to the fast development and implementation of security patches. The organization no longer relies on the vendor of the operating system to identify and rectify security issues.
- **Linux provides for efficiency through customization.** The user can make the Linux system into exactly what the organization needs, thus making most efficient use of equipment (Dalheimer et al. 11–19).

Conclusion

The Microsoft Windows desktop has established a dominant presence in enterprise desktop marketplace. Nonetheless, as the IT manager at Design International, I must be alert for opportunities to improve service delivery that is safe, secure, and cost effective. Among the options that are currently available in the marketplace is the range of Linux-based desktop solutions. While the Linux desktop solution is a viable candidate for many enterprise environments, as the IT manager I must carefully weigh the strengths and weaknesses of each available system to assure that the best solution for the individual circumstances is chosen.

Recommendation

We should consider migrating to a Linux-based desktop system at Design International. A further study should be conducted to consider the economic impact of migration on our budget, resources, and personnel.

Glossary

Term, Acronym, or Abbreviation	Definition
Beta testing	The process used to identify the correctness, completeness, security, and quality of computer software.
Browser	A user interface that allows navigation. For example, Web browser provides access to the World Wide Web; Wiki browser lets you read articles from Wikipedia; a file browser manages files and related objects; a code browser allows you to access the source code of a computer program.
Drivers	A driver is an electronic component used to control another electronic component. The term refers to a specialized chip that controls high-power transistors.

Use a glossary to define high-tech terms, abbreviations, and acronyms. By using a glossary, you do not have to burden the report's text with definitions or slow your readers down. Readers unfamiliar with terms can look for definitions in the glossary, while readers familiar with the terms can continue to read without interference.

15

(Continued)

Term, Acronym, or Abbreviation	Definition
GNOME	GNOME is an international project to create an easy-to-use computing system built from free software.
GPL	The General Public License allows anyone to use or distribute source code.
GUI	Graphical User Interface. GUI, pronounced like "gooey," uses graphical images to represent information or actions available to the user.
I/O	Input/Output. Input allows a computer system user to manipulate the system; Output allows the system to produce what the user has asked for.
KDE	K Desktop Environment is a free desktop environment that uses GUIs.
Kernel	The kernel is the central component of most operating systems (OSs). It manages the system's resources and allows for communication between hardware and software.
Open Source Software	Open source software is computer software that permits users to change and to redistribute the software in the revised form.
OS	A computer's Operating System manages the computer's hardware and software.
RPC	Remote Procedure Call—a situation where one computer initiates a call on another computer or network.
Source Code	A program's source code is its programming language.
Utilities	Utility software helps manage the computer's hardware and operating system.

16

FIGURE 17.11 Continued

Documentation

See Chapter 5 for more discussion of documenting sources.

Works Cited

"About Us." *Free Standards Group*. Web. 20 Mar. 2005.

Almond, Chris, et al. "Linux Client Migration Cookbook." *IBM Redbooks*. 2004. Web. 20 Apr. 2005.

Brown, Bob. "Centeris Eyes Windows/Linux Simplicit." *Network World*. 7 Aug. 2006: 22. Print.

Dalheimer, Matthias, et al. *Running Linux*. Sebastapol, CA: O'Reilly Safari Books, 2003. Print.

"Desktop Linux Capabilities 1.0." *Open Source Development Lab*. Web. 21 Mar. 2005.

"Desktop Linux Position Statement." *2003 Open Source Development Lab*. Web. 15 Mar. 2005.

Eckert, Jason C., and John M. Schitka. *Linux + Guide to Linux Certification*. Boston, MA: Course Technology, 2003. Print.

Figueiredo, Daniel. "GNOME Desktop Environment." *Faulkner Information Services 2005*. Web. 22 Mar. 2005.

Freedman, David H. "Tech geeks have long praised open-source software. Now's the time to see what the fuss is about." Inc. Jun. 2006: 69–71. Print.

Galli, Peter. "OSDL Answers Microsoft Claims; Study: Linux is cost-effective alternative to Windows." *eWeek*. 13 Feb. 2006: 16. Print.

"GNU General Public License Version 2." *Free Software Foundation*. 1991. Web. 10 Mar. 2005.

Kenston, Geoff. "Software Licensing Issues." *Faulkner Information Services 2004*. Web. 14 Feb. 2005.

Khanna, Poonam. "Deal Isn't About the Customer." *Computing Canada*. 17 Nov. 2006: 10. Print.

Lai, Eric. "Microsoft Follows Oracle In Tighter Linux Embrace." Computerworld. 6 Nov. 2006: 6. Print.

Mundie, Craig. "Prepared Text of Remarks by Craig Mundie." *Microsoft*. 3 May 2001; Web. 4 Mar. 2005.

Murphy, Pete. "Linux Enterprise Market Trends." *Faulkner Information Services* 2004. Web. 22 Mar. 2005.

Raymond, Eric. *The Cathedral and the Bazaar*. Sebastapol, CA: O'Reilly Safari Books, 1999. Print.

Petreley, Nicholas. "Security Report: Windows vs. Linux." *The Register*. 22 Oct. 2004: 20–28. Print.

Sturdevant, Cameron. "Can't Windows and Linux Just Get Along?" *eWeek*. 28 Aug. 2006: 47. Print.

"Work Groups." *Free Standards Group*. Web. 20 Mar. 2005.

Yaghmour, Karim. *Building Embedded Linux Systems*. Sebastapol, CA: O'Reilly Safari Books, 2003. Print.

17

(Continued)

CHAPTER HIGHLIGHTS

1. Long, formal reports require time, resources of people and money, and have far-reaching effects.

2. Long, formal reports often deal with topics that consider serious and complex issues.

3. Long reports can provide information, be analytical, and/or recommend a course of action.

4. Long, formal reports include front matter (title page, cover letter, table of contents, list of illustrations, and abstract); text (introduction, discussion, conclusion/recommendation); and back matter (glossary, works cited, and optional appendix).

5. To help your readers navigate the report, create a clear hierarchy of headings.

6. When you write your long report, consider your audience and enhance your discussion with figures and tables for clarity, conciseness, and cosmetic appeal.

7. Primary research is research performed and generated by you.

8. Surveys or questionnaires are primary sources of research.

9. When you conduct secondary research, you rely on already printed and published information.

10. Secondary research is from books, periodicals, databases, and online searches.

APPLY YOUR KNOWLEDGE

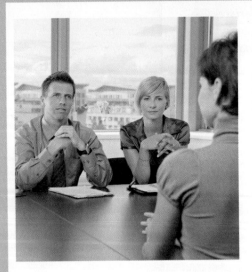

CASE STUDY

Alpha/Beta Consulting (A/B/C) solves customer problems related to hardware, software, and Web-based applications. The consulting company helps clients assess their computing needs, provides training, installs required peripherals to expand computer capabilities, and builds e-commerce applications, such as Web sites and corporate blogs. Their client base includes international companies and academic institutions. A/B/C works extensively with FIRE industries (Finance, Insurance, and Real Estate).

A new client, Home and Hearth Security Insurance Company, has asked A/B/C to build its e-commerce opportunities. Home and Hearth's current Web site is outdated. Since the Web site was created, Home and Hearth has added new services, including insurance coverage for electronic commerce, employment-related practices liability, financial institutions, management protection, and medical professional liability. In addition, Home and Hearth has expanded globally. It now has a presence in the Far East (Japan, Taiwan, and Singapore), the Mid East (Dubai, Jordan, and Israel), and South Africa. Home and Hearth's new services and new service locations are not evident on the current Web site.

An additional challenge for Home and Hearth is client contact. Currently, Home and Hearth depends on a hard-copy newsletter. However, corporate communications at Home and Hearth believes that a corporate blog would be a more effective means of connecting with its modern client base.

A/B/C can meet Home and Hearth's needs by building a new Web site and by creating a corporate blog.

Assignments

1. Write the abstract that will preface A/B/C's long, formal report to Home and Hearth Security Insurance Company.
2. Write the introduction that will preface A/B/C's long, formal report to Home and Hearth Security Insurance Company.
3. Create a survey questionnaire asking Home and Hearth employees what they would like to see on the company's Web site and corporate blog.
4. Outline what you believe Home and Hearth's new Web site should include to cover its new international focus. This constitutes the discussion section of the report.
5. Conduct any research you need to find more information about either Web design or the development of a corporate blog.

INDIVIDUAL AND TEAM PROJECTS

1. Read a long, formal report. It might be one your instructor provides you or one you obtain from a business or online (see the links provided in this chapter's Web Workshop and Dot. com Updates). If the report contains an abstract or executive summary, is it successful, based on the criteria provided in this chapter? Explain your decision. If the abstract or executive summary could be improved upon, revise it to make it more successful. If the report does not have an abstract or executive summary, write one.
2. Read the following abstract from Alpha/Beta Consulting, written for its client Nitrous Solutions. It needs revision. How would you improve its layout and content?

◀ EXAMPLE

> Nitrous Solutions, a multidiscipline architectural design, is a start-up business with plans to begin operation in October 2012. The company currently has no information technology network for internal and external communication. Alpha/Beta Consulting will recommend a server to meet Nitrous Solutions' unique needs, a backup solution for archiving architectural materials and critical communications, software for digital asset management, a proposed network design. After implementing these suggestions, Nitrous Solutions will have an IT network that will be safe, secure, dependable, and prompt. The system that Alpha/Beta Consulting is recommending also can be restructured with simplicity as the company's needs grow. By installing the IT network recommended in this report, Nitrous Solutions will be able to fulfill their company's mission statement.

3. Alpha/Beta Consulting, as agreed upon in a meeting with its client Nitrous Solutions, determined that its research for a contracted long, formal recommendation report would take no less than 350 hours and no more than 400 hours. As detailed in the following time sheet, A/B/C's four employees assigned to the project worked 376 hours. Using the information provided in the time sheet, create a line graph that tracks the employees' hours and a pie chart showing the percentage of time each employee worked.

TABLE 1 Time Sheet

Staff	Hours Worked Each Week														Total Hours for Project
	wk 1	wk 2	wk 3	wk 4	wk 5	wk 6	wk 7	wk 8	wk 9	wk 10	wk 11	wk 12	wk 13	wk 14	
Kenyon Patel	6.0	10.5	7.0	7.5	7.0	8.5	6.0	11.0	8.0	8.0	7.0	6.5	12.0	8.0	**113.00**
Randy Butler	6.5	5.5	5.5	3.0	5.5	8.0	4.0	5.5	6.0	8.0	7.0	9.5	15.0	8.0	**97.00**
Maria Villatega	0.0	0.0	3.5	4.0	4.5	5.0	3.0	1.5	10.0	17.0	8.0	9.0	8.0	4.0	**77.50**
James Soto	4.5	3.5	4.0	5.5	5.0	5.0	3.0	2.0	14.0	12.0	8.5	10.5	7.5	3.5	**88.50**
Total	17.00	19.50	20.00	20.00	22.00	26.50	16.00	20.00	38.00	45.00	30.50	35.50	42.50	23.50	376.00

4. At the conclusion of their long, formal report, the employees at A/B/C wrote a recommendation. Their recommendation, though excellent in terms of its content, needed to be reformatted for easier access and better emphasis of key points. Read the employees' recommendation and improve its layout.

EXAMPLE ▶

> ### Recommendations to Meet Nitrous Solutions' Network Needs
>
> To ensure a secure and dependable network that can be restructured to meet Nitrous Solutions' developing needs, Alpha/Beta Consulting recommends the following:
>
> You can improve your computing system's "Application Layer." The applications software programs that Alpha/Beta Consulting recommends consist of the Adobe Creative Suite 2, for imaging, editing, illustrations, file sharing, web designs, and digital processing; Microsoft Office, for spreadsheet and Excel documents, word processing files, project timelines, and maintenance; and Internet communication utilities, such as web browsers and FTP clients, to guarantee rapid deployment of company communications and contract sales. All these applications will be run on both the current Macintosh and Microsoft operating systems.
>
> You also need to add a "Data Layer" to your new computing system. The data transferred on the new network will consist of large graphic files and fonts being accessed from a centralized server, postscript printing, voice, and messaging. TCP/IP will be the main communication protocol, but Appletalk may also be used.
>
> To meet Nitrous Solutions' network infrastructure requirements, Alpha/Beta Consulting recommends Category 6 (Cat 6) cables connected to 1Gbps switches. These constitute your "Network Layer." The logical star topology network will enable a higher degree of performance for all users associated with the production of architectural designs, meet all organizational duties and communications requirements, and provide a centralized server as the main point of access for all the client machines.
>
> Finally, once installed, the physical network must have a "Technology Layer." This will enable the company to perform at a competitive level. This communication system will be a vital piece of the overall success of the company. The server will provide quick and reliable data storage, meet Nitrous Solutions' data backup needs, and allow for streamlined work flow.

PROBLEM-SOLVING THINK PIECES

Read the following text from a long, formal report written by Alpha/Beta Consulting. Based on the explanation in this chapter, decide whether the text informs, analyzes, and/or recommends. Explain your decisions.

EXAMPLE ▶

> ### Purpose Statement
>
> This report will recommend the design parameters, hardware specifications, and estimated expenses needed to build and install external and internal computer network technology to meet Nitrous Solutions' communication needs.
>
> This report's recommendation will focus on the following key areas:
>
> - A solution for Nitrous Solutions' server needs
> - Recommendations for a backup solution
> - Possible software for digital asset management
> - A proposed network design
>
> ### Needs Analysis
>
> **1.** *Company Background*
>
> Nitrous Solutions, a multidiscipline architectural design firm located in Raleigh, NC, is a start-up company. Nitrous Solutions will occupy a 2,000 square foot office space.

They plan to begin operations in October 2012. The company will provide industrial, interior, landscape, and green architecture services for an international client base. Communication is a major component of the business plan. Thus, a reliable, efficient, creative communications/IT technology system is mandatory.

2. *Basic IT Requirements*

Currently, Nitrous Solutions has no network. The company needs an infrastructure plan that allows for the following:

- An IT platform that will be compatible with Macintosh and Windows.
- Eight Windows work stations and four Macintosh platforms.
- A network based on WiMAX impact.
- An internal and external network that is safe, secure, reliable, and fast.
- An internal and external network that can store, execute, and transmit architectural materials.
- A network that will allow for successful communications to clients, partners, and vendors.
- A data backup solution.
- Digital asset management software.

3. *Detailed IT Requirements*

In addition to the basic network requirements for Nitrous Solutions, the company also has asked for recommendations to meet the following micro network needs:

- Information about the most efficient server operating system for reliable communication in a cross-platform environment.
- Requirements for different file and storage server hardware and software.
- An analysis of which FTP server software would transfer secure data reliably to external clients.
- The need for scripts to scan recently modified files and back them up to a server.
- Reliable backup solutions and backup media types.
- A comparison of different software applications that would prevent data duplication in the permanent backup archives.
- A comparison of software that could catalog backup storage.
- A software solution allowing Microsoft Word documents to be converted and catalogued into one PDF file.

WEB WORKSHOP

You can find many examples of long, formal reports online. Compare the format of the online reports with the criteria for long, formal, researched reports discussed in this chapter. Explain where the online reports are similar to or different from this chapter's criteria. Do the online reports successfully communicate information and analyze issues and problems? Check out these sites, for example:

- International Narcotics Control Strategy Report, Released by the Bureau for International Narcotics and Law Enforcement Affairs (http://www.state.gov/p/inl/rls/nrcrpt/2003/vol2/html/29910.htm)
- The Use of Radio Frequency Identification by the Department of Homeland Security to identify and track suspicious individuals (http://www.dhs.gov/xlibrary/assets/privacy/privacy_advcom_rpt_rfid_draft.pdf)
- Reports from the District Department of the Environment (http://dceo.dc.gov/dceo/lib/dceo/Green_DC_Week.pdf)
- The United States' Government Accountability Offices' Report on "Women's Participation in the Sciences" (http://www.gao.gov/new.items/d04639.pdf)

QUIZ QUESTIONS

1. What are reasons for writing long, formal reports?
2. Explain informative formal reports.
3. Explain analytical formal reports.
4. Explain recommendation formal reports.
5. What are the components of a long, formal report?
6. What do you include in an abstract?
7. What are the reasons for using research in a long, formal report?
8. Explain primary research.
9. Explain secondary research.
10. What is the purpose for using a hierarchy of headings in a long, formal report?

Proposals

COMMUNICATION AT WORK

In this scenario, Bellaire Educational Supplies/Technologies uses proposals to sell its products.

BEST (Bellaire Educational Supplies/Technologies) manufactures and markets school supplies for all levels of education (K–12, public or parochial, and college).

These supplies include the following:

- Desks and chairs
- Modular computer workstations
- Blackboards and whiteboards
- Markers, chalk, and erasers
- Overhead projectors (both mobile and fixed computer projection systems)
- Pull-down screens
- Pens and pencils
- Grade books

BEST produces catalogs and mails fliers and brochures, but the majority of its large-ticket sales (to entire school systems) come from external proposals. These proposals, which

1. Format proposals effectively using components such as a title page, cover letter, table of contents, list of illustrations, abstract, introduction, discussion, conclusion/recommendation, glossary, works cited or references, and appendices

2. Write effective internal proposals to persuade corporate decision makers to address issues and provide resources

3. Write effective external proposals to sell a new service or product to a potential customer

4. Distinguish among common proposal terms including RFP, T & C, SOW, boilerplate, solicited proposals, and unsolicited proposals

5. Apply research techniques to gather information for proposals

6. Evaluate proposals using the proposal checklist

usually are written in response to requests for proposals (RFPs) or as follow-ups to person-to-person sales meetings, focus on the following components:

- Cover letters prefacing the proposal to personalize the enclosed text as well as direct the audience to key parts of the proposal
- Table of contents itemizing the major headings and subheadings within the text
- A directory of tables and figures
- An executive summary, geared toward low-tech readers with limited time to read the entire proposal
- An introduction highlighting the reader's unique educational supplies problems and explaining how BEST will meet the client's needs
- Text that details costs, services, delivery, maintenance, guarantees, and product quality
- A conclusion summing up the proposal's key points
- A glossary to define unfamiliar words, acronyms, and abbreviations

Many of BEST's proposals are actually hard-copy printouts of PowerPoint slides. BEST's sales force and technical writers have concluded that most readers want less text and easy-to-read documentation, complete with ample graphics. PowerPoint is the solution to this need. Not only can BEST's sales staff use PowerPoint to make oral presentations about its products, but also they can print out the PowerPoint slides as $11 \times 8\frac{1}{2}$-inch landscape pages and bind them as handouts. The hard-copy pages are easy to create, easy to read, and meet the audience's need for conciseness.

BEST's products, once used, sell themselves. However, the proposals open the door. Through proposals, BEST has found a perfect way to present clear, concise, and thorough documentation of the company's assets.

Check out our quarterly newsletters at www.pearsonhighered.com/gerson for dot.com updates, new case studies, insights from business professionals, grammar exercises, and facts about technical communication.

Why Write a Proposal?

When you write a proposal, your goal is to sell an idea persuasively. Consider this scenario: Your company is growing rapidly. As business increases, several changes must occur to accommodate this growth. For example, the company needs a larger facility. This new building could be located in your city's vibrant new downtown expansion corridor, in a suburban setting, or entail the expansion of your current site. A new building or expansion should include amenities to improve recruitment of new employees, such as workout facilities, daycare, restaurant options, and even a gaming room. Finally, as part of new employee recruitment, the company must increase its diversity hiring practices.

Internal Proposals

How will you convey these ideas to upper-level management? The topic is large and will require extensive financial obligations, time for planning, and a commitment to new staffing. A short, informal report will not suffice. In contrast, you will have to write a type of longer, formal report—an *internal proposal* for your company's management.

Additional examples of internal proposals include the following:

- Your company needs to improve its mobile communication abilities for employees who work at diverse locations. To accomplish this goal, you write an internal proposal requesting the purchase of WiFi-compatible laptops, MiFi-capabilities, handheld computers, and smart phones with Internet access.
- Your company's insurance coverage is skyrocketing. As a member of the human resources staff, you have researched insurance carriers and now will propose insurance options or opt-out options to upper-level management.
- Your company is migrating to a new software platform. Employees will need training to use the software. You propose consulting companies which can offer the training, optional schedules, funding sources, and post-training certification.

External Proposals

Whereas *internal proposals* are written to management within your company, *external proposals* are written to sell a new service or product to an audience outside your company. Your biotechnology company, for example, has developed new software for running virtual cell cultures. The software simulates cell runs and displays synchronous strip charts for sterile monitoring. Data from the runs are graphed for comparison purposes. Not only will your company sell the software, but also the company provides consulting services to train clients in the software use. Your responsibility is to write an *external proposal* selling the benefits of this new corporate offering to a prospective client.

Requests for Proposals

Many external proposals are written in response to *requests for proposals* (RFPs). Often, companies, city councils, and state or federal agencies need to procure services from other corporations. A city, for example, might need extensive road repairs. A governmental agency needs Internet security systems for its offices. A hospital asks engineering companies to submit proposals about facility improvements. An insurance company needs to buy a fleet of cars for its adjusters. To receive bids and analyses of services, the city will write an RFP, specifying the scope of its needs. Competing companies will respond to this RFP with an external proposal.

In each of these instances, you ask your readers to make significant commitments regarding employees, schedules, equipment, training, facilities, and finances. Only a proposal, complete with research, will convey your content sufficiently and successfully.

Reports

See Chapter 16 for more discussion of short, informal reports. See Chapter 17 for more discussion of long, formal reports.

Research

See Chapter 5 for more discussion of research and documenting sources.

Web Resources

www.pearsonhighered.com/gerson See our companion website for more sample proposals.

Q: When I read about proposals, I see terms like *RFP, T & C, SOW, boilerplate,* and *solicited* and *unsolicited.* What do these words mean?

A: Here's a table defining these common proposal terms:

Proposal Terms	Definitions
Boilerplate	Any content (text or graphics) that can be used in many proposals
RFP	**Request for Proposals**—means by which external companies and agencies ask for proposals
Solicited Proposal	A proposal written in response to a request
SOW	**Scope of Work** *or* **Statement of Work**—a summary of the costs, dates, deliverables, personnel certifications, and/or company history
T & C	**Terms and Conditions**—the exact parameters of the request and expected responses
Unsolicited Proposal	A proposal written on your own initiative

Criteria for Proposals

To guide your readers through a proposal, provide the following:

- Title page
- Cover letter (or cover e-mail message for electronic submission of proposals)
- Table of contents
- List of illustrations
- Abstract
- Introduction
- Discussion (the body of the proposal)
- Conclusion/recommendation
- Glossary
- Works cited (or references) page
- Appendix

For information about title pages, cover letters, table of contents, list of illustrations, discussion, glossaries, works cited, and appendices, see Chapter 16. Each of these components, typical of long, formal reports, is thoroughly covered in that chapter. Following is information specifically related to your proposal's abstract, introduction, and conclusion/recommendation.

Abstract

Your audience for the proposal will be diverse. Accountants might read your information about costs and pricing, technicians might read your technical descriptions and process analyses, human resources personnel might read your employee biographies, and shipping/delivery might read your text to determine deadlines. One group of readers will be management—supervisors, managers, and highly placed executives. How do these readers' needs differ from others? Because these readers are busy with management concerns and might have little technical knowledge, they need your help in

Abstracts and Executive Summaries

See Chapter 17 for more discussion of abstracts and summaries.

two ways: They need information quickly, and they need it presented in low-tech terminology. You can achieve both these objectives through an abstract or executive summary.

The abstract, limited to approximately 3 to 10 sentences, presents the *problems* leading to your proposal, the suggested *solutions*, and the *benefits* your audience will derive. The following is an example of a brief, low-tech abstract for an internal proposal.

EXAMPLE ▶

An effective abstract highlights the problem, possible solutions, and benefits in the proposal.

Abstract

Due to deregulation and the recent economic recession, we must reduce our workforce by 12%.

Our plan for doing so involves

- Freezing new hires
- Promoting early retirement
- Reassigning second-shift supervisors to our Desoto plant
- Temporarily laying off third-shift line technicians

Achieving the above will allow us to maintain production during the current economic difficulties.

Introduction

Your introduction should include two primary sections: (1) purpose and (2) problem.

Purpose. In one to three sentences, tell your readers the purpose of your proposal. Your purpose statement clarifies the proposal's context. The following is an effective purpose statement.

EXAMPLE ▶

Purpose Statement

The purpose of this report is to propose the immediate installation of the 102473 Numerical Control Optical Scanner. This installation will ensure continued quality checks and allow us to meet agency specifications.

Problem (Needs Analysis). To clarify for the audience why this proposal is important, explain the problems leading to your suggestions. For example, computer viruses are attacking your company's workstations. This is leading to a decline in productivity, compromised security, and corrupted documents. Your proposal highlights these problems to explain why new computer security measures are needed. One way to help your readers understand the problem is through the use of highlighting techniques, especially headings and subheadings. See Figure 18.1 for a sample introduction.

Discussion

When writing the text for your proposal, sell your ideas persuasively, develop your ideas thoroughly through research, observe ethical technical communication standards, organize your content so the audience can follow your thoughts easily, and use graphics.

Communicating Persuasively. A successful proposal will make your audience act. Writing persuasively is especially important in an *unsolicited* proposal since your audience has not

FIGURE 18.1   Introduction with Purpose Statement and Needs Analysis

1.0 Introduction

1.1 Purpose Statement

This is a proposal for a storm sewer survey for Yakima, Washington. First, the survey will identify storm sewers needing repair and renovation. Then it will recommend public works projects that would control residential basement flooding in Yakima.

1.2 Needs Analysis

1.2.1. Increased Flooding

Residential basement flooding in Yakima has been increasing. Fourteen basements were reported flooded in 2011, whereas 83 residents reported flooded basements in 2012.

1.2.2. Property Damage

Basement flooding in Yakima results in thousands of dollars in property damage. The following are commonly reported as damaged property:

- Washers
- Dryers
- Freezers
- Furniture
- Furnaces

Major appliances cannot be repaired after water damage. Flooding also can result in expensive foundation repairs.

1.2.3. Indirect Costs

Flooding in Yakima is receiving increased publicity. Flood areas, including Yakima, have been identified in newspapers and on local newscasts. Until flooding problems have been corrected, potential residents and businesses may be reluctant to locate in Yakima.

1.2.4. Special-Interest Groups

Citizens over 55 years old represent 40 percent of the Yakima population. In city council meetings, senior citizens with limited incomes expressed their distress over property damage. Residents are unable to obtain federal flood insurance and must bear the financial burden of replacing flood-damaged personal and real property. Senior citizens (and other Yakima residents) look to city officials to resolve this financial dilemma.

> Provide specific details to explain the problem. Doing so shows that you understand the reader's need and highlights the proposal's importance.

asked for your report. A *solicited* proposal, perhaps in response to an RFP, is written to meet an audience's specific request. Your audience wants you to help them meet a need or solve a problem. In contrast, when you write an unsolicited proposal, your audience has not asked for your assistance. Therefore, in this type of proposal, you must convincingly persuade the audience that a need exists and that your proposed recommendations will benefit the reader.

To write persuasively, accomplish the following:

- Arouse audience involvement—focus on your audience's needs that generated this proposal.

> **Persuasion**
>
> See Chapter 11 for more discussion of persuasive writing.

- Refute opposing points of view in the body of your proposal.
- Give proof to develop your content, through research and proper documentation.
- Urge action—motivate your audience to act upon your proposal by either buying the product or service or adopting your suggestions or solutions.

Researching Content for Proposals.   As in any long, formal report, consider developing your content through research. This can include primary and secondary sources such as the following:

- Interviewing customers, clients, vendors, and staff members
- Creating a survey and distributing it electronically or as hard-copy text
- Visiting job sites to determine your audience's needs
- Using the Internet
- Reading journals, books, newspapers, and other hard-copy text

Communicating Ethically in Proposals.   When you write a proposal, your audience will make decisions based on your content. They will decide what amounts of money to budget, how to allocate time, what personnel will be needed to complete a task, and if additional equipment or facilities will be required. Therefore, your proposal must be accurate and honest. You cannot provide information in the proposal that dishonestly affects your decision makers. To write an ethical proposal, provide accurate information about credentials, pricing, competitors, needs assessment, and sources of information and research. When using research, for example, cite sources accurately to avoid plagiarism, as discussed in Chapter 5.

Organizing Your Content.   Your proposal will be long and complex. To help your audience understand the content, use modes of organization. These can include the following:

- **Comparison/contrast.** Rely on this mode when offering options for vendors, software, equipment, facilities, and more.
- **Cause/effect.** Use this method to show what created a problem or caused the need for your proposed solution.
- **Chronology.** Show the timeline for implementation of your proposal, reporting deadlines to meet, steps to follow, and payment schedules.
- **Analysis.** Subdivide the topic into smaller parts to aid understanding.

See Table 18.1 for organization and key components of a proposal's discussion section.

Using Graphics.   Graphics, including tables and figures, can help you emphasize and clarify key points. For example, note how the following graphics can be used in your proposal's discussion section:

- **Tables.** Your analysis of costs lends itself to tables.
- **Figures.** The proposal's main text sections could profit from the following figures:
  - **Line charts.** Excellent for showing upward and downward movement over a period of time. A line chart could be used to show how a company's profits have decreased, for example.
  - **Bar charts.** Effective for comparisons. Through a bar or grouped bar chart, you could reveal visually how one product, service, or approach is superior to another.
  - **Pie charts.** Excellent for showing percentages. A pie chart could help you show either the amount of time spent or amount of money allocated for an activity.

TABLE 18.1 Key Components of the Proposal's Discussion Section

Analyses of the existing situation, your suggested solutions, and the benefits your audience will derive	Spatial descriptions of mechanisms, tools, facilities, or products	Process analyses explaining how the product or service works	Chronological instructions explaining how to complete a task
Comparative approaches to solving a problem	Comparing and contrasting purchase options	Managerial chains of command	Chronological schedules for implementation, reporting, maintenance, delivery, payment, or completion
Corporate and employee credentials	Years in business	Testimonials from satisfied clients	Certifications
Analyses of previous accomplishments	Biographical sketches of personnel	Chronological listing of projected milestones (forecasts)	Comparative cost charts

- **Line drawings.** Effective for technical descriptions and process analyses.
- **Photographs.** Effective for technical descriptions and process analyses.
- **Flowcharts.** A successful way to help readers understand procedures.
- **Organizational charts.** Excellent for giving an overview of managerial chains of command.

Conclusion/Recommendations

Sum up your proposal, providing your readers closure. The conclusion can restate the problem, your solutions, and the benefits to be derived. Your recommendation will suggest the next course of action. Specify when this action will or should occur and why that date is important. The following example is a conclusion/recommendation from an internal proposal.

Web Resources

www.pearsonhighered.com/ gerson For more information about research and documentation, visit our companion website.

3.0 Solutions for Problem

Our line capability between San Marcos and LaGrange is insufficient. Presently, we are 23% under our desired goal. Using the vacated fiber cables will not solve this problem because the current configuration does not meet our standards. Upgrading the current configuration will improve our capacity by only 9% and still present us the risk of service outages.

4.0 Recommended Actions

We suggest laying new fiber cables for the following reasons. They will

- Provide 63% more capacity than the current system
- Reduce the risk of service outages
- Allow for forecasted demands when current capacity is exceeded
- Meet standard configurations

If these new cables are laid by September 1, 2012, we will predate state tariff plans to be implemented by the new fiscal year.

◄ **EXAMPLE**

Summarize the key elements of the proposal.

Recommend follow-up action and show the benefits derived.

PROPOSAL CHECKLIST

Have you included the following in your proposal?

_____ 1. Title page (listing title, audience, author or authors, and date)

_____ 2. Cover letter or e-mail cover message (stating why you're writing and what you're writing about; what exactly you're providing the readers; what's next—follow-up action)

_____ 3. Table of contents (listing all major headings, subheadings, and page numbers)

_____ 4. List of illustrations (listing all figures and tables, including their numbers and titles, and page numbers)

_____ 5. Abstract (stating in low-tech terms the problem, solution, and benefits)

_____ 6. Introduction (providing a statement of purpose and a lengthy analysis of the problem)

_____ 7. Discussion (solving the readers' problem by discussing topics such as procedures, specifications, timetables, materials/equipment, personnel, credentials, facilities, options, and costs)

_____ 8. Conclusion (restating the benefits and recommendation for action)

_____ 9. Glossary (defining terminology)

_____ 10. Appendix (optional additional information)

SPOTLIGHT

How to Write an Effective Proposal

In an interview, **Mary Woltkamp**, president of Effective Technical Communication, Inc., made the following comments about the importance of proposals in her job.

Q: Tell me about proposals.

A: All of the proposals that I write are for an external audience: potential clients. Thus, the tone is very formal and businesslike. One thing I always keep in mind when I'm writing is word choice. It's very easy to slip into the jargon of our industry, but more often than not, those terms are Greek to our clients. When I have to use industry-specific words, I define them. I'm also very conscious of the length of my words, sentences, and paragraphs. In general, the people who read my proposals are very busy. They don't have time to wade through a lot of unnecessary verbiage. I'm a big fan of headings, subheadings, and bulleted lists, and I always use lots of white space.

Q: What are some components in your proposals?

A: I include the following headings in a project proposal:

- Contact Information—for both the client and my company—always the first page of the document.

- Situation—What is the client's need? This assures the client that we understand the dilemma.

- Business Objectives—How will this project positively affect the client's business and return on investment?

- Project Objectives—What behaviors will be changed as a result of the project or training?

- Scope—A very detailed list of all the tasks that we think will need to be done in order to complete the project.

- Deliverables—What the client gets when we're all done, such as paper materials, electronic files, etc.

- Project Timeline—The milestones that have been identified at this point in the process.

- Time and Cost Estimates—A table that outlines our time and cost estimates for each task listed in the Scope.

- Project Team and Company Background—A short biography on all team members, plus our expertise and references (tailored to the prospective client's industry).

Q: Who constitutes the audiences for your proposals?

A: It varies. Sometimes the proposal is addressed to someone pretty high up the food chain, such as the company's president and CEO. Other times, we're dealing with a department manager or project manager who is responsible for the project.

Q: Do you use boilerplates?

A: Absolutely. I will cut and paste from multiple documents if it saves me time and makes the proposal stronger. I always have to make changes to whatever I paste into the proposal, but at least I don't have to start from scratch on every paragraph. I'm a firm believer in NOT reinventing the wheel—or retyping content when it can be avoided.

Q: Do you follow a writing process (routing, approvals, team-written, etc.), and if so, what might it be?

A: I follow a process every time I write. As for proposals, once I have completed a rough draft, I pass it off to our sales and marketing manager for input. After I make any changes my colleagues suggest, the document goes to our editor/business manager, who goes over it with a fine-tooth comb—both for grammar and for detail. Bottom line: A proposal never leaves this office without at least two pairs of eyes looking at it very closely. Our proposal is the first sample of our work that most clients see; therefore, we make sure that it is as clean, concise, and user-friendly as we can possibly make it. It's our first opportunity to impress the client, and I'm happy to say that we often do.

The Writing Process

Prewriting	Writing	Rewriting

• Determine whether you are writing an internal or external proposal.
• Determine whether you are writing a solicited or unsolicited proposal.
• Conduct primary and secondary research to gather information.
• Use a prewriting technique, such as brainstorming, outlining, or answering reporter's questions, to organize your thoughts.
	• Organize your content using modes such as problem/solution, cause/effect, comparison, argument/persuasion, chronology, etc.
	• Write persuasively by arousing your audience's interest, refuting opposing points of view, giving proof to support your contentions, and urging action in the proposal's recommendation.
	• Use figures and tables to clarify content.
		• Use headings and talking headings for access.
		• Revise your draft by
		  • Adding details
		  • Deleting wordiness
		  • Simplifying words
		  • Enhancing the tone
		  • Reformatting your text
		  • Proofreading and correcting errors

The Writing Process at Work

For your proposal, to persuade your audience to act, you will gather data, organize information, and revise text. To help you accomplish these tasks, prewrite, write, and rewrite.

Prewriting

You first have to prewrite by considering the goals of your proposal. Mary Woltkamp had to write a proposal to show how she was going to solve a potential client's problem. For her prewriting, she created a questionnaire. In the questionnaire, she interviewed employees about challenges they faced when using a manual to complete a job-related task. With this primary research, she was able to gather information from the client. See Figure 18.2 for Mary's questionnaire.

Writing

After you gather information, you can then write a draft of a proposal. Below is Mary's draft of part of the Discussion section of her proposal (Figure 18.3).

Rewriting

After drafting the document, Mary revised the proposal excerpt based on input from her colleagues. Figure 18.4 is Mary's revised proposal excerpt.

FIGURE 18.2   Questionnaire for Planning a Proposal

About the Manual
- Are instructions for completing tasks set off clearly from the other text and written in a numbered, step-by-step format?
- Are graphics and screen shots used when appropriate to enhance the instructions?
- Does the material have a table of contents and/or index to help readers find the information they need?
- Are terms and acronyms clearly defined?

From Interviews with Employees
- What tasks do you perform without making mistakes? How often do you perform those tasks?
- What tasks do you struggle with? How often do you perform those tasks?
- Do you refer to the training manual when you are having trouble?
- If you do, can you find the information you need easily? Or at all?

From Interview with Manager
- What mistakes are causing the most problems?
- How often are these mistakes made?

FIGURE 18.3   Partial Draft of "Findings" Section with Colleague Suggestions

Mary, you need to include more details about goals and recommendations. That way we can clarify what we hope to achieve in the proposal.

Try using bullets and subheadings to break the information into smaller chunks for easy access.

The content is good, but we need to make this more persuasive. Let's highlight the errors that employees were making. Doing so will remind the readers exactly what prompted this proposal and how important our solutions are.

Findings:

Employees perform 80% of the tasks in Application X with 100% accuracy. Each of these tasks is performed at least once a week if not daily. Employees also consistently identified five tasks that they have trouble completing. Three of these tasks are complicated, multistep tasks with costly consequences when errors are made. Two of the tasks are performed only once a quarter.

Next, the instructions in the existing training manual are well written. Graphics and screen shots are used often to enhance the clarity of the instructions. Because of this, employees found the manual to be very thorough and helpful in the classroom training sessions.

However, the manual does not have an index, and the table of contents is skeletal. Employees do not use the manual as a reference when they are having trouble with a task because the manual is cumbersome to handle and instructions are hard to find.

FIGURE 18.4   Proposal Excerpt

Client Request

Company ABC (Client) has asked Effective Technical Communication, Inc. (ETC) to submit a proposal for the redesign of an existing training manual. The manual is used to train new Finance Department employees on Application X, the company's expense reporting application.

Employees are currently making costly mistakes, and the Client believes that the manual is failing to communicate what employees need to know or do. The Client's request for proposal (RFP) does not indicate what information the Client relied on when deciding to have the manual revised.

Project Goal

The goal of the project is to eliminate the mistakes employees make when using Application X.

Findings

In response to the RFP, ETC asked for and received permission to conduct a front-end analysis to confirm the cause of the mistakes being made in the application.

The findings include the following:

- Employees perform 80% of the tasks in Application X with 100% accuracy. Each of these tasks is performed at least once a week if not daily.
- Employees consistently identified five tasks that they have trouble completing. Three of these tasks are complicated, multistep tasks with costly consequences when errors are made. Two of the tasks are performed only once a quarter.
- The instructions in the existing training manual are well written. Graphics and screen shots are used often to enhance the clarity of the instructions.
- Employees found the manual to be very thorough and helpful in the classroom training sessions.
- The manual does not have an index, and the table of contents is skeletal.
- Employees do not use the manual as a reference when they are having trouble with a task because the manual is cumbersome to handle and instructions are hard to find.

Recommendation

Although the manual would benefit from the addition of an index and a more complete table of contents, a complete revision of the manual is unnecessary. Instead, ETC recommends the creation of job aids for those tasks that are complicated, multistep processes or for those tasks that are performed infrequently.

Sample Internal and External Proposals

Figure 18.5 illustrates a sample external proposal that you can use as a model for your writing. Figure 18.6 shows a sample internal proposal.

FIGURE 18.5 External Proposal Sample

Proposal Title

Proposal
for
Reducing Oxygen Expenses

Proposal Audience

Prepared for
Dr. Richard Davis, Director
La Habra Retirement Center

Proposal Author

by
Robert Maxwell
Marketing Director
Orson Medical Supplies

Date of Submission

December 8, 2012

FIGURE 18.5 (Continued)

Orson Medical Supplies
"Improving the Quality of Life"
12345 College Blvd.
Overland Park, OR 90091
976-988-2000

December 8, 2012

Dr. Richard Davis, Director
La Habra Retirement Center
220 Cypress
La Habra, CA 90631

Dear Dr. Davis:

Submitted for your review is our proposal regarding the Electronic Demand Cannula oxygen-saving system. This document is in response to your September 25, 2012, letter and our subsequent discussions.

Within our report, you will find the following supporting materials geared toward your requests:

- Product specifications . . . . pages 3–4
- Operating instructions . . . . page 4
- Qualifications and experience . . . . page 5
- Cost . . . . page 6
- Warranty . . . . page 7

Thank you for your interest in our product. We look forward to serving you and will call within two weeks to finalize arrangements.

Sincerely,

Robert Maxwell

Robert Maxwell
Marketing Director

Enclosure: Proposal

Itemizing the cover letter body focuses on key points within the proposal and provides the appropriate page numbers for reference. Maintaining a pleasant tone and being persuasive will appeal to your reader.

(Continued)

FIGURE 18.5 External Proposal Sample (Continued)

Headings, subheadings, and page numbers help the audience find information and navigate the text.

Table of Contents

List of Illustrations

Figure and table numbers plus titles allow the reader to find the visuals quickly.

iii

FIGURE 18.5 (Continued)

Abstract

Expenses for medical oxygen have increased steadily for several years. Now the federal government is reducing the amount of coverage that Medicare allows for prescription oxygen.

These cost increases can be reduced through the use of our new Electronic Demand Cannula (EDC). The EDC delivers oxygen to the patient only when the patient inhales. Oxygen does not flow during the exhalation phase. Therefore, oxygen is conserved.

This oxygen-saving feature can reduce your oxygen expenses by as much as 50 percent. Patients who use portable oxygen supplies can enjoy prolonged intervals between refilling, thus providing more freedom and mobility.

Emphasize problems generating the proposal.

Show how the proposal can solve the problems for the client.

Highlight the benefits derived from accepting the proposal. Stating the problem, solution, and benefits gives focus to the proposal.

iv

(Continued)

FIGURE 18.5   External Proposal Sample (Continued)

1.0 Introduction

1.1 Purpose

This is a proposal to sell the new Electronic Demand Cannula (EDC)* to the La Habra Retirement Center, La Habra, California. This bid to sell offers you a special discount when you purchase our EDCs in the quantities suggested in this proposal.

1.2 Problems

1.2.1 *High Costs*

Since 2000, the price of medical-grade oxygen has skyrocketed. It cost $10 per 1,000 cubic feet (cu ft) in 2000. Today, medical-grade oxygen costs $26 per 1,000 cu ft. In fact, you can expect next year's oxygen expenses to double the amount you spent this year.

1.2.2 *Governmental/Insurance Involvement*

Many factors have contributed to this soaring cost, including demand, product liability, and inflation. However, two factors contributed the most. First, legislation reduced the amount that Medicare pays for prescription oxygen. Second, few insurance companies offer programs covering long-term prescription oxygen. Therefore, you, or your patients, must pay the additional expenses.

1.2.3 *Decreased Quality of Service*

Because prescription oxygen has risen in cost so dramatically, few medical service companies can produce affordable EDCs and stay competitive. Since 2001, according to *Medical Digest Bulletin,* 80 percent of medical service vendors have gone out of business. Your ability to receive quality service at an affordable price has diminished.

*This and subsequent terms marked by an asterisk (*) are defined in the glossary.

1

The IEEE numbering system (1.2, 1.2.1) aids document design and allows the writer to organize detailed content.

Headings and subheadings are used to help the reader navigate the text. The writer places these headings and subheadings on the Table of Contents thus giving readers quick access to all parts of the proposal.

FIGURE 18.5   (Continued)

2.0 Discussion

2.1 Implementation of Electronic Demand Cannula

Because the price of oxygen will not go down, you must try to use less while obtaining the same clinical benefits.

Orson Medical Supplies, a leader in oxygen-administering technology, proposes the implementation of our new EDC*. Using state-of-the-art electronics, the EDC senses the patient's inspiratory effort.* When a breath is detected, the EDC dispenses oxygen through the patient's cannula.* The patient receives oxygen only when he or she needs it.

Continuously flowing cannulas waste gas during exhalation and rest. Clinical studies have proved that 50 percent of the oxygen used by cannula patients is wasted during that phase. These same tests also revealed that blood oxygen saturation* does not significantly vary between continuous and intermittent flow cannulas. The patient receives the same benefit from less oxygen. Table 1 explains this in greater detail.

Explanations and details address the audience's need to be persuaded by seeing the benefits of the proposal.

Table 1
BLOOD OXYGEN SATURATION
USING THE ELECTRONIC DEMAND CANNULA
VERSUS
CONTINUOUS FLOW CANNULAS

Prescribed Flowrate (L/min)*	Breaths per Minute (bpm)*	Blood Oxygen Saturation %	
		Intermittent	Continuous
0.5	12	96%	98%
1	12	98%	99%
2	12	99%	100%
3	12	100%	100%
4	12	100%	100%

A table adds visual appeal, makes complex information easier to understand, and enhances the readability of numbers.

We have included a technical description of the EDC to help explain how this system will benefit oxygen cannula users.

2

(Continued)

FIGURE 18.5   External Proposal Sample (Continued)

2.2 Technical Description

The EDC* is an oxygen-administering device that is designed to conserve oxygen. The EDC is composed of six main parts: oxygen inlet connector, visual display indicators (LEDs),* power switch, patient connector, AC adapter connector, and high-impact plastic case (see Figures 1 and 2).

Line drawings combined with dimensions help the reader to visualize the equipment. In this way, the writer anticipates questions that the reader might have about the equipment.

Figure 1 EDC Front View

Oxygen Inlet Connector:
The oxygen inlet connector is a DISS No. 1240 (diameter index safety system)* and is made of chrome-plated brass.

Visual Display: Two LEDs* provide visual indications of important functions. Alarm functions are monitored by a red LED, Motorola No. R32454. An indication of each delivered breath is given by the pulse display, which is a yellow LED, Motorola No. Y32454.

Power Switch: The power switch is an ALCO No. A72-3 slide switch. The dimensions are $0.5'' \times 0.30''$: button height is $0.20''$. Electrical Specifications: Dry contact rating is 1 amp, contract resistance is 20 milliohms, and the life expectancy is 100,000 actuations.

3

FIGURE 18.5 (Continued)

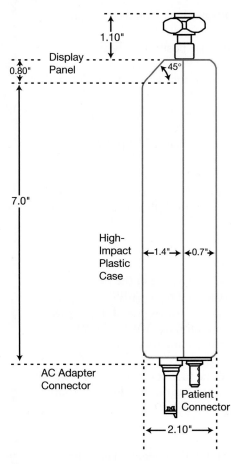

Patient Connector: Attachment of the patient cannula system is made at the patient connector, which is located at the bottom of the case. The white nylon connector, Air Logic No. F-3120-85, is a 10-32, UNF male threaded, straight barbed connector for 1/8″ ID flexible tubing.

AC Adapter Connector: An optional AC adapter* and battery charger assembly, part number PA-32, plugs into the AC adapter connector, which is located at the bottom left-hand side of the case. The connector is a male, D-subminiature, 12-pin flush insert supplied by Dupont Connector Systems. The part number is DCS: 68237009.

High-Impact Plastic Case: The case housing is made from an impact-resistant, flame-retardant, oxygen-compatible ABS plastic*.

Using technical descriptions creates a "word picture" of the equipment.

Figure 2 EDC Side View

2.3 Operating Instructions

The EDC is an oxygen-saving and administering device (see Figures 1 and 2). By following these five easy steps, you will be able to enjoy the benefits of intermittent demand oxygen.

Warning: Federal law prohibits the sale or use of this device without the order of a physician.

4

(Continued)

FIGURE 18.5   External Proposal Sample (Continued)

1. Attach your oxygen supply to the Oxygen Inlet Connector located at the top of the case.
2. Move the Pulse-Steady Switch to the Pulse position to begin intermittent demand flow.
3. Connect your nasal cannula to the Patient Outlet Connector located at the bottom of the case.
4. Adjust your oxygen supply to the oxygen flow prescribed by your physician.
5. Put on your nasal cannula and breathe normally. The pulse light will turn on when a breath is delivered.

You are now ready to conserve oxygen by as much as 50 percent. Should you have the need to go back to continuous flow, just push the Pulse-Steady Switch to the "steady" position.

2.4 Qualifications and Experience

Qualifications and experience help to persuade the reader of your company's expertise, reputation, and ability to do the job.

Orson Medical Supplies has been an international leader in the field of respiratory therapy since 1975. Pure Air introduced the first IPPB* respirator on the market. In 1985, responding to the needs of doctors and therapists, we produced the first life support volume ventilator, the VV-1. The VV-1 became the industry standard by which all other ventilators were measured.

In 1990, Orson Medical Supplies introduced the first computer-controlled life support system, the VV-2. Technology developed for this product has found application in other areas as well. Recently, we introduced one such product, the Electronic Demand Cannula.

Orson Medical Supplies is located in Overland Park, Oregon. The main manufacturing and engineering facility employs 450 people. Regional sales and service branch offices are located throughout the United States.

2.5 Personnel

Each of our engineering facilities is staffed by trained technicians ready to answer your questions. The following individuals have been assigned to La Habra Retirement Center:

5

FIGURE 18.5 (Continued)

Randy Draper
Randy (BS, Electrical Engineering, South Central Texas University, 1992) has worked at Orson Medical Supplies since 1992. In 16 years at Orson, Randy has been promoted from service technician to supervisor. Randy specializes in developing new medical equipment. He has supervised the development teams that worked on the X29 respirator, the Z284-00 ventilator, and the Omega R-449 sphygmomanometer. Randy was lead development specialist for the Pure Air EDC.

Randy will be in charge of your account. Please contact him directly regarding any questions you might have about the EDC.

Ruth Bressette
Ruth (BS, Mechanical Engineering, Pittsburgh State University, 1997) has worked at Orson Medical Supplies since 1998. She has risen in our company from service technician to manager of troubleshooting/maintenance. Ruth has received the highest-level certification (Master Technician) offered by the IEEE for service on every piece of equipment developed, manufactured, and sold by Orson.

Ruth will be the manager of your Orson equipment maintenance and troubleshooting crew. Her responsibility is to ensure that your equipment is kept in outstanding working condition. She will schedule maintenance checks and promptly assign technicians to troubleshoot potential malfunctions.

Douglas Loeb
Doug (AA, Electrical Engineering Technology, Plainview Community College, 1992) is one of our most accomplished troubleshooters. Having worked at Orson for 16 years, Doug is commended annually for his speed, accuracy, and skill. Your equipment is in good hands with Doug. He will be your primary troubleshooter and maintenance person.

2.6 Cost

Orson Medical Supplies is pleased to offer our Pure Air EDC at cost-effective pricing. Table 2 explains the benefits you'll derive when purchasing in quantity.

Provide detailed credentials and emphasize the qualifications of the team members for the audience. Such details can help persuade the audience to accept the proposal.

6

(Continued)

FIGURE 18.5   External Proposal Sample (Continued)

Table 2   LIST PRICE VS. DISCOUNT PRICE FOR EDC

List Price	Quantity Warranty*	Extended *per EDC*	Total Cost
$350	1–9 units	$75	$425
Discount Price			
$310	10–24 units	$60	$370
$300	25+ units	$50	$350

As you can see, Orson is happy to offer you substantial savings when you purchase our Pure Air EDC in volume. At these prices, and assuming normal use, the oxygen cost savings will exceed your initial investment in less than one year, as shown in Figure 3.

Figure 3 Return on Investment

Visuals, such as tables and figures, break up potentially monotonous text and help to persuade the audience through accessible content.

2.7 Warranty

Orson Medical Supplies warrants this product to be free of manufacturing defects for a one-year period after the original date of consumer purchase. This warranty does not include damage done to the product due to accident, misuse, improper installation or operation, or unauthorized repair. This warranty also does not include replacement of parts due to normal wear.

*Orson's extended warranty is discussed in Section 2.8.

7

FIGURE 18.5 (Continued)

If the product becomes defective within the warranty period, we will replace or repair it free of charge.

This warranty gives you special legal rights. You may also have other rights that vary from state to state. Some states do not allow exclusions or limitations of incidental damage, so the above limitations may not apply to you.

2.8 Extended Warranty

In addition to the coverage provided in our unconditional warranty, you might want to take advantage of our extended warranty package. For the prices provided in Table 2, Orson Medical Supplies will extend the warranty to cover a three-year period after the original date of consumer purchase. With this three-year extended warranty, Orson not only covers manufacturing defects but also replaces worn parts free of charge.

The extended warranty does not cover damage to the product resulting from accident, misuse, improper installation or operation, or unauthorized repair.

For answers to any of your questions regarding repair, replacement, warranty, or extended warranty, please call 1-800-555-ORSN, or write to Manager, Customer Relations, Orson Medical Supplies, 12345 College Blvd., Overland Park, OR 90091.

3.0 Conclusion

3.1 Major Concern

Prescription oxygen expenses are escalating whereas government support has been reduced. The cost increase to the patient and the health care facility will be enormous.

3.2 Recommendation

To offset the inevitable rise of oxygen expenses, we recommend the use of the Electronic Demand Cannula.

8

Offering the customer an extended warranty could persuade the reader of the benefits of the proposal.

(Continued)

FIGURE 18.5   External Proposal Sample (Continued)

4.0 Glossary

ABS plastic	Acrylonitrile butadiene styrene—a durable and long-lasting plastic
AC adapter	A remote power supply used to convert alternating current to direct current
Blood oxygen saturation	The partial pressure of oxygen in alveolar blood recorded in percent
Cannula	A small tube inserted into the nose, specifically for administering oxygen
Electronic Demand Cannula (EDC)	An electronically-controlled device that dispenses oxygen only when triggered by an inspiratory effort
Inspiratory effort	The act of inhaling
Light-emitting diode (LED)	A solid-state semiconductor device that produces light when current flows in the forward direction

List of Acronyms/Abbreviations

bpm	Breaths per minute
DISS	Diameter index safety system
EDC	Electronic Demand Cannula
IPPB	Intermittent positive pressure breathing
L/min	Liters per minute

9

Alphabetized glossary defining jargon and technical terms helps both a low-tech and a lay audience to understand the text.

FIGURE 18.6   Sample Internal Proposal Written
in PowerPoint

*B*io
*S*taffing

Your One-Stop Shop for Biomedical Needs

Submitted to
Leann Towner
Chief Financial Officer

By
Jonathan Bacon
Manager, Information Technology Department

June 12, 2009

Table of Contents

1.0 **Abstract**
   1.1 Problem
   1.2 Solution
   1.3 Benefits
2.0 **Introduction**
   2.1 Purpose
   2.2 Problem
      2.2.1 Technology Policies
      2.2.2 Outdated Hardware
3.0 **Discussion**
   3.1 Revised Technology Replacement Policy
      3.1.1 Biannual Rotating Replacements
   3.2 Hardware Purchases
      3.2.1 Hardware Allocation Analysis
      3.2.2 Hardware Costs
4.0 **Conclusion**
   4.1 Conclusion
   4.2 Recommendation
   4.3. Benefits
5.0 **Glossary**

Date: June 12, 2009
To: Leann Towner, Chief Financial Officer
From: Jonathan Bacon, Information Technology Manager
Subject: Proposal for New Corporate Technology Support

*B*io
*S*taffing

Leann, in response to your request, the IT Department is happy to propose
a new technology support system.

Among detailed analyses of our current system, this proposal presents the
following:
   ✓ BioStaffing's current technology challenges
   ✓ Hardware purchasing suggestions
   ✓ Technology vendor recommendations

We are confident that our suggestions will maintain BioStaffing's
competitive edge and increase employee satisfaction. If I can answer any
questions, either call me (ext. 3625) or e-mail me at jbacon@biostaff.com.

(Continued)

FIGURE 18.6   Sample Internal Proposal Written
in PowerPoint (Continued)

Both figure/table
numbers and titles
are given for easy
reference and
clarity.

List of Illustrations

Figure 1: Increase in Costs Due to Repairs and Retrofitting per
        Workstation

Figure 2: Time Spent Printing Based on PPM

Figure 3: Vendor Evaluation Matrix

Table 1:   Biannual Rotating Technology Replacements

Table 2:   Business Application Hardware Requirements

An abstract often
presents the problem
necessitating the
proposal, provides the
proposed solutions,
and shows the
potential benefits.

1.0 Abstract

1.1 Problem
BioStaffing's current technology policies are not responsive to our evolving needs, our
hardware is outdated, and our current vendor is changing ownership.

1.2 Solution
To solve these problems, the Information Technology (IT) Department suggests the
following:
- Upgrading our technology needs biannually rather than every five years
- Purchasing new computers, printers, scanners, and digital cameras
- Hiring a new vendor to supply and repair our hardware

1.3 Benefits
A biannual replacement schedule will allow us to stay current with hardware
advancements. Purchasing new technologies will help our staff more effectively meet
client needs. Hiring a new vendor is key to these goals. Our current vendor agreement
ends June 30, 2009. A new vendor will provide better turnaround, pricing, maintenance,
and merchandise.

Pronouns such as
"our" and "we"
are appropriate
for an internal
proposal written
to coworkers.

2.0 Introduction

2.1 Purpose

The purpose of this proposal is to improve BioStaffing's current
technology. By revising our policies, purchasing new hardware, and
hiring a new technology vendor, we can increase both customer and
worker satisfaction.

2.2 Problem

BioStaffing faces three key challenges:

2.2.1 *Technology Policies*
2.2.2 *Outdated Hardware*
2.2.3 *Vendor Changes*

FIGURE 18.6   (Continued)

2.0 Introduction cont.
Problems

2.2.1 *Technology Policies*
Since 1987, we have replaced hardware and software on a five-year, rotating basis. This was an effective policy initially since new-computer costs were expensive.

However, the current policy is no longer responsive to our needs, for the following reasons:

- *Prices have gone down*. Today's hardware costs are more affordable. In addition, vendors will offer BioStaffing buyer-incentives. We can buy hardware on an as-needed basis by taking advantage of special sales pricing.

  In contrast, our current five-year hardware replacement policy is not responsive. It disregards today's changes in technology pricing, and it disallows us from taking advantage of dealer incentives.

A variety of highlighting techniques, such as boldface headings, underlined subheadings, and bullets, make the text easier to read.

2.0 Introduction cont.
Problems

- *Repair costs have gone up*. By replacing technology only every five years, BioStaffing has resorted to repairing and retrofitting outdated equipment. Costs for these repairs have increased more than five-fold over the last ten years, as noted in the following line graph (Figure 1).

Figure 1: Increase in Costs Due to Repairs and Retrofitting Per Workstation

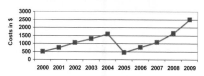

The line graph adds clarity and provides visual appeal.

2.0 Introduction cont.
Problems

2.2.2 *Outdated Hardware*
We need to purchase new technology items because our current hardware and software are outdated.

- *Our computers have insufficient memory*. Most of our computers have only 2.0 GHz processors with approximately 256 MB of RAM versus the current minimal standard 3.0 GHZ and 512 MB. This negatively impacts speed of document retrieval.*

(*Continued*)

FIGURE 18.6   Sample Internal Proposal Written in PowerPoint (Continued)

The proposal uses specific details to persuade the audience.

2.0 Introduction cont.
Problems

- *Our ink jet printers are slow and produce poor quality documents*. Our current black and white printers produce only 7 ppm with a dpi resolution of 1440 x 720.* In contrast, new black and white laser printers will produce up to 20 ppm at a dpi resolution of 5760 x 1440.

> NOTE: At 7 ppm (times 60 minutes per hour), BioStaffing personnel can print 420 pages per hour. Current printing requires **9.5 hours**.
>
> In contrast, at 20 ppm (times 60 minutes per hour), improved printers can produce 1200 pages per hour, requiring only **3.3 hours**.
>
> <u>**Laser printers can save our company over 6 hours of lost time each day.**</u>

2.0 Introduction cont.
Problems

To clarify the importance of ppm, look at Figure 2.

Figure 2: Time Spent Printing Based on PPM

3.0 Discussion

3.1 Revised Technology Replacement Policy
We must revise BioStaffing's technology replacement policy. It is not responsive to our growing technology needs. We suggest this approach:

3.1.1 Biannual Rotating Replacements

We propose that BioStaffing divide its 12 departments into two categories. The "Red" team would receive technology upgrades one year (even-numbered years), while the "Blue" team would receive technology upgrades the next year (odd-numbered years).

IT has divided the departments into teams alphabetically for fairness, as shown in Table 1:

FIGURE 18.6 (Continued)

3.0 Discussion cont.

Table 1: Biannual Rotating Technology Replacements	
Red Team (*even*-numbered years)	**Blue Team** (*odd*-numbered years)
Accounting	Manufacturing
Administrative Services	Personnel
Corporate Communication	Sales
Information Technology	Shipping and Receiving

The table, color, and white space show how formatting aids readability.

3.0 Discussion cont.

3.2. Hardware Purchases

After researching department requirements, IT determined that BioStaffing needs to upgrade business application hardware as follows:

3.2.1 Hardware Allocation Analysis

Based on survey results from each department, IT suggests that the hardware listed in Table 2 be allocated as follows:

- **Accounting**—2 desktops, 1 laptop, 5 handhelds, and 1 laser printer
- **Administrative Services**—3 desktops, 1 laptop, 1 handheld, and 1 laser printer
- **Corporate Communication**—1 laptop, 7 handhelds, 1 scanner, 1 laser printer, and 2 digital cameras
- **Information Technology**—1 laptop, 7 handhelds, 1 scanner, 1 laser printer, and 2 digital cameras
- **Manufacturing**—2 desktops, 1 laptop, 2 handheld, and 1 laser printer
- **Personnel**—2 desktops, 1 laptop, 4 handhelds
- **Sales**—2 desktops, 1 laptop, 10 handhelds, and 1 digital camera
- **Shipping and Receiving**—1 desktop, 1 laptop, 2 handhelds

Providing specific amounts shows how you will meet the reader's needs.

3.0 Discussion cont.

3.2.2 Hardware Costs

Table 2: Business Application Hardware Costs		
Hardware	Individual Costs	Total Costs per Item
12 Desktop Computers	$1,000 each	$12,000
8 Laptop Computers	$700 each	$5,600
38 Handheld computers	$150 each	$5,700
5 Laser printers	$400 each	$2,000
3 Scanners	$170 each	$510
5 digital cameras	$200 each	$1,000
Total Costs for all items		$26,810

(Continued)

FIGURE 18.6   Sample Internal Proposal Written in PowerPoint (Continued)

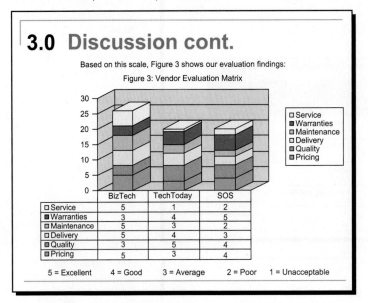

3.0 Discussion cont.

Based on this scale, Figure 3 shows our evaluation findings:

Figure 3: Vendor Evaluation Matrix

	BizTech	TechToday	SOS
☐ Service	5	1	2
■ Warranties	3	4	5
■ Maintenance	5	3	2
☐ Delivery	5	4	3
▨ Quality	3	5	4
■ Pricing	5	3	4

5 = Excellent   4 = Good   3 = Average   2 = Poor   1 = Unacceptable

4.0 Conclusion

4.1 Conclusion

BioStaffing's technology needs are critical.

- _Repair costs_ due to old hardware have risen five-fold per workstation, from as low as $500 to as high as $2,500.

- _Outdated printers_, producing only 7 pages per minute, require up to 9.5 hours a day to print BioStaffing's 4,000 pages of documentation.

- _Expired technology vendor contracts_ need to be renegotiated.

Key to the above challenges is our current 5-year technology replacement policy. Technology advances demand a more responsive policy.

4.0 Conclusion cont.

4.2 Recommendation

The Information Technology Department, based on research and surveys, proposes the following:

- _A biannual technology replacement policy_—this would allow BioStaffing to meet employee technology needs more responsively as well as benefit from vendor pricing incentives.

- _New communication hardware_—this would include desktop, laptop, and handheld computers; laser printers; scanners; and digital cameras.

- _A new technology vendor_—based on our research, we suggest that BizTech Warehouse will best meet our technology needs.

FIGURE 18.6 (Continued)

4.0 Conclusion cont.

4.3 Benefits

These changes must occur before June 30, 2009, when our current vendor contract ends. By acting now, BioStaffing will benefit in several ways.

- We can save up to 20% on hardware and maintenance costs.

- We will maintain our competitive edge in the marketplace.

- Most importantly, our employee satisfaction and productivity will increase as they work with the latest technology and software upgrades.

Positive words and phrases such as "benefit," "Save," "competitive edge," "Provide employee satisfaction," and "increase productivity" help sway the audience toward the writer's stance.

5.0 Glossary

Acronym/Abbreviation	Definition
dpi	Dots per inch
GHz	Gigahertz
GB	Gigabyte
HD	Hard drive
MB	Megabyte
MHz	Megahertz
ppm	Pages per minute

Though all readers might not need terms defined, a glossary provides clarity to a less specialized audience.

CHAPTER HIGHLIGHTS

1. You might have multiple readers for a proposal, including internal and external audiences. Consider the needs of your audience. To communicate with different levels of readers, include abstracts, glossaries, and definitions.
2. A proposal could include the following:
   - Title page
   - Cover letter or cover e-mail message
   - List of illustrations
   - Abstract
   - Introduction
   - Discussion
   - Conclusion
   - Recommendation
   - Glossary
   - Works cited (or references)
   - Appendix

3. Subheadings and visual aids will make your proposal more accessible.

4. Use primary and secondary research to develop your content.

5. Write persuasively to convince your audience to act. To accomplish this goal, arouse reader interest, refute opposing points of view, gather details to support your argument, and urge action.

6. Write ethically by documenting sources and making sure your content (prices, timelines, credentials, etc.) are accurate.

APPLY YOUR KNOWLEDGE

CASE STUDIES

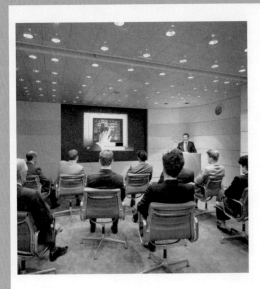

1. The technical communication department at Bellaire Educational Supplies/Technologies (BEST) needs new computer equipment. Currently, the department has outdated hardware, outdated word processing software, an outdated printer, and limited graphics capabilities. Specifically, the department is using computers with 15-inch monitors, hard drives with only 256 KB of memory, and one, 10 MB hard disk drive. The word processing package used is WordPro 3.0, a version created in 2000. Since then, WordPro has been updated four times; the latest version is 6.5. The department printer is a black-and-white Amniprint machine. To create art, the department must go off-site to a part-time graphic artist who charges $35 an hour, so the department uses very few graphics.

   Because of these problems, the company's user manuals, reports, and sales brochures are being poorly reviewed by customers. Further, BEST's Web site doesn't have an option for online purchases or social media links for client contacts. The bottom line: BEST is falling behind the curve, and profits are off 27 percent from last year.

   As department manager, you have consulted with five staff members (Jim Nguyen, Mario Lozano, Mike Thurmand, Amber Badger, and Maya Liu) to correct these problems. As a team, you have decided the company needs to purchase new equipment:

   - **Six new personal computers.** Each computer must have a 19-inch color monitor, 5 GB of memory, a 500 GB hard drive, a 16x DVD+/-RW drive for burning DVDs, and a VGA graphics card.

   - **Two laser printers.** These must have a print speed of 24 ppm, resolution of 600 × 600 dpi, and 4 MB of memory, expandable to 132 MB.

   - **Word processing software.** WordPro 6.5 with these capabilities: voice-activated annotations, typing, and correcting; automatic footnoting and endnoting; envelope labeling; grammar and spell checking; thesaurus; help options; automatic index generating; 50 or more scalable fonts.

   - **Graphics software.** For professional-quality newsletters and brochures, BEST needs the capability for quick demonstrations and design tips; a layout checker with at least 10 online views; 200 true type fonts, each scalable; a table creation toolbar; 2,000 clip art images; and a logo creator with 50 border design options.

   - **Scanner.** To increase your graphics potential, BEST also needs a flatbed scanner with these specifications: 300 dpi image resolution; 155 ppm gray-scale scanning capability; 8-bit, 256 gray-scale color support; and approximately 8 ½ × 12-inch bed size.

   Using the criteria provided in this chapter, write an internal proposal to BEST's CEO, Jim McWard. In this proposal, explain the problem, discuss the solution to this problem, and then highlight the benefits derived once the solution has been implemented. These benefits will include increased productivity, better public relations, increased profits, and less employee stress. Develop these points thoroughly, and provide Mr. McWard the names of vendors for the required hardware and software. To find these vendors, you could search the Internet.

2. You own Buzz Electronics Co., 4256 Blue Mountain Blvd., Raleigh, North Carolina 65221. Mr. and Mrs. Allan Thibodeux, 3876 Spanish Moss Drive, Bayside, North Carolina 65223, have asked you to give them a bid on electrical work for a new family room they are adding to their home.

   You and Mr. and Mrs. Thibodeux have gone over the couple's electrical needs, including the following. The room, which will measure 18 feet (east to west) by 15 feet (north to south), should have four 110 V outlets for 3 lamps, a clock, a radio, a CD and DVD player, and a high-definition television with 48-inch screen. The family wants the four 110 V outlets to be placed equidistant throughout the room.

   The client wants two 220 V outlets. One 220 V outlet will go by the southwest window on the west wall where the family plans to put a window air conditioning unit. The window will be located 3 feet in from the south wall. There will be another window on the west wall, located 3 feet in from the north wall. A third picture window, measuring 6 feet wide by 4 feet high, will be centered on the south wall. The other 220 V outlet must be placed on the east wall, where the family plans to put home office equipment (computer, printer, scanner, and fax machine). Their office desk will sit 5 feet from the door leading into the room. The door will be built on the east wall where it comes to a corner meeting the south wall.

   Centered in the ceiling, the family wants electrical wiring for a fan with a light package. In addition to this light, the family also wants a light mounted on the east wall above the desk area, so wiring is needed there, approximately 5 to 6 feet up from the floor.

   The family wants two light switches in the room: one by the door and one on the north wall, approximately 6 feet in from where the north and west walls meet. The Thibodeuxs plan to have a couch and lamp in that area for reading. They want both to be double switches, one to control the fan and ceiling lights; the other to control additional floor and ceiling lights in the room. All light switches need rheostats for dimming. Finally, the Thibodeuxs plan to have a whole-house vacuum system installed in the walls, and they have asked you if you can provide this service.

   Buzz Electronics has been in business since 1995. The company has worked with thousands of satisfied customers, including both residential and business owners. Buzz has long-standing contracts for service with Acme Construction, J&L Builders, Food-to-Go Groceries, the City of Piedmont, North Carolina, and Ross and Reed Auto Showroom.

   As owner of Buzz Electronics, you have an associate's degree in electronics from Sandy Shoal Community College, Sandy Shoal, North Carolina. You are ETA-I (Electronics Technicians Association International) Certified; NASTec (National Appliance Service Technician) Certified; and a Certified Industrial Journeyman. You have eight employees, all of whom also are Certified Industrial Journeymen.

Assignment

Write a short (three- to five-page) external proposal—bid for contract. To do so, study the Thibodeux's electrical needs, list the parts you will need to complete the job, estimate the time for your labor—including setup, work performed, and cleanup. Then, provide a price quote. You might need to research the wiring and equipment needed for this job, either online, in technical journals, or in parts catalogues. Follow the guidelines provided in the textbook for short reports and proposals.

INDIVIDUAL AND TEAM PROJECTS

1. Write an external proposal. To do so, create a product or a service and sell it through a proposal. Your product can be an improved radon detection unit, a new MP3 player, safety glasses for construction work, bar codes for pricing or inventory control, a piece of biotechnology equipment for monitoring blood work, computer graphics for an advertising

agency, and so on. Your service may involve dog grooming, automobile servicing, computer maintenance, home construction (refinishing basements, building decks, room additions, and so on), freelance technical writing, at-home occupational therapy, or home theater installation and maintenance. The topic is your choice. Draw from your job experience, college coursework, or hobbies. To write this proposal, follow the process provided in this chapter.

2. Write an internal proposal. You can select a topic from either work or school. For example, your company or department is considering a new venture. Research the prospect by reading relevant information. Interview involved participants or survey a large group of people. Once you have gathered your data, document your findings and propose to management the next course of action. If you choose a topic from school, you could propose a daycare center, on-campus bus service, improved computer facilities, tutoring services, coed dormitories, pass/fail options, and so on. Write an internal proposal to improve your company's Web site; expand or improve the security of your company's parking lot; create or improve your company's physical site security in light of post 9/11 issues; improve policies for overtime work; improve policies for hiring diversity; or improve your company's policies for promotion. Research your topic by reading relevant information or by interviewing or surveying students, faculty, staff, and administration. Once you have gathered your data, document your findings and recommend a course of action.

PROBLEM-SOLVING THINK PIECES

1. Stinson, Heinlein, and Brown Accounting, LLC, employs over 2,000 workers, including accountants, computer information specialists, legal staff, paralegals, and office managers. The company requires a great deal of written and oral communication with customers, vendors, governmental agencies, and co-workers. For example, a sample of their technical communication includes the following:

   - Written reports to judges and lawyers
   - Letters and reports to customers
   - E-mail and memos to co-workers
   - Oral communication in face-to-face meetings, videoconferences, and sales presentations
   - Maintaining the company's Facebook and blog sites

   Unfortunately, not all employees communicate effectively. The writing companywide is uneven. Discrepancies in style, grammar, content, and format hurt the company's professionalism. The same problems occur with oral communication.

   George Hunt, a mid-level manager, plans to write an internal, unsolicited proposal to the company's principal owners, highlighting the problems and suggesting solutions. What must Mr. Hunt include in his proposal—beyond the obvious proposal components (a title page, table of contents, abstract, introduction, and so forth)—to persuade the owners to accept his suggestions? Suggest ways in which the problem can be solved.

2. Toby Hebert is sales manager at Crab Bayou Industries (Crab Bayou, Louisiana). In his position, Toby manages a sales staff of 12 employees who travel throughout Louisiana, Texas, Arkansas, and Mississippi. Currently, the sales staff members use their own cars to make sales calls, and CBI pays them 31 cents per mile for travel expenses. Each staff member currently travels approximately 2,000 miles a month, with cars getting 20 miles per gallon.

   Gasoline prices, at the moment, are over $3 a gallon. With gasoline and car maintenance costs higher than ever, the current rate of 31 cents per mile means that CBI's sales employees are losing money. Something must be done to solve this problem. Toby has met with his staff, and they have decided to write a proposal to Andre Boussaint, CBI's CEO.

   What must Toby and his staff include in the proposal—beyond the obvious proposal components (a title page, table of contents, abstract, introduction, and so forth)—to persuade the CEO to accept the suggestions? Suggest ways in which the problem can be solved.

WEB WORKSHOP

By typing "RFP," "proposal," "online proposal," or "online RFP" in an Internet search engine, you can find tips for writing proposals and requests for proposals (RFPs), software products offered to automatically generate e-proposals and winning RFPs, articles on how to write proposals, samples of RFPs and proposals, and online RFP and proposal forms.

To perform a more limited search, type in phrases like "automotive service RFP," "computer maintenance RFP," "desktop publishing RFP," "Web design RFP," and many more topics. You will find examples of both proposals and RFPs from businesses, school systems, city governments, and various industries.

To enhance your understanding of business and industry's focus on proposal writing, search the Web for information on RFPs and proposals. Using the criteria in this chapter and your knowledge of effective technical writing techniques, analyze your findings.

- How do the online proposals or RFPs compare to those discussed in this textbook, in terms of content, tone, layout, and so on?
- What information provided in the textbook is missing in the online discussions?
- What are some of the industries that are requesting proposals, and what types of products or services are they interested in?
  a. Report your findings, either in an oral presentation or an e-mail message.
  b. Imagine that you are requesting a proposal for a product or service. Create your own online form to meet this need.
  c. Rewrite any of the proposals that you think can be improved, using the criteria in this chapter as your guide.
  d. Respond to an online RFP by writing a proposal. To complete this assignment, go online to research any information you need for your content.

QUIZ QUESTIONS

1. Explain the purpose of an internal proposal.
2. Explain the purpose of an external proposal.
3. What content do you include in the introduction?
4. What is an abstract in a proposal?
5. What do you include in a purpose statement?
6. What are the typical components of the discussion section?
7. What do you include in the recommendation?
8. Explain why you include a glossary.
9. What can you include in an appendix?
10. What type of visuals can you include in a proposal?
11. Explain techniques for organizing a proposal.
12. How can you format a proposal for reader-friendly ease of access?
13. In what ways can you enhance the tone of your proposal?
14. In what ways can you write persuasively in a proposal?
15. What are RFPs, SOWs, and boilerplates?

Oral Communication

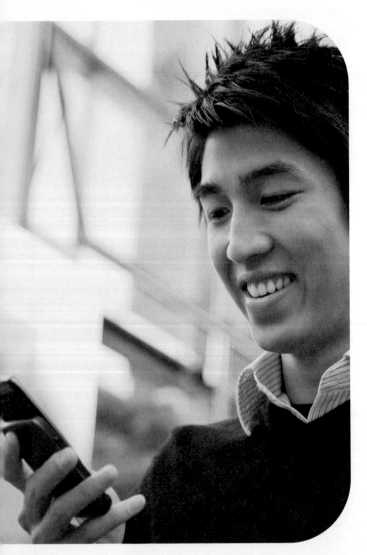

COMMUNICATION AT WORK

In the TechStop scenario, a customer, Carolyn Jensen, complains about malfunctioning equipment. Her complaints are not handled successfully by the sales help, so Shuan Wang, the vice president of customer service, has decided to make an oral presentation to his staff to address this issue.

TechStop, with 12 locations throughout the state, sells DVDs, high-definition televisions, audio components, computers, and computer peripherals. Lately, Shuan Wang has been receiving complaints from customers about poor service. For example, one customer, Carolyn Jensen, owns a large electronics company. She has done business with TechStop for six years. During that time, Carolyn's company has purchased over 1,000 pieces of equipment, including computers, printers, paper items, and cell phones.

Carolyn called a local TechStop to complain about problems her company was having with over a dozen printers. Her company's text looks fine on screen but hard copies print differently than they look. The company cannot print three

Objectives

When you complete this chapter, you will be able to

1. Communicate effectively with customers, vendors, and co-workers

2. Understand the importance of verbal and nonverbal communication

3. Develop effective listening skills

4. Use the telephone to communicate successfully

5. Use voicemail effectively

6. Deliver effective informal oral presentations on the job

7. Participate in teleconferences

8. Communicate effectively in a videoconference or webconference

9. Deliver formal oral presentations

10. Use a variety of visual aids to enhance your oral presentations, including techniques for effective PowerPoint presentations

11. Use the writing process—prewriting, writing, and rewriting—for informal and formal oral presentations

different sized envelopes (as had been promised by the salesperson when Carolyn purchased the printers). Graphics are not printing in color, though the color ink cartridges are full. Finally, printers are stopping their print jobs before all pages of a document have been printed. She purchased the printers 12 months and two weeks ago. The printer's in-store warranty was for 12 months, guaranteeing full replacement of parts and labor coverage.

Carolyn realizes that the warranty expired two weeks ago. However, she contends that several facts should negate this deadline. The deadline expired over a Christmas weekend, a severe snowstorm left many homes and businesses without power for several days, and other, more pressing business activities required her attention. When she informed the sales help of her situation and her company's long-standing patronage of TechStop, the salesperson said, "Sorry, lady. The warranty's just no good anymore. Hey, we've

all got problems. Anyway, there's no way my boss will even listen to you with this complaint. He told me to leave him alone when he's busy."

Unfortunately, Shuan has heard of other such complaints regarding rude and unresponsive sales personnel. Shuan fears that TechStop's staff is developing an overall corporate disregard for customer satisfaction. To address this issue for the entire sales staff at TechStop's numerous locations throughout the state, Shuan plans to give a formal oral presentation on the following:

- Sales etiquette and customer interaction
- Store policy regarding customer satisfaction
- The impact of poor customer relations
- The consequences of failing to handle customers correctly

Shuan only hopes that his actions aren't too late.

> Check out our quarterly newsletters at www.pearsonhighered.com/gerson for dot.com updates, new case studies, insights from business professionals, grammar exercises, and facts about technical communication.

585

The Importance of Oral Communication

Many people, even the seemingly most confident, are afraid to speak in front of others. A recent Monster.com poll asked, "What is your biggest career-related phobia?" Table 19.1 provides the results of the poll.

This chapter offers techniques to make your oral communication experience rewarding rather than frightening. You may have to communicate orally with your peers, your subordinates, your supervisors, and the public. Oral communication is an important component of your business success. You will be required to speak formally and informally on an everyday basis, as follows:

- 40 percent of employees use oral communication face-to-face and one-on-one.
- 30 percent of employees talk on the phone.
- 36 percent of employees attend group meetings.
- 8 percent of employees participate in group teleconferences. (Miller)

Everyday Oral Communication

"Hi. My name is Bill. How may I help you?" Think about how often you have spoken to someone today or this week at your job. You constantly speak to customers, vendors, and co-workers face-to-face, on the telephone, or by leaving messages on voicemail:

- If you work an 800 hotline, your primary job responsibility is oral communication.
- When you return the dozens of calls you receive or leave voicemail messages, each instance reveals your communication abilities.
- As an employee, you must achieve rapport with your co-workers. Much of your communication to them will be verbal. What you say impacts your working relationships.

Every time you communicate orally, you reflect something about yourself and your company. The goal of effective oral communication is to ensure that your verbal and nonverbal skills make a good impression and communicate your messages effectively.

What Is Verbal Communication?

What you say is either positively or negatively impacted by *how* you say it. In fact, "after your physical appearance, your voice is the first thing people notice about you" (Clarke). Verbal communication includes your pace (rate of speech) and modulation.

Pace.   A key concern is pace, the speed with which you speak. Generally, we speak about 150 to 200 words a minute. Speeding up this rate can have several meanings. You could imply that you are happy, fearful, angry, or surprised. In addition, if you talk too fast, not

TABLE 19.1 Career-Related Phobias

Percentage	Phobia
42%	Giving a speech or presentation
32%	Confronting a co-worker or boss
15%	Networking
11%	Writing a report or proposal

Source: (Monster Career Advice Newsletter)

only will your co-workers have trouble following your train of thought, but also they might think you're nervous or impatient. If you are in a meeting, for instance, and you speak too rapidly, your voice might be telling your co-workers it's time to move on to the next topic. You are saying, "I don't have time for you" or "I want to be doing something else (anything else)."

Talking too slowly causes problem also. Slowing down your pace could convey sadness, boredom, or lack of interest. People also might feel you are talking down to them. A very slow delivery style could imply condescension—"I assume that you can't keep up with my thoughts, so I'll speak *Reallllly Slowlllly*."

Modulation. Modulation is the loudness, tone, and pitch of your voice as you speak. A very loud voice will be perceived as dominating, while a very quiet voice could make you sound meek. Mumbling can destroy your credibility. A monotone voice, one with no change in pitch, will put your listener to sleep. Modulation, varying your pace and pitch, adds interest to your comments.

What Is Nonverbal Communication?

Nonverbal communication is another phrase for body language. Basically, nonverbals entail "any conscious or unconscious movement of a part of all of the body that communicates an emotional message" (Tilton 15). Nonverbal cues can include any of the following.

Eye Contact. Good eye contact suggests openness, confidence, and interest in your audience. However, too much of anything is potentially bad. Do not overdo eye contact; staring and glaring will make any audience uncomfortable. Many people believe that "duration of the eye contact should last between 3 to 5 seconds" (Tilton 15). Any more than that might be unsettling.

Facial Expressions. Smiling, frowning, yawning, wide-eyed bewilderment, open-mouthed shock, lip-curling angry glares—each of your facial expressions conveys a message. For example, if you are listening to a colleague speak and you are simultaneously wrinkling your nose, arching your eyebrows, scrunching your face, or furrowing your forehead, you are conveying a very negative message. You are showing your co-worker your uncertainty or disbelief. In contrast, a pleasant smile shows your involvement and support. Figure 19.1 illustrates facial expressions.

Posture. Imagine you are in a meeting. A co-worker is speaking, but you are fidgeting, slouching, placing your hands behind your head, or leaning to the side. You are reflecting boredom and disinterest. On the other hand, if you lean forward, you show you are interested and actively participating in a team meeting. Nodding your head up and down shows your affirmation. Figure 19.2 demonstrates how posture reveals attitude.

Proximity. What does proximity to your audience show? If you stand very close to a spouse or child, you show your love. However, if you stand very close to a co-worker, you will encroach upon his or her personal space. This can suggest a threatening or domineering behavior. In *The Silent Language*, Edward T. Hall suggests that we have four main proximity zones of personal space (see Table 19.2).

Figure 19.3 shows the effect of proximity.

Multicultural Concerns. Sometimes, proximity is a cultural consideration. Studies suggest that people from many Middle Eastern cultures stand closer while speaking than North Americans are used to. Middle Eastern men might greet each other by grasping both hands and placing kisses on each cheek. There is more physical contact, and conversations often involve touching. These "intimate" and "personal" behaviors in Western culture are acceptable "public" behaviors in the Mideast. In contrast, touching is not common in the Far East (China and Japan) where more formality and distance from the speaker are the norm.

FIGURE 19.1 Facial Expressions

Smiling

Frowning

Angry

Surprised

Bored

Gestures. We speak a lot with our hands. A "thumbs up" gesture in the United States signifies that everything is "A-OK." A "thumbs down" gesture denotes a negative opinion. These two gestures are commonly understood in Western culture. Other gestures, however, are subtler. Making a fist shows anger, while rubbing the back of your neck or your forehead shows stress. If you place your hands in front of your mouth, you might be showing secretiveness and deception (Tilton 4). Finger pointing might help you make a point, but it is also very assertive, even threatening. We all know how annoying a person's rhythmic drumming can be. This suggests that the individual is impatient and bored.

Multicultural Concerns. Many of our gestures are culturally based. Finger pointing in the Far East is considered to be rude. Worse still are the "thumbs up" and "OK" signs. In Italy and Australia, a "thumbs up" motion can be considered obscene. The "OK" gesture can be considered obscene in Brazil and Germany.

Listening Skills

In addition to controlling verbal and nonverbal reactions, you must also learn to listen effectively. If you are not listening while others are speaking, no communication can

FIGURE 19.2 Attitude Revealed Through Posture

Look at the people sitting in the picture. What does each person's posture tell you about his or her attitude?

TABLE 19.2 Proximity Zones of Personal Space

Zone	Distance	Who Gets Access	What Occurs Within the Zone
Intimate	6–18 inches (15–45 cm)	Spouse, children, and significant others	Intimate relationships, such as touching and hugging
Personal	1 1/2–4 feet (45–125 cm)	Family members and close personal friends	Within an arm's length, less intimate but still very close encounters
Social	4–12 feet (1.25–3.6 m)	Friends and colleagues	Social and business relationships
Public	Over 12 feet (+3.6 m)	Strangers/general public	Little interaction

occur. Collaboration demands that all voices be heard. Successful workplaces value everyone's input.

Barriers to Active Listening. What gets in the way of your active listening abilities? Following are barriers to active listening.

Multitasking. Envision this scenario: You take your seat at the meeting table. You are still on your cell phone, speaking to a customer. You are also checking your PDA to see when you can meet with this client. You send a tweet to a vendor about a proposed meeting. Meanwhile, you are thinking about the deadline you have for submitting a proposal. You are wondering if your graphic artist has completed the sketch you need for the upcoming meeting. You are juggling many tasks at the same time. Now, the meeting you are currently in has begun, and your co-worker is speaking. How can you also pay attention to this colleague, when you have so many other tasks to complete and so many other conflicting responsibilities?

FIGURE 19.3 Proximity Zones of Personal Space

Personal

Public

Social

Intimate

Preconceived Notions.   If you believe that you already know what a co-worker will say, then you will not hear his or her comments objectively. Assumptions like this are a barrier to successful teamwork. Preconceived notions will diminish the colleague's comments and hinder your contributions to the team effort.

Focusing on Your Response.   While your colleague is speaking, you do not wait to hear all he or she says. You only focus on your follow-up response. You can barely wait to jump in with a clever retort, a witty comeback, an oppositional point of view, or debunking doubt. This does your colleague and your team a disservice. By thinking only of what you might say in response, you risk not hearing all that is said in the meeting.

Interrupting the Speaker.   Interrupting a speaker is rude. A person should have the right to his or her say. If you interject your comments abruptly, then your co-worker cannot complete his or her thoughts. Waiting to respond is not only good manners; it is also good business.

External Distractions. Think of all the possibilities:

- The room is too hot or too cold.
- It is 11:30 a.m. and you are hungry.
- You receive a tweet and want to read and respond to it immediately.
- It is Friday and you have exciting plans for the weekend.
- A road crew is repairing the street outside your office complex, complete with the sounds of a jackhammer and the annoying beeping of a truck backing up.

These external distractions might disallow you from focusing on what your co-workers are saying.

Keys to Effective Listening. Effective listening is critical to a company's goal of continuous improvement. Employees can best represent a company by actively listening to vendors, customers, and co-workers. When you hear what others say, then you also can contribute to discussions more effectively and provide useful feedback. Following are some keys to effective listening.

Stop What You Are Doing; Concentrate on the Task at Hand. Turn off and put away your smart phone. Don't text or tweet while you are in a meeting. Leave your unfinished work in your office. Focus on the team's project and on the speaker.

Do Not Talk. It is hard to speak and listen at the same time. Sure, you might have valuable things to add, but wait your turn. Let the speaker have his or her say. Once that person has spoken, you can.

Make Eye Contact. When someone in the group is speaking, look at this person. Do not look out the window, look at your watch, stare at your shoes, check your smart phone for messages, or look at your fingernails. Making eye contact with the speaker will help you focus your attention.

Take Notes. Another way to stay focused is by taking notes. Write down key ideas for future reference. You also can jot down any thoughts you might have for follow-up discussion. This way, you will not feel compelled to interrupt the speaker. The notes will help you remember what you had wanted to say, once the speaker has completed his or her thoughts.

Be Objective. Without a doubt, a colleague might say something with which you disagree. That's okay. We all have different points of view. However, remember that these differences are exactly what make cross-functional teams successful. Be open to alternatives; be willing to accept differences in opinion. Remember to judge the content of the comment instead of the speaker. Often, you will be asked to work with someone you might not like. Maybe you do not like his or her politics, work ethic, or personality. Still, that individual might make valuable comments.

Ask Questions. Once someone has spoken, then review your notes and ask questions. Be sure your questions are sincere and not oppositional. The goal of a question should be clarity. Then, after he or she answers your question, repeat what the speaker has said to confirm your understanding.

Control Your Reactions. If you hear something you do not agree with, avoid just shouting out your opposition. Your goal is to *respond*, not *react*. When you *react*, you tend to do so defensively. Immediate reactions are based on impatience, maybe even aggression. A more effective approach to successful listening is to calmly *respond*, after a brief delay— 5 to 10 seconds. This delay acts as a filter or a buffer. Even a short delay will allow you to digest the issue, consider the person's point of view, organize your thoughts, and then respond more professionally.

Communicating on the Job

Telephone and Voicemail

You speak on the telephone dozens of times each week. When speaking on a telephone, make sure that you do not waste either your time or your listener's time. Follow these 10 tips for telephone and voicemail etiquette:

1. Know what you are going to say before you call.
2. Speak clearly and enunciate each syllable.
3. Avoid rambling conversationally.
4. Avoid lengthy pauses.
5. Leave brief messages.
6. Avoid communicating bad news.
7. Repeat your phone number twice including the area code.
8. Offer your e-mail address as an option.
9. Sound pleasant, friendly, and polite.
10. If a return call is unnecessary, say so.

DOT-COM UPDATES

For more information about using voicemail effectively, check out the following link:

- "Top 10 Tips for Using Voicemail Effectively": http://www.sideroad.com/ Business_Etiquette/ voicemail-greeting.html

Informal Oral Presentations

As a team member, manager, supervisor, employee, or job applicant, you often will speak to a co-worker, a group of colleagues, or a hiring committee. You will need to communicate orally in an informal setting for several reasons:

- Your boss needs your help preparing a presentation. You conduct research, interview appropriate sources, and prepare reports. When you have concluded your research, you might be asked to share your findings with your boss in a brief, informal oral presentation.
- Your company is planning corporate changes (staff layoffs, mergers, relocations, or increases in personnel). As a supervisor, you want to provide your input in an oral briefing to a corporate decision maker.
- You are applying for a job. Your interview, though not a formal, rehearsed presentation, requires that you speak effectively before a hiring committee.
- You are working from a remote location but need to participate in company-sponsored training. You cannot physically attend the workshop, but through a Web conference (Webinar) or teleconference, you and other employees from around the country can be trained simultaneously on your computers.

Video and Teleconferences.   If you are communicating with a group and you want to hear what everyone else is saying simultaneously, video or teleconferences are an answer. Consider using a video or teleconference when three or more people at separate locations need to talk.

In a video or teleconference, you want all participants to feel as if they are in the same room facing each other. With expensive technology in place, such as cameras, audio components, coder/decoders, display monitors, and user interfaces, avoid wasting time and money with poor communication. Follow these 10 tips for successful video and teleconferences:

1. Inform participants of conference date, time, time zone, and expected duration.
2. Make sure participants have printed materials before the teleconference.
3. Ensure that equipment has good audio quality.

4. Choose your room location carefully for quiet and privacy.
5. Consider arrangements for hearing impaired participants. You might need a text telephone (TTY) system or simultaneous transcription in a chat room.
6. Introduce all participants.
7. Direct questions and comments to specific individuals.
8. Do not talk too loudly, too softly, or too rapidly.
9. Turn off cell phones and pagers.
10. Limit side conversations.

Webconferencing.   Due to costs of airline tickets, time-consuming travel, and the need to complete projects or communicate information quickly, many companies and organizations are using Webconferencing. Webconferencing, sometimes called webinars, e-seminars, webcasts, teleclasses, or virtual meetings, include Web-based seminars, lectures, presentations, or workshops transmitted over the Internet (Coyner; Murray). You can achieve successful Webconferences by following these 10 tips:

1. Limit Webconferences to 60 to 90 minutes.
2. Limit a Webconference's focus to three or four important ideas.
3. Start quickly by limiting introductory comments.
4. Keep it simple. Instead of using too many Internet tools, limit yourself to simple and important features like polling and messaging.
5. Plan ahead. Make sure that all Webconference participants have the correct Internet hardware and software requirements; know the correct date, time, agenda, and Web login information for your Webconference; and have the correct Web URL or password.
6. Before the Webconference, test your equipment, hypertext links, and PowerPoint slide controls.
7. Use both presenter and participant views. One way to ensure that all links and slides work is by setting up two computer stations. Have a computer set on the presenter's view and another computer logged in as a guest. This will allow you to view accurately what the audience sees and how long displays take to load.
8. Involve the audience interactively through questions and/or text messaging.
9. Personalize the presentation. Introduce yourself, other individuals involved in the presentation, and audience members.
10. Archive the presentation.

Formal Oral Presentations

You might need to make a formal presentation for the following reasons:

- Your company asks you to visit a civic club meeting and provide an oral presentation to maintain good corporate or community relations.
- Your company asks you to represent it at a city council meeting. You will give an oral presentation explaining your company's desired course of action or justifying activities already performed.
- Your company asks you to represent it at a local, regional, national, or international conference by giving a speech.
- A customer has requested a proposal. In addition to writing this long report, you need to make an oral presentation promoting your service or product to the potential customer.

Types of Formal Oral Presentations

Three types of formal oral presentations include the memorized speech, manuscript speech, and extemporaneous speech.

Memorized Speech.   The least effective type of oral presentation is the memorized speech. This is a well-prepared speech which has been committed to memory. Although such preparation might make you feel less anxious, too often these speeches sound mechanical and impersonal. They are often stiff and formal, and allow no speaker-audience interaction.

Manuscript Speech.   In a manuscript speech, you read from a carefully prepared manuscript. The entire speech is written on paper. This may lessen speaker anxiety and help you to present information accurately, but such a speech can seem monotonous, wooden, and boring to the listeners.

Extemporaneous Speech.   Extemporaneous speeches are probably the best and most widely used method of oral communication. You carefully prepare your oral presentation by conducting necessary research, and then you create a detailed outline. However, you avoid writing out the complete presentation. When you make your presentation, you rely on notes or PowerPoint slides with the major and minor headings for reference. This type of presentation helps you avoid seeming dull and mechanical, allows you to interact with the audience, and still ensures that you correctly present complex information. Figure 19.4 shows the oral presentation process.

Parts of a Formal Oral Presentation

A formal oral presentation consists of an introduction, a discussion (or body for development), and a conclusion.

Introduction

The introduction should welcome your audience, clarify your intent by providing a verbal road map, and capture your audience's attention and interest. This is the point in the

FIGURE 19.4   The Oral Presentation Process

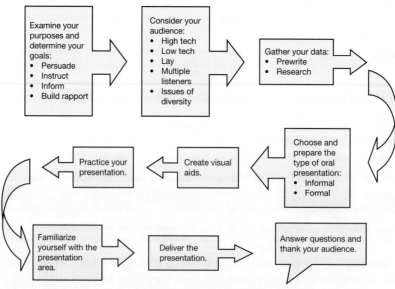

Source: (Adapted from "Communications: Oral Presentations")

presentation where you are drawing in your listener, hoping to create enthusiasm and a positive impression.

To create a positive impression, set the table. Address your audience politely by saying "Good morning" or "Good afternoon." Tell the audience your name and the names of others who might be speaking. Welcome the audience and thank them for inviting you to speak.

- Table setters for goodwill

◀ EXAMPLE

"Good evening. Thank you for allowing us to speak to you tonight. I'm (name and title). You'll also hear from (name and title)."

Next, provide a road map, clarifying what points you'll discuss and laying out the order of the topics.

- Road map (thesis statement)

◀ EXAMPLE

"What we are going to talk about, and in this order, are the following key points:

- The issues that led us to consider road improvements
- Challenges to this construction task
- Optional approaches
- Costs
- A timeframe for your consideration."

You can use a variety of openings to capture your audience's attention, such as the following:

- Word pictures—anecdotes (specific in *time, place, person,* and *action*), quotes, and data (facts and figures)

◀ EXAMPLE

Word Pictures to Arouse Your Audience's Interest

"From November 2011 to February 2012, our police department received over 75 reports of problems regarding Elm Street. These ranged from injuries related to hill jumping, cars sliding into the street's ditches, increased rush hour traffic, and tight turn lanes. One stretch of the road, from grid 39 to grid 47, is too narrow for snow removal crews to clean effectively. And these issues promise to get worse with residential growth anticipated to increase by 39%. As Sgt. Smith of the police department stated, 'Elm Street is a disaster waiting to happen.'"

- Question or series of questions—to ask questions involves the audience immediately. A training facilitator could begin a workshop as follows:

◀ EXAMPLE

Questions to Involve Your Audience

"How many reports do you write each week or month? How often do you receive and send e-mail messages to customers and colleagues? How much time do you spend on the telephone? Face it; technical communication is a larger part of your engineering job than you ever imagined."

These three questions are both personalized and pertinent. Through the use of the pronoun *you,* the facilitator speaks directly to each individual in the audience. By focusing on the listeners' job-related activities, the questions directly lead into the topic of conversation—technical communication.

- Quotation from a famous person—the training facilitator in the previous example could have begun his speech with a quote from Warren Buffett, a famous businessperson.

EXAMPLE ▶

A Quote to Arouse Audience Interest

"How important is effective technical communication? Just listen to what Warren Buffett has to say on the topic:

'For more than forty years, I've studied the documents that public companies file. Too often, I've been unable to decipher just what is being said or, worse yet, had to conclude that nothing was being said. . . . Perhaps the most common problem . . . is that a well-intentioned and informed writer simply fails to get the message across to an intelligent, interested reader. In that case, stilted jargon and complex constructions are usually the villain. . . . When writing Berkshire Hathaway annual reports, I pretend that I'm talking to my sisters. I have no trouble picturing them: Though highly intelligent, they are not experts in accounting or finance. They will understand plain English, but jargon may puzzle them. My goal is simply to give them the information I would wish them to supply me if our positions were reversed. To succeed, I don't need to be Shakespeare; I must, though have a sincere desire to inform.'

That's what I want to impart to you today: Good writing is communication that is easy to understand. If simple language is good enough for Mr. Buffett, then that should be your goal."

(Buffett 1–2)

Discussion (or Body)

After you have aroused your listeners' attention and clarified your goals, you have to prove your assertions. In the *discussion* section of your formal oral presentation, provide details to support your thesis statement. You can develop your content in a variety of ways, including the following:

Research

See Chapter 5 for more discussion of research and documentation.

- **Quotes, testimony, anecdotes.** You can find this type of information through primary and secondary research. For example, primary research, such as a survey or questionnaire, will help you substantiate content. By using people's direct comments, you help your audience to relate to the content. You also can find quotes, testimony, or anecdotes in secondary research, such as periodicals, newspapers, books, or online. This type of information validates your comments. As an ethical oral communicator, cite sources of information where appropriate or provide the audience with a references page.
- **Data.** Facts and figures, again found through research or interviews, develop and support your content. Stating exactly how often your company's computer system has been attacked by viruses will support the need for improved firewalls. A statistical analysis of the increase in insurance premiums will clarify the need for a new insurance carrier. Showing the exponential increase in the cost of gasoline over the last 10 years will show why your company should consider purchasing

hybrids. To help your audience follow your oral communication, present these details using any of the following modes of organization:

Organization

See Chapter 3 for more discussion of types of organization.

Comparison/contrast.   In your presentation, you could compare different makes of office equipment, employees you are considering for promotion, different locations for a new office site, vendors to supply and maintain your computers, different employee benefit providers, and so forth. Comparison/contrast is a great way to make value judgments and provide your audience options.

Problem/solution.   You might develop your formal oral presentation by using a problem-to-solution analysis. For example, you might need to explain to your audience why your division needs to downsize. Your division has faced problems with unhappy customers, increased insurance premiums, decreased revenues, and several early retirements of top producers. In your speech, you can then suggest ways to solve these problems ("We need to downsize to lower outgo and ultimately increase morale"; "Let's create a 24-hour, 1-800 hotline to answer customer concerns"; "We should compare and contrast new employee benefits packages to find creative ways to lower our insurance costs").

Persuasion

See Chapter 11 for more discussion of persuasive communication.

Argument/persuasion.   Almost every oral presentation has an element of argument/persuasion to it—as does all good written communication. You will usually be persuading your audience to do something based on the information you share with them in the presentation.

Importance.   Prioritizing the information you present from least to most important (or most important to least) will help your listeners follow your reasoning more easily. To ensure the audience understands that you are prioritizing, provide verbal cues. These include simple words like *first, next, more important,* and *most important.* Do not assume that these cues are remedial or obvious. Remember, sometimes it is hard to follow a speaker's train of thought. Good speakers realize this and give the audience verbal signposts, reminding the listeners exactly where they are in the oral presentation and where the speaker is leading them.

Chronology.   A chronological oral presentation can outline for your audience the order of the actions they need to follow. For example, you might need to prepare a yearly evaluation of all sales activities. Provide your audience with target deadlines and with the specific steps they must follow in their reports each quarter.

Maintaining coherence.   To maintain coherence, guide your audience through your speech as follows:

- **Use clear topic sentences.** Let your listeners know when you are beginning a new, key point: "Next, let's talk about the importance of conciseness in your technical communication."
- **Restate your topic often.** Constant restating of the topic is required because listeners have difficulty retaining spoken ideas. A reader can refer to a previously discussed point by turning back a page or two. Listeners do not have this option.

  Furthermore, a listener is easily distracted from a speech by noises, room temperature, uncomfortable chairs, or movement inside and outside the meeting site. Restating your topic helps your reader maintain focus.
- **Use transitional words and phrases.** This helps your listeners follow your speech. Transitional words and phrases, like those shown in Table 19.3, aid reader comprehension, emphasize key points, and highlight your speech's organization.

TABLE 19.3 Transitional Words

Purpose	Examples
Cause and effect	because, since, thus, therefore, for this reason, due to this, as a result of, consequently, in order to
Example	for instance, for example, another
Interpreting jargon	that is, in other words, more commonly called
Sequencing ideas	first, second, next, last, following, finally, above, below
Adding a point	furthermore, next, in addition, besides, not only . . . but also, similarly, likewise
Restating	in other words, that is, again, to clarify
Contrasting	but, instead, yet, however, on the other hand, nevertheless, in contrast, on the contrary, whereas, still
Emphasizing	in fact, more importantly, clearly
Summarizing	to summarize, therefore, in summary, to sum up, consequently, therefore

Conclusion

Conclude your speech by restating the main points, by recommending a future course of action, or by asking for questions or comments. A polite speaker leaves time for a few follow-up questions from the audience. Gauge your time well, however. You do not want to bore people with a lengthy discussion after a lengthy speech. You also do not want to cut short an important question-and-answer (Q and A) session. If you have given a controversial speech that you know will trouble some members of the audience, you owe them a chance to express their concerns.

Presentation Delivery Skills

Effective oral communicators interact with and establish a dynamic relationship with their audiences. The most thorough research will be wasted if you are unable to create rapport and sustain your audience's interest. Although smaller audiences are usually easier to connect with, you can also establish a connection with much larger audiences through a variety of delivery techniques.

Eye Contact.   Avoid keeping your eyes glued to your notes. You will find it easy to speak to one individual because you will naturally look him or her in the eye. The person will respond by looking back at you.

With a larger audience, whether the audience has 20 or 200-plus people, keeping eye contact is more difficult. Try looking into different people's eyes as you move through your presentation (or look slightly above their heads if that makes you more comfortable). Most of the audience has been in your position before and can sympathize.

Rate.   Vary your rate of speech to keep the audience interested. Slow down for the most important and most interesting parts. Speed up your delivery when you reach a section of less interesting facts. Match your rate of speaking to the content of your speech, just as actors vary their speech rate to reflect emotion and changes in content.

Enunciation.   Speak each syllable of every word clearly and distinctly. Rarely will an audience ask you to repeat something even if they could not understand you the first time. It is up to you to avoid mumbling. Remember to speak more clearly than you might in a more conversational setting.

Pitch. When you speak, your voice creates high and low sounds. That's *pitch*. In your presentation, capitalize on this fact. Vary your pitch by using even more high and low sounds than you do in your normal, day-to-day conversations. Modulate to stress certain keywords or major points in your oral presentations.

Pauses. One way to achieve a successful pace is to pause within the oral presentation. Pause to ask for and to answer questions, allow ideas to sink in, and use visual aids or give the audience handouts. These pauses will not lengthen your speech; they will only improve it.

A well-prepared speech will allow for pauses and will have budgeted time effectively. Know in advance if your speech is to be 5 minutes, 10 minutes, or an hour long. Then plan your speech according to time constraints, building pauses into your presentation. Practice the speech beforehand so you can determine when to pause and how often.

Emphasis. You will not be able to underline or boldface comments you make in oral presentations. However, just as in written communication, you will want to emphasize key ideas. Your body language, pitch, gestures, and enunciation will enable you to highlight words, phrases, or even entire sentences.

Conflict Resolution

You might be confronted with a hostile listener who either disagrees with you or does not want to be in attendance. You need to be prepared to deal with such a person. If someone disagrees with you or takes issue with a comment you make, try these responses:

- "That is an interesting perspective."
- "Thanks for your input."
- "Let me think about that some more and get back to you."
- "I have got several more ideas to share. We could talk about that point later, during a break."

The important point to remember is to not allow a challenging person to take charge of your presentation. Be pleasant but firm and maintain control of the situation. You will be unable to please all of your listeners all of the time. However, you should not let one unhappy listener destroy the effect of your presentation for the rest of the audience.

Visual Aids

Most speakers find that visual aids enhance their oral communication. "Visuals have the greatest, longest lasting impact—show as much or more visually as what you say. Use pictures; use color. Use diagrams and models" (O'Brien).

Although PowerPoint slide shows, graphs, tables, and flip charts are powerful means of communication, you must be the judge of whether visual aids will enhance your presentation. Avoid using them if you think they will distract from your presentation or if you lack confidence in your ability to create them and integrate them effectively. However, with practice, you probably will find that visuals add immeasurably to the success of most presentations.

Table 19.4 lists the advantages, disadvantages, and helpful hints for using visual aids. For all types of visual aids, practice using them before you actually make your presentation. When you practice your speech, incorporate the visual aids you plan to share with the audience.

TABLE 19.4 Advantages and Disadvantages of Visual Aids

Type	Advantages	Disadvantages	Helpful Hints
Chalkless Whiteboards	Help audiences take notes. Allow you to emphasize a point. Allow audiences to focus on a statement. Help you be spontaneous. Break up monotonous speeches.	Require unique, erasable pens. Can stain clothing. Some pens can be hard to erase if left on the board too long. Pens that run low on ink create light, unreadable impressions.	Use blue, black, or red ink. Cap pens to avoid drying out. Use pens made especially for these boards. Erase soon after use.
Flip Charts	Can be prepared in advance. Are neat and clean. Can be reused. Are inexpensive. Are portable. Help you avoid a nonstop presentation. Allow for spontaneity. Help audiences take notes. Allow you to emphasize key points. Encourage audience participation. Allow easy reference by turning back to prior pages. Allow highlighting with different colors.	Are limited by small size. Require an easel. Require neat handwriting. Won't work well with large groups. You can run out of paper. Markers can run out of ink.	Have two pads. Have numerous markers. Use different colors for effect. Print in large letters. Turn pages when through with an idea so audience will not be distracted. Don't write on the back of pages where print bleeds through.
DVDs	Are easy to use. Are entertaining. Have many to choose from. Can be used for large groups.	Require equipment and outlets. Make note taking difficult. Deny speaker–audience interaction. Can malfunction and become dated.	Use up-to-date films. Avoid long films. Provide discussion time. Use to supplement the speech, not replace it. Practice with the equipment.
PowerPoint Presentations	Are entertaining. Offer flexibility, allowing you to move from topics with a mouse click. Can be customized and updated. Can be used for large groups. Allow for speaker–audience interaction. Can be supplemented with handouts easily generated by PPt. Can be prepared in advance. Can be reused. Allow you to return to a prior slide. Can include animation and hyperlinks.	Require computers, screens, and outlets. Work better with dark rooms. Computers can malfunction. Can be too small for viewing. Can distance the speaker from the audience.	Practice with the equipment. Bring spare computer cables. Be prepared with a backup plan if the system crashes. Have the correct computer equipment (cables, monitors, screens, etc.). Make backup handouts. Practice your presentation.

Powerpoint Presentations

One of the most powerful oral communication tools is visual—Microsoft PowerPoint (PPt). Whether you are giving an informal or formal oral presentation, your communication will be enhanced by PowerPoint slides.

Today, you will attend very few meetings where the speakers do not use PowerPoint slides. PowerPoint slides are used frequently because they are simple to use, economical,

EFFECTIVE ORAL PRESENTATION CHECKLIST

_____ 1. Does your speech have an introduction,
- Arousing the audience's attention?
- Clearly stating the topic of the presentation?

_____ 2. Does your speech have a body,
- Explaining what exactly you want to say?
- Developing your points thoroughly?

_____ 3. Does your speech have a conclusion, suggesting
- What is next?
- Explaining when (due date) a follow-up should occur?
- Stating why that date is important?

_____ 4. Does your presentation provide visual aids to help you make and explain your points?

_____ 5. Do you modulate your pace and pitch?

_____ 6. Do you enunciate clearly so the audience will understand you?

_____ 7. Have you used verbal and nonverbal communication skills effectively?

_____ 8. Do you speak slowly and remember to pause so the audience can think?

_____ 9. Have you practiced with any equipment you might use?

_____ 10. Have you left time for Q/A?

FAQs: Using PowerPoint

Q: Why is the use of PowerPoint so important in the workplace?

A: Widely used by businesspeople, educators, and trainers, PowerPoint is among the most prevalent forms of presentation technology. "According to Microsoft, 30 million PowerPoint presentations take place every day: 1.25 million every hour" (Mahin). Employees use PowerPoint not only for oral presentations but also as hard-copy text.

Q: Does everyone like PowerPoint? Aren't there any negative attitudes toward this technology?

A: Not everyone likes PowerPoint. Opposition to the use of this technology, however, usually stems from the following problems:

- Dull PowerPoint slides, lacking in variety and interest
- The use of Microsoft's standardized templates
- "Death by Bullet Point," caused by an excessive dependence on bullets
- Excessive content on a PowerPoint slide
- People who read the slides to an audience

However, these challenges can be overcome easily through techniques discussed in this chapter.

DOT.COM UPDATES

For more information about PowerPoint, both techniques for using this technology effectively and potential problems, check out the following links:

Sites Related to Effective PowerPoint Usage

Finklestein, Ellen. "PowerPoint Tips Blog." http://www.ellenfinkelstein.com/powerpoint_tip.html.

"The PowerPoint FAQ." http://www.pptfaq.com/.

Paradi, Dave. "How to Avoid Death by PowerPoint." http://www.communicateusingtechnology.com/articles/avoid_death_by_ppt.htm.

Russell, Wendy. "10 PowerPoint Tutorials for Beginners—How to Use PowerPoint." http://presentationsoft.about.com/od/powerpoint101/a/begin_guide.htm.

Wuorio, Jeff. "Presenting with PowerPoint: 10 dos and don'ts." http://www.microsoft.com/smallbusiness/resources/technology/business_software/presenting_with_powerpoint_10_dos_and_donts.mspx#EDD.

Opposing Points of View Regarding PowerPoint

Garber, Angela. "Death By Powerpoint." http://www.smallbusinesscomputing.com/biztools/article.php/684871.

Tufte, Edward. "PowerPoint Is Evil. Power Corrupts. PowerPoint Corrupts Absolutely." http://www.wired.com/wired/archive/11.09/ppt2.html.

and transportable. Even if you have never created a slide show before, you can use the templates in the software and the auto layouts to develop your own slide show easily. An added benefit of PowerPoint slides is that you can print them and create handouts for audience members.

Benefits of PowerPoint

When you become familiar with Microsoft PowerPoint, you will be able to achieve the following benefits:

1. Choose from many different presentation layouts and designs.
2. Create your own designs and layouts, changing colors and color schemes from preselected designs.
3. Add, delete, or rearrange slides as needed. By left-clicking on any slide, you can copy, paste, or delete it. By left-clicking between any of the slides, you can add a new blank slide for additional information.

4. Insert art from the Web, add images, or create your own drawings.

5. Add hyperlinks either to slides within your PowerPoint presentation or to external Web links.

Tips for Using PowerPoint

To make it easy for you to add PowerPoint slides to your presentations, consider the following hints:

1. **Create optimal contrast.** Use dark backgrounds for light text or light backgrounds for dark text. Avoid using red or green text (individuals who are color blind cannot see these colors). You should use color for emphasis only.

2. **Choose an easy-to-read font size and style.** Arial is considered to be the best to use because sans serif fonts (those without feet) show up best in PowerPoint. Use no more than three font sizes per slide. Use at least a 24-point font size for text and 36-point font for headings.

3. **Limit the text to six or seven lines per slide and six or seven words per line.** Think 6 × 6. Two or more short, simple lines of text are better than one slide with many words. Also, use no more than 40 characters per line (a character is any letter, punctuation mark, or space). You can accomplish these goals by creating a screen for each major point discussed in your oral presentation.

4. **Use headings for readability.** To create a hierarchy of headings, use larger fonts for a first-level heading and smaller fonts for second-level headings. Each slide should have at least one heading to help the audience follow your thoughts.

5. **Use emphasis techniques.** To call attention to a word, phrase, or idea, use color (sparingly), boldface, all caps, or arrows. Use a layout that includes white space. Include figures, graphs, pictures from the Web, or other line drawings.

6. **End with an obvious concluding screen.** Often, if speakers do not have a final screen that *obviously* ends the presentation, the speakers will click to a blank screen and say, "I didn't realize I was through," or "I guess that's it." In contrast, an obvious ending screen will let you as the speaker end graciously—and without surprise.

7. **Prepare handouts.** Give every audience member a handout, and leave room on the handouts for note taking.

8. **Avoid reading your screens to your audience.** Remember they can read and will become quickly bored if you read slides to them. Speakers lose their dynamism when they resort to reading slides rather than speaking to the audience.

9. **Elaborate on each screen.** PowerPoint should not replace you as the speaker. In contrast, PowerPoint should add visual appeal, while you elaborate on the details. Give examples to explain fully the points in your presentation.

10. **Leave enough time for questions and comments.** Instead of rushing through each slide, leave sufficient time for the audience to consider what they have seen and heard. Both during and after the PowerPoint presentation, give the audience an opportunity for input.

Adding Hypertext Links to a PowerPoint Presentation

To enhance your oral presentation, you can add hypertext links to PowerPoint (PPt). Doing so will allow you to accomplish two goals: moving seamlessly between slides within the PPt presentation and linking to external sites.

To create hypertext links, follow these steps:

1. Highlight the word or words that you want to make a hypertext link.
2. Right click on that word or those words. A box will pop up, as shown in the figure below.

3. Scroll to **Hyperlink** and right click. You will see the following pop up.

4. To create a hypertext link to screens within your PPt presentation, left click on **Place in This Document**.
5. To create a hypertext link to external Web sites, left click on **Existing File or Web Page**. Then, type the URL to the existing site in the Address line.

How is PowerPoint used in business?

Mark Uhlig is chairperson and CEO of Uhlig Communications, Inc. This advanced-technology publishing company specializes in the management and creation of marketing and business documents for a diverse clientele, ranging from apartment and condominium communities to individual businesses.

Mark has to speak formally and informally every day to employees, customers, investors, and strategic partners. In many instances, Mark says, "Oral communication is superior to written correspondence. It's quicker, it requires less preparation, and it's more persuasive since you get immediate feedback from your audience and, thus, can react and adjust your message. The challenge," notes Mark, "is to speak with the same deliberation and structure as you would when writing."

That's where PowerPoint comes in. According to Mark, PPt is the perfect hybrid, fusing the benefits of written and oral communication. With PPt, "you can talk from an outline or an agenda. You can prepare talking points to help you stay on message. Thus, you have the power of oral persuasion enhanced by the structure of written communication." In addition, PPt slides "allow the audience to use the printed slides as a record of the conversation, ensuring that their recollections are accurate." Finally, people happily will read PowerPoint slides "because they are short and can be accessed at a glance," Mark says.

Because PPt has become "the standard currency" for oral presentations, Mark uses PPt slides when speaking to his employees about their 401k plans, to investors about Uhlig Communications financial performance, and to customers about the benefits of his products.

Mark sketched the following table to clarify how PPt is the perfect hybrid between oral and written communication:

Oral Communication		PowerPoint	Written Communication	
Positives	Negatives	Positives	Positives	Negatives
• Fast • Easy • Persuasive—allowing for give and take	• Hard for audience to recall key ideas • Heard in real time • Not good for complex facts/figures	Easy Structured Combines oral persuasion and written record Printed notes help recall	• Allows for unlimited detail • Allows for random access	• Takes time and effort to read • Disallows give and take

PPt is just one of the many tools Mark employs in his publishing company to meet the needs of customers and other stakeholders. Clear communication is his goal, and PPt addresses this need.

The Writing Process at Work

Nurani Singh is a recruitment specialist in the human resources department for Omaha, Nebraska. She uses PowerPoint to make new-hire orientation presentations to city employees.

POWERPOINT SLIDES CHECKLIST

_____ 1. Does the presentation include headings for each slide?

_____ 2. Have you used an appropriate font size for readability?

_____ 3. Did you choose an appropriate font type for readability?

_____ 4. Did you limit yourself to no more than three different font sizes per screen?

_____ 5. Has color been used effectively for readability and emphasis, including font color and slide background?

_____ 6. Have you avoided overusing special effects?

_____ 7. Have you limited text on each screen (remembering the 6 × 6 rule)?

_____ 8. Did you size your graphics correctly for readability, avoiding ones that are too small or too complex?

_____ 9. Have you used highlighting techniques (arrows, color, white space) to emphasize key points?

_____ 10. Have you edited for spelling and grammatical errors?

The Writing Process

Prewriting	Writing	Rewriting

• Decide whether your presentations will be formal or informal.
• Decide whether you are writing to inform, instruct, persuade, or build rapport.
• Determine whether your audience is high tech, low tech, or lay.
• Gather information through primary and secondary research.
	• Prepare outlines or note cards.
	• Organize your content using modes such as problem/solution, cause/effect, comparison, argument/persuasion, analysis, chronology, etc.
	• Use visual aids to emphasize key points.
		• Consider all aspects of style, delivery, appearance, and body language and gestures.
		• Practice in front of peers or colleagues for their input.
		• Revise your content by
		  • Adding details
		  • Deleting wordiness
		  • Simplifying words
		  • Enhancing the tone
		  • Reformatting your text
		  • Proofreading and correcting errors

Prewriting

To plan a presentation, Nurani considered the following:

- Goal—to communicate human resources information and regulations
- Audience—new city employees
- Channels—oral presentation, PowerPoint slides, and handouts
- Data—details about governmental regulations illustrated by real-life experiences of other employees

Figure 19.5 shows how Nurani used brainstorming to plan her communication.

Writing

Nurani drafted PowerPoint slides, such as the one shown in Figure 19.6.

FIGURE 19.5 Nurani's Brainstorming List

Before the interview: Be prepared, find the best place and time for the interview, make up the interview questions, invite the panel of interviewers, get water and glasses, paper and pens.

    During the interview: Create a comfortable environment, ask questions correctly (check with our HR legal staff to find out what we can and can't ask), use correct body language when speaking to the applicant, and be sure to manage time (don't go too long, so we have enough time for the next interview).

    At the end of the interview: Thank the applicant, walk her or him out of the building (or to meet management), meet with the panel to discuss the interview, rate the applicant, and prepare documentation for records.

Figure 19.7 shows Nurani's revised PowerPoint slide.

FIGURE 19.6 Rough Draft of a PowerPoint Slide for Nurani's Presentation on "Conducting Effective Interviews"

Nurani says, "After my first unsuccessful presentation, I realized I was too slide dependent. My slides had too much content and were way too boring and I read them to my audience! No wonder I scored poorly on the effectiveness of the presentation."

Create a Comfortable Environment

Not every applicant will be comfortable in an interview. It's your job to make them comfortable. Conduct the interview in a pleasant, informal, and conversational manner. Here's what to do:

Welcome applicants and make them feel comfortable by creating a climate that shows trust. Introduce any panel members and explain how the interviews will be conducted. Tell the applicant how many questions will be asked, who will ask them, the sequence of the questions, and let them know about note-taking resources available during the interview, like pen, paper, or reference materials. Let applicants know that they will be allowed to ask questions at the end of the interview if there's time and encourage them to organize their thoughts before responding to questions.

Create a positive body language by making eye contact and facing the applicants so that you seem caring and interested. Avoid negative body language. Don't frown, cross your arms, look at your watch, take phone calls, or yawn during the interview.

FIGURE 19.7 Revised PowerPoint Slide

Large heading to help reader navigation

Subheadings using Arial font for readability

Graphics and white space

Four lines of text, bulleted, with no more than six words per line

Create a Comfortable Environment

In the interview, be pleasant, informal, and conversational.

- Welcome applicants by using their name.
- Introduce the interviewers.
- Allow applicants to ask questions.
- Avoid negative body language.

CHAPTER HIGHLIGHTS

1. Effective oral communication ensures that your message is conveyed successfully and that your verbal skills make a good impression.
2. You will communicate verbally on an everyday basis, informally and formally.
3. Developing listening skills will enhance your oral communication.
4. A Webinar is a way to offer online training for employees at remote locations.
5. An effective oral presentation introduction might include an anecdote, question, quote, facts, or a "table setter."
6. In the discussion section of a formal oral presentation, organize content according to comparison/contrast, problem/solution, argument/persuasion, importance, and chronology.
7. Conclude your formal speech by restating main points, recommending a future course of action, or by asking for questions and comments.
8. One of the most powerful oral communication tools is Microsoft PowerPoint.
9. Many visual aids can enhance your oral presentation. Choose from chalkless whiteboards, flip charts, DVDs, and PowerPoint.
10. Be prepared for questions and discussion after you conclude your oral presentation.

CASE STUDIES

1. Read the TechStop scenario that begins this chapter. In this scenario, a customer is treated poorly by a salesperson. The vice president of customer service, Shuan Wang, has received reports of other problems with customer and staff interaction. Shuan needs to make a formal oral presentation to all sales personnel to improve customer service.

Assignment

Based on the scenario at the beginning of the chapter, outline the content for Shuan's oral presentation as follows:

- Determine what kind of introduction should be used to arouse the audience's interest.
- Provide a thesis statement.
- Develop information to teach employees sales etiquette and customer interaction, store policy regarding customer satisfaction, the impact on poor customer relations, and the consequences of failing to handle customers correctly. Research these topics (either online, in a library, or through interviews) to find your content.
- Organize the speech's content with appropriate transitions to aid coherence.
- Determine which types of visual aids would work best for this presentation.

2. Halfmoon is an online wholesaler of hiking, biking, and boating gear. Their staff is located in four northeastern states. Halfmoon's CEO, Montana Wildhack, wants his employees to excel in oral and written communication. He has hired a consultant to train the employees. To accomplish this training, the consultant will provide a teleconference. Fifty employees will meet with the consultant in a training room, while another 150 Halfmoon staff members view the training session from their different work locations.

During the presentation, the consultant highlights his comments by writing on a chalkless whiteboard, using green and yellow markers. He attempts humor by suggesting that Halfmoon's thermal stocking caps would be a perfect gift for one of the seminar's balding participants, Joe. In addition, the consultant has participants break into small groups to role play. They take mock customer orders, field complaints, and interact with vendors, using a prepared checklist of do's and don'ts.

At one point, when a seminar participant calls in a question, the consultant says, "No, that would be wrong, wrong, wrong! Why would you ever respond to a customer like that? Common sense would dictate a different response." During a break, the consultant does not mute his clip-on microphone. To conclude the session, the consultant asks for questions and comments. When he cannot provide a good answer, he asks the participants to suggest options, and he says he will research the question and place an answer on his corporate Web site.

Assignment

Using the guidelines in the chapter, discuss the speaker's oral communication skills. Explain what went wrong and why. Explain what was effective.

INDIVIDUAL AND TEAM PROJECTS

Evaluating an Oral Presentation

Listen to a speech. You could do so on television; at a student union; at a church, synagogue, or mosque; at a civic event, city hall meeting, or community organization; at a company activity; or in your classroom. Answer the following questions:

1. What was the speaker's goal? Did the speaker try to persuade, instruct, inform, or build trust? Explain your answer by giving examples.
2. What type of introduction did the speaker use to arouse listener interest (anecdote, question, quote, or facts)? Give examples to support your decision.
3. What visual aids were used in the presentation? Were these visual aids effective? Explain your answers.
4. How did the speaker develop the assertions? Did the speaker use analysis, comparison/contrast, argument/persuasion, problem/solution, or chronology? Give examples to prove your point.
5. Was the speaker's delivery effective? Use the Presentation Delivery Rating Sheet to assess the speaker's performance by placing check marks in the appropriate columns.

PRESENTATION DELIVERY RATING SHEET			
Delivery Techniques	Good	Bad	Explanation
Eye contact			
Rate of speech			
Enunciation			
Pitch of voice			
Use of pauses			
Emphasis			
Interaction with listeners			
Conflict resolution			

Giving an Oral Presentation

1. Find an advertised job opening in your field of expertise or degree program. You could look either online, in the newspaper, at your school's career placement center, or at a company's human resource office. Perform a mock interview for this position. To do so, follow this procedure:
   - Designate one person as the applicant.
   - He or she should prepare for the interview by making a list of potential questions (ones that the applicant believes he or she will be asked, as well as questions to ask the search committee).
   - Designate others in the class to represent the search committee.
   - The mock search committee should prepare a list of questions to ask the applicant.
2. Research a topic in your field of expertise or degree program. The topic could include a legal issue, a governmental regulation, a news item, an innovation in the field, or a published article in a professional journal or public magazine. Make an oral presentation about your findings.

Creating a PowerPoint Slide Presentation

For assignments 1 and 2 above ("Giving an Oral Presentation"), create PowerPoint slides to enhance your oral communication. To do so, follow the guidelines provided in this chapter.

Assessing PowerPoint Slides

After reviewing the following Microsoft PowerPoint slides, determine which are successful, which are unsuccessful, and explain your answers, based on the guidelines provided in this chapter.

A Library Perspective on Distance Learning

1. Library online research sites are rapidly evolving.
2. Instructors engaged in online course development should visit with library staff for updates on resources.
3. Many traditional library services are available to serve distance learning students, including course reserves, interlibrary loan, reference, reciprocal borrowing policies, document delivery, access to online databases and collections
4. Many new services, electronic books and journals, are available from libraries.
5. The Internet is not the online equivalent of an academic library.

A Technology Perspective on Online Education

<u>Students</u>—24-hour Call Centers answer student hardware/software needs

<u>Faculty</u>—The Tech Center helps faculty with course creations and tech resources

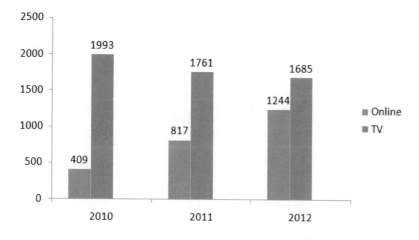

As DISTANCE LEARNING classes have risen in popularity, many *diverse areas* of interest across campuses have converged.

These include counseling, library services, technology, faculty, administration, continuing education, etc.

To ensure student success, *Distance Learning Coordinating Committees* need to draw from these multiple disciplines and collaborate on solutions to arising technology challenges.

Comparison of Enrollment: Online vs. TV Classes

Enrollment data by year (Online and TV):
- 2010: Online 409, TV 1993
- 2011: Online 817, TV 1761
- 2012: Online 1244, TV 1685

PROBLEM-SOLVING THINK PIECES

After reading the following scenarios, answer these questions:

- Would your oral presentation be everyday, informal, or formal?
- Would you use a videoconference, teleconference, or face-to-face meeting?
- Which visual aids would work best for your presentation?
- Is your oral presentation goal to persuade, instruct, inform, or build trust?

In addition, complete any or all of these assignments:

- Write an outline for the presentation.
- Write a brief, introductory lead-in to arouse your listeners' interest.
- Create a questionnaire.

1. You work at FlashCom Electric. Your company has created a new interface for modems. Your boss has asked you to make a presentation to sales representatives from 20 potential vendors in the city. The oral presentation will explain to the vendors the benefits of your product, the sales breaks you will offer, and how the vendors can increase their sales.
2. After working for Friendly's, a major discount computer hardware and software retailer, for over a year, you have created a new organizational plan for their vast inventory. In an oral presentation, you plan to show the CEO and board of directors why your plan is cost effective and efficient.
3. You are the manager of an automotive parts supply company, Plugs, Lugs, & More. Your staff of 10 in-store employees lacks knowledge of the store's new merchandise, has not been meeting sales goals, and does not always treat customers with respect and care. It's time to address these concerns.
4. As CEO of your Engineering/Architectural firm (Levin, Lisk, and Lamb), you must downsize. Business is decreasing and costs are rising. To ensure third-quarter profitability, 10 percent of the staff must be laid off. That will amount to a layoff of over 500 employees. You now must make an oral presentation to your stockholders and the entire workforce (located in three states) at your company's annual meeting to report the situation.
5. Your Quality Circle Team is composed of account representatives, management, engineers, and technicians. You have been asked to research new ways to improve product delivery. You have concluded your research. Now, you need to share your suggestions with your 15 team members.

WEB WORKSHOP

PowerPoint has proven to be a useful tool in business presentations. However, PowerPoint must be used correctly, as discussed in the chapter, to be effective. Access an online search engine and type in phrases such as follows (include any additional ones you create):

- "using powerpoint in business"
- "effective powerpoint use"
- "tips for using powerpoint"
- "powerpoint + business presentations"
- "powerpoint + pros and cons"

Once you have found articles discussing this topic, summarize your findings and make an oral presentation (using PowerPoint slides).

QUIZ QUESTIONS

1. What are four nonverbal cues that communicate through your body language?
2. What are three ways you can achieve successful telephone conversations?
3. Why would you participate in a teleconference?
4. How can you create an effective teleconference or videoconference?
5. How many parts are in a formal oral presentation?
6. In what ways can you arouse audience attention in an oral presentation?
7. What does a thesis statement in your formal oral presentation achieve?
8. How can you organize the details in the discussion section of a formal oral presentation?
9. How can you achieve coherence in a formal oral presentation?
10. How can you conclude a formal oral presentation?

APPENDIX A
GRAMMAR, PUNCTUATION, MECHANICS, AND SPELLING

Correct organization and development of your memos, letters, or reports are important for the success of your technical communication. However, no one will be impressed with the quality of your work, or with you, if your writing is riddled with errors in sentence construction or punctuation. Your written correspondence is often your first contact with business associates. Many people mistakenly believe that only English teachers notice grammatical errors and wield red pens, but businesspeople as well take note of such errors and may see the writer as less competent.

We were working recently with a young executive who is employed by a branch of the federal government. This executive told us that whenever his supervisor found a spelling error in a subordinate's report, this report was paraded around the office. Everyone was shown the mistake and had a good laugh over it, and the report was then returned to the writer for correction. Our acquaintance assured us that all of this was in good-natured fun. However, he also said that employees quickly learned to edit and proofread their written communication to avoid such public displays of their errors. He went on to say that his dictionary was well thumbed and always on his desk.

Your writing at work may not be exposed to such scrutiny by co-workers. Instead, your writing may go directly to another firm, and those readers will see your mistakes. To avoid this problem, you must evaluate your writing for grammar, punctuation, and spelling errors. If you don't, your customers, bosses, and colleagues will.

Grammar Rules

To understand the fundamentals of grammar, you must first understand the basic components of a sentence.

A correctly constructed sentence consists of a subject and a predicate (some sentences also include a phrase or phrases).

The meeting	began	at 4:00 A.M.
subject	predicate	phrase

◀ EXAMPLE

Subject: The *doer* of the action; the subject usually precedes the predicate.
Predicate: The *action* in the sentence

He	ran	to the office to avoid being late.
doer	action	phrases

◀ EXAMPLE

If the subject and the predicate (a) express a complete thought and (b) can stand alone, you have an *independent clause*.

The meeting began	at 4:00 P.M.
independent clause	phrase

◀ EXAMPLE

A *phrase* is a group of related words that does not contain a subject and a predicate and cannot stand alone or be punctuated as a sentence. The following are examples of phrases:

> at the house
> in the box
> on the job
> during the interview

If a clause is dependent, it cannot stand alone.

EXAMPLE ▶

Although he tried to hurry, he was late for the meeting.
dependent clause   independent clause

EXAMPLE ▶

He was late for the meeting although he tried to hurry.
independent clause   dependent clause

NOTE: When a dependent clause begins a sentence, use a comma before the independent clause. However, when an independent clause begins a sentence, do not place a comma before the dependent clause.

Agreement Between Pronoun and Antecedent (Referent)

A pronoun has to agree in gender and number with its antecedent.

> *Susan* went on *her* vacation yesterday.
> The *people* who quit said that *they* deserved raises.

1. Problems often arise when a singular indefinite pronoun is the antecedent. The following antecedents require singular pronouns: *anybody, each, everybody, everyone, somebody,* and *someone.*

incorrect

Anyone can pick up *their* applications at the job placement center.

correct

Anyone can pick up *his* or *her* applications at the job placement center.

2. Problems also arise when the antecedent is separated from the pronoun by numerous words.

incorrect

Even when the best *employee* is considered for a raise, *they* often do not receive it.

correct

Even when the best *employee* is considered for a raise, *he or she* often does not receive it.

Agreement Between Subject and Verb

Writers sometimes create disagreement between subjects and verbs, especially if other words separate the subject from the verb. To ensure agreement, ignore the words that come between the subject and verb.

incorrect

Her *boss* undoubtedly *think* that all the employees want promotions.

correct

Her *boss* undoubtedly *thinks* that all the employees want promotions.

incorrect

The *employees* who sell the most equipment *is* going to Hawaii for a week.

correct

The *employees* who sell the most equipment *are* going to Hawaii for a week.

If a sentence contains two subjects (a compound subject) connected by *and*, use a plural verb.

incorrect

Joe and Tiffany *was* both selected employee of the year.

correct

Joe and Tiffany *were* both selected employee of the year.

incorrect

The bench workers and their supervisor *is* going to work closely to complete this project.

correct

The bench workers and their supervisor *are* going to work closely to complete this project.

Add a final *s* or *es* to create most plural subjects or singular verbs, as follows:

PLURAL SUBJECTS	SINGULAR VERBS
bosses hire	a boss hires
employees demand	an employee demands
experiments work	an experiment works
attitudes change	the attitude changes

If a sentence has two subjects connected by *either . . . or, neither . . . nor,* or *not only . . . but also,* the verb should agree with the closest subject. This also makes the sentence less awkward.

Either the salespeople or the warehouse worker deserv**es** raises.
Not only the warehouse worker but also the salespeople deserv**e** raises.
Neither the salespeople nor the warehouse worker deserv**es** raises.

◄ **EXAMPLE**

Singular verbs are used after most indefinite pronouns such as the following:

another	everything
anybody	neither
anyone	nobody
anything	no one
each	nothing
either	somebody
everybody	someone
everyone	something

> *Anyone* who works here *is* guaranteed maternity leave.
> *Everybody* wants the company to declare a profit this quarter.

Singular verbs often follow collective nouns such as the following:

class	organization
corporation	platoon
department	staff
group	team

> The *staff is* sending the boss a bouquet of roses.

Comma Splice

A *comma splice* occurs when two independent clauses are joined by a comma rather than separated by a period or semicolon.

incorrect

Sue was an excellent employee, she got a promotion.

Several remedies will correct this error.

1. Separate the two independent clauses with a semicolon.

correct

Sue was an excellent employee; she got a promotion.

2. Separate the two independent clauses with a period.

correct

Sue was an excellent employee. She got a promotion.

3. Separate the two independent clauses with a comma and a *coordinating conjunction (and, but, or, for, so, yet).*

correct

Sue was an excellent employee, *so* she got a promotion.

4. Separate the two independent clauses with a semicolon (or a period), a conjunctive adverb, and a comma. *Conjunctive adverbs* include *also, additionally, consequently, furthermore, however, instead, moreover, nevertheless, therefore,* and *thus.*

correct

Sue was an excellent employee; *therefore,* she got a promotion.

or

Sue was an excellent employee. *Therefore,* she got a promotion.

5. Use a *subordinating conjunction* to make one of the independent clauses into a dependent clause. Subordinating conjunctions include *after, although, as, because, before, even though, if, once, since, so that, though, unless, until, when, where,* and *whether.*

correct

Because Sue was an excellent employee, she got a promotion.

Faulty or Vague Pronoun Reference

A pronoun must refer to a specific noun (its antecedent). Problems arise when (a) there is an excessive number of pronouns (causing vague pronoun reference) and (b) there is no specific noun as an antecedent. Notice that there seems to be an excessive number of pronouns in the following passage, and the antecedents are unclear.

◀ **EXAMPLE**

Although Bob had been hired over two years ago, *he* found that his boss did not approve *his* raise. In fact, *he* was also passed over for *his* promotion. The boss appears to have concluded that *he* had not exhibited zeal in *his* endeavors for their business. Instead of being a highly valued employee, *he* was not viewed with pleasure by those in authority. Perhaps it would be best if *he* considered *his* options and moved to some other company where *he* might be considered in a new light.

The excessive and vague use of *he* and *his* causes problems for readers. Do these words refer to Bob or to his boss? You are never completely sure. To avoid this problem, limit pronoun usage, as in the following revision.

◀ **EXAMPLE**

Although Bob had been hired over two years ago, he found that his boss, Joe, did not approve his raise. In fact, Bob was also passed over for promotion. Joe appears to have concluded that Bob had not exhibited zeal in his endeavors for their business. Instead of being a highly valued employee, Bob was not viewed with pleasure by those in authority. Perhaps it would be best if Bob considered his options and moved to some other company where he might be considered in a new light.

To make the preceding paragraph more precise, we have replaced vague pronouns (*he* and *his*) with exact names (*Bob* and *Joe*).

Fragments

A *fragment* occurs when a group of words is incorrectly used as an independent clause. Often the group of words begins with a capital letter and has end punctuation but is missing either a subject or a predicate.

Working with computers.

(lacks a predicate and does not express a complete thought)

The group of words may have a subject and a predicate but be a dependent clause.

Although he enjoyed working with computers.

(has a subject, *he*, and a predicate, *enjoyed*, but is a dependent clause because it is introduced by the subordinate conjunction *although*)

It is easy to remedy a fragment by doing one of the following:

- Add a subject.
- Add a predicate.
- Add both a subject and a predicate.
- Add an independent clause to a dependent clause.

Joe found that working with computers used his training.

(subject, *Joe*, and predicate, *found*, have been added)

Although he enjoyed working with computers, he could not find a job in a computer-related field.

(independent clause, *he could not find a job*, added to the dependent clause, *Although he enjoyed working with computers*)

Fused Sentence

A *fused sentence* occurs when two independent clauses are connected with no punctuation.

The company performed well last quarter its stock rose several points.

There are several ways to correct this error.

1. Write two sentences separated by a period.

The company performed well last quarter. Its stock rose several points.

2. Use a comma and a coordinating conjunction to separate the two independent clauses.

The company performed well last quarter, *so* its stock rose several points.

3. Use a subordinating conjunction to create a dependent clause.

correct

Because the company performed well last quarter, its stock rose several points.

4. Use a semicolon to separate the two independent clauses.

correct

The company performed well last quarter; its stock rose several points.

5. Separate the two independent clauses with a semicolon, a conjunctive adverb or a transitional word or phrase, and a comma.

correct

The company performed well last quarter; *therefore*, its stock rose several points.

correct

The company performed well last quarter; *for example*, its stock rose several points.

The following are transitional words and phrases, listed according to their use:

TO ADD

again	in addition
also	moreover
besides	next
first	second
furthermore	still

TO COMPARE/CONTRAST

also	nevertheless
but	on the contrary
conversely	still
in contrast	

TO PROVIDE EXAMPLES

for example	of course
for instance	put another way
in fact	to illustrate

TO SHOW PLACE

above	here
adjacent to	nearby
below	on the other side
elsewhere	there
further on	

afterward	second
first	shortly
meanwhile	subsequently
presently	thereafter

TO SUMMARIZE

all in all	last
finally	on the whole
in conclusion	therefore
in summary	thus

Modification

A *modifier* is a word, phrase, or clause that explains or adds details about other words, phrases, or clauses.

Misplaced Modifiers.   A *misplaced modifier* is one that is not placed next to the word it modifies.

incorrect

He had a heart attack *almost* every time he was reviewed by his supervisor.

correct

He almost had a heart attack every time he was reviewed by his supervisor.

incorrect

The worker had to *frequently* miss work.

correct

The worker frequently had to miss work.

Dangling Modifiers.   A *dangling modifier* is a modifier that is not placed next to the word or phrase it modifies. To avoid confusing your readers, place modifiers next to the word(s) they refer to. Don't expect your readers to guess at your meaning.

incorrect

While working, tiredness overcame them.
(Who was working? Who was overcome by tiredness?)

correct

While working, the staff became tired.

incorrect

After soldering for two hours, the equipment was ready for shipping. (Who had been soldering for two hours? Not the equipment!)

correct

After soldering for two hours, the technicians prepared the equipment for shipping.

Parallelism

All items in a list should be parallel in grammatical form. Avoid mixing phrases and sentences (independent clauses).

We will discuss the following at the department meeting:

1. Entering mileage in logs (phrase)
2. All employees have to enroll in a training seminar. (sentence)
3. Purpose of quarterly reviews (phrase)
4. Some data processors will travel to job sites. (sentence)

We will discuss the following at the department meeting:

1. Entering mileage in logs
2. Enrolling in training seminars
3. Reviewing employee performance quarterly
4. Traveling to job sites

} phrases

At the department meeting, you will learn how to

1. Enter mileage in logs
2. Enroll in training seminars
3. Review employee performance quarterly
4. Travel to job sites

} phrases

Punctuation

Apostrophe (')

Place an *apostrophe* before the final *s* in a singular word to indicate possession.

Jim's tool chest is next to the furnace.

◀ EXAMPLE

Place the apostrophe after the final *s* if the word is plural.

The employees' reception will be held next week.

◀ EXAMPLE

Don't use an apostrophe to make singular abbreviations plural.

The EXT's will be shipped today.

correct

The EXTs will be shipped today.

Colon (:)

Use a *colon* after a salutation.

EXAMPLE ▶

> Dear Mr. Harken:

In addition, use a colon after an emphatic or cautionary word if explanations follow.

EXAMPLE ▶

> Note: Hand-tighten the nuts.
> Caution: Wash thoroughly if any mixture touches your skin.

Finally, use a colon after an independent clause to precede a quotation, list, or example.

EXAMPLE ▶

> She said the following: "No comment."
> These supplies for the experiment are on order: plastic hose, two batteries, and several chemicals.
> The problem has two possible solutions: hire four more workers, or simply give everyone a raise.

NOTE: In the preceding examples, the colon follows an independent clause.

A common mistake is to place a colon after an incomplete sentence. Except for salutations, headings, and cautionary notes, whatever precedes a colon *must* be an independent clause.

incorrect

The two keys to success are: earning money and spending wisely.

correct

The two keys to success are earning money and spending wisely.

> or

The two keys to success are as follows: earning money and spending wisely.

> or

The two keys to success are as follows:

1. Earning money
2. Spending wisely

Comma (,)

Writers often get in trouble with *commas* when they employ one of two common "words of wisdom."

- When in doubt, leave it out.
- Use a comma when there is a pause.

Both rules are inexact. Writers use the first rule to justify the complete avoidance of commas; they use the second rule to sprinkle commas randomly throughout their writing. On the contrary, commas have several specific conventions that determine usage:

1. Place a comma before a coordinating conjunction *(and, but, or, for, so, yet)* linking two independent clauses.

> You are the best person for the job, *so* I will hire you.
> We spent several hours discussing solutions to the problem, *but* we failed to decide on a course of action.

◀ EXAMPLE

2. Use commas to set off introductory comments.

> First, she soldered the components.
> In business, people often have to work long hours.
> To work well, you need to get along with your co-workers.
> If you want to test equipment, do so by 5:25 P.M.

◀ EXAMPLE

3. Use commas to set off sentence interrupters.

> The company, started by my father, did not survive the last recession.
> Mrs. Patel, the proprietor of the store, purchased a wide array of merchandise.

◀ EXAMPLE

4. Set off parenthetical expressions with commas.

> A worker, it seems, should be willing to try new techniques.
> The highway, by the way, needs repairs.

◀ EXAMPLE

5. Use commas after each item in a series of three or more.

> Prakash, Mirren, and Justin were chosen as employees of the year.
> We found the following problems: corrosion, excessive machinery breakdowns, and power failures.

◀ EXAMPLE

6. Use commas to set off long numbers.

> She earns $100,000 before taxes.

◀ EXAMPLE

NOTE: Very large numbers are often written as words.

> Our business netted over $2 million in 2012.

◀ EXAMPLE

7. Use commas to set off the day and year when they are part of a sentence.

EXAMPLE ▶

> The company hired her on September 7, 2012, to be its bookkeeper.

NOTE: If the year is used as an adjective, do not follow it with a comma.

EXAMPLE ▶

> The 2012 corporate report came out today.

8. Use commas to set off the city from the state and the state from the rest of the sentence.

EXAMPLE ▶

> The new warehouse in Austin, Texas, will promote increased revenues.

NOTE: If you omit either the city or the state, you do not need commas.

EXAMPLE ▶

> The new warehouse in Austin will promote increased revenues.

Dash (—)

A *dash,* typed as two consecutive hyphens with no spaces before or after, is a versatile punctuation mark used in the following ways:

1. After a heading and before an explanation.

EXAMPLE ▶

> Forecasting—Joe and Joan will be in charge of researching fourth-quarter production quotas.

2. To indicate an emphatic pause.

EXAMPLE ▶

> You will be fired—unless you obey company rules.

3. To highlight a new idea.

EXAMPLE ▶

> Here's what we can do to improve production quality—provide on-the-job training, salary incentives, and quality controls.

4. Before and after an explanatory or appositive series.

EXAMPLE ▶

> Three people—Sue, Luci, and Tom—are essential to the smooth functioning of our office.

Ellipses (. . .)

Ellipses (three spaced periods) indicate omission of words within quoted materials.

> "Six years ago, prior to incorporating, the company had to pay extremely high federal taxes."
>
> "Six years ago, . . . the company had to pay extremely high federal taxes."

◀ EXAMPLE

Exclamation Point (!)

Use an *exclamation point* after strong statements, commands, or interjections.

> You must work harder!
> Do not use the machine!
> Danger!

◀ EXAMPLE

Hyphen (-)

A *hyphen* is used in the following ways:

1. To indicate the division of a word at the end of a typed line. Remember, this division must occur between syllables.

2. To create a compound adjective.

> He is a well-known engineer.
> Until her death in 2012, she was a world-renowned chemist.
> Tom is a 24-hour-a-day student.

◀ EXAMPLE

3. To join the numerator and denominator of fractions.

> Four-fifths of the company want to initiate profit sharing.

◀ EXAMPLE

4. To write out two-word numbers.

> Twenty-six people attended the conference.

◀ EXAMPLE

Parentheses ( )

Parentheses enclose abbreviations, numbers, words, or sentences for the following reasons:

1. To define a term or provide an abbreviation for later use.

> We belong to the Society for Technical Communication (STC).

◀ EXAMPLE

2. To clarify preceding information in a sentence.

EXAMPLE ▶

> The people in attendance (all regional sales managers) were proud of their accomplishments.

3. To number items in a series.

EXAMPLE ▶

> The company should initiate (1) new personnel practices, (2) a probationary review board, and (3) biannual raises.

Period (.)

A *period* must end a declarative sentence (independent clause).

EXAMPLE ▶

> I found the business trip rewarding.

Periods are often used with abbreviations:

D.C.	Mrs.	A.M. or a.m.
e.g.	Ms.	P.M. or p.m.
Mr.		

It is incorrect to use periods with abbreviations for organizations and associations.

incorrect

S.T.C. (Society for Technical Communication)

correct

STC (Society for Technical Communication)

State abbreviations do not require periods if you use two capital letters.

incorrect

KS. (Kansas)

MO. (Missouri)

TX. (Texas)

correct

KS

MO

TX

Question Mark (. . .)

Use a *question mark* after direct questions.

◀ EXAMPLE

> Do the lab results support your theory?
> Will you work at the main office or at the branch?

Quotation Marks (" ")

Quotation marks are used in the following ways:

1. When citing direct quotations.

◀ EXAMPLE

> He said, "Your division sold the most compressors last year."

NOTE: When you are citing a quotation within a quotation, use double quotation marks (" ") and single quotation marks (' ').

◀ EXAMPLE

> Kim's supervisor, quoting the CEO, said the following to explain the new policy regarding raises: "'Only employees who deserve them will receive merit raises.'"

2. To note the title of an article or a subdivision of a report.

◀ EXAMPLE

> The article "Robotics in Industry Today" was an excellent choice as the basis of your speech.
> Section III, "Waste Water in District 9," is pertinent to our discussion.

When using quotation marks, abide by the following punctuation conventions:

- Commas and periods always go *inside* quotation marks.

◀ EXAMPLE

> She said, "Our percentages are fixed."

- Colons and semicolons always go *outside* quotation marks.

◀ EXAMPLE

> He said, "The supervisor hasn't decided yet"; however, he added that the decision would be made soon.

- Exclamation points and question marks go inside the quotation marks if the quoted material is either exclamatory or a question. However, if the quoted material is not exclamatory or a question, then these punctuation marks go outside the quotation marks.

EXAMPLE ▶

John said, "Don't touch that liquid. It's boiling!"

(Although the sentence isn't exclamatory, the quotation is. Thus, the exclamation point goes inside the quotation marks.)

EXAMPLE ▶

How could she say, "We haven't purchased the equipment yet"?

(Although the quotation isn't a question, the sentence is. Thus, the question mark goes outside the quotation marks.)

Semicolon (;)

Semicolons are used in the following instances:

1. Between two independent clauses *not* joined by a coordinating conjunction.

EXAMPLE ▶

The light source was unusual; it emanated from a crack in the plastic surrounding the cathode.

2. To separate items in a series containing internal commas.

EXAMPLE ▶

When the meeting was called to order, all members were present, including Susan Johnson, the president; Ruth Schneider, the vice president; Harold Holbert, the treasurer; and Linda Hamilton, the secretary.

Mechanics

Abbreviations

Never use an abbreviation that your reader will not understand. A key to clear technical writing is to write on a level appropriate to your reader. You may use the following familiar abbreviations without explanation: *Mrs., Dr., Mr., Ms.,* and *Jr.*

A common mistake is to abbreviate inappropriately. For example, some writers abbreviate *and* as follows:

EXAMPLE ▶

I quit my job & planned to retire young.

This is too colloquial for professional technical writing. Spell out *and* when you write.

The majority of abbreviation errors occur when writers incorrectly abbreviate states and technical terms.

States. Writers often abbreviate the names of states incorrectly. Use the U.S. Postal Service abbreviations in addresses.

Abbreviations for States

AL	Alabama	MT	Montana
AK	Alaska	NC	North Carolina
AZ	Arizona	ND	North Dakota
AR	Arkansas	NE	Nebraska
CA	California	NV	Nevada
CO	Colorado	NH	New Hampshire
CT	Connecticut	NJ	New Jersey
DE	Delaware	NM	New Mexico
FL	Florida	NY	New York
GA	Georgia	OH	Ohio
HI	Hawaii	OK	Oklahoma
IN	Indiana	OR	Oregon
IA	Iowa	PA	Pennsylvania
ID	Idaho	RI	Rhode Island
IL	Illinois	SC	South Carolina
KS	Kansas	SD	South Dakota
KY	Kentucky	TN	Tennessee
LA	Louisiana	TX	Texas
ME	Maine	UT	Utah
MD	Maryland	VT	Vermont
MA	Massachusetts	VA	Virginia
MI	Michigan	WA	Washington
MN	Minnesota	WV	West Virginia
MS	Mississippi	WI	Wisconsin
MO	Missouri	WY	Wyoming

Technical Terms. Units of measurement and scientific terms must be abbreviated accurately to ensure that they will be understood. Writers often use such abbreviations incorrectly. For example, "The unit measured 7.9 cent." is inaccurate. The correct abbreviation for centimeter is *cm*, not *cent*. Use the following abbreviation conventions.

Technical Abbreviations for Units of Measurement and Scientific Terms

absolute	abs	audio frequency	AF
alternating current	AC	azimuth	az
American wire gauge	AWG	barometer	bar.
ampere	amp	barrel, barrels	bbl
ampere-hour	amp-hr	billion electron volts	BeV
amplitude modulation	AM	biochemical oxygen demand	BOD
angstrom unit	Å	board foot	bd ft
atmosphere	atm	Brinell hardness number	BHN
atomic weight	at wt	British thermal unit	Btu

(Continued)

bushel	bu	gallon	gal
calorie	cal	gallons per day	GPD
candela	cd	gallons per minute	GPM
Celsius	C	grain	gr
center of gravity	cg	grams	g (or gm)
centimeter	cm	gravitational acceleration	g
circumference	cir	hectare	ha
cologarithm	colog	hectoliter	hl
continuous wave	CW	hectometer	hm
cosine	cos	henry	H
cotangent	cot	hertz	Hz
cubic centimeter	cc	high frequency	HF
cubic foot	cu ft (or ft^2)	horsepower	hp
cubic feet per second	cfs	horsepower-hours	hp-hr
cubic inch	cu in. (or in.^{3})	hour	hr
cubic meter	cu m (or m^3)	hundredweight	cwt
cubic yard	cu yd (or yd^3)	inch	in.
current (electric)	I	inch-pounds	in.-lb
cycles per second	CPS	infrared	IR
decibel	dB	inner diameter or inside dimensions	ID
decigram	dg		
deciliter	dl	intermediate frequency	IF
decimeter	dm	international unit	IU
degree	deg	joule	J
dekagram	dkg	Kelvin	K
dekaliter	dkl	kilocalorie	kcal
dekameter	dkm	kilocycle	kc
dewpoint	DP	kilocycles per second	kc/sec
diameter	dia	kilogram	kg
direct current	DC	kilohertz	kHz
dozen	doz (or dz)	kilojoule	kJ
dram	dr	kiloliter	kl
electromagnetic force	emf	kilometer	km
electron volt	eV	kilovolt	kV
elevation	el (or elev)	kilovolt-amperes	kVa
equivalent	equiv	kilowatt-hours	kWH
Fahrenheit	F	lambert	L
farad	F	latitude	lat
faraday	f	length	l
feet, foot	ft	linear	lin
feet per second	ft/sec	linear foot	lin ft
fluid ounce	fl oz	liter	l
foot board measure	fbm	logarithm	log.
foot-candle	ft-c	longitude	long.
foot-pound	ft lb	low frequency	LF
frequency modulation	FM	lumen	lm

lumen-hour	lm-hr	radio frequency	RF
maximum	max	radian	rad
megacycle	mc	radius	r
megahertz	MHz	resistance	r
megawatt	MW	revolution	rev
meter	m	revolutions per minute	rpm
microampere	µamp	second	s (or sec)
microinch	µin.	secant	sec
microsecond	µsec	specific gravity	sp gr (or SG)
microwatt	µw	square foot	ft^2
miles per gallon	mpg	square inch	$in.^2$
milliampere	mA	square meter	m^2
millibar	mb	square mile	mi^2
millifarad	mF	tablespoon	tbs (or tbsp)
milligram	mg	tangent	tan
milliliter	ml	teaspoon	tsp
millimeter	mm	temperature	t
millivolt	mV	tensile strength	ts
milliwatt	mW	thousand	m
minute	min	ton	t
nautical mile	NM	ultra high frequency	UHF
negative	neg or −	vacuum	vac
number	no.	very high frequency	VHF
octance	oct	volt-ampere	VA
ounce	oz	volt	V
outside diameter	OD	volts per meter	V/m
parts per billion	ppb	volume	vol
parts per million	ppm	watt-hour	whr
pascal	pas	watt	W
positive	pos or +	wavelength	WL
pound	lb	weight	wt
pounds per square inch	psi	yards	y (or yd)
pounds per square inch absolute	psia	years	y (or yr)
pounds per square inch gauge	psig		
quart	qt		

Capital Letters

Capitalize the following:

1. Proper nouns.

people	cities	countries	companies	schools	buildings
Susan	Houston	Italy	Bendix	Harvard	Oak Park Mall

 EXAMPLE

2. People's titles (only when they precede the name).

EXAMPLE ▶

Governor Sally Renfro

or

Sally Renfro, governor
Technical Supervisor Todd Blackman

or

Wes Schneider, the technical supervisor

3. Titles of books, magazines, plays, movies, television programs, and CDs (excluding the prepositions and all articles after the first article in the title).

EXAMPLE ▶

(Italicize titles of books, television series, and movies.)
Mad Men
The Colbert Report
The Catcher in the Rye
New Moon
American Idol

4. Names of organizations.

EXAMPLE ▶

Girl Scouts
Phoenix, AZ, Regional Home Care Association
Kansas City Regional Council for Higher Education
Programs for Technical and Scientific Communication
American Civil Liberties Union
Students for a Democratic Society

5. Days of the week, months, and holidays.

EXAMPLE ▶

Monday
December
Thanksgiving

6. Races, religions, and nationalities.

EXAMPLE ▶

American Indian
Jewish
Polish

7. Events or eras in history.

EXAMPLE ▶

the Gulf War
the Vietnam War
World War II

8. North, South, East, and West (when used to indicate geographic locations).

◀ EXAMPLE

> They moved from the North.
> People are moving to the Southwest.

NOTE: Don't capitalize these words when giving directions.

◀ EXAMPLE

> We were told to drive south three blocks and then to turn west.

9. The first word of a sentence.
10. Don't capitalize any of the following:

◀ EXAMPLE

> Seasons—spring, fall, summer, winter
> Names of classes—sophomore, senior
> General groups—middle management, infielders, surgeons

Numbers

Write out numbers one through nine. Use numerals for numbers 10 and above:

10	12
104	2,093
536	5,550,286

Although the preceding rules cover most situations, there are exceptions:

1. Use numerals for all percentages.

◀ EXAMPLE

> 2 percent   18 percent   25 percent

2. Use numerals for addresses.

◀ EXAMPLE

> 12 Elm   935 W. Harding

3. Use numerals for miles per hour.

◀ EXAMPLE

> 5 mph   225 mph

4. Use numerals for time.

◀ EXAMPLE

> 3:15 A.M.

5. Use numerals for dates.

◀ EXAMPLE

> May 31, 2012

6. Use numerals for monetary values.

EXAMPLE ▶

$45 $.95 $2 million

7. Use numerals for units of measurement.

EXAMPLE ▶

14′   6 3/4″   16 mm   10 V

8. Do not use numerals to begin sentences.

incorrect

568 people were fired last August.

correct

Five hundred sixty-eight people were fired last August.

9. Do not mix numerals and words when writing numbers. When two or more numbers appear in a sentence and one of them is 10 or more, figures are used.

EXAMPLE ▶

We attended 4 meetings over a 16-day period.

10. Use numerals and words in a compound number adjective to avoid confusion.

EXAMPLE ▶

The worker needed six 2-inch nails.

Spelling

The following is a list of commonly misspelled or misused words. You can avoid many common spelling errors if you familiarize yourself with these words. Remember to run spell check; also remember that spell check will not understand context. The incorrect word contextually could be spelled correctly.

accept, except	council, counsel	personal, personnel
addition, edition	desert, dessert	principal, principle
access, excess	disburse, disperse	quiet, quite
advise, advice	fiscal, physical	rite, right, write
affect, effect	forth, fourth	stationery, stationary
all ready, already	incite, insight	their, there, they're
assistants, assistance	its, it's	to, too, two
bare, bear	loose, lose	whose, who's
brake, break	miner, minor	your, you're
coarse, course	passed, past	
cite, site, sight	patients, patience	

Proofreader's Marks

The following table illustrates proofreader's marks and how to use them.

PROOFREADER'S MARKS			
Symbol	Meaning	Mark on Copy	Revision
e	delete	hire better peoples*e*	hire better people
(tr) or ⌒	transpose	computer systme	computer system
⌒	close space	we ne ed 40	we need 40
#	insert space	three#mistakes	three mistakes
¶	begin paragraph	¶The first year	The first year
(RUN IN)	no paragraph	financial. We earned twelve dollars.	financial. We earned twelve dollars.
⊏	move left	Next year	Next year
⊐	move right	Next year	Next year
∧	insert period	happy employee ∧	happy employee.
∧	insert comma	For two years ∧	For two years,
∧	insert semicolon	We need you ∧ come to work.	We need you; come to work.
∧	insert colon	dogs∧brown, black, and white.	dogs: brown, black, and white.
∧	insert hyphen	first∧rate	first-rate
∨	insert apostrophe	Marys car	Mary's car
∨/∨	insert quotation marks	∨Like a Rolling Stone∨	"Like a Rolling Stone"
(SP)	spell out	4 chips	four chips
(cap) or ≡	capitalize	the meeting	The meeting
(lc)	lower case	the Boss	the boss
∧	insert	the left∧ margin	the left margin

Apply your Knowledge

Spelling

In the following sentences, circle the correctly spelled words within the parentheses.

1. Each of the employees attended the meeting (accept except) the line supervisor, who was out of town for job-related travel.
2. The (advise advice) he gave will help us all do a better job.
3. Management must (affect effect) a change in employees' attitudes toward absenteeism.
4. Let me (site cite sight) this most recent case as an example.
5. (Its It's) too early to tell if our personnel changes will help create a better office environment.
6. If we (lose loose) another good employee to our competitor, our production capabilities will suffer.
7. I'm not (quite quiet) sure what she meant by that comment.
8. (Their There They're) budget has gotten too large to ensure a successful profit margin.
9. We had wanted to attend the conference (to too two), but our tight schedule prevented us from doing so.
10. (You're Your) best chance for landing this contract is to manufacture a better product.

In the letter below, correct the misspelled words.

March 5, 2012

Joanna Freeman
Personel Director
United Teletype
1111 E. Street
Kansas City, MO 68114

Dear Ms. Freeman:

Your advertizmemt in the Febuary 18, 2012, Kansas City Star is just the opening I have been looking for. I would like to submit my quallifications.

As you will note in the inclosed resum, I recieved an Enginneering degree from the Missouri Institute of Technology in 2005 and have worked in the electronic enginneering department of General Accounts for three years. I have worked a great deal in design electronics for microprocesors, controll systems, ect.

Because your company has invented many extrordinary design projects, working at your company would give me more chances to use my knowlege aquired in school and through my expirences. If you are interested in my quallifications, I would be happy to discuss them futher with you. I look foreward to hearing from you.

Sincerly,

Bob Cottrell

Bob Cottrell

Fragments and Comma Splices

In the following sentences, correct the fragments and comma splices by inserting the appropriate punctuation or adding any necessary words.

1. She kept her appointment with the salesperson, however, the rest of her staff came late.
2. When the CEO presented his fiscal year projections, he tried to motivate his employees, many were not excited about the proposed cuts.
3. Even though the company's sales were up 25 percent.
4. The supervisor wanted the staff members to make suggestions for improving their work environment, the employees, however, felt that any grievances should be taken directly to their union representatives.
5. Which he decided was an excellent idea.
6. Because their machinery was prone to malfunctions and often caused hazards to the workers.
7. They needed the equipment to complete their job responsibilities, further delays would cause production slowdowns.
8. Their client who was a major distributor of high-tech machinery.
9. Robotics should help us maintain schedules, we'll need to avoid equipment malfunctions, though.
10. The company, careful not to make false promises, advertising their product in media releases.

In the letter below, correct the fragments and comma splices.

May 12, 2012

Maurene Pierce
Dean of Residence Life
Mann College
Mannsville, NY 10012

Subject: Report on Dormitory Damage Systems

Here is the report you authorized on April 5 for an analysis of the current dormitory damage system used in this college.

The purpose of the report was to determine the effectiveness of the system. And to offer any concrete recommendations for improvement. To do this, I analyzed in detail the damage cost figures for the past three years, I also did an extensive study of dormitory conditions. Although I had limited manpower. I gathered information on all seven dormitories, focusing specifically on the men's athletic dorm, located at 1201 Chester. In this dorm, bathroom facilities, carpeting, and air conditioning are most susceptible to damage. Along with closet doors.

Nonetheless, my immediate findings indicate that the system is functioning well, however, improvements in the physical characteristics of the dormitories, such as new carpeting and paint, would make the system even more efficient.

I have enjoyed conducting this study, I hope my findings help you make your final decision. Please contact me. If I can be of further assistance.

Rob Harken

Rob Harken

Punctuation

In the following sentences, circle the correct punctuation marks. If no punctuation is needed, draw a slash mark through both options.

1. John took an hour for lunch (, ;) but Joan stayed at her desk to eat so she could complete the project.
2. Sally wrote the specifications (, ;) Randy was responsible for adding any needed graphics.
3. Manufacturing maintained a 93.5 percent production rating in July (, ;) therefore, the department earned the Golden Circle Award at the quarterly meeting.
4. In their year-end requests to management (, ;) supervisors asked for new office equipment (, ;) and a 10 percent budget increase for staffing.
5. The following employees attended the training session on stress management (, :) Steve Janasz, purchasing agent (, ;) Jeremy Kreisler, personnel director (, ;) and Prakash Patel, staff supervisor.
6. Promotions were given to all sales personnel (, ;) secretaries, however, received only cost-of-living raises.
7. The technicians voted for better work benefits (, ;) as an incentive to improve morale.
8. Although the salespeople were happy with their salary increases (, ;) the technicians felt slighted.
9. First (, ;) let's remember that meeting schedules should be a priority (, ;) and not an afterthought.
10. The employee (, ;) who achieves the highest rating this month (, ;) will earn 10 bonus points (, ;) therefore (, ;) competition should be intense.

In the letter below, no punctuation has been added. Instead, there are blanks where punctuation might be inserted. First, decide whether any punctuation is needed (not every blank requires punctuation). Then, insert the correct punctuation—a comma, colon, period, semicolon, or question mark.

January 8_ 2012

Mr_ Ron Schaefer
1324 Homes
Carbondale_ IL_ 34198

Dear Mr_ Schaefer_

Yesterday_ my partners_ and I read about your invention in the Herald Tribune_ and we want to congratulate you on this new idea_ and ask you to work with us on a similar project_

We cannot wait to begin our project_ however_ before we can do so_ I would like you to answer the following questions_

• Has your invention been tested in salt water_
• What is the cost of replacement parts_
• What is your fee for consulting_

Once_ I receive your answers to these questions_ my partners and I will contact you regarding a schedule for operations_ We appreciate your design concept_ and know it will help our business tremendously_ We look forward to hearing from you_

Sincerely_

Elias Agamenyon

Elias Agamenyon

Agreement (Subject/Verb and Pronoun/Antecedent)

In the following sentences, circle the correct choice to achieve agreement between subject and verb or pronoun and antecedent.

1. The employees, though encouraged by the possibility of increased overtime, (was were) still dissatisfied with their current salaries.
2. The supervisor wants to manufacture better products, but (they he) doesn't know how to motivate the technicians to improve their work habits.
3. The staff (was were) happy when the new manager canceled the proposed meeting.
4. Anyone who wants (his or her their) vote recorded must attend the annual board meeting.
5. According to the printed work schedule, Susan and Tom (work works) today on the manufacturing line.
6. According to the printed work schedule, either Susan or Tom (work works) today on the manufacturing line.
7. Although Tamara is responsible for distributing all monthly activity reports, (she they) failed to mail them.
8. Every one of the engineers asked if (he or she they) could be assigned to the project.
9. Either the supervisor or the technicians (is are) at fault.
10. The CEO, known for her generosity to employees and their families, (has have) been nominated for the humanitarian award.

In the following memo, find and correct the errors in agreement.

MEMO

DATE: October 30, 2012
TO: Tammy West
FROM: Susan Lisk
SUBJECT: REPORT ON AIR HANDLING UNIT

There has been several incidents involving the unit which has resulted in water damage to the computer systems located below the air handler.

The occurrences yesterday was caused when a valve was closed creating condensation to be forced through a humidifier element into the supply air duct. Water then leaked from the duct into the room below causing substantial damage to four disc-drive units.

To prevent recurrence of this type of damage, the following actions has been initiated by maintenance supervision:

- Each supervisor must ensure that their subordinates remove condensation valves to avoid unauthorized operation.
- Everyone must be made aware that they are responsible for closing condensation valves.
- The supply air duct, modified to carry away harmful sediments, are to be drained monthly.

Maintenance supervision recommend that air handlers not be installed above critical equipment. This will avoid the possibility of coil failure and water damage.

Capitalization

In the following memo, nothing has been capitalized. Capitalize those words requiring capitalization.

---

date: december 5, 2012
to: jordan cottrell
from: richard davis
subject: self-contained breathing apparatus (scba) and negative pressure respirator
          evaluation and fit-testing report

the evaluation and fit-testing have been accomplished. the attached list identifies the following:

• supervisors and electronic technicians who have used the scba successfully.
• the negative pressure respirators used for testing in an isoamyl acetate atmosphere.

fit-testing of waste management personnel will be accomplished annually, according to president chuck carlson. new supervisors and technicians will be fit-tested when hired.

all apex corporation personnel located in the new york district (12304 parkview lane) must submit a request form when requesting a respirator or scba for use. any waste management personnel in the north and south facilities not identified on the attached list will be fit-tested when use of scba is required. if you have any questions, contact chuck carlson or me (richard davis, district manager) at ext. 4036.

Grammar Quiz

The following sentences contain errors in spelling, punctuation, verb and pronoun agreement, sentence structure (fragments and run-ons), and modification. Circle the letter corresponding to the section of the sentence containing the error.

> When you recieve the salary increase, your family will celebrate the occasion.
>   A   B   C

◀ **EXAMPLE**

1. Each department manager should tell his subordinates to advise of any negative
                                  A   B
occurrences regarding in-house training.
   C   D

2. The lawyers' new offices were similar to their former ones _____ the offices
   A   B   C
were on a quiet street.
         D

3. New York City is divided into five boroughs ; Manhattan, the Bronx, Queens,
                 A   B   C
Brooklyn, and Staten Island.
       D

4. Fashion consultants explain that clothes create a strong impression _____ so
                        A   B
they advise executives _____ to choose wardrobes carefully.
                    C   D

5. Everyone should make sure that they are well represented in union meetings ;
                           A   B
otherwise , management could become too powerful.
       C   D

6. The employment agency , too busy to return the telephone calls from
                    A B
prospective clients , are harming business opportunities.
          C D

7. The supervisory staff are making decisions based on scheduling , but all
                A   B
employees want to ensure quality control.
   C   D

8. Because the price of cars has risen dramatically _____ most people keep their
                                    A   B
cars longer _____ to save capital expenditures.
       C   D

9. The <u>department's</u> manager heard that a merger was possible _____ but he
    A   B

 decided <u>to</u> keep the news <u>quiet</u>.
    C   D

10. The manager of the department believed that her <u>employee's</u> were excellent , but
                   A   B

 she decided that no raises could be given _____ because the price of stocks
                  C

 <u>was</u> falling.
  D

11. The detailed report from the audit _____ of the department gave good <u>advise</u>
             A   B

 about how to restructure , so we are <u>all ready</u> to do so.
        C   D

12. When my boss and <u>I</u> looked at the books , we found these problems <u>;</u> lost
        A   B   C

 invoices, unpaid bills , and late payments.
        D

13. The most dedicated staff members <u>are</u> <u>accepted</u> for the on-site training sessions
              A B

 because <u>:</u> they are responsive to criticism, represent the <u>company's</u> future, and
    C   D

 strive for improvements.

14. Despite her assurances to the contrary , there <u>is</u> still three unanswered
                A B

 questions <u>:</u> who will make the payments, when will these payments occur, and
      C

 why is <u>there</u> a delay?
     D

15. The reputation of many <u>companies</u> often <u>depend</u> on one employee <u>who</u>
         A   B   C

 <u>represents</u> that company.
  D

16. Either my monthly activity report or my <u>year-end</u> report <u>are</u> due today , but my
                A   B   C

 computer is broken <u>;</u> therefore, I need to use yours.
        D

17. Today's American <u>manpower</u> , according to many <u>foreign</u> governments, <u>suffers</u>
      A B   C   D

 from lack of discipline.

18. Rates are increasing next year because <u>:</u> fuel, <u>maintenance,</u> and insurance <u>are</u> all
                 A   B   C

 higher <u>than</u> last year.
    D

19. Many colleges have <u>long-standing</u> football rivalries , one of the most famous
                       A   B
    ones <u>is</u> between KU and KSU <u>(two universities in Kansas)</u>.
         C   D

20. <u>Everyone</u> who <u>wants</u> to enroll in the business school should do so before June if
    A   B
    <u>they</u> can to <u>ensure</u> getting the best classes.
    C   D

21. Because John wanted high visibility for his two <u>businesses</u> , he paid top dollar
                                                    A   B
    _____ and spent long hours looking for appropriate <u>cites.</u>
         C   D

22. Many people apply for jobs at Apex , however , only a few <u>are</u> <u>accepted.</u>
                                       A   B   C   D

23. Because most cars break down <u>occasionally</u> , all drivers should know how to
                                 A   B
    change a flat tire , and how to signal for <u>assistance.</u>
                    C   D

24. Arriving on time, working <u>diligently,</u> and closing the office securely — everyone
                              A   B
    <u>needs</u> to know that <u>they are</u> responsible for these job duties.
    C   D

25. Mark McGwire not only hit the ball <u>further</u> <u>than</u> other players, but also he hit
                                        A   B
    more dingers than other players <u>who</u> hit <u>fewer</u> homers.
                                    C   D

To document research correctly, you must provide parenthetical source citations following the quote or paraphrase within the text. At the end of your document, supply a references page (American Psychological Association and Council of Science Editors) or a works cited page (Modern Language Association).

Parenthetical Source Citations

The American Psychological Association (APA), the Council of Science Editors (CSE), and the Modern Language Association (MLA) use a parenthetical form of source citations. If your boss or instructor requests footnotes or endnotes, you should still use these forms. However, the most modern approach to source citations according to MLA and APA requires only that you cite the source of your information parenthetically after the quotation or paraphrase. The CSE style manual uses a number after the quote or paraphrase to reference the source to the references page at the end of the document. CSE also uses an author-year sequence similar to APA.

APA Format

One Author. If you do not state the author's name or the year of the publication in the lead-in to the quotation, include the author's name, year of publication, and page number in parenthesis, after the quotation.

> "Social media has helped companies quickly answer customer complaints" (Cottrell, 2011, p. 118).

(Page numbers are included for quoted material. The writer determines whether page numbers are included for source citations of summaries and paraphrases.)

Two Authors. When you cite a source with two authors, always use both last names with an ampersand (&).

> "Line charts reveal relationships between sets of figures" (Gerson & Gerson, 2011, p. 158).

Three or More Authors. When your citation has more than two authors but fewer than six, use all the last names in the first parenthetical source citation. For subsequent citations, list the first author's last name followed by *et al.* (Latin for "and others"), the year of publication, and for a quotation, the page number.

> "Employees require instantaneous access to crisis communication in the workplace" (Conners et al., 2011, p. 2).

Anonymous Works. When no author's name is listed, include in the source citation the title or part of a long title and the year. Book titles are underlined or italicized, and periodical titles are placed in quotation marks.

> Flash drives have revolutionized data storage (*Electronic Databases*, 2011).

> Effective e-mail messages can be organized in three paragraphs ("Using Templates," 2011).

CSE Format

In-text citations for quoted or paraphrased material are in the form of superscript numbers. Sometimes, editors prefer that numbers are placed in parentheses or in brackets. The number refers to the numbered source citations on the references list at the end of the document. Many editors and publishers believe that a numbered form of citation is less intrusive to the reader than the method used by APA or MLA.

> "Social media has helped companies quickly answer customer complaints."[1]

> "Line charts reveal relationships between sets of figures" (2).

> "Employees require instantaneous access to crisis communication in the workplace" [3].

MLA Format

One Author. After the quotation or paraphrase, parenthetically cite the author's last name and the page number of the information.

> "Viewing the molecular activity required state-of-the-art electron microscopes" (Heinlein 193).

Note that the period follows the parenthesis, not the quotation. Also note that no comma separates the name from the page number and that no lowercase *p* precedes the number.

Two Authors. After the quotation or paraphrase, parenthetically cite the authors' last names and the page number of the information.

> "Twitter has dramatically changed the way we write on the job" (Crider and Berry 292).

Three or More Authors. Writing a series of names can be cumbersome. To avoid this, if you have a source of information written by three or more authors, parenthetically cite one author's name, followed by *et al.* and the page number.

> "The development of gaming software is a growing industry" (Norwood et al. 93).

Anonymous Works. If your source has no author, parenthetically cite the shortened title and page number.

> "Robots are more accurate and less prone to errors caused by long hours of operation than humans" ("Useful Robots" 81).

Documentation of Sources

Parenthetical source citations are an abbreviated form of documentation. In parentheses, you tell your readers only the names of your authors and the page numbers on which the information can be found, or you provide a number that parallels the numbered source at the end of the document. Such documentation alone would be insufficient. Your readers would not know the names of the authors (in CSE numerical-sequence format), the titles of the books, the names of the periodicals, or the dates, volumes, or publishing companies. This more thorough information is found on the references page (APA) or works cited page (MLA), a listing of research sources alphabetized either by author's name or title (if anonymous). On the references page (CSE), you organize the citations numerically by the order in which the quote or paraphrase appeared in the text. This is the last page(s) of your research report.

Your entries should follow APA, CSE, and MLA standards. (Additional style manuals are available for many professions.) MLA no longer requires the use of URLs in source citations. Because Web addresses can change and documents sometimes appear in several different databases, MLA says that most readers can find electronic sources using title or author searches in Internet search engines. If you do include a URL, MLA says to put the URL in angle brackets after the date of access. Use slash marks to break a URL.

APA References

The APA style is commonly used in both engineering and scientific fields. The following are sample entries for the reference page which is placed at the end of the document. Include on the reference page only sources from which you cited in the document. For a comprehensive illustration of reference page entries, use the *Publication Manual of the American Psychological Association* (2009) and the *APA Style Guide to Electronic References* (2007).

A book with one author

Cottrell, R. C. (2006). *Smoke jumpers of the civilian public service in World War II.* London: McFarland and Co., Inc.

A book with two authors

Heath, C. & Heath, D. (2007). *Made to stick: why some ideas survive and others die.* New York: Random House.

A book with three or more authors

Nadell, J., McNeniman, L., & Langan, J. (1997). *The Macmillan writer.* Boston: Allyn & Bacon.

A book with a corporate authorship

Corporate Credit Union Network. (2007). *A review of the credit union financial system: history, structure, and status and financial trends.* Kansas City, MO: U.S. Central.

A translated book

Phelps, R. (Ed.). (1983). *The collected stories of Colette* (M. Ward, Trans.). New York: Farrar, Straus & Giroux.

An entry in a collection or anthology

Hamilton, K. (2005). What's in a Name? In R. Atwan (Ed.), *America now: short readings from recent periodicals* (pp. 12–20). New York: Bedford/St. Martin's.

A signed article in a journal

Davis, R. (2007, April). Getting—and keeping good clients. *Intercom*, 8–12.

A signed article in a magazine

Rawe, J. (2007, May 28). A question of honor. *Time*, 59–60.

A signed article in a newspaper

Gertzen, J. (2007, March 29). University to go wireless. *The Kansas City Star*, p. C3.

An unsigned article

Effective communication with clients. (2009, September 23). *Technical Communication*, 22.

Encyclopedias and almanacs

Internet. (2000). *The world book encyclopedia.* Chicago: World Book.

Computer software

Drivers and Utilities [Computer software]. (2002–2004). Dell, Inc.

An article from an online database (or other electronic subscription service)

Pascal, J. (2005). Top ten qualities/skills employers want. *Job Outlook 2006 Student Version*. National Association of Colleges and Employers, 5 (12–16). Retrieved from ProQuest database.

E-mail

According to APA, do not include e-mail messages in the list of references. You should cite the message parenthetically in your text. (R. Schneider, personal communication, April 2, 2011).

Blog

McWard, J. (2011, May 31). Graphics on-line. Message posted to http://www.jmcward .net.

Personal Web site

Mohr, E. (2011, Dec. 29). Home page of Ellen Mohr's Web Site. Retrieved from http://emohr.edu.

Professional Web site

Johnson County Community College Writing Center. (2011, Jan. 5). Johnson County Community College. Retrieved from http://jccc.edu.

Posting to a discussion listserv

Tsui, P. (2011, Sep. 15). Questionnaire [Msg.16]. Message posted to http://groups.stc.com ./html.

APA References Page

Place the references page at the end of the document or in an appendix. The entries on the reference page are alphabetized by author's last name or title.

References

Corporate Credit Union Network. (2007). *A review of the credit union financial system: History, structure, and status and financial trends.* Kansas City, MO: U.S. Central.

Effective communication with clients. (2009, September 23). *Technical Communication,* 22.

Gertzen, J. (2007, March 29). University to go wireless. *The Kansas City Star,* p. C3.

Pascal, J. (2005). Top ten qualities/skills employers want. *Job Outlook 2006 Student Version.* National Association of Colleges and Employers, 5 (12–16). Retrieved from ProQuest database.

Tsui, P. (2011, Sep. 15). Questionnaire [Msg.16]. Message posted to http://groups.stc.com./html

---

CSE References

The CSE style guide shows two systems for organizing references at the end of the document. First, you can use the citation-sequence system that lists the numbered references in the order cited within the text (illustrated below). Second, you can follow the name-year system that lists references in alphabetical order by author's last name.

Personal e-mail messages, blog entries, personal Web sites, and entries to listservs should not automatically be included on a references list. According to the CSE style manual, the decision to include such references is left to publishers and editors.

The CSE style of documentation is used in the fields of biology and medicine. Following are sample entries using the numerical system for the references list. For a comprehensive illustration of entries on the references list, use the *Scientific Style and Format: The CSE Manual for Authors, Editors, and Publishers* (2006).

A book with one author

6. Cottrell RC. Smoke jumpers of the civilian public service in World War II. London: McFarland and Co., Inc.; 2006. p 27–28.

A book with two authors

2. Heath C., Heath D. Made to stick: why some ideas survive and others die. New York: Random House; 2007. p 217–24.

A book with three or more authors

17. Nadell J., McNeniman L., Langan J. The Macmillan writer. Boston: Allyn and Bacon; 1997. p 224.

A book with a corporate authorship

22. Corporate Credit Union Network. A review of the credit union financial system: History, structure, and status and financial trends. Kansas City, MO: U.S. Central; 2007.

A translated book

3. Patel J. Technical communication and globalization. McWard J, translator. New York: Bedford/St. Martin's; 2006. p 15.

An entry in a collection or anthology

8. Hamilton K. What's in a name? R. Atwan, editor. America now: short readings from recent periodicals. New York: Bedford/St. Martin's; 2005. p 12–20.

A signed article in a journal

1. Davis R. Getting—and keeping good clients. Intercom 2007 April: 8–12.

A signed article in a magazine

15. Rawe J. A question of honor. Time 2007 May 28: 59–60.

A signed article in a newspaper

5. Gertzen J. University to go wireless. The Kansas City Star 2007 Mar 16; Sect C: 3.

An unsigned article

32. Effective communication with clients. Technical Communication 2009 Sep 23; 22.

An article from an online database (or other electronic subscription service)

4. Pascal J. Re: top ten qualities/skills employers want. In: Job outlook 2006 student version. 2005. National Association of Colleges and Employers; 2006 May 5 [cited 2006 Apr 14]. Available from: ProQuest database.

Professional web site

41. Johnson County Community College Writing Center. Johnson County Community College. [cited 2011 Jan 5]. Available from: http://jccc.edu.

CSE References Page

The references page in CSE style is placed either at the end of the document or in an appendix. List the sources in the order in which they appeared in the document.

References

1. Davis R. Getting—and keeping good clients. Intercom 2007; April: 8–12.
2. Heath C., Heath D. Made to stick: why some ideas survive and others die. New York: Random House; 2007. p 217–24.
3. Patel J. Technical communication and globalization. McWard J, translator. New York: Bedford/St. Martin's; 2006. p 15.
4. Pascal J. Re: top ten qualities/skills employers want. In: Job outlook 2006 student version. 2005. National Association of Colleges and Employers; 2006 May 5 [cited 2006 Apr 14]. Available from: ProQuest database.
5. Gertzen J. University to go wireless. The Kansas City Star 2007 Mar 16; Sect C: 3.

MLA Works Cited

MLA documentation format is used in the arts and humanities fields. Following are examples of entries on the works cited page. For a comprehensive illustration of MLA format for the works cited page, refer to the *MLA Style Manual and Guide to Scholarly Publishing* (2008). MLA also recently provided an abbreviated, updated style guide, *A Guide to MLA Style 2009 Update*.

A book with one author

Cottrell, Robert C. *Smoke Jumpers of the Civilian Public Service in World War II*. London: McFarland and Co., Inc., 2006. Print.

A book with two or three authors

Heath, Chip, and Dan Heath. *Made to Stick: Why Some Ideas Survive and Others Die*. New York: Random House, 2007. Print.

A book with four or more authors

Nadell, Judith, et al. *The Macmillan Writer*. Boston: Allyn and Bacon, 1997. Print.

A book with a corporate authorship

Corporate Credit Union Network. *A Review of the Credit Union Financial System: History, Structure, and Status and Financial Trends*. Kansas City: U.S. Central, 2007. Print.

A translated book

Phelps, Robert, ed. *The Collected Stories of Colette*. Trans. Matthew Ward. New York: Farrar, Straus Giroux, 1983. Print.

An entry in a collection or anthology

Hamilton, Kendra. "What's in a Name?" *America Now: Short Readings from Recent Periodicals*. Ed. Robert Atwan. New York: Bedford/St. Martin's, 2005. 12–20. Print.

A signed article in a journal

Davis, Rachel. "Getting—and Keeping Good Clients." *Intercom* (April 2007): 8–12. Print.

A signed article in a magazine

Rawe, Julie. "A Question of Honor." *Time* 28 May 2007: 59–60. Print.

A signed article in a newspaper

Gertzen, Jason. "University to go wireless." *The Kansas City Star* 29 Mar. 2007: C3. Print.

An unsigned article

"Effective Communication with Clients." *Technical Communication* 23 Sep. 2009: 22. Print.

Encyclopedias and almanacs

"Internet." *The World Book Encyclopedia*. 2000 ed. Chicago: World Book. Print.

Computer software

Drivers and Utilities. Computer software. Dell, Inc., 2002–2004. Print.

An article from an online database (or other electronic subscription service)

Pascal, Janet. "Top Ten Qualities/Skills Employers Want." *Job Outlook 2006 Student Version*. National Association of Colleges and Employers 5 (2005): 12–16. ProQuest. Web. 25 May 2009.

E-mail

Schneider, Ruth. "Re: Teaching Technical Communication." Message to Sharon Gerson. 2 Apr. 2011. Email.

Blog

McWardj [Jim McWard]. "Graphics On-line." *Myinfosite*. 31 May 2009. Web. 7 Sep. 2011.

Personal Web site

Mohr, Ellen. Home page. 29 Dec. 2009. Web. 30 December 2011.

Professional Web site

Johnson County Community College Writing Center. Johnson County Community College. 5 Jan. 2009. Web. 14 Feb. 2011.

Posting to a discussion listserv

Ptsui [Peter Tsui]. "Questionnaire." *Society for Technical Communication Listserv*. 15 Sep. 2011. Web. 17 Sep. 2011.

MLA Works Cited Page

At the end of the document, include a works cited page or place this page in an appendix.

Works Cited

Corporate Credit Union Network. *A Review of the Credit Union Financial System: History, Structure, and Status and Financial Trends*. Kansas City: U.S. Central, 2007. Print.

"Effective Communication with Clients." *Technical Communication* 23 Sep. 2009: 22. Print.

McWardj [Jim McWard]. "Graphics On-line." *Myinfosite*. 31 May 2009. Web. 7 Sep. 2011.

Pascal, Janet. "Top Ten Qualities/Skills Employers Want." *Job Outlook 2006 Student Version*. National Association of Colleges and Employers 5 (2005): 12–16. ProQuest. Web. 25 May 2009.

Rawe, Julie. "A Question of Honor." *Time* 28 May 2007: 59–60. Print.

REFERENCES

CHAPTER 1

"About Google Wave." *Google Wave Preview.* 2009. Web. 22 Nov. 2009.

"Achieving the Promise of the Mobile Enterprise." *Motorola: Position Paper.* Aug. 2005. Web. 31 Aug. 2007.

"Business Identity." *Crane's.* 2007. Web. 28 Jul. 2007.

Costas, Bob. "Importance of Writing." *The National Commission on Writing.* 2007. Web. 15 May 2007.

Fisher, Lori, and Lindsay Bennion. "Organizational Implications of the Future Development of Technical Communication: Fostering Communities of Practice in the Workplace." *Technical Communication* 52 Aug. 2005: 277–288. Print.

Hughes, Michael A. "Managers: Move from Silos to Channels." *Intercom* (Mar. 2003): 9–11. Print.

"Individual's and Teams' Roles and Responsibilities." GOAL/QPC. 2003. Web. 24 Feb. 2003.

Mader, Stewart. "Using Wiki in Education." *The Science of Spectroscopy.* 14 Dec. 2005. Web. 7 Apr. 2006.

Mader, Stewart. "Your Wiki Isn't Wikipedia: How to Use It for Technical Communication." *Intercom* (Jan. 2009): 14–15. Print.

"Mashup." *Wikipedia.* Web. 22 Nov. 2009.

McNeill, Angie E. "Self-Directed Work Teams." *Intercom* (Sep./Oct. 2000): 15–17. Print.

Miller, Carolyn R. "Communication in the Workplace: A Collaborative Teacher-Student Research Project." Oct. 1996. Web. Aug. 2002.

Nesbitt, Pamela, and Elizabeth Bagley-Woodward. "Practical Tips for Working with Global Teams." *Intercom* (Jun. 2006): 25–30. Print.

"Planning Job Choices." NaceWeb. 2004. Web. 15 Jan. 2004.

Reddick, Thomas. E-mail message to Steve Gerson. 18 Nov. 2009. E-mail.

Robidoux, Charlotte, and Beth Hewett. "Is There a Write Way to Collaborate?" *Intercom* (Feb. 2009): 4–9. Print.

"Top Ten Qualities/Skills Employers Want." *Job Outlook 2006 Student Version.* National Association of Colleges and Employers. Web. 16 Nov. 2011.

Trapani, Gina. "Google Wave's Best Use Cases." *Lifehacker.* 14 Oct. 2009. Web. 22 Nov. 2009.

"TWiki—Enterprise Wiki & Collaboration Platform." 2007.

TWiki. Web. 4 Mar. 2007.

"We Are Smarter Than Me." 2006. Web. 18 Dec. 2006.

"What Is Wikipedia." *Wikipedia.* Web. 9 Jan. 2007.

"Wiki." *Wikipedia.* Web. 9 Jan. 2007.

"Writing: A Ticket to Work . . . Or A Ticket Out, A Survey of Business Leaders." *National Commission on Writing.* Sep. 2004. Web. 2007.

"Writing Skills Necessary for Employment, Says Big Businesses." *National Commission on Writing.* 2007. Web. 1 Aug. 2007.

CHAPTER 2

Albers, Michael J. "Single Sourcing and the Technical Communication Career Path." *Technical Communication* 50 Aug. 2003: 336–343. Print.

Campbell, Kim S., et al. "Leader-Member Relations as a Function of Rapport Management." *Journal of Business Communication* 40 (2003): 170–194. Print.

Cargile-Cook, Kelli. "Usability Testing in the Technical Communication Classroom." *Intercom* Sep./Oct. 1999: 36–37. Print.

Carter, Locke. "The Implications of Single Sourcing for Writers and Writing." *Technical Communication* 50 Aug. 2003: 317–320. Print.

Dorazio, Pat. Interview. Jun. 2000.

Rizzo, Tony. "Dropped Decimal Point Becomes a Costly Point of Contention." *The Kansas City Star* 19 Feb. 2005: A1, A6. Print.

"Writing: A Powerful Message from State Government." *Report of the National Commission on Writing.* College Board, 2005. Print.

CHAPTER 3

Adams, Rae, et al. "Ethics and the Internet." *Proceedings: 42nd Annual Technical Communication Conference.* 23–26 Apr. 1995: 328. Print.

Anderson-Hancock, Shirley A. "Society Unveils New Ethical Guidelines." *Intercom* 42 Jul./Aug. 1995: 6–7. Print.

Barker, Thomas, et al. "Coming into the Workplace: What Every Technical Communicator Should Know—Besides Writing." *Proceedings: 42nd Annual Technical Communication Conference*. 23–26 Apr. 1995: 38–39. Print.

Bowman, George, and Arthur E. Walzer. "Ethics and Technical Communication." *Proceedings: 34th Annual Technical Communication Conference*. 10–13 May 1987: MPD–93. Print.

Bremer, Otto A., et al. "Ethics and Values in Management Thought." *Business Environment and Business Ethics: The Social, Moral, and Political Dimensions of Management*. Ed. Karen Paul. Cambridge, MA: Ballinger, 1987: 61–86. Print.

"Circular 66, Copyright Registration for Online Works." *U.S. Copyright Office*. 2009. Web. 15 Sep. 2010.

"Code for Communicators." *Rocky Mountain Chapter, Society for Technical Communication*. 2007. Web. 1 Aug. 2007.

"College Degree Nearly Doubles Annual Earnings, Census Bureau Reports." *U.S. Census Bureau News*. 28 Mar. 2005. Web. 5 Mar. 2007.

Dillon, Sam. "Literacy Falls for Graduates from College, Testing Finds." *The New York Times*. 16 Dec. 2005. Web. 16 Dec. 2005.

Gerson, Steven M., and Sharon J. Gerson. "A Survey of Technical Writing Practitioners and Professors: Are We on the Same Page?" *Proceedings: 42nd Annual Technical Communication Conference*. 23–26 Apr. 1995: 44–47. Print.

Girill, T. R. "Technical Communications and Ethics." *Technical Communication* 34 Aug. 1987: 178–179. Print.

Guy, Mary E. *Ethical Decision Making in Everyday Work Situations*. New York: Quorum Books, 1990. Print.

Harkness, Holly E. "Sarbanes-Oxley & New Opportunities." *Intercom* (Jan. 2004): 16–18. Print.

Hartman, Diane B., and Karen S. Nantz. "Send the Right Messages About E-Mail." *Training & Development* May 1995: 60–65. Print.

Johnson, Dana R. "Copyright Issues on the Internet." *Intercom* (Jun. 1999): 17. Print.

LeVie, Donald S. "Internet Technology and Intellectual Property." *Intercom* (Jan. 2000): 20–23. Print.

Mollison, Andrew. "U.S. Loses Its Lead in High School and College Grads." *Cox Newspapers* 6 Nov. 2001: 1–3. Print.

Moore, Linda E. "Serving the Electronic Reader." *Intercom* (Apr. 2003): 17. Print.

Perlin, Neil. "Technical Communication: The Next Wave." *Intercom* (Jan. 2001): 4–8. Print.

Reyman, Jessica. "Rethinking Plagiarism for Technical Communication." *Technical Communication* (Feb. 2008): 61–67. Print.

Self, Tony. "What If Readers Can't Read?" *Intercom* (Feb. 2009): 10–14. Print.

"STC Ethical Principles for Technical Communicators." *The Society for Technical Communication*. 1998. Web. 1 Sep. 2007.

"Summary of Sarbanes-Oxley Act of 2002." *American Institute of Certified Public Accountants*. 2002. Web. 8 Jan. 2004.

Turner, John R. "Ethics Online: Looking Toward the Future." *Proceedings: 42nd Annual Technical Communication Conference*. 23 Apr. 1995: 59–62. Print.

Wilson, Catherine Mason. "Product Liability and User Manuals." *Proceedings: 34th International Technical Communication Conference*. 10 May 1987: WE-68–71. Print.

"Writing: A Ticket to Work . . . Or A Ticket Out, A Survey of Business Leaders." *National Commission on Writing*. Sep. 2004. Web. 14 May 2007.

CHAPTER 4

Cardarella, Toni. "Business." *The Kansas City Star*. 30 Sep. 2003. Web. 30 Sep. 2003.

"Civilian Labor Force by Age, Sex, Race, and Hispanic Origin, 1994, 2004, and Projected 2014." *The United States Department of Labor*. 7 Dec. 2005. Web. 6 Mar. 2007.

"Coca-Cola Company." 11 Aug. 2010. Web. 11 Aug. 2010.

Courtis, John K., and Salleh Hassan. "Reading Ease of Bilingual Annual Reports." *Journal of Business Communication* 39 Oct. 2002: 394–413. Print.

"50 Million Americans Speak Languages Other Than English at Home." *Progressive Policy Institute*. 16 Apr. 2008. Web. 4 Jun. 2009.

Flint, Patricia, et al. "Helping Technical Communicators Help Translators." *Technical Communication* 46 May 1999: 238–248. Print.

Gardner, Amanda. "Language a Widening Barrier to Health Care." *Health on the Net Foundation*. 9 Jan. 2007. Web. 6 Mar. 2007.

Grimes, Diane Susan, and Orlando C. Richard. "Could Communication Form Impact Organizations' Experience with Diversity?" *Journal of Business Communication* 40 Jan. 2003: 7–27. Print.

Horton, William. "The Almost Universal Language: Graphics for International Documents." *Technical Communication* 40 Nov. 1993: 682–693. Print.

Hussey, Tim, and Mark Homnack. "Foreign Language Software Localization." *Proceedings: 37th International Technical Communication Conference*. 20–23 May 1990: RT-44–47. Print.

Jordan, Katrina. "Diversity Training: What Works, What Doesn't, and Why?" *Civil Right Journal* Fall 1999: 1. Print.

King, Janice M. "The Challenge of Communicating in the Global Marketplace." *Intercom* 42.5 May/Jun. 1995: 1, 28. Print.

McInnes, R. "Workforce Diversity: Changing the Way You Do Business." *Diversity World*. 2003. Web. 12 Feb. 2003.

Melgoza, Cesar M. "Geoscape Findings Confirm Hispanic Market Growth." *Hispanic MPR.com*. 2007. Web. 6 Mar. 2007.

Miller, Carolyn R. "Communication in the Workplace: A Collaborative Teacher-Student Research Project." 2002. Web. 1 Aug. 2002.

"Multilingual Features in Windows XP Professional." 2001. Web. 24 Aug. 2001.

Nethery, Kent. "Let's Talk Business." 2003. Web. 16 Feb. 2003.

"Occupational Employment in Private Industry by Race/Ethnic Group/Sex and by Industry, United States 2005." *The U.S. Equal Employment Opportunity Commission*. 9 May 2005. Web. 6 Mar. 2007.

"Population Projections." *U.S. Census Bureau*. 31 Jul. 2008. Web. 4 Jun. 2009.

Rains, Nancy E. "Prepare Your Documents for Better Translation." *Intercom* 41.5 Dec. 1994: 12. Print.

Sanchez, Mary. "KC Hospitals Seek to Overcome Language Barriers." *The Kansas City Star* 7 Jan. 2003: A1, A4. Print.

Scott, Julie S. "When English Isn't English." *Intercom* (May 2000): 20–21. Print.

Skettino, Michele. "America in the 21st Century: A Marketer's Guide to the Nation." *Interep Research Division*. 2004. Web. 24 Aug. 2004.

St. Amant, Kirk R. "Communication in International Virtual Offices." *Intercom* (Apr. 2003): 27–28. Print.

St. Amant, Kirk R. "Designing Web Sites for International Audiences." *Intercom* (May 2003): 15–18. Print.

St. Amant, Kirk R. "International Integers and Intercultural Expectations." *Intercom* (May 1999): 25–28. Print.

Swenson, Lynne V. "How to Make (American) English Documents Easy to Translate." *Proceedings: 34th International Technical Communication Conference*. 10 May 1987: WE-193–195. Print.

"Swiss Fight Encroachment of English." *The Kansas City Star* 7 Dec. 2002: A16. Print.

U.S. Census Bureau. 2009. Web. 4 Jun. 2009.

Walmer, Daphne. "One Company's Efforts to Improve Translation and Localization." *Technical Communication* 46 May 1999: 230–237. Print.

Weiss, Edmund H. "Twenty-five Tactics to 'Internationalize' Your English." *Intercom* (May 1998): 11–15. Print.

"What Is the 'Business Case' For Diversity?" *Society for Human Resource Management*. 2003. Web. 12 Feb. 2003.

CHAPTER 5

Reyman, Jessica. "Rethinking Plagiarism for Technical Communication." *Technical Communication* (Feb. 2008): 61–67. Print.

CHAPTER 6

Bradley, Tony. "Policing Instant Messaging: Create A Policy To Govern IM Applications." *Processor* 5 Aug. 2005. Web. 12 Apr. 2008.

Clark, Robert. Interview. June 2010.

Gerson, Steven M., et al. "Core Competencies." Survey. Prentice Hall Publishing Company, 2004. Print.

Goforth, Alan. "Texting Can Be Ad Vehicle." *The Kansas City Star* 7 Jul. 2009: D14–D15. Print.

Hoffman, Jeff. "Instant Messaging in the Workplace." *Intercom* (Feb. 2004): 16–17. Print.

"How To: Instant Messaging Security." *PC World* 14 Jul. 2006. Web. 12 Apr. 2008.

Miller, Carolyn, et al. "Communication in the Workplace." *Center for Communication in Science, Technology, and Management*. Oct. 1996. Print.

Munter, Mary, et al. "Business E-mail: Guidelines for Users." *Business Communication Quarterly* 66 March 2003: 29. Print.

Ollman, Gunter. "Instant Message Security." *Technical Info*. Mar. 2004. Web. 12 Apr. 2008.

Reardon, Marguerite. "Americans Text More Than They Talk." *CNET.com*. 22 Sep. 2008. Web. 13 Jul. 2009.

Shinder, Deb. "Instant Messaging: Does It Have a Place in Business Networks?" *WindowSecurity.com*. 2005. Web. 8 Jun. 2005.

"The Transition to General Management." *Harvard Business School*. 1998. Web. 12 Nov. 2002.

"Writing: A Ticket to Work . . . Or A Ticket Out, A Survey of Business Leaders." *National Commission on Writing*. Sep. 2004. Web. 12 Mar. 2007.

CHAPTER 7

"About." Twitter.com. 25 Aug. 2010. Web. 19 Oct. 2010.

"Bosses 'Should Embrace Facebook.'" *BBC News*. 29 Oct. 2008. Web. 21 May 2009.

Braud, Gerard. "Crafting a Crisis Communication Plan." *International Association of Business Communicators Online Newsletter*. 2008. Web. 10 Apr. 2008.

Brewer, Melissa. "YouTube Marketing Basics—Know and Follow the Rules." 17 Dec. 2007 Web. 17 Sep. 2010.

Brito, Michael. Presentation. Social Networking Conference. 25 Jun. 2009.

Brodkin, Jon. "IBM Opens Sales Center in Second Life." *Network World*. 15 May 2007. Web. 15 Sep. 2010.

Brogan, Chriss. "50 Ideas on Using Twitter for Business." 20 Aug. 2008. Web. 4 Feb. 2009.

Brooks, Rich. "How to Use Twitter for Business." *Flyte Newsletters*. 2009. Web. 4 Feb. 2009.

Carr, David F. "Second Life: Is Business Ready For Virtual Worlds?" *Baseline*. 1 Mar. 2007. Web. 15 Sep. 2010.

Ciortea, Marius. Presentation. Social Networking Conference. 25 Jun. 2009.

Clifford, Stephanie. "Video Prank at Domino's Taints Brand." *New York Times*. 15 Apr. 2009. Web. 25 Aug. 2010.

"ComScore: YouTube Now 25 Percent Of All Google Searches." *Tech Blips*. 19 Dec. 2008. Web. 17 Jun. 2009.

"Facebook Surpasses 175 Million Users, Continuing to Grow by 600k Users/Day." *InsideFace-book.com*. 14 Feb. 2009. Web. 18 May 2009.

"Facing Up to Facebook." *British Trade Union*. Aug. 2007. Web. 21 May 2009.

"Fastest Growing Demographic on Facebook: Women Over 55." *Insidefacebook.com*. 2 Feb. 2009. Web. May 2009.

Foremski, Tom. "IBM is preparing to launch a massive corporate wide blogging initiative." *Silicon Valley Watcher*. 13 May 2005. Web. 23 May 2005.

"Gates Backs Blogs for Businesses." *BBC News*. 21 May 2004. Web. 23 May 2005.

Goforth, Alan. "Firms Find Friend in Facebook." *The Kansas City Star* 19 May 2009: D:15, D:17. Print.

Griffin, Chip. "Using New Media to Tame a Crisis." *International Association of Business Communicators Online Newsletter*. 2007. Web. 10 Apr. 2008.

Jana, Reena. "Breathing Second Life Into Business." *Business Week*. 16 Oct. 2006. Web. 3 Feb. 2009.

Jantsch, John. "Using Twitter for Business." *Duct Tape Marketing*. 2009. Web. Feb. 2009.

Kazi, Sumaya. Presentation. Social Networking Conference. 25 Jun. 2009.

Kharif, Olga. "Blogging for Business." *Business Week Online*. 9 Aug. 2004. Web. 23 May 2005.

King, Rachael. "How Companies Use Twitter to Bolster Their Brands." *Business Week*. 6 Sep. 2008. Web. 15Sep. 2010.

Kirkpatrick, Marshall. "Study: Fastest Growing US Companies Rapidly Adopting Social Media." *Read Write Web*. 15 Aug. 2008. Web. May 2009.

Kyzer, Lindy. "US Army Enlists Facebook, Twitter." 27 Apr. 2009. Web. 21 May 2009.

Li, Charlene. "Blogging: Bubble or Big Deal?" *Forrester*. 5 Nov. 2004. Web. 23 May 2005.

Martel, David. "The Threat of Social Networks' Rise." *Northern Light*. 29 Jan. 2009. Web. 17 Oct. 2009.

Mathews, Whitney. Interview. 4 May 2009.

Miller, Michael. "Using YouTube to Promote Your Business." *InformIT*. 28 Sep. 2007. Web. 4 Feb. 2009.

Ochman, B.L. "The Beauty, Secrets and Utility of Twitter for Business." *Whatsnextblog*. 8 Feb. 2009. Web. 8 Feb. 2009.

Pogue, David. "Twittering Tips for Beginners." *New York Times*. 15 Jan. 2009. Web. 15 Sep. 2010.

"Promoting Your Facebook Page Outside Facebook." *Facebook*. 21 May 2009. Web. 12 May 2010.

Ramel, David. "YouTube for Your Business." *PCWorld*. 23 Jun. 2007. Web. 4 Feb. 2009.

Ranii, David. "More Companies Use Social Media for Marketing." *The News & Observer*. 12 Dec. 2008. Web. 4 Feb. 2009.

Ray, Ramon. "Blogging for Business." *Inc.com*. Sep. 2004. Web. 23 May 2005.

Reddick, Paul. Interview. 15 Oct. 2009.

"Respect the YouTube Community." Youtube.com. May 2009. Web. May 2009.

Rife, Martine Courant. "Technical Communicators and Digital Writing Risk Assessment." *Technical Communication* (May 2007): 157–170. Print.

Science Friday. Web. May 2009.

Secondlife Grid. Web. 7 Jul. 2009.

Smith, Aaron. "New Numbers for Blogging and Blog Readership." *Pew Internet*. 22 Jul. 2008. Web. 21 May 2009.

Sohn, Judi. "12 Ways to Use Facebook Professionally." *Web Worker Daily*. 24 Jul. 2007. Web. 21 May 2009.

Stafford, Diane. "What Happens at Work Doesn't Stay at Work." *Kansas City Star* 9 Jun. 2009: A1, A6. Print.

"Statement of Rights and Responsibilities." *Facebook*. May 2009. Web. 1 May 2009.

Swartz, Jon. "Businesses Get Cheap Help from a Little Birdie." *USA Today* 26 Jun. 2009: B1– B2. Print.

"Tracking Facebook and the Facebook Platform for Developers and Marketers." *Inside Facebook*. 2009. Web. 21 May 2009.

"Twitter." *Wikipedia*. 7 Jun. 2009.

Vascellaro, Jessica E. "Why Email No Longer Rules." *The Wall Street Journal* 12 Oct. 2009: R1, R3. Print.

Wagner, Mitch. "Using Second Life as a Business-To-Business Tool." *Information Week*. 26 Apr. 2007. Web. 15 Oct. 2009.

Webber, Sheila. "Second Life for Business." *Fumsi*. Oct. 2008. Web. 10 May 2009.

"Welcome to Technorati." *Technorati.* 10 Apr. 2008. Web.7 Sep. 2009.

Wuorio, Jeff. "Blogging for business: 7 tips for getting started." *Microsoft.* 23 May 2005. Web. 10 Aug. 2010.

Wuorio, Jeff. "5 Ways Blogging Can Help Your Business." *Microsoft.* 23 May 2005. Web. 5 Jun. 2005.

Zahorsky, Darrel. "YouTube Your Business With Online Video." About.com. 20 Apr. 2009. Web. 23 Jan. 2010.

CHAPTER 8

Bloch, Janel M. "Online Job Searching: Clicking Your Way to Employment." *Intercom* (Sep./Oct. 2003): 11–14. Print.

Dikel, Margaret F. "The Online Job Application: Preparing Your Resume for the Internet." 1999. Web. 6 Sep. 1999.

Dixson, Kirsten. "Crafting an E-mail Resume." *BusinessWeek Online* 13 Nov. 2001. Web. 15 Sep. 2010.

Drakeley, Caroline A. "Viral Networking: Tactics in Today's Job Market." *Intercom* (Sep./Oct. 2003): 5–7. Print.

Hartman, Peter J. "You Got the Interview. Now Get the Job!" *Intercom* (Sep./Oct. 2003): 23–25. Print.

Isaacs, Kim. "Resume Critique Checklist." *MonsterTRAK Career Advice Archives.* 5 Sep. 2006. Web. 15 Sep. 2010.

Kallick, Rob. "Research Pays Off During Interview." *The Kansas City Star* 23 Mar. 2003: D1. Print.

Kendall, Pat. "Electronic Resumes." 9 Dec. 2003. Web. 15 Sep. 2010.

LeVie, Donald S., Jr. "Resumes: You Can't Escape." *Intercom* (Apr. 2000): 8–11. Print.

Mulvaney, Mary Kay. "The Information Interview: Bridging College and Beyond." *Business Communication Quarterly* 66 Sep. 2003: 66–72. Print.

Nale, Mike. "10 Things That Make Up a Good Video Resume." *ere.net.* 8 Feb. 2008. Web. 20 Jul. 2009.

Ralston, Steven M., et al. "Helping Interviewees Tell Their Stories." *Business Communication Quarterly* 66 Sep. 2003: 8–22. Print.

"Resume Guidelines." Career Services Center. Overland Park, KS: Johnson County Community College, 2003. Print.

Robart, Kay. "Submitting Resumes via E-Mail." *Intercom* (Jul./Aug. 1998): 12–14. Print.

Schwabel, Dan. "7 Secrets to Getting Your Next Job Using Social Media." *Mashable.* 5 Jan. 2009. Web. 20 Jul. 2009.

Stafford, Diane. "Show Up Armed with Answers." *The Kansas City Star* 3 Aug. 2003: L1. Print.

Stern, Linda. "New Rules of the Hunt." *Newsweek* 17 Nov. 2003: 67. Print.

"Top Ten Qualities/Skills Employers Want." *Job Outlook 2005 Student Version.* National Association of Colleges and Employers. 2005. Web. 6 Nov. 2007.

Vogt, Peter. "Your Resume's Look Is as Important as Its Content." *MonsterTRAK Career Advice Archives.* 6 Mar. 2007. Web.15 Sep. 2010.

"Writing: A Powerful Message from State Government." *Report of the National Commission on Writing.* College Board, 2005. Print.

CHAPTER 9

Schriver, Karen A. "Quality in Document Design." *Technical Communication* (1993): 250–251. Print.

CHAPTER 10

Horton, William. "The Almost Universal Language: Graphics for International Documents." *Technical Communication* (Nov. 1993): 682–693. Print.

Johnson, Dana R. "Copyright Issues on the Internet." *Intercom* (Jun. 1999): 17. Print.

LeVie, Donald S. "Internet Technology and Intellectual Property." *Intercom* (Jan. 2000): 20–23. Print.

Reynolds, Michael, and Liz Marchetta. "Color for Technical Documents." *Intercom* (Apr. 1998): 5–7. Print.

"Time Spent on Facebook up 700%, but MySpace Still Tops for Video." *Nielsen Wire.* 2 Jun. 2009. Web. 10 Aug. 2009.

CHAPTER 11

"CDC and CPSC Warn of Winter Home Heating Hazards." *Centers for Disease Control and Prevention.* 26 Oct. 2006. Web. 6 Mar. 2007.

Dorazio, Pat. Interview. June 2000.

CHAPTER 13

Sharp, Roger A. "Incorporating Animation into Help Files." *Intercom* (Nov. 2007): 8–12. Print.

Tietjen, Phil. "Adventures in Screencasting." *Intercom* (Jul./Aug. 2008): 8–10. Print.

CHAPTER 14

Berst, Jesse. "Seven Deadly Web Site Sins." *ZDNet.* 30 Jan. 1998. Web. 15 Dec. 2005.

Blumberg, Max. "Blogging Statistics." *Max Blumberg Positioning Game.* 2005. Web. 3 Jan. 2005.

Bruner, Rick. "Blogging Statistics." *Idiotprogrammer.* 23 Apr. 2004. Web. 6 Jun. 2005.

Cross, Jay. "Blogging for Business." *Learning Circuits.* Aug. 2003. Web. 23 May 2005.

Dorazio, Pat. Interview. June 2000.

Eddings, Earl, Kim Buckley, Sharon Coleman Bock, and Nathaniel Williams. Software Documentation Specialists at PDA. Interview. 12 Jan. 1998.

"'.eu': A New Internet Top Level Domain." *The Official Journal of the European Union* 24 May 2003. Web. 4 Jun. 2004.

Fallows, Deborah. "China's Online Population Explosion." *Pew/Internet.* 12 Jul. 2007. Web. 29 Jul. 2007.

"508 Law." *Section 508.* 15 Aug. 2002. Web. 10 Jun. 2004.

Fogg, B.J., et al. "How Do People Evaluate a Web Site's Credibility?" *Consumer Reports* 29 Oct. 2002. Web. 1 Aug. 2007.

Foremski, Tom. "IBM Is Preparing to Launch a Massive Corporate Wide Blogging Initiative." *Silicon Valley Watcher.* 13 May 2005. Web. 23 May 2005.

Gard, Lauren. "The Business of Blogging." *BusinessWeek Online.* 13 Dec. 2004. Web. 6 Jun. 2005.

"Gates Backs Blogs for Businesses." *BBC News.* 21 May 2004. Web. 23 May 2005.

Goldenbaum, Don, and George Calvert. Applied Communications Group. Interview. 10 Jan. 1998.

Hemmi, Jane A. "Differentiating Online Help from Printed Documentation." *Intercom* (Jul./Aug. 2002): 10–12. Print.

"Human-Computer Interaction." Rensselaer Polytechnic Institute: Academic Programs. 1998. Web. 21 Apr. 1998.

"Internet Accessibility." *STC AccessAbility SIG.* 9 Jun. 2004. Web. 15 Jun. 2004.

"Internet Growth: Today's Road to Business and Trade." *Internet World Stats.* 2004. Web. 30 May 2004.

Kharif, Olga. "Blogging for Business." *BusinessWeek Online.* 9 Aug. 2004. Web. 23 May 2005.

Knight, Fred. S. "The Internet and Web Go Mainstream." *Business Communications Review* (Oct. 1998): 42–46. Print.

LeVie, Donald S. "Internet Technology and Intellectual Property." *Intercom* (Jan. 2000): 20–23. Print.

Li, Charlene. "Blogging: Bubble or Big Deal?" *Forrester.* 5 Nov. 2004. Web. 23 May 2005.

McAlpine, Rachel. "Passing the Ten-Second Test." *Wise-Women.com.* 22 Apr. 2002. Web. 15 Sep. 2010.

McGowan, Kevin S. "The Leap Online." *Intercom* (Sep./Oct. 2000): 22–25. Print.

Moore, Linda E. "Serving the Electronic Reader." *Intercom* (Apr. 2003): 16–17. Print.

Olive, Eric G. "Usability: Making the Web Work." *Intercom* (Nov. 2002): 8–10. Print.

OnGuard Online. 2007. Web. 1 Aug. 2007.

Pratt, Jean A. "Where Is the Instruction in Online Help Systems." *Technical Communication* (Feb. 1998): 33–37. Print.

"Privacy Initiatives." *Federal Trade Commission* 2007. Web. 1 Aug. 2007.

"Protecting Personal Information." *Federal Trade Commission* 2007. Web. 1 Aug. 2007.

Ray, Ramon. "Blogging for Business." *Inc.com* Sep. 2004. Web. 23 May 2005.

Redish, Janice C. "Writing for the Web: Letting Go of the Words." *Intercom* (Jun. 2004): 4–10. Print.

Shearer, Richard. "Internet Growth Reinforces Reasons for Online Branding." *The National Business Review.* 30 May 2004. Web. 3 Jun. 2004.

Spyridakis, Jan H. "Guidelines for Authoring Comprehensible Web Pages and Evaluating Their Success." *Technical Communication* 47 Aug. 2000: 359–382. Print.

Timpone, Donna. "Help! Six Fixes to Improve the Usability of Your Online Help." *Society for Technical Communication: 44th Annual Conference, 1997 Proceedings.* May 1997: 306. Print.

Wagner, Carol A. "Using HCI Skills to Create Online Message Help." *Society for Technical Communication: 44th Annual Conference, 1997 Proceedings.* May 1997: 35. Print.

Williams, Thomas R. "Guidelines for Designing and Evaluating the Display of Information on the Web." *Technical Communication* 47 Aug. 2000: 383–396. Print.

"World Internet Usage." 2007. Web. 1 Aug. 2007.

Wuorio, Jeff. "Blogging for Business: 7 Tips for Getting started." *Microsoft.* 2005. Web. 23 May 2005.

Wuorio, Jeff. "5 Ways Blogging Can Help Your Business." *Microsoft.* 2005. Web. 23 May 2005.

Yeo, Sarah C. "Designing Web Pages That Bring Them Back." *Intercom* (Mar. 1996): 12–14. Print.

CHAPTER 16

"Robert's Rules of Order Revised." *Constitution Society.* 2004. Web. 15 Sep. 2004.

CHAPTER 19

Buffett, Warren. "Preface." In *A Plain English Handbook: How to Create Clear SEC Disclosure Documents,* Office of Investor Education and Assistance. Washington, D.C.: U.S. Securities and Exchange Commission, Aug. 1998: 1–2. Print.

Clarke, Robyn D. "Put Your Best Voice Forward." *Black Enterprise.* Jun. 2000. Web. 14 Sep. 2010.

"Communications: Oral Presentations." 2004. Web. 2004.

Coyner, Dale. "Webconferencing 101." Sep. 2001. Web. 7 Mar. 2007.

Mahin, Linda. "PowerPoint Pedagogy." *Business Communication Quarterly* 67 Jun. 2004: 219–222. Print.

Miller, Carolyn R. "Communication in the Workplace: A Collaborative Teacher-Student Research Project." 2003. Web. 15 Sep. 2010.

Monster Career Advice Newsletter. 29 Aug. 2006. Print.

Murray, Krysta. "Web Conferencing Tips from a Pro." 1 Jul. 2004. Web. 7 Mar. 2007.

O'Brien, Tim. "In Front of a Group." *The Kansas City Star* 15 Jul. 2003: C5. Print.

Tilton, James E. "Adventures in Public Speaking: A Guide for the Beginning Instructor or Public Speaker." *The FBI Law Enforcement Bulletin.* Feb. 2002. Web. 16 Sep. 2010.

INDEX